AF606466

VOLUME SIXTY FOUR

ADVANCES IN BOTANICAL RESEARCH

ADVANCES IN BOTANICAL RESEARCH

Series Editors

Jean-Pierre Jacquot
Professeur, Membre de L'Institut Universitaire de France, Unité Mixte de Recherche INRA, UHP 1136 "Interaction Arbres Microorganismes", Université de Lorraine, Faculté des Sciences, Vandoeuvre, France

Pierre Gadal
Professor honoraire, Université Paris-Sud XI, Institut Biologie des Plantes, Orsay, France

VOLUME SIXTY FOUR

GENOMIC INSIGHTS INTO THE BIOLOGY OF ALGAE

Volume Editor

Gwenaël Piganeau

CNRS Research Fellow
Evolutionary and Environmental Genomics of Phytoplankton
Laboratoire de Biologie Integrative des Organismes Marins
UMR CNRS-UPMC 7232
France

ELSEVIER

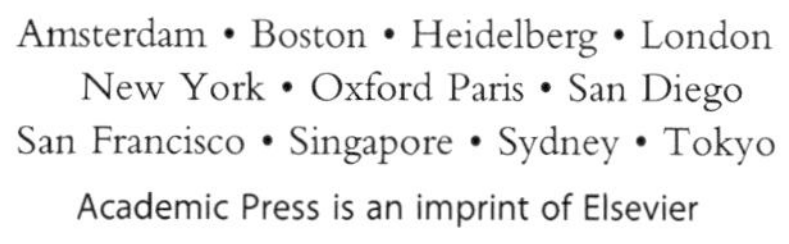

Amsterdam • Boston • Heidelberg • London
New York • Oxford Paris • San Diego
San Francisco • Singapore • Sydney • Tokyo
Academic Press is an imprint of Elsevier

Academic Press is an imprint of Elsevier
32 Jamestown Road, London NW17BY, UK
225 Wyman Street, Waltham, MA02451, USA
525 B Street, Suite 1900, San Diego, CA92101-4495, USA
Radarweg 29, PO Box 211, 1000 AE Amsterdam, The Netherlands

First edition 2012

ISBN: 978-0-12-391499-6
ISSN: 0065-2296

For information on all Academic Press publications
visit our Web site at store.elsevier.com

Printed and bound in USA
12 13 14 15 11 10 9 8 7 6 5 4 3 2 1

CONTENTS

CONTRIBUTORS

The Ectocarpus Genome Consortium
See complete list page 141

John M. Archibald
Canadian Institute for Advanced Research, Program in Integrated Microbial Biodiversity, Department of Biochemistry and Molecular Biology, Dalhousie University, Halifax NS B3H 4R2, Canada

Romain Blanc-Mathieu
UPMC Univ Paris 06, UMR 7232, Observatoire Océanologique, Avenue du Fontaulé, BP44, 66651 Banyuls-sur-Mer, France; CNRS, UMR 7232, Observatoire Océanologique, Avenue du Fontaulé, BP44, 66651 Banyuls-sur-Mer, France

Kenny A. Bogaert
Phycology Research Group, Biology Department, Ghent University, Krijgslaan 281 S8, 9000 Ghent, Belgium

Jean-Paul Cadoret
Ifremer, Laboratoire Physiologie et Biotechnologie des Algues, rue de l'île d'Yeu BP 21105 44311 Nantes cedex 3, France

Olivier De Clerck
Phycology Research Group, Biology Department, Ghent University, Krijgslaan 281 S8, 9000 Ghent, Belgium

J. Mark Cock
Algal Genetics Group, UMR 7139 CNRS-UPMC Marine Plants and Biomolecules Station Biologique, Place Georges Teissier, 29682 Roscoff, France

Erwan Corre
Computer and Genomics Resource Centre, FR 2424, Station Biologique de Roscoff, Place Georges Teissier, BP74, 29682 Roscoff Cedex, France

Yves Desdevises
CNRS, UMR7232, University Pierre et Marie Curie Paris 06, Laboratoire de Biologie Intégrative des Organisms Marins, Observatoire Océanologique, Banyuls-sur-Mer, France

Matthieu Garnier
Ifremer, Laboratoire Physiologie et Biotechnologie des Algues, rue de l'île d'Yeu BP 21105 44311 Nantes cedex 3, France

Sébastien Gourbière
UMR 5244 CNRS-UPVD, Ecologie et Evolution des Interactions, Université de Perpignan via Domitia, 66860 Perpignan, France

Nigel H. Grimsley
CNRS, UMR7232, University Pierre et Marie Curie Paris 06, Laboratoire de Biologie Intégrative des Organisms Marins, Observatoire Océanologique, Banyuls-sur-Mer, France

Stéphan Jacquet
INRA, Stationd' Hydrobiologie Lacustre, 74203 Thonon-les-bains cedex, France

Jessica U. Kegel
Alfred Wegener Institute for Polar and Marine Research, Am Handelshafen 12, D-27570 Bremerhaven, Germany

Wiebe H.C.F. Kooistra
SZN, Stazione Zoologica Anton Dohrn, Villa Comunale, 80121 Naples, Italy

Frederik Leliaert
Phycology Research Group, Biology Department, Ghent University, Krijgslaan 281 S8, 9000 Ghent, Belgium

Monica Medina
Molecular and Cell Biology, University of California, Merced, Merced, California 95343, USA

Linda K. Medlin
University of Pierre and Marie Curie, CNRS, Observatoire Océanologique, LOMIC, UMR 7621, BP 44, Banyuls sur Mer 66651, France

Thomas Mock
School of Environmental Sciences, University of East Anglia, Norwich Research Park, Norwich NR47TJ, UK

Hervé Moreau
CNRS, UMR7232, University Pierre et Marie Curie Paris 06, Laboratoire de Biologie Intégrative des Organisms Marins, Observatoire Océanologique, Banyuls-sur-Mer, France

Fabrice Not
UPMC University Paris 06, UMR 7144, Station Biologique de Roscoff, 29680 Roscoff, France; CNRS, UMR 7144, Station Biologique de Roscoff, 29680 Roscoff, France

Bradley J.S.C. Olson
Molecular Cellular and Developmental Biology, Ecological Genomics Institute, Division of Biology, Kansas State University, Manhattan, Kansas, USA

Gwenaël Piganeau
UPMC Univ Paris 06, UMR 7232, Observatoire Océanologique, Avenue du Fontaulé, BP44, 66651 Banyuls-sur-Mer, France; CNRS, UMR 7232, Observatoire Océanologique, Avenue du Fontaulé, BP44, 66651 Banyuls-sur-Mer, France

Ian Probert
UPMC University Paris 06, FR 2424, Station Biologique de Roscoff, 29680 Roscoff, France

Bruno Saint-Jean
Ifremer, Laboratoire Physiologie et Biotechnologie des Algues, rue de l'île d'Yeu BP 21105 44311 Nantes cedex 3, France

Raffaele Siano
IFREMER, Centre de Brest, DYNECO/Pelagos, BP 70 29280 Plouzané, France

Nathalie Simon
UPMC University Paris 06, UMR 7144, Station Biologique de Roscoff, 29680 Roscoff, France; CNRS, UMR 7144, Station Biologique de Roscoff, 29680 Roscoff, France

Rozenn Thomas
CNRS, UMR7232, University Pierre et Marie Curie Paris 06, Laboratoire de Biologie Intégrative des Organisms Marins, Observatoire Océanologique, Banyuls-sur-Mer, France

Eve Toulza
UPMC Univ Paris 06, UMR 7232, Observatoire Océanologique, Avenue du Fontaulé, BP44, 66651 Banyuls-sur-Mer, France; CNRS, UMR 7232, Observatoire Océanologique, Avenue du Fontaulé, BP44, 66651 Banyuls-sur-Mer, France

James G. Umen
Donald Danforth Plant Science Center, St. Louis, Missouri, USA

Daniel Vaulot
UPMC University Paris 06, UMR 7144, Station Biologique de Roscoff, 29680 Roscoff, France; CNRS, UMR 7144, Station Biologique de Roscoff, 29680 Roscoff, France

Michele X. Weber
Molecular and Cell Biology, University of California, Merced, Merced, California 95343, USA

PREFACE

Algae range in size from 1-μm single-celled organisms to 60-m long giant kelps and have colonized virtually every single aquatic habitat on Earth. Algae comprise diverse and numerous oxygenic photosynthetic eukaryotes with representatives all over the eukaryotic tree of life, with the exception of the non-photosynthetic Unikont superphylum that includes all animals and fungi. Cellular and molecular studies of the last century provided the evidence that all algae derive from a major transition in the evolution of life, the endosymbiosis between one eukaryotic host cell and a photosynthetic bacterium, that subsequently evolved into an organelle, the plastid. Thus, despite their phylogenetic diversity, all algae are 'bound by plastids' (Delwiche, 2007).

In the last decades, the use of sequence data from a handful of genes has shed light on the huge diversity and the complex evolutionary history of algae. The recent development of large-scale sequencing studies provides unprecedented access to both the metabolic potential and the ultimate record of the evolution of a cell, its genome. Genomics, defined here in its broad sense as large-scale DNA sequencing projects, has revolutionized the study of algae in several ways. These data enable us to scale up our knowledge on algal diversity and evolution and open the book to reading about the underlying molecular mechanisms; the rise and fall of gene families, the transfer of genes from the endosymbiont to the nuclear genomes, as well as the hallmark of selection on gene content, gene amino acid structure and non-coding regions.

In this volume, we address some of the genomic insights gained into the ecology, the evolution and the biology of algae.

The first chapter (Not et al., 2012) provides an overview of the diversity and ecology of algae in the largest ecosystem on Earth: the ocean's surface. A bucket of seawater may contain hundreds of algal species invisible to the eye, marine phytoplankton, and the authors report how early morphological studies and DNA sequences from environmental samples, metagenomics, fostered the discovery of their diversity.

The following chapters review the evolutionary scenarios leading to this highly diverse group of eukaryotes. Chapter 2 (De Clerck, Bogaret, &

Leliaert 2012) reviews the diversity of the descendants of the primary endosymbiosis event, the Archaeplastida, that comprise the green algae, the red algae and the glaucophytes. Chapter 3 (Archibald, 2012) summarizes the complex history of algae that have evolved from the engulfment of some Archaeplastida algae by a eukaryote. These secondary endosymbioses of green or red algae have occurred several times independently in the course of evolution. Some algal groups are descendants of tertiary endosymbiosis; the engulfment of an algae that acquired its photosynthetic metabolism from a secondary endosymbiotic event. This complex evolutionary story of serial cell capture and enslavement was unveiled by the comparative analysis of the genome sequences of both the chloroplast, the nuclear and the nucleomorph genomes of these algae.

Chapter 4 (Weber & Medina 2012) gives an overview of the genomics of one of the many symbioses between algae and non-algae – the corals and anemones with their dinoflagellate *Symbiodinium* algal partners.

The following chapters are dedicated to case studies of genome projects within the green algae *Volvox* and *Chlamydomonas* (chapter 6 (Umen & Olson 2012)), the brown alga *Ectocarpus* (chapter 5 (The Ectocarpus Genome Consortium 2012)) and the Diatoms (chapter 7 (Mock & Medlin 2012)). Chapter 8 (Cadoret, Garnier, & Saint-Jean 2012) reports some of the genomic insights gained into the physiology of algae and their biotechnological potentials as nutritional complements, medicines or biofuels.

The last two chapters review how genomics enables a glimpse into the molecular basis of the interactions between algae and their environment. Chapter 9 (Grimsley et al. 2012) concerns recent genomic insights of algal viruses, providing clues about host-viral evolutionary and functional relationships. Chapter 10 (Toulza et al. 2012) reviews the power and challenges of metagenomics, the scaling up of genomes to the community level, for the microbial algae.

I would like to thank all contributors for joining in to produce this publication and for their effort to fit into a tight time schedule.

I would also like to thank Cecile Meunier, Keith Cornelius, Adam Eyre-Walker, Falk Hildebrand, Jonathan Green, Pascal Hingamp, Jean-François Gout, Hiro Ogata, Philippe Deschamps, David Moreira, Michael Guarnieri and Sophie Sanchez-Ferrandin, for discussion and constructive comments on specific sections of this volume.

Gwenaël Piganeau

REFERENCES

Archibald, J. (2012). The evolution of algae by secondary and tertiary endosymbiosis. *Advances in Botanical Research, 64*, 87–118.

Cadoret, J., Garnier, M., & Saint-Jean, B. (2012). Microalgae, functional genomics and biotechnology. *Advances in Botanical Research, 64*, 285–341.

De Clerck, O., Bogaret, K., & Leliaert, F. (2012). Diversity and evolution of algae: primary endosymbiosis. *Advances in Botanical Research, 64*, 56–86.

Delwiche, C. (2007). Algae in the warp and weave of life, bound by plastids. In J. Brodie, & J. Lewis (Eds.), *Unravelling the algae, the past, present and future of algal systematics* (pp. 7–20). CRC Press.

Grimsley, N., Thomas, R., Kegel, J., Jacquet, S., Moreau, H., & Desdevises, Y. (2012). Genomics of algal host–virus interactions. *Advances in Botanical Research, 64*, 343–378.

Mock, T., & Medlin, L. K. (2012). Genomics and genetics of diatoms. *Advances in Botanical Research, 64*, 245–284.

Not, F., Siano, R., Kooistra, W. H. C. F., Simon, N., Vaulot, D., & Probert, I. (2012). Diversity and ecology of eukaryotic marine phytoplankton. *Advances in Botanical Research, 64*, 1–53.

The Ectocarpus Genome Consortium. (2012). The Ectocarpus genome and brown algal genomics. *Advances in Botanical Research, 64*, 141–184.

Toulza, E., Blanc-Mathieu, R., Gourbière, S., & Piganeau, G. (2012). Environmental genomics of microbial algae: power and challenges of metagenomics. *Advances in Botanical Research, 64*, 379–423.

Umen, J., & Olson, B. (2012). Genomics of volvocine algae. *Advances in Botanical Research, 64*, 185–243.

Weber, M., & Medina, M. (2012). The role of microalgal symbionts (Symbiodinium) in holobiont physiology. *Advances in Botanical Research, 64*, 119–140.

CONTENTS OF VOLUMES 35–63

Series Editor (Volumes 35–44)

J.A. CALLOW
School of Biosciences, University of Birmingham, Birmingham, United Kingdom

Contents of Volume 35

Contents of Volume 36

Contents of Volume 37

ANTHOCYANINS IN LEAVES

Edited by K. S. Gould and D. W. Lee

Contents of Volume 38

Contents of Volume 39

Contents of Volume 40

Contents of Volume 41

Contents of Volume 42

Contents of Volume 43

Contents of Volume 44

Series Editors (Volume 45–60)

JEAN-CLAUDE KADER
Laboratoire Physiologie Cellulaire et Moléculaire des Plantes, CNRS, Université de Paris, Paris, France

MICHEL DELSENY
Laboratoire Génome et Développement des Plantes, CNRS IRD UP, Université de Perpignan, Perpignan, France

Contents of Volume 45

RAPESEED BREEDING

Contents of Volume 46

Contents of Volume 47

Contents of Volume 48

Contents of Volume 49

Contents of Volume 50

Contents of Volume 51

Contents of Volume 52

Contents of Volume 53

Contents of Volume 54

Contents of Volume 55

Contents of Volume 56

Contents of Volume 57

Contents of Volume 58

Contents of Volume 59

Contents of Volume 60

Contents of Volume 61

Contents of Volume 62

Contents of Volume 63

CHAPTER ONE

Diversity and Ecology of Eukaryotic Marine Phytoplankton

Fabrice Not[*,†,1] **Raffaele Siano**[‡], **Wiebe H.C.F. Kooistra**[§], **Nathalie Simon**[*,†], **Daniel Vaulot**[*,†], **and Ian Probert**[¶]

[*]UPMC University Paris 06, UMR 7144, Station Biologique de Roscoff, 29680 Roscoff, France
[†]CNRS, UMR 7144, Station Biologique de Roscoff, 29680 Roscoff, France
[‡]IFREMER, Centre de Brest, DYNECO/Pelagos, BP 70 29280 Plouzané, France
[§]SZN, Stazione Zoologica Anton Dohrn, Villa Comunale, 80121 Naples, Italy
[¶]UPMC University Paris 06, FR 2424, Station Biologique de Roscoff, 29680 Roscoff, France
[1]Corresponding author: Fabrice Not, not@sb-roscoff.fr

Contents

Advances in Botanical Research, Volume 64
ISSN 0065-2296,
http://dx.doi.org/10.1016/B978-0-12-391499-6.00001-3

Abstract

Marine phytoplankton, the photosynthetic microorganisms drifting in the illuminated waters of our planet, are extremely diverse, being distributed across major eukaryotic lineages. About 5000 eukaryotic species have been described with traditional morphological methods, but recent environmental molecular surveys are unveiling an ever-increasing diversity, including entirely new lineages with no described representatives. Eukaryotic marine phytoplankton are significant contributors to major global processes (such as oxygen production, carbon fixation and CO_2 sequestration, nutrient recycling), thereby sustaining the life of most other aquatic organisms. In modern oceans, the most diverse and ecologically significant eukaryotic phytoplankton taxa are the diatoms, the dinoflagellates, the haptophytes and the small prasinophytes, some of which periodically form massive blooms visible in satellite images. Evidence is now accumulating that many phytoplankton taxa are actually mixotrophs, exhibiting alternate feeding strategies depending on environmental conditions (e.g. grazing on prey or containing symbiotic organisms), thus blurring the boundary between autotrophs and heterotrophs in the ocean.

1. PHYTOPLANKTON FEATURES

1.1. Diversity of Phytoplankton

This chapter provides an overview of current knowledge on the diversity and ecology of the phytoplankton that drift in the illuminated waters of seas and oceans. The term phytoplankton here corresponds to the functional grouping of single-celled organisms (prokaryotes and eukaryotes) that have the capacity to perform oxygenic photosynthesis. Marine phytoplanktonic prokaryotes all belong to the phylum Cyanobacteria within the domain Bacteria. In contrast, eukaryotic phytoplankton, the focus of the present chapter, is taxonomically very diverse, having representatives in all but one lineage of the eukaryotic tree of life (Fig. 1.1). The early evolutionary history of eukaryotic phytoplankton (and more generally of all plastid bearing eukaryotes) was shaped by series of endosymbiotic events, involving the engulfment of a cyanobacterium by a eukaryote (Chapter II of this volume, De Clerck, Bogaret, & Leliaert, 2012) or the engulfment of a photosynthetic eukaryote by another eukaryote (Chapter III of this volume, Archibald

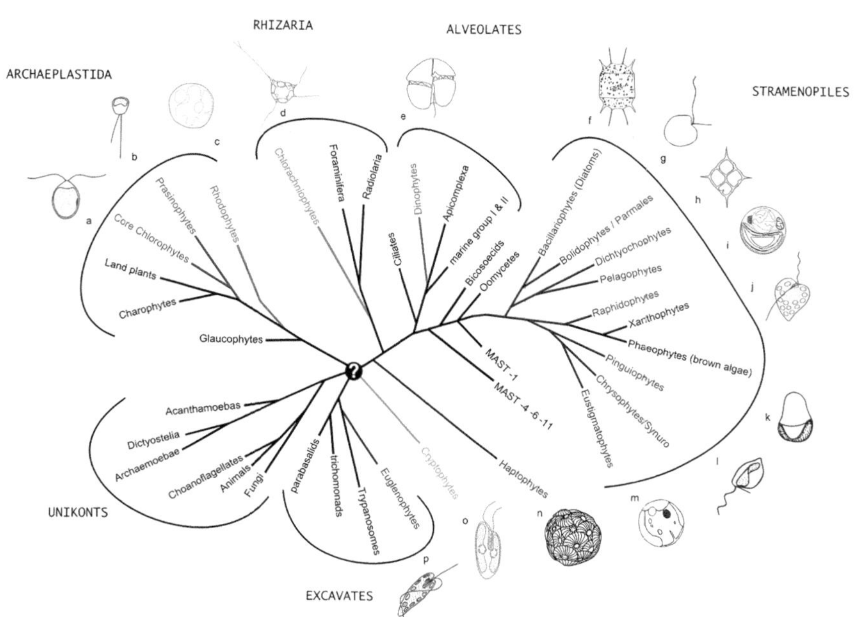

Figure 1.1 Schematic phylogenetic tree representing the distribution of phytoplanktonic taxa across eukaryote lineages (in color). Illustrations of (a) Chlorophyceae, (b) *Pseudoscourfieldia* sp., (c) *Porphyridium cruentum*, (d) *Gymnochlora dimorpha*, (e) Dinoflagellates, (f) *Odontella* sp. (g) *Bolidomonas pacifica*, (h) *Dictyocha* sp., (i) *Aureococcus anophagefferens*, (j) *Heterosigma akashiwa*, (k) *Pinguiochrysis pyriformis*, (l) *Ochromonas* sp., (m) *Nannochloropsis salina*, (n) *Calcidiscus* sp., (o) *Cryptomonas* sp., (p) Euglenids; 'a, b, e, f, h, j, l, n, o and p' are adapted from Tomas (1997), 'c' adapted from Lee (1999), 'd' from Ota, Kudo, and Ishida (2011), 'g' adapted from http://tolweb.org/Bolidomonas/142186, 'i' from Andersen and Preisig (2000), 'k' from Kawachi *et al.* (2002) and 'm' from Van Den Hoek, Mann, and Jahns (1995). For colour version of this figure, the reader is referred to the online version of this book.

2012). These endosymbiotic events were accompanied by massive gene transfers from the genomes of the endosymbionts to the genome of the host, traces of which can be detected in modern eukaryotic primary producers.

Historically, the diversity of eukaryotic phytoplankton has been assessed by microscope-based comparison of morphological features. Based on these observations, less than 5000 species have been described to date (Simon, Cras, Foulon, & Lemée, 2009; Sournia, Chretiennot-Dinet, & Ricard, 1991; Tett & Barton, 1995), but there is a general agreement that this number largely underestimates the real extent of phytoplankton diversity. In the last decade, evaluation of environmental diversity using molecular approaches has highlighted massive undescribed diversity, including whole lineages without any cultured representatives and for which only environmental sequences are available (Massana & Pedros-Alio, 2008; Vaulot, Eikrem, Viprey, & Moreau, 2008). Some of these environmental lineages are so distantly related to all other groups that they may represent new phyla, one example being the picobiliphyte (Not *et al.*, 2007). Combination of molecular phylogenetic and morphological analyses has repeatedly demonstrated the existence of cryptic (or pseudocryptic) species, even in supposedly well-known groups such as diatoms and coccolithophores, fuelling the debate concerning species delineation in protists (Amato *et al.*, 2007; Saez *et al.*, 2003). In addition, detailed studies comparing well-defined species complexes demonstrate that commonly used molecular markers (e.g. the 18S rRNA gene) often underestimate diversity, particularly for organisms like phytoplankton that have huge population sizes and high turnover rates (Piganeau, Eyre-walker, Grimsley, & Moreau, 2011a). Information on the genetic diversity of phytoplankton is likely to significantly increase in the future with the advent of environmental meta-barcoding surveys (Bik *et al.*, 2012; Toulza, Blanc-Mathieu, Gourbiere, & Piganeau, 2012). This is particularly true for small-sized phytoplankton for which very few distinguishing morphological characters are available.

1.2. Size Matters

Eukaryotic phytoplankton cells are not only taxonomically very diverse but also span an exceptionally wide size range both between and within taxonomic groups. Size spectra can even vary temporally and/or spatially in response to varying environmental conditions or succession of life cycle stages. Phytoplankton cells span more than three orders of magnitude in size, ranging from picoplankton (0.2–2 μm) up to mesoplankton (0.2–2 mm). Individual cells of most species are solitary, but many species (e.g. most species within the diatom genera *Chaetoceros* and *Thalassiosira*, the dinoflagellate *Alexandrium catenella*, or the haptophyte *Phaeocystis*) also have the ability to form chains or colonies. Although exceptions exist, the largest size classes of marine phytoplankton are generally dominated by 'golden brown' groups, notably diatoms and dinoflagellates, whereas smallest size classes essentially consist of green algae from the prasinophyte lineage (e.g. *Ostreococcus tauri*, which has a cell diameter less than 1 μm; Chrétiennot-Dinet *et al.*, 1995). In practice, this wide range of cell sizes requires the deployment of various collecting devices (plankton nets and filtration on various mesh sizes) and observation methodologies (optical and electronic microscopy) to characterize phytoplankton diversity. Cell size also affects numerous functional characteristics of phytoplankton. For instance, because of their large surface to volume ratio which facilitates passive nutrient uptake, small cells are particularly well adapted to stable and oligotrophic (nutrient poor) waters, whereas larger cells typically perform better in mixed and eutrophic (nutrient rich) settings (Finkel *et al.*, 2010; Marañón *et al.*, 2001). Because the marine environment exhibits heterogeneous physicochemical structures across space and time, cell size is an important feature to consider from an ecological point of view.

1.3. Global Ecological Patterns

Phytoplankton plays a significant role in global ecology and ecosystem functioning. First and foremost, phytoplankton species are primary producers and contribute to about half of the primary production on the planet, of which one forth is estimated to occur in oligotrophic waters (essentially performed by the cyanobacteria *Prochlorococcus*), one forth in eutrophic waters and half in mesotrophic regions (Field, C. B., Behrenfeld, M. J., Randerson, J. T., & Falkowski, P. 1998). Phytoplankton participates to the global carbon cycle through the so-called biological pump, by fixing carbon, a portion of which is subsequently sequestered at depth. Carbon is ultimately buried at the sea floor for centuries or longer (Falkowski, 2012). Phytoplankton is also at the base of virtually all marine food webs. Under specific light and nutrient conditions, some phytoplankton taxa can form large blooms, particularly in coastal waters of temperate seas. Some bloom-forming phytoplankton produce toxins that affect higher trophic levels (*H*armful *A*lgal *B*looms or HABs), thus having significant ecological and economic impacts (Hinder *et al.*, 2011; Imai & Yamaguchi, 2012). Classically, study of the ecology of phytoplankton communities involves one or a combination of microscope-based morphological studies, flow cytometric cell counting, molecular surveys and/or measurement of the presence of specific photosynthetic pigments (Jeffrey, 1997). Although each approach has its inherent limitations, general ecological patterns can be drawn from the literature. Eutrophic coastal and continental shelf waters are classically dominated by diatoms, dinoflagellates and calcifying haptophytes (coccolithophores), groups that contain species that have the capacity to form large blooms, while other groups such as the euglenophytes, cryptophytes and raphidophytes produce more localized blooms (Assmy & Smetacek, 2009). Open oceans tend to be dominated by groups such as green algal prasinophytes, *Chrysochromulina*-like haptophytes and small stramenopiles like pelagophytes and chrysophytes (Not *et al.*, 2008; Reynolds, 2006).

Since phytoplankton are primary producers living in a dispersive environment, abiotic physico-chemical factors exert a strong control on the composition and dynamics of phytoplankton communities. Several bloom-forming phytoplankton taxa have the ability to bio-mineralize silica or calcium, which, among other biogeochemical impacts, drives long-term carbon sequestration by accentuating sinking to the sea floor after bloom events. Valuable fossil records exist for these bio-mineralizing taxa and these are extensively used for paleo-stratigraphy and paleo-climatology. Bio-mineralization is also probably involved in mechanical defense and probably explains, at least in part, why diatoms and coccolithophores are ubiquitous in spring blooms in temperate and boreal systems (Smetacek, 2001).

Besides the control exerted by the zooplankton, which feed upon phytoplankton, the impact of biotic parameters on the global ecology of phytoplankton has generally not been studied in great detail. There is now a growing awareness of the impact of viruses and of parasitic and mutualistic symbiotic interactions on phytoplankton community structure (Brussaard *et al.*, 2008; Siano *et al.*, 2011) and ultimately on global biogeochemical cycles (Strom, 2008). We refer to Chapter IX of this volume for a review on genomic insights into the diversity of algal viruses (Grimsley *et al.* 2012).

Each phytoplankton lineage employs diverse trophic strategies. Indeed, although phytoplanktonic organisms are primarily photosynthetic, many exhibit mixotrophic behavior, feeding on prokaryotes or other small phytoplankton in addition to conducting photosynthesis. This has been well characterized for certain dinoflagellates (e.g. species within the genera *Gymnodinium* or *Amphidinium*; Lee, 1999) and haptophytes (e.g. *Chrysochromulina* sp.; Kawachi, Inouye, Maeda, & Chihara, 1991). Phytoplankton can also live in symbiosis with larger heterotrophic protists such as foraminifers or radiolarians and also with metazoans (e.g. in coral reefs; (Weber & Medina, 2012). Recently, several lines of evidence (e.g. stable isotope labelling) promote the conclusion that such mixotrophic strategies are more frequent than previously thought in the marine environment (Frias-Lopez, Thompson, Waldbauer, & Chisholm, 2009; Liu *et al.*, 2009; Stoecker, Johnson, de Vargas, & Not, 2009). While exogenous abiotic and biotic factors exert key controls on phytoplankton growth and mortality, internal factors such as life cycle traits (D'alelio *et al.*, 2010) or control of cell death (Biddle & Falkowski, 2004) fine-tune the regulation of population dynamics.

1.4. Current Conceptual Challenges

Studies of phytoplankton diversity and ecology, and more generally of microbial ecology and evolution, are driven by a number of major unresolved conceptual challenges, perhaps the foremost of which is the definition of what is a species. The species stands as a key concept, a basic unit and a common currency for studies of diversity and ecology in any environment; yet, there is no consensus on how to define a species. Phytoplankton species are traditionally defined according to morphological features, but (1) comparisons of morphological and molecular data often provide evidence for cryptic diversity within 'morphospecies' (Amato *et al.*, 2007), (2) morphological traits are prone to change under varying environmental conditions (Pizay *et al.*, 2009), and (3) the smallest phytoplanktonic cells usually lack distinctive features (Potter, Lajeunesse, Saunders, & Anderson, 1997). As for prokaryotes, the classical biological species concept defined by E. Mayr in 1969 (i.e. members of an interbreeding population reproductively isolated from other such groups and capable of producing fertile descendants; Mayr, 1969) cannot be applied to most microbial eukaryotes due to the lack of knowledge on sexual reproduction (Silva, 2008). Other species concepts have been proposed (e.g. ecological, phylogenetic, morphological) (De Queiroz, 2007). While progress has been made towards the proposition of a unified species concept (De Queiroz, 2007; Samadi & Barberousse, 2006), operationally applicable non-subjective criteria are lacking to circumscribe phytoplankton, and more generally microbial, species.

Another major scientific puzzle that has its historical roots in the nineteenth century (O'malley, 2007) and is currently at the centre of an intense debate in the field of microbial ecology concerns the conceptual principle of 'everything is everywhere, but the environment selects', postulating that the abundance of individuals in microbial species is so large that dispersal is never restricted by geographical barriers (Finlay, 2002). Intuitively, this might be thought to be particularly true for oceanic phytoplankton, unicellular eukaryotes drifting in a dispersive environment. This statement still structures the ecological and evolutionary understanding of microbial distribution (De Wit & Bouvier, 2006). However, microbial ecology no longer relies on culture-based studies. With the advent of molecular tools, evidence is accumulating that tends to show that physico-chemical barriers do exist for marine plankton and that species are not globally distributed (Casteleyn *et al.*, 2010) and can occupy distinct niches (Foulon *et al.*, 2008). In the near future, the use of relevant molecular markers coupled to massive sequencing depth provided by high throughput technologies will probably allow this question to be fully addressed.

Finally, another unresolved question is that of the paradox of the plankton formulated by G.E. Hutchinson in 1961, who asked 'why do so many planktonic species co-exist in a supposedly homogeneous habitat? (i.e. under the competitive exclusion principle of Gause, given the limited range of resources required for their growth)'. For specific ecosystems, proposed mechanisms to explain the extreme diversity of phytoplankton include spatial and temporal heterogeneity in physical and biological environments at different scales, oscillation and chaos generated by internal and external causes and self limitation by toxin-producing phytoplankton. A general and well-accepted theory to explain environmental plankton diversity is still, however, lacking (Roy & Chattopadhyay, 2007). This question is extremely challenging in the context of the uncertainties mentioned above concerning species delineation and enumeration in natural phytoplankton assemblages.

2. THE GREEN PHYTOPLANKTON: THE CHLOROPHYTES

2.1. General Considerations

The Chlorophyta together with the land plants form the green lineage (Viridiplantae). This group arose after an endosymbiotic event between a cyanobacterium-related organism and a heterotrophic eukaryote that was at the origin of the Plantae, also named Archaeplastida, a super group of eukaryotes that also includes the red algae and glaucophytes (Leliaert, Verbruggen, & Zechman, 2011). The extant Streptophyta include the land plants as well as diverse freshwater algal lineages, while the Chlorophyta include some freshwater algae and all marine representatives (De Clerck *et al.*, 2012). The Chlorophyta and Streptophyta possess the following common unique features: a double membrane bound plastid containing chlorophyll *b* as the main accessory pigment and starch as well as a unique stellate structure linking pairs of microtubules in the flagellar base. The Chlorophyta form a strongly supported group in molecular phylogenies and are characterized by unique biochemical and ultrastructural features (Leliaert *et al.*, 2011).

In marine waters, Chlorophyta are especially important within the smallest size classes, in particular the picoplankton and nanoplankton, which are formally defined as cells between 0.2–2 and 2–20 μm, respectively.

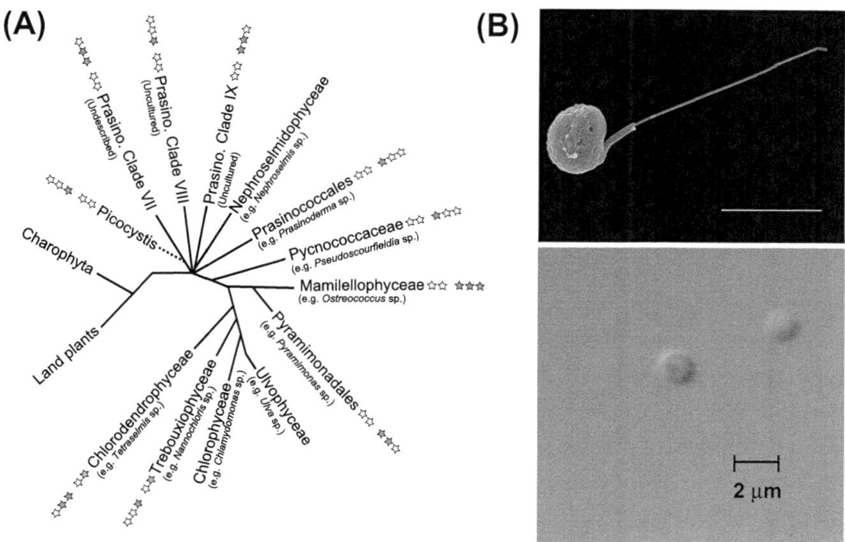

Figure 1.2 A) Schematic phylogenetic tree of the green algae and land plants lineages showing the relationships among major phytoplanktonic taxa and an estimation of their ecological significance. Typical representative of each lineage is indicated in brackets. The overall ecological significance (illustrated by a five-star ranking) is subjective and has been established based on parameters such as abundance, distribution, bloom formation, trophic strategies, toxicity, etc. (color code is 1 blue star = having freshwater members, 1 red star = important toxic or harmful species, 1–3 green stars range = other relevant ecological parameters, no stars means multi-cellular or no marine species). (B) Illustration of two important prasinophytes belonging to the Mamiellophyceae. Top: scanning electron microscopy of the common and abundant *Micromonas* sp. (E. Foulon). Bottom: the smallest photosynthetic eukaryote *Ostreococcus* sp. (D. Vaulot). See the colour plate.

A rather small proportion of the lineage belongs to the Trebouxiophyceae (that are mostly freshwater or terrestrial species), but most of the species described from marine isolates (Vaulot *et al.*, 2008) and most 18S rRNA gene sequences recovered from the oceanic environment correspond to prasinophytes. Prasinophytes form a polyphyletic assemblage (Fig. 1.2) with very few common characters and taxonomists are slowly re-organizing this group by creating new classes for each of the existing clades (Guillou *et al.*, 2004; Marin & Melkonian, 2010).

2.2. The Mamiellophyceae

The recently defined class Mamiellophyceae (Marin & Melkonian, 2010) encompasses three orders (Mamiellales, Dolichomastigales, and Monomastigales).

From an ecological point of view, the Mamiellales is the most important order (Fig. 1.2), containing three key genera: *Micromonas* (Butcher, 1952), with the first described picoplanktonic species *Micromonas pusilla, Ostreococcus* (Chrétiennot-Dinet *et al.*, 1995), containing the smallest known photosynthetic algal species (0.8-μm cell diameter), *Ostreococcus tauri*, and *Bathycoccus* (Eikrem & Throndsen, 1990), with a single scale-bearing coccoid species, *Bathycoccus prasinos*. These three related genera, which share few morphological features, are typical of coastal waters (Cheung *et al.* 2010; Collado-Fabri, Ulloa, & Vaulot 2011; Medlin, Metfies, Wiltshire, Mehl, & Valentin, 2006; Not *et al.* 2004) but can also bloom under specific conditions in oceanic waters (Treusch *et al.*, 2011) or be dominant in Arctic ecosystems (Lovejoy *et al.*, 2007). Members of these three genera can relatively easily be isolated into pure laboratory culture, facilitating their adoption as biological and ecological models. Full genome sequences for the three representative genera cited above are now available (Derelle *et al.*, 2006; Moreau *et al.*, 2012; Worden *et al.*, 2009), and their analysis has started to reveal genes that are relevant to studies of ecology and speciation (Piganeau, Grimsley, & Moreau, 2011b).

The single described species of the genus *Micromonas, M. pusilla*, is characterized by naked cells with a short flagellum with a characteristic hair-point and is genetically differentiated into at least three (but probably more) clades (Foulon *et al.*, 2008; Guillou *et al.*, 2004; Slapeta, Lopez-Garcia, & Moreira, 2006). Two of the major clades (A and B; sensu Guillou *et al.* 2004) are found in coastal waters, while clade C is typically oceanic (Foulon *et al.*, 2008). Within clade B, a specific lineage seems to be restricted to Arctic waters (Lovejoy *et al.*, 2007), where it can completely dominate the picophytoplankton size fraction (Balzano, Marie, Gourvil, & Vaulot, 2012).

Ostreococcus is characterized by small naked coccoid cells with no specific morphological features except a very salient starch grain in the pyrenoid (Ral *et al.*, 2004). As in the case of *Micromonas*, it can be subdivided into at least four clades based on phenotypic, genetic and genomic traits (Rodriguez *et al.*, 2005). While clade C is mostly restricted to environments where it was initially discovered (coastal lagoons), clade A is typical of surface coastal waters and clade B appears to be associated with deeper layers of the euphotic zone, displaying specific photoacclimation strategies (Six *et al.*, 2008). However, analyses of the distribution of the 18S rRNA gene of *Ostreococcus* clades in the Pacific Ocean as well as in the subtropical and tropical North Atlantic indicate that the ecophysiological parameters influencing clade distribution are more complex than irradiance alone, with factors such as temperature and nutrients also being involved in the control of the distribution of ecotypes (Demir-Hilton *et al.*, 2011). *Ostreococcus* can form localized blooms not only in coastal waters (O'Kelly, Sieracki, Thier, & Hobson, 2003) but also in open ocean regions (Treusch *et al.*, 2011). In certain ecosystems, such as the coastal upwelling off Chile, it is the most abundant picophytoplankton species (Collado-Fabri *et al.*, 2011).

The third member of the Mamiellales, *B. prasinos*, is characterized by spider-like scales covering the cell surface (Eikrem & Throndsen, 1990). In contrast to the two other genera, there is little evidence as yet for the existence of distinct clades with the genus *Bathycoccus* (Guillou *et al.*, 2004). Although initially described from the bottom of the euphotic zone (hence the prefix 'Bathy'; Eikrem & Throndsen 1990), *Bathycoccus* is typical of surface coastal waters (Collado-Fabri *et al.*, 2011; Not *et al.* 2004). The analysis of metagenomes obtained from sorted cells from coastal and pelagic deep chlorophyll maximum waters suggests that there may indeed be distinct *Bathycoccus* ecotypes or species adapted to these different environments (Monier *et al.*, 2012; Vaulot *et al.*, 2012). Two other nanoplanktonic genera, *Mantionella* and *Mamiella*, also belong to Mamiellales. The order Dolichomastigales contains two genera *Dolichomastix* and *Crustomastix* with nanosized cells possessing two very long flagella. The 18S rRNA gene sequences related to these four genera have been found in the Mediterranean Sea (Viprey, Guillou, Férréol, & Vaulot, 2008), in the Atlantic, and even associated to deep sediment samples (Marin & Melkonian, 2010), but very little information is available on their global distribution and ecology.

2.3. Other Prasinophytes

The Pyramimonadales (prasinophyte clade I) encompasses more than 35 species (Guiry & Guiry, 2012) within the main genus *Pyramimonas*, characterized by nanosized cells typically possessing four flagella. This order can be ecologically important in coastal areas (Bergesch, Odebrecht, & Moestrup, 2008) as well as in polar waters (Balzano *et al.*, 2012; Rodriguez, Varela, & Zapata, 2002).

The Chlorodendrophyceae (prasinophyte clade IV) is a recently established class (Massjuk, 2006), which contains one major genus, *Tetraselmis*, with around 30 species (Guiry & Guiry, 2012). Cells possess four equal flagella arranged in two opposite pairs and a theca composed of aggregated scales. This group does not appear to be ecologically important in marine waters, although related sequences have been found in the Mediterranean

Sea (Viprey *et al.*, 2008). Cultured strains are widely used for applications such as aquaculture (Mohammady, 2004).

The Pycnococcaceae (prasinophyte clade V) contains only two major species: *Pseudoscourfieldia marina*, a flagellate, and *Pycnococcus provasolii*, a coccoid cell. The two species share 100% 18S rRNA gene identity and could actually be the two forms of a single life cycle (Fawley, Yun, & Qin, 1999; Guillou *et al.*, 2004). *Pycnococcus* has been found to be abundant in specific ecosystems such as the Magellan Straits (Zingone, Sarno, Siano, & Marino, 2011). These species are easily isolated from oceanic waters and similar sequences have been found, for example, in the Mediterranean Sea (Viprey *et al.*, 2008), suggesting that this group may be widespread.

As in the case of the Pycnococcaceae, the order Prasinococcales (prasinophyte clade VI) contains only two genera, *Prasinoderma* and *Prasinococcus*, both falling in the picoplankton size range and containing in total three species (Guiry & Guiry, 2012). All three species produce some kind of gelatinous matrix, which, in the case of *Prasinococcus capsulatus*, has been identified as consisting of a sulfated and carboxylated polyanionic polysaccharide named capsulan (Sieburth, Keller, Johnson, & Myklestad, 1999). They are easily isolated from marine waters (Le Gall *et al.*, 2008), but few 18S rRNA gene sequences are recovered from planktonic environmental clone libraries (Viprey *et al.*, 2008), suggesting that they may be associated to specific marine habitats such as marine particles.

Prasinophyte clade VII has not yet been formerly described, despite the fact that it contains cultured strains, all of which are picosized and coccoid (Vaulot *et al.*, 2008). It is divided into two well-supported subclades (A and B) and, depending on phylogenetic analyses, can include *Picocystis salinarum*, a small species found in inland saline lakes (Lewin, Krienitz, Goericke, Takeda, & Hepperle, 2000). The 18S rRNA gene sequences from this clade have been recovered from moderately oligotrophic areas from the Pacific Ocean and Mediterranean Sea (Shi, Marie, Jardillier, Scanlan, & Vaulot, 2009; Viprey *et al.*, 2008) as well as from coastal waters (Romari & Vaulot, 2004).

In contrast to clade VII, no cultures have yet been isolated from prasinophyte clades VIII and IX that were first discovered from 18S rRNA gene sequences in the Mediterranean Sea (Viprey *et al.*, 2008). Sequences from clade IX (but not VIII) have also been found in the very oligotrophic waters of the South East Pacific gyre (Shi, Lepère, Scanlan, & Vaulot, 2009). These clades appear to be extremely diversified and are probably an important component of the photosynthetic picoplankton in the central oceanic gyres.

2.4. Trebouxiophyceae

Trebouxiophyceae are mostly terrestrial algae, in particular associated with lichens. However, several genera, including *Picochlorum* (erected to regroup salt-tolerant *Nannochloris* species; Henley *et al.*, 2004), *Stichococcus* and *Chlorella*, can be isolated from marine waters and have been found in environmental 18S rRNA gene clone analyses from coastal waters (Medlin *et al.*, 2006).

3. THE PHYTOPLANKTON WITH CALCAREOUS REPRESENTATIVES: THE HAPTOPHYTES

3.1. Origins of the Haptophytes

The haptophytes are a distinct and almost exclusively photosynthetic protistan lineage that is widespread and often very abundant in diverse marine settings. Haptophytes are characterized by the presence of a unique organelle called a haptonema (from the Greek *hapsis*, touch, and *nema*, thread), which is superficially similar to a flagellum but differs in the arrangement of microtubules and in function, being implicated in attachment or capture of prey. The group includes some well-known taxa, such as *Phaeocystis*, *Prymnesium* and *Chrysochromulina*, that form periodic harmful or nuisance blooms in coastal environments, the calcifying species (coccolithophore) *Emiliania huxleyi* that produces massive 'white-water' blooms in high latitude coastal and shelf ecosystems (Fig. 1.3), and *Pavlova lutheri* and *Isochrysis galbana* that are species extensively used as feedstock in aquaculture. *E. huxleyi* has become a model species, notably for studies of the effects of ocean acidification on coccolithophore calcification (Beaufort *et al.*, 2011; Iglesias-Rodriguez *et al.*, 2008; Riebesell *et al.*, 2000), and is the only haptophyte for which extensive genomic data (including full genome sequences) are currently available.

The origin and evolutionary affiliations of the Haptophyta remain contentious. Haptophytes were tentatively grouped within stramenopiles (Cavalier-Smith, 1981) since in both lineages plastids contain chlorophylls *a* and *c* as well as various carotenoids, typically giving them a golden or brown colour, and the photosynthetic carbohydrate storage product is a β-1,3-linked glucan. Haptophytes possess a network of endoplasmic reticulum immediately below the cell membrane that was suggested to be homologous to alveoli of ciliates, amphiesmal vesicles of dinoflagellates, the inner membrane complex of apicomplexans, the periplast of

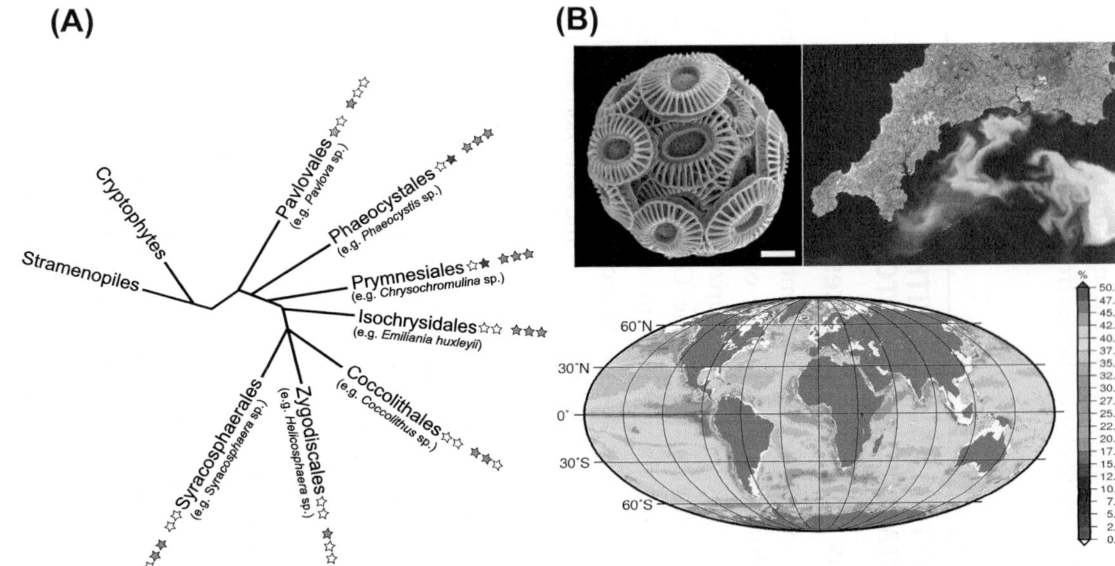

Figure 1.3 A) Legend as Figure 1.2 A but for Haptophyta. (B) Scanning electronic microscopy illustration of the coccolithophore *Emiliania huxleyi*, scale bar 1 μm (top left), satellite image showing a coccolithophore bloom off south-western England (image source: http://ina.tmsoc.org/galleries/photodujour/source/cornwall-bloom_ehux.htm) (top right) and Haptophyta pigment concentration estimates across the world oceans. *(adapted from Liu* et al., *2009)*. See the colour plate.

cryptophytes and possibly mucosal structures of heterokont algae (Andersen, 2004; Cavalier-Smith, 2002; Daugbjerg & Andersen, 1997), supporting the hypothesis that these lineages form a supergroup termed the chromalveolates, with plastids originating from a single secondary endosymbiosis event (Cavalier-Smith, 1999). Data from plastid genes have generally supported the monophyly of chromalveolate lineages (e.g. Fast, Kissinger, Roos, & Keeling, 2001; Harper & Keeling, 2003), including evidence from a lateral gene transfer common to the plastids of haptophytes and cryptophytes (Rice & Palmer, 2006). However, the chromalveolate hypothesis also implies that host nuclear lineages are monophyletic, which has not been confirmed despite the use of substantial genetic data sets. Nuclear-based phylogenomics have consistently shown that the heterokonts (Stramenopiles) and Alveolates are closely related, forming a strongly supported group with Rhizaria, together constituting the so-called SAR (*S*tramenopiles, *A*lveolates, *R*hizaria) group (Burki *et al.*, 2007). Haptophytes generally branch together with cryptophytes, picobiliphytes and several heterotrophic groups (telonemids, centrohelids and katablepharids) in these analyses (e.g. Burki *et al.*, 2009). Based on congruent plastid and nuclear data, haptophytes and cryptophytes were proposed to be a distinct chromalveolate lineage, the Hacrobia (Okamoto, Chantangsi, Horak, Leander, & Keeling, 2009). A recent phylogenomic study based on alignment of 258 genes provided strong support for the hypothesis that haptophytes are sister to the SAR group, possibly together with telonemids and centrohelids, but that cryptophytes and katablepharids have a common origin and are not related to other hacrobians rather than branching with plants (Burki, Okamoto, & Keeling, 2011).

3.2. Haptophytes Diversity

The known diversity of extant haptophytes is relatively low compared to other ecologically predominant microalgal groups, with only ca. 400 extant species having been described (Jordan, Cros, & Young, 2004). Most haptophytes occur as solitary planktonic cells possessing two smooth flagella (i.e. completely lacking mastigonemes) in addition to the haptonema, but solitary non-motile planktonic or benthic cells, pseudofilamentous forms and colonies also exist. Most described haptophytes fall into the nanoplankton size class (cells 2–20 μm in diameter), but results of environmental molecular surveys indicate the existence of numerous taxa of very small (<2–3 µm) undescribed pico-haptophytes (Liu *et al.*, 2009; Moon-van der Staay *et al.*,

2000). Common ultrastructural characters of the group include the presence of plastids surrounded by four membranes, chloroplast lamellae consisting of three thylakoids without girdle (interconnecting) lamellae, tubular mitochondria and characteristic distension of Golgi vesicles in which organic scales are produced prior to being exported onto the cell surface via exocytosis. A number of haptophytes are known to undergo a haplodiplontic life cycle, with alternation between haploid and diploid phases both capable of independent asexual division, each phase characterized by distinct scale morphology.

Both morphological and molecular evidence support the division of the Haptophyta into two classes, the Pavlovophyceae and the Prymnesiophyceae (Edvardsen *et al.*, 2000). The likely existence of one or more haptophyte lineages that occupy an intermediate phylogenetic position between the two described classes has been revealed by analysis of molecular data from environmental surveys in marine (Shi *et al.*, 2009) and freshwater (Shalchian-Tabrizi, Reier-Roberg, Ree, Klaveness, & Brate, 2011; Slapeta, Moreira, & Lopez-Garcia, 2005).

The Pavlovophyceae contains only 13 described species classified in a single order, the Pavlovales (Fig. 1.3). Structural features common to all or most members of the Pavlovophyceae that distinguish them from the Prymnesiophyceae include the markedly anisokont (i.e. unequal in length) nature of the flagella and the relatively simple arrangement of microtubular and fibrous roots of the pavlovophycean flagellar-haptonematal basal complex (Hori & Green, 1994). The Pavlovophyceae are also known to synthesise certain specific sterols and conjugates, the pavlovols (Véron, Dauguet, & Billard, 1996; Volkman, Farmer, Barrett, & Sikes, 1997) and a unique photosynthetic pigment signature with unknown xanthophyll and two polar chlorophyll *c* forms (Van Lenning *et al.*, 2003). Intriguingly, some pavlovophytes possess an eyespot (Lee, 1999). The organic scales of Pavlovophyceae, when present, consist of small dense bodies ('knob scales') in contrast to the plate scales of the Prymnesiophyceae. The phylogenetic relationships between known Pavlovales were recently elucidated, leading to a taxonomic revision of the group (Bendif *et al.*, 2011).

The vast majority of the known diversity of haptophytes occurs in the Prymnesiophyceae, which comprises two orders of non-calcifying taxa, the Phaeocystales and the Prymnesiales, together with the coccolithophores making up a monophyletic sublineage (the subclass Calcihaptophycidae) containing four orders (Isochrysidales, Coccolithales, Syracosphaerales and Zygodiscales; Fig. 1.3).

The Phaeocystales contains a single genus, *Phaeocystis*, with less than 10 non-calcifying species, several of which have complex life histories involving solitary cells and colonies. The Prymnesiales contains two families of non-calcifying taxa, the Prymnesiaceae and the Chrysochromulinaceae (Edvardsen *et al.*, 2011), the organic plate scales of which are often relatively highly elaborated. The two families contain roughly equivalent numbers of species (30–40), but the Chrysochromulinaceae, which are known to be able to catch prey using their characteristically long haptonema (Kawachi *et al.*, 1991), appear to contain a massive undescribed diversity (possibly hundreds of genotypes) of extremely small taxa (Liu *et al.*, 2009). One recently described Prymnesiacean species produces siliceous scales (Yoshida, Noel, Nakayama, Naganuma, & Inouye, 2006).

The coccolithophores possess an exoskeleton of calcareous plates called coccoliths. Calcification occurs intracellularly in Golgi-derived vesicles, using organic base-plate scales as the substrate for crystal nucleation. Two main types of coccoliths exist: heterococcoliths, formed of a radial array of complex-shaped interlocking crystals units, and holococcoliths, constructed of numerous small, similar sized and simple-shaped calcite elements. Heterococcolith- and holococcolith-bearing taxa were originally thought to be morphologically and phylogenetically distinct species, but they are now known to be different phenotypes exhibited within the life cycle of coccolithophore species (typically, diploid life cycle stages bear heterococcoliths and haploid stages bear holococcoliths). Although the underlying structures of coccoliths are universal (Young, Didymus, Bown, Prins, & Mann, 1992), coccolith morphology is extremely diverse, with several morphological categories recognized (Young *et al.*, 2003). Coccolithophore taxonomy is almost exclusively based on comparison of coccolith morphology; however, molecular studies have demonstrated significant cryptic genetic diversity within several coccolithophore species (Saez *et al.*, 2003). Despite the existence of numerous theories (such as involvement in intracellular supply of CO_2 for photosynthesis, protection from predators or concentration of light towards plastids; see Young, 1994), the function of coccoliths remains unknown.

3.3. Haptophytes Evolution

Coccoliths are extremely abundant as microfossils, providing an outstanding tool for biostratigraphic dating and studies of evolution. The earliest appearance of coccoliths in the fossil record (corresponding to the origin of the calcareous haptophytes) is dated at ca. 220 Mya and several clear evolutionary transitions through the subsequent evolutionary history of coccolithophores have been accurately dated (Bown, 1998). This allows calibration of multiple nodes on phylogenetic trees and hence molecular clock analysis of the evolution of the group as a whole. A recent molecular clock study based on multigene analysis (nuclear 18S rRNA gene or SSU, 28S rRNA gene or LSU and plastid tufA and rbcL genes) estimated that the haptophytes diverged from other chromists in the Neoproterozoic Era ca. 824 Mya (1031–637 Mya) around the time of the onset of the Cryogenian 'snowball Earth' (Liu, Aris-Brossou, Probert, & de Vargas, 2010). In the same study, the divergence of the two extant haptophyte classes was estimated to have occurred 543 Mya, early in the Cambrian period that witnessed the most rapid and widespread diversification of life in Earth's history. The primary radiation within the Prymnesiophyceae (the divergence of Phaeocystales from other prymnesiophytes) was estimated to have occurred 329 Mya in the Carboniferous period, which was characterized by the presence of widespread shallow epicontinental seas. The timing of the next two divergences within the Prymnesiophyceae, that of the Prymnesiales and the primary radiation of the Calcihaptophycideae, both apparently followed important Earth system transitions early in the Permian and the Triassic, respectively (Liu *et al.*, 2010). Within the Calcihaptophycidae, extant coccolithophores appear to have diversified from a few lineages that survived the major extinction at the Cretaceous/Tertiary (K/T) boundary ca. 65 Mya, whereas non-calcifying haptophytes were not affected by the K/T extinction (Medlin, Saez, & Young, 2008). The adaptation of non-calcifying haptophytes to eutrophic coastal environments and their ability to switch nutrition modes from autotrophy to mixotrophy were posited as possible explanations for their survival during this abrupt global change event. Members of the Isochrysidalean family Noelaerhabdaceae have numerically dominated coccolithophore communities for the last 20 million years and continue to do so in modern oceans. The noelaerhadacean *E. huxleyi* dominates modern coccolithophore assemblages despite the fact that it is a very young species in geological terms, having originated only 220 Kya (Thierstein, Geitzenauer, Molfino, 1977).

3.4. Distribution and Ecology of Haptophytes

Haptophytes adopt diverse ecological strategies, being present in open-ocean, shelf, upwelling, coastal, littoral, brackish and freshwater environments. In addition to undertaking photosynthesis, many haptophytes are known to be capable of using particulate and/or dissolved organic food sources, the group probably therefore being predominantly mixotrophic (de Vargas, Aubry, Probert, & Young, 2007). Life cycle transitions, with each phase adapted to distinct ecological niches (Noel, Kawachi, & Inouye, 2004), may also be an integral part of the ecological strategy for different haptophytes.

Pavlovophytes have been described mostly from cultures isolated from littoral, brackish water and in some cases freshwater environments, and they are common components of near-shore planktonic and benthic microalgal communities in widespread locations. Due to the difficulty in identifying pavlovophytes in the light microscope, it is not clear whether they commonly occur in open-ocean environments. *Phaeocystis* is ubiquitous from poles to tropics and from coastal to open ocean waters and certain species regularly produce extensive blooms, notably in Arctic, Antarctic and North Sea waters. *Phaeocystis* blooms can be detrimental to the growth and reproduction of shellfish and zooplankton, and hemolytic and toxic effects have been reported on fish (Schoemann, Becquevort, Stefels, Rousseau, & Lancelot, 2005). Some Prymnesiales species also form periodic blooms in coastal waters, occasionally harmful to fish and other biota. The Prymnesiaceae tend to be restricted to coastal waters, whereas the Chrysochromulinaceae also thrive in oligotrophic open ocean regions (Liu *et al.*, 2009) where mixotrophic nutrition is likely an important strategy. Through the process of calcification and export of calcite to the deep ocean following cell death, coccolithophores play a key role in the 'biological pump' that contributes significantly to global carbon cycling (Rost & Riebesell, 2004). Massive annual blooms of *E. huxleyi* in temperate and subpolar coastal and shelf environments are visible in satellite images (Fig. 1.3). Among the factors that may contribute to the ecological success of *E. huxleyi* are high affinities for inorganic nutrient uptake (Paasche, 2002) and physiological mechanisms for maintaining growth under high irradiance (Loebl, Cockshutt, Campbell, & Finkel, 2010; Ragni, Aris, Leonardos, & Geider, 2008). Specific viruses play an important role in regulation of *E. huxleyi* blooms (Wilson *et al.*, 2002). Although it does not form bloom populations in subtropical and tropical environments, *E. huxleyi* (and the closely related species *Geophyrocapsa oceanica*) is widespread and relatively abundant in these ecosystems. The Coccolithales genus *Coccolithus* includes relatively large cells (ca. 20 μm) that, together with *E. huxleyi*, dominate coccolithophore communities in subpolar and temperate regions of the North Atlantic as well as occurring in western-margin upwelling zones around the globe. The Coccolithales, Syracosphaerales and Zygodiscales include species that contribute significantly to communities in warm water mesotrophic settings, including *Calcidiscus*, *Umbilicosphaera*, *Syracosphaera*, *Helicosphaera*, *Discosphaera*, *Rhabdosphaera* and *Scyphosphaera*. The genus *Syracosphaera* is notably very diverse in mesotrophic to oligotrophic coccolithophore communities. *Umbellosphaera*, a genus of uncertain phylogenetic affinities, is abundant in oligotrophic surface layer communities. The coccolithophores *Florisphaera* and *Gladiolithus* are important members of deep-photic zone communities, but the adaptations that allow them to thrive in conditions of extremely low light are unknown.

4. THE MULTIFACETED PHYTOPLANKTON: THE DINOFLAGELLATES

4.1. Dinoflagellates as Members of the Alveolate Lineage

The Alveolata constitutes a diverse group of single-celled eukaryotes present in both marine and terrestrial ecosystems, the principal shared morphological feature of which is the presence of flattened vesicles (cortical alveoli) packed into a continuous layer supporting the cell membrane (Cavalier-Smith & Chao, 2004). These structures have been associated by immunolocalization to a family of proteins, named alveolins, common to all alveolates (Gould, Tham, Cowman, Mcfadden, & Waller, 2008). Alveolates exhibit extremely diverse trophic strategies, including predation, photo-autotrophy and intracellular parasitism. Most alveolates fall into one of three main subgroups (or phyla): ciliates, dinoflagellates and Apicomplexa, all sharing a common ancestor (Leander & Keeling, 2003). Apicomplexans are obligate parasites of animal cells, including humans (e.g. *Plasmodinium* which causes malaria). Ciliates are mainly aquatic predators that perform essential roles as consumers in microbial food webs, although some taxa can be parasitic or may contain sequestered plastids (e.g. *Myrionecta rubra*). Dinoflagellates are either photo-autotrophs, free-living predators (heterotrophs) or both, either simultaneously or alternatively (mixotrophs), while some also live as parasites or symbionts (Taylor, Hoppenrath, & Saldarriaga, 2008). Other lineages of marine alveolates (MALV) have been identified in culture independent surveys and associated to the class Syndinea (order Syndiniales). These are divided into two main groups, MALV-I and MALV-II, the phylogenetic placement of which is still uncertain (Massana & Pedros-Alio, 2008). Both groups have been proposed to correspond to heterotrophic parasites of marine organisms (Guillou *et al.*, 2008; Gunderson, John, Boman, & Coats, 2002).

Dinoflagellates and Apicomplexa are the only groups of alveolates that possess plastids. These structures are actively involved in photosynthesis only in dinoflagellates, while they are vestigial in apicomplexans (Mcfadden, Reith, Munholland, 1996). This common character between these two groups suggests that their common ancestor was photosynthetic (Leander & Keeling, 2003). In contrast to other alveolates, dinoflagellates can develop cellulose-like polysaccharide plates within the cortical alveoli, forming a theca. The arrangement and ornamentation of these plates leads to an astonishing variety of shapes, the description and comparison of which forms the traditional basis of species classification for dinoflagellates.

4.2. Dinoflagellates Diversity

The majority of dinoflagellates (perhaps 80%) are free-living marine planktonic or benthic flagellates, the remainder inhabiting equivalent freshwater habitats (Taylor *et al.*, 2008). Half of the estimated 2000 extant species of dinoflagellates are considered photosynthetic. Dinoflagellates contain a characteristic nucleus with permanently condensed chromosomes (the dinokaryon). The haploid motile stage within dinoflagellate life cycles typically possesses two dissimilar flagella: a ribbon-like flagellum with multiple waves situated in a transverse groove (cingulum) and a more conventional flagellum emerging from a ventral furrow (sulcus) which beats posteriorly to the cell. After sexual recombination, dinoflagellates produce resistant diploid benthic stages (hypnozygotes, also termed resting cysts) to escape predation and adverse environmental conditions as well as to colonize ecosystems. Some dinoflagellates are noxious to humans and other marine organisms due to the production of potent toxins, while others cause hypoxia, anoxia or mechanical damage to marine fauna. Finally, certain dinoflagellates, including some toxin-producing species, can form large blooms ('red tides') that can negatively affect economic activities in coastal areas (Hallegraeff, 2003, 2010) (Fig. 1.4).

Nine major orders (Gonyaulacales, Peridiniales, Gymnodiniales, Suessiales, Prorocentrales, Dinophysiales, Blastodiniales, Phytodiniales and Noctilucales) are recognized within the dinoflagellates (Fig. 1.4), the order Thoracosphaerales being suspected to belong to the Peridiniales (Saldarriaga, Taylor, Cavalier-Smith, Menden-Deuer, & Keeling, 2004). Dinoflagellate orders can be distinguished on the basis of major morphological characters

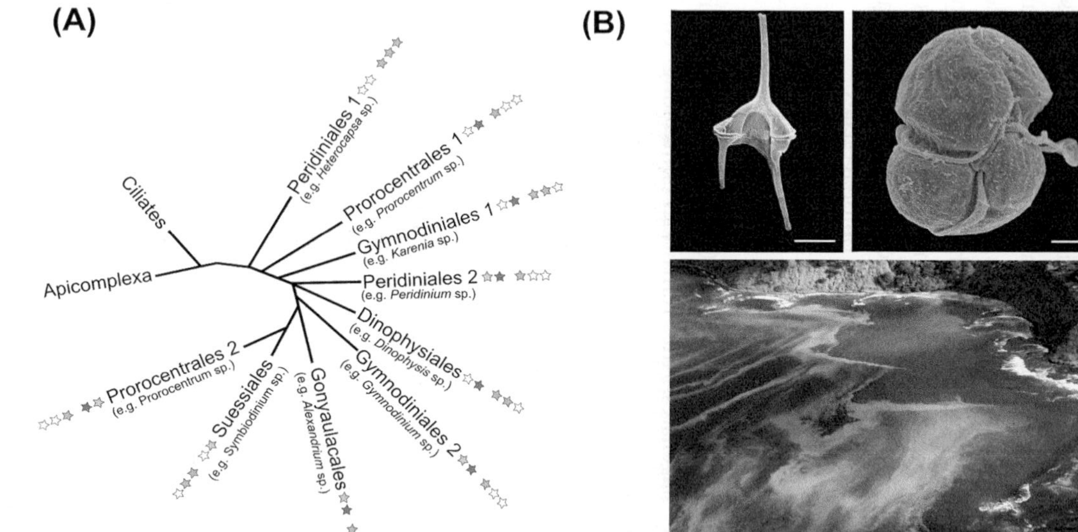

Figure 1.4 A) Legend as Figure 1.2 A, but for dinoflagellates. (B). Scanning Electron Microscopy (SEM) micrographs of *Neoceratium candelabrum*, a non-toxic thecate micro-dinoflagellate (scale bar = 30 μm), a species of the Gonyaulacales lineage (top left) and *Karlodinium veneficum*, belonging to lineage Gymnodiniales 1, an athecate nano-dinoflagellate (scale bar 2 μm) that is toxic for a range of marine invertebrates and fish (top right). Red discoloration caused by the dinoflagellate *Noctiluca scintillans* off Waiheke Island (New Zealand) (http://www.TeAra.govt.nz/en/plankton/1/4) (bottom). See the colour plate.

and life cycle features of their members (Table 1.1). The Gonyaulacales and Peridiniales are characterized by the presence of cellulose-like thecal plates within the cortical alveoli. The thecae of both groups are constituted of five latitudinal series of plates (apical, anterior intercalary, precingular, postcingular and antapical) plus the cingular and sulcal series. The two orders were separated by Taylor (1980). Members of the order Blastodiniales have

Table 1.1 Main Morphological Features for Major Orders of Dinoflagellates

Order	Main morphological features
Gonyaulacales	Cellulose-like thecal plates within the cortical alveoli
	Left-handed torsion of the epitheca
	Small anterior intercalary plates (often absent)
	Apical pore complex often with a hook-like groove
	Asymmetric antapical plates
Peridiniales	Cellulose-like thecal plates within the cortical alveoli
	Bilateral symmetry
	Anterior intercalary plates
	Small apical pore
	Two subequal antapical plates
Gymnodiniales	Cortical alveoli without thecal plates (athecate, unarmored, naked dinoflagellates)
	Apical furrow on the cell apex
Suessiales	Cortical alveoli without thecal plates (athecate, unarmored, naked dinoflagellates)
	Generally 7—10 longitudinal series of cortical alveoli (less than in Gymnodiniales)
	Elongated apical vesicle on the cell apex
Prorocentrales	Division of theca into lateral halves joined by a sagittal suture
	Lack of the sulcus and the cingulum
	Two pores, the flagella emerging from the larger one
	Tiny periflagellar platelets
Dinophysiales	Division of theca into lateral halves joined by a sagittal suture
	Two pores, the flagella emerging from the larger one
Blastodiniales	Temporary dinokaryon
	Parasitic lifestyle
	Dinospores with peridinioid plate tabulation
Phytodiniales	Shift from a non-calcareous coccoid cell or continuous-walled colonial stage to a vegetative stage
Noctilucales	Highly mobile ventral tentacle
	Lack (at least in some life stages) of the ribbon-like transverse flagellum or the condensed chromosomes of the dinokaryon

a parasitic lifestyle and a temporary dinokaryon, which fostered the hypothesis that Blastodiniales diverged early from the dynokaryotic dinoflagellate lineage (Saldarriaga *et al.*, 2004). However, dinospores of the genus *Blastodinium* have been shown to have peridinioid plate tabulation, suggesting affiliation to the Peridiniales. Although the genus *Blastodinium* has been demonstrated to be late-branching, consistent with Peridiniales evolution, the affiliation of the genus and of the whole order of Blastodiniales to Peridiniales is still to be proved (Skovgaard, Massana, & Saiz, 2007). The Prorocentrales and the Dinophysiales share a major synapomorphic feature, unique within dinoflagellates: the division of theca into lateral halves joined by a sagittal suture. The order Phytodiniales includes species characterized by a shift from a non-calcareous coccoid cell or continuous-walled colonial stage to a vegetative stage. Similar life shifts have, however, also been observed in genera of other orders (e.g. Suessiales [*Symbiodinium*], Gonyaulacales [*Pyrocystis*]). The Phytodiniales presently includes poorly understood genera for which little molecular data are available. The Noctilucales is an early-diverging order that includes aberrant dinoflagellates characterized by a highly mobile ventral tentacle, which is missing in typical dinoflagellates and other alveolates. The tentacle does not play a role in keeping the cell in suspension but seems rather related to food capture. The Noctilucales have the ability to incorporate, replace or lose chloroplasts, a rare phenomenon in other alveolate groups. Whether these chloroplasts are kleptoplastids or derive from ancient endosymbiosis, as in other dinoflagellate families, remain to be demonstrated (Gomez, Moreira, & Lopez-Garcia, 2010).

4.3. Dinoflagellates Evolution

To date, comparison of morphological, cytological and nuclear genetic markers (18S rRNA gene, 28S rRNA gene, ITS rDNA) has not clearly resolved relationships between dinoflagellate orders. This raises the question as to whether all dinoflagellate orders emerged about the same time during a major radiation period (Hoppenrath & Leander, 2010). Phylogenetic studies carried out using plastid genes (*rbcl*, *atpB*) are limited since many dinoflagellates are heterotrophic, and mitochondrial genes (*cob* and *cox* 1) are only useful in combination with other genetic markers and for specific groups of dinoflagellates (Zhang, Bhattacharya, & Lin, 2005). Some major phylogenetic traits can, however, be identified for dinoflagellates (Fig. 1.4). The genus *Oxyrrhis* is often considered a predinoflagellate lineage (Saldarriaga, Mcewan, Fast, Taylor, & Keeling, 2003), followed by *Noctiluca scintillans* (order Noctilucales), which now appears more closely related to Syndinea (Gomez *et al.*, 2010). The Dinophysiales (Gomez, Lopez-Garcia, & Moreira, 2011), Suessiales (Siano, Montresor, Probert, Not, & De Vargas, 2010) and Gonyaulacales (Saldarriaga *et al.*, 2004) are strongly supported holophyletic groups. The Prorocentrales splits into two branches in both 18S rRNA and 28S rRNA genes phylogenies (Saldarriaga *et al.*, 2004), but not in mitochondrial *cox*-1 topology (Murray, Ip, Moore, Nagahama, & Fukuyo, 2009). *Cox*-1 demonstrates the monophyly of the morphologically very cohesive group of prorocentroid dinoflagellates, but it separates the group into two clades that include species present, respectively, in the two branches of the 28S rRNA gene phylogeny. The Gymnodiniales is a polyphyletic order and together with the Peridiniales are the most evolutionary complex groups of dinoflagellates. The *Karenia/Karlodinium* clade separates from other gymnodinioid clades (e.g. *Akashiwo*, *Gymnodinium*, *Amphidinium* clades) in both 18S and 28S rRNA gene phylogenies (Murray, Jorgensen, Ho, Patterson, & Jermilin, 2005; Saldarriaga *et al.*, 2004). The Peridiniales appears to be a complex paraphyletic group of dinoflagellates, that is, its phylogeny comprises non-Peridiniales branches. The *Heterocapsa* clade often has a basal position to other peridinioids and in general to other thecate dinoflagellates in phylogenies inferred from the 18S and 28S rRNA genes, despite having mediocre branching support. Other Peridiniales branches (*Peridinium*, *Scrippsiella* and *Protoperidium*) branch later in phylogenetic trees. General dinoflagellate phylogenies still require molecular data for many dinoflagellate genera (especially for heterotrophic species), taxonomic revision of some species and identification of species likely to correspond to missing branches of phylogenetic trees (Fig. 1.4).

Considerable diversity of chloroplast types and pigment composition occurs in photosynthetic dinoflagellates, acquired through secondary and tertiary endosymbioses (Cavalier-Smith, 1999) with in some cases multiple losses and replacement of plastids, as revealed by molecular phylogenetic analysis (Saldarriaga, Taylor, Keeling, & Cavalier-Smith, 2001). Some dinoflagellates harbor foreign plastids that are periodically lost and gained during their life cycle (kleptoplastidy, from the Greek 'klepto' – stealing), or bear photosynthetic endosymbionts that are kept for longer periods but not fully integrated (Moestrup & Daugbjerg, 2007). Beyond chlorophyll *a*, *c2* and β-carotene, dinoflagellates with permanent plastids can have three other accessory pigments: peridinin (Gonyaulacales, Peridiniales, Prorocentrales, Suessiales and some Gymnodiniales), fucoxanthin or fucoxanthin derivatives (the family Kareniaceae of the order Gymnodiniales) and chlorophyll *b* (the genus *Lepidodinium* of the order Gymnodiniales). The toxin-producing genus *Dinophysis* (Dinophysiales) is a peculiar case. It is still debated whether *Dinophysis acuminata* has permanent plastids of cryptophyte origin (Garcia-Cuetos, Moestrup, Hansen, & Daugbjerg, 2010) or whether it maintains a temporary plastid (kleptoplast) acquired from prey (Wisecaver & Hackett, 2010). The latter hypothesis is supported by the fact that *Dinophysis* species can only be cultured using the ciliate *Myrionecta rubra* as prey, which itself feeds on cryptophytes and would thus be the source for *Dinophysis* of the cryptophytes and their chloroplasts (Park *et al.*, 2006). The fact that *Dinophysis* needs the ciliate as an intermediary to acquire cryptophyte plastids suggests that it probably lacks critical enzymes to initially process the prey or maintain their plastids (Wisecaver & Hackett, 2010). Future transcriptomic and genomic sequencing would help up to understand if this dinoflagellate possesses a photosynthetic machinery and how it regulates genes that coordinate photosynthetic activity and/or prey capture.

4.4. Ecology of Dinoflagellates

Photosynthetic dinoflagellates are common and abundant in pelagic and benthic habitats of both marine and freshwater ecosystems. Typically, they reach their highest abundances in estuaries and coastal marine waters, in concomitance with high nutrient supply from land sources and/or deep water upwelling. Blooms of noxious species (HABs) are more common under these conditions (Fig. 1.4). Using their two perpendicular flagella, dinoflagellates exhibit directed movement in response to chemical stimuli, physical variations, gravity and light. Due to this motility, dinoflagellates are able to find optimal conditions for growth and survival under high physical disturbance (turbulence and shear forces), intense light stress and nutrient limitation. Because dinoflagellates have various habitat preferences, multiple life strategies and are nutritionally versatile, they are very competitive with other groups of protists for resource acquisition. Although they can sometimes be important in terms of biomass, both micro- and nano-dinoflagellates are rarely reported to dominate in terms of abundance within the phototrophic fraction of plankton communities (Siokou-Frangou *et al.*, 2010). This is probably partly due to the fact that quantitative information is very fragmentary for dinoflagellates because of inadequate identification methodologies, notably for athecate species.

In coastal waters, bloom initiation can be due to germination of vegetative cells from hypnozygotes. The biological and physical factors that trigger bloom initiation are poorly known for most dinoflagellates, including harmful species (Burkholder, Azanza, & Sako, 2006). During bloom development, many dinoflagellate species are capable of rapid growth, attaining abundances up to 10^9 cells/L (up to 400–500 μg Chl*a*/L) (Taylor & Pollingher, 1987). Dinoflagellates can grow at rates of up to 3.5 divisions per day, but only 15% of the larger free-living harmful species have growth rates greater than 1.0 division per day (Smayda, 1997). Photosynthetic dinoflagellates are primarily limited by phosphorous and nitrogen, although they can store these nutrients in a species-specific way that in some cases allows one species to outcompete others (Graham & Wilcox, 2000; Labry *et al.*, 2008). As for other phytoplankton taxa, micro-nutrients, including forms of selenium and iron, have been shown to influence blooms of some harmful phototrophic dinoflagellates (Boyer & Brand, 1998; Doblin, Blackburn, & Hallegraeff, 2000). Apart from nutrient limitation, bloom termination can be caused by water dispersion and dilution, zooplankton grazing and biological endogenous cycles, as well as parasite (Chambouvet, Morin, Marie, & Guillou, 2008) and viral (Nagasaki, Tomaru, Shirai, Takao, & Mizumoto, 2006) infections.

Field observations indicate that bloom-forming dinoflagellate species have neither strict habitat preferences nor uniform responses. Smayda and Reynolds (2001) recognized nine different pelagic habitats where dinoflagellates bloom, arranged along an onshore–offshore gradient of decreasing nutrients, reduced mixing and deepening of the photic water layer. Each of the nine types of habitat is characterized by a specific dinoflagellate life-form, which suggests that dinoflagellates have evolved multiple adaptive strategies, rather than a common ecological strategy. Subsequently, these authors introduced five rules of assembly for marine dinoflagellate communities, which state that specific habitat conditions correspond to specific life forms that are mainly selected on the basis of abiotic factors (turbulence and nutrient availability). Within the species pool of a given habitat, seasonal succession is stochastic and characterized by a high degree of unpredictability (Smayda & Reynolds, 2003). Conversely, for cyst-forming dinoflagellates, the existence of endogenous or exogenous factors determining the presence and succession of species has been hypothesized, suggesting that the apparent random succession of species within a pool of species is understandable and predictable (Anderson & Rengefors, 2006).

The ecological role of dinoflagellates in the functioning of marine ecosystems and in the marine food web can be significant. Some heterotrophic species of the genera *Ornithocercus*, *Histioneis* and *Citharistes* can host photosynthetic endosymbionts. In autumn in oligotrophic subtropical waters of the Gulf of Aqaba (Israel), peaks of these species coincided with extended nitrogen limitation and high abundances of free-living N-fixing cyanobacteria. It has been proposed that heterotrophic dinoflagellate hosts may provide cyanobacterial symbionts with the anaerobic microenvironment necessary for efficient N fixation, which would in turn determine the ecological success of the hosts (Gordon, Angel, Neori, Kress, & Kimor, 1994). This symbiotic relationship as well as the capacity of N fixation by the cyanobacterial symbionts has been suggested (Foster, Carpenter, & Bergman, 2006). Alternatively, some dinoflagellates are symbionts of benthic organisms (*Symbiodinium*) (Freudenthal, 1962) or of pelagic protists (*Pelagodinium*) (Siano *et al.*, 2010). Many, if not most, photosynthetic dinoflagellates are considered mixotrophic (Smalley & Coats, 2002). It has been suggested that the larger size and lower cell surface-to-volume ratios of dinoflagellates generally result in lower affinities for dissolved nutrients than smaller protists, therefore positively selecting for mixotrophy among photosynthetic species (Smayda, 1997). However, mixotrophy is often difficult to assess clearly because of low feeding rates, intermittent feeding dependent on conditions poorly simulated in cultures, specificity of prey or the fact that organelles can obscure food vacuoles (Stoecker, 1999).

5. THE SILICEOUS PHYTOPLANKTON: THE DIATOMS

Diatoms, also called Bacillariophyceae, are unicellular photoautotrophic stramenopiles, the defining feature of which is the compound silica cell wall, called a frustule. The diatoms constitute one of the most diverse lineages of eukaryotes with possibly over 100,000 extant species (Mann & Droop, 1996). They are ubiquitous in marine and freshwater habitats and in damp terrestrial environments. It is therefore not surprising that diatoms have been studied intensively ever since microscopes became available. We review here the diversity and ecology of diatoms and refer to Chapter VII of this volume for a review on diatom genomics (Mock & Medlin 2012).

5.1. The Hallmark of the Diatom: The Silica Cell Wall

The diatom frustule is composed of two overlapping thecae (the larger called the epitheca and the smaller the hypotheca), each of which consists of a valve and an accompanying series of girdle bands. These frustule elements contain rows of pores, called interstriae, with ribs between them, called striae. The pores in the interstriae form the principal conduits for the uptake of nutrients and exudation of metabolites. Girdle bands are architecturally relatively simple and always consist of a single layer, whereas valves are more elaborate, with a flat area, called the valve face, and a rim, called the mantle. Valves are either composed of a single layer or two layers. In the latter case, the internal layer has larger pores and the two layers are connected via perpendicular cross walls in a rectangular or honeycomb pattern, giving rise to chambered valves (see Round, Crawford, & Mann, 1990).

Construction of new silica cell wall elements proceeds in silica deposition vesicles (SDVs) near the plasma membrane. Dissolved silicic acid is actively taken up from water and concentrated in the cytoplasm far above the level at which silica would normally polymerize. Precipitation in the cytoplasm is avoided because silicic acid transporter proteins bind the silica and shepherd it through the cytoplasm into the SDV. Other classes of peptides, including silaffins, silacidins and long-chain polyamines, are produced in the endoplasmic reticulum, transported in vesicles to the Golgi apparatus where they are modified and activated, then transported into the SDV where they form a matrix onto which the super-saturated silica precipitates in an amorphous form (Hildebrand, 2008; Kroger & Poulsen, 2008).

During vegetative cell division, new thecae are laid down in such a way that the cell is covered completely by silica cell wall elements throughout the division phase. Newly formed thecae must therefore be formed within the confines of existing thecae. The daughter cell inheriting the parental epitheca (the larger half of the frustule) forms a new hypotheca of the same size as that of the parental cell and hence this daughter cell is of the same size as the parent. By contrast, the daughter cell inheriting the parental hypotheca (the smaller half of the frustule) uses it as an epitheca within which a new hypotheca is formed; hence, this daughter cell is slightly smaller than its parent. Consequently, average cell diameter diminishes with ongoing mitotic division. The only escape from this continual miniaturization is sexual reproduction (see Section 5.3).

5.2. Diatom Diversity

Diatoms are categorized into centrics and pennates based on characteristics of valves, including shape, ultrastructure, ornamentation and types of processes (tubes, slits) (Fig. 1.5). The principal difference is the way in which the interstriae are organized. Centric diatoms possess radially organized valves with striae radiating from a central region or ring, whereas pennates possess elongated valves with striae oriented perpendicular to a midrib (called

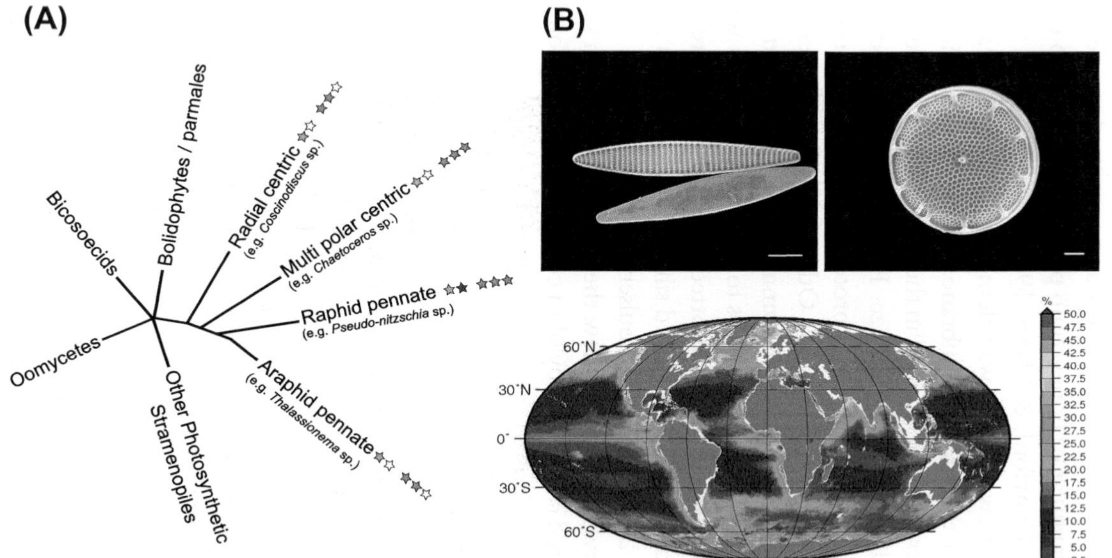

Figure 1.5 A) Legend as Figure 1.2 A, but for diatoms. (B) Pennate diatom *Fragilariopsis kerguelensis*, top valve: interior side, bottom valve: exterior side, scale bar 10 μm (by Marina Montresor) (top left), centric diatom *Thalassiosira tealata* showing valve exterior with 10 peripheral strutted processes, a central strutted process and a peripheral labiate process, scale bar 1 μm (by Diana Sarno) (top right), and diatom pigment concentration estimates across the world oceans (*adapted from Liu* et al., *2009*). See the colour plate.

a sternum), like in a feather (Round *et al.*, 1990). Electron microscope illustrations of ultrastructural details of diatoms can be found in Round *et al.*

One group of centric diatoms, the radial centrics, possesses valves shaped like Petri dishes. Radial centrics usually possess a ring of so-called labiate processes around their valve mantle, forming tubes that may enable intake or secretion of organic material. Important planktonic representatives of radial centrics include *Aulacoseira*, *Corethron*, *Coscinodiscus*, *Leptocylindrus*, *Melosira* and *Rhizosolenia*. A second group of centrics, the so-called multipolar centrics (bi-, tri-, etc., referring to polarity of shape) also exhibit a radial pore organization, but their cell form is usually elongate, triangular or starlike; that is, exhibits polarity. Labiate processes, if present, are located on the central area of valves. Many species possess fields of densely packed pores at their valve apices. Mucilage is exuded through these apical pore fields, enabling benthic or epiphytic species to attach to the substratum or to form chains. Planktonic representatives such as *Eucampia* also form chains this way. Secondarily radial centric genera such as *Lauderia*, *Porosira*, *Skeletonema* and *Thalassiosira* (Order Thalassiosirales) possess specialized tubes in their valves, called strutted processes, through which chitin filaments are exuded. These filaments extend into the surrounding medium and link cells into chains. *Chaetoceros* and *Bacteriastrum* (Order Chaetocerotales) form chains by means of setae; thin hollow silica tubes that are formed following cell division. Setae probably evolved principally as a grazer deterrent, but they have acquired additional functions. In some species, plastids migrate up and down setae. Chaetocerotales and Thalassiosirales are highly diverse and their species form important constituents of phytoplankton blooms.

Pennate diatoms are elongate, their defining feature being the midrib from which striae and interstriae extend perpendicularly. The pennates are subdivided into raphid pennates, which possess a slit-shaped process, called a raphe, and araphid pennates, which lack a raphe, but have apical pore fields and apical labiate processes. The raphe slit enables raphid pennates to move actively by means of a cost-effective system of traction. Long-chain organic compounds protrude partially through one end of the slit and connect to the substratum. Subsequently, the chain is pulled along the slit and then exuded. Consequently, the diatom moves in the opposite direction. Most raphid pennates are benthic, but a few, including *Fragilariopsis* and *Pseudo-nitzschia*, have secondarily acquired a planktonic lifestyle. The araphid pennates are also typically benthic, but there are also several lineages that have acquired a planktonic lifestyle, including *Asterionella*, *Asterionellopsis*, *Lioloma*, *Thalassionema* and *Thalassiothrix*.

Arguably most diatom diversity is not planktonic, but benthic or epiphytic. Many species occur unattached, but their frustules are robust and way too heavy for a planktonic existence, thus abounding drifting over sandy bottoms. Diatom diversity shows several clades and grades containing planktonic species, but these are most often rooted in benthic ancestry. This versatility between planktonic and benthic existence in the evolutionary history of the diatoms means that traits acquired in a benthic setting may later have provided a benefit in the plankton. For example, raphid pennate diatoms are typically found in benthic habitats where they use their raphe to move actively over the substratum. *Pseudo-nitzschia*, a planktonic representative that evolved from benthic ancestry, use their raphe to enable daughter cells to slide along each other adjacent valve faces to assume a position in which they are just attached at their valve apices. In this way, *Pseudo-nitzschia* has acquired a novel way to form chains in the plankton.

5.3. Diatom Life Cycle

The diatoms possess a diplontic life cycle consisting of a long period (up to several years) during which diploid cells divide mitotically and a brief period (a few days) during which sexual reproduction takes place. Gametes fuse to form a zygote, which inflates to generate an auxospore. The new frustule elements of the resultant vegetative cell are formed within the confines of the auxospore wall and once this is completed the cell emerges from the auxospore (Round *et al.*, 1990). Centric diatoms form non-motile macrogametes and flagellated microgametes. In radial centrics and Thalassiosirales, the zygote inflates isometrically and forms an organic wall in which, in some species, small silica elements are embedded. In multipolar centrics, gamete formation and conjugation proceed as in radial centrics, but zygote inflation is anisometric; in this case, a series of silica bands, together called a properizonium, is laid down in sequence to mould the expanding auxospore into a bi- or multipolar shape. Pennate diatoms are isogamous (= produce gametes of equal size, though generally not of equal behaviour) and usually dioecious. Sexual reproduction involves alignment of cells of opposite mating types, gamete formation, migration in an amoeboid fashion to the partner gametes and zygote formation. Zygote inflation is constrained by a double series of perizonial bands into a cigar-like auxospore. The new frustule is formed within the confines of this perizonium. For pennates, it is unclear what provokes sexual reproduction, how potential mates detect each other and how and in what order they initiate the process of gametogenesis.

Several centric diatoms form resting spores (Ishii, Iwataki, Matsuoka, & Imai, 2011; Kooistra *et al.*, 2010). Spores are formed usually in response to deteriorating conditions. Resting spores can remain viable for extended periods (Härnström, Ellegaard, Andersen, & Godhe, 2011) and are often so robust that they exist as fossil markers (Suto, 2006). Auxospores are not resting cysts; their silica elements are flimsy and disintegrate and dissolve soon after the initial cell hatches.

5.4. Diatom Evolution

Results of molecular phylogenetic studies have generally revealed that centrics are the most ancient group of diatoms, forming a grade (= paraphyletic group). Multipolar centrics evolved from radial centric ancestry, but they form a grade because all pennates form a clade within this group. Within the pennates, the raphid pennates form a clade inside a grade of araphid pennates (Fig. 1.5).

This phylogenetic pattern suggests that the properizonial bands of multipolar centric auxospores and the perizonial bands of pennate auxospores are homologous structures and that these bands were acquired in the last common ancestor of the clade containing all multipolar centrics and pennates. The formation of these bands during auxospore formation enabled diatoms to generate shapes diverging from those resembling Petri dishes and tubes. Moreover, the results suggest that isogametogenesis is a derived trait of all pennates and that it evolved from oogenesis of their multipolar ancestor, possibly because it is a more efficient mode of sexual reproduction in benthic environments. The organization of the pennate valve with its midrib evolved from a radial organization mode. The advantage of such a midrib may have been the structural reinforcement it provides in lightweight but elongated diatom valves. Finally, the raphe of raphid pennates evolved from the apical labiate processes of their araphid pennate ancestry (Kooistra, Gersonde, Medlin, & Mann, 2007).

Each of these acquisitions apparently resulted in rapid diversification of the lineage exhibiting the novelty. Radial centrics appear to consist of a restricted number of remnant lineages, as do most multipolar centrics. Within the latter, the Thalassiosirales with their chitin threads and the Chaetocerotales with their setae constitute two highly diverse clades. The pennates as a whole form a huge clade with a poorly resolved basal structure and the same is true for raphid pennates that constitute a relatively novel lineage, but by far the most diverse group. Nonetheless, radial centrics do not constitute an evolutionary dead-end. Many radial centrics, such as *Leptocylindrus*, are important bloom formers in coastal regions all over the world.

Diatoms possess an extensive fossil record. Frustule elements of vegetative cells and spores are preserved, often in exquisite detail, over millions of years (e.g. Gersonde & Hardwood, 1990; Ishii *et al.*, 2011; Suto, 2006). In addition, biomarkers specific for particular diatom lineages, for instance *Rhizosolenia*, are detectable in petroleum of definable age (Sinninghe-Damsté *et al.*, 2004). These sources of information reveal that radial centrics first appeared in the Jurassic, multipolar centrics in the Early Cretaceous, pennates in the Late Cretaceous and the first raphid pennates at 55 Mya in the Paleogene. This order of appearance corroborates the above-mentioned phylogenetic patterns in extant diversity. The diatoms apparently traversed the K/T boundary relatively unscathed, with diversity subsequently increasing (Harwood, Chang, & Nikolaev, 2004; Stoermer & Smol, 1999).

5.5. Diatom Ecology

Diatoms are ubiquitous in the plankton and benthos of marine and freshwater habitats. Centric diatoms are typically marine, but a few exclusively freshwater genera exist and a few large marine genera have freshwater representatives (e.g. *Thalassiosira*). Araphid pennates are predominantly marine, but there are also many freshwater genera and some marine genera have freshwater representatives. Raphid pennates seem to be well represented in both freshwater and marine environments and many genera have representatives in both. Because of their abundance in shallow coastal seas, it has been estimated that planktonic diatoms account for as much as 20% of global photosynthetic fixation of carbon (~20 Pg carbon fixed per year; Mann, 1999), which is more than all the world's tropical rainforests, and that their contribution to nutrient cycling is significant (Boyd *et al.*, 2000). Planktonic diatoms are particularly important bloom-formers in nutrient-rich coastal regions, especially in temperate zones and upwelling zones in warm-temperate and tropical regions. Two characteristics render them particularly well suited to such environments. Firstly, they are well adapted to growth in deeply mixed turbulent water, where cells are only intermittently exposed to high light levels (Falkowski & Raven, 1997). Secondly, they are well adapted to pulsed availability of nutrients because they can use their often large central vacuole for nutrient storage. Diatoms are heavily grazed upon by a plethora of herbivores, in particular copepods. Diatoms seem to have few viral adversaries, but they are subject to attack from a range of protistan parasites such as oomycetes and thraustochytrids (Hanic, Sekimoto, & Bates, 2009). Defense mechanisms against biological pressures include structural reinforcement of frustule elements with double layers and internal struts and buttresses, such as in the pennate diatom *Fragilariopsis* and in many radial centric diatoms, setae in *Chaetoceros* and *Bacteriastrum*, barbed spines in *Corethron*, or barbed spine-like extensions of the valve as in *Asterionellopsis*. Other strategies include biochemical defense by means of production of metabolites that become bioactive when the diatom cell is disrupted by grazers. Biological attack results in a large part of diatom production being rapidly remineralized in surface waters. Nonetheless, a considerable part of the primary production of coastal diatom blooms sinks. The bulk of this organic material is re-mineralized in deeper water, but given high sedimentation rates in coastal regions, a small part is trapped in sediments. In this way, coastal planktonic diatoms contribute to about half of total long-term organic carbon sequestration in the marine environment. This is why petroleum is found in present and past river plume sediments along so-called passive continental fringes (Brazil, the North Sea, Caspian Sea, Venezuela, Gulf of Mexico and the Middle East).

Diatoms are also important constituents of phytoplankton communities in the Southern ocean. In this region, diatom cell densities are typically low, mainly because of restricted availability of iron, but population sizes are nevertheless considerable given the extent of the region. Southern ocean diatoms possess well-developed physical grazer defenses in the form of spines (e.g. *Corethron*, *Chaetoceros*) or chambered and structurally reinforced valve elements (e.g. *Fragilariopsis*). Diatoms also abound (but do not dominate) in the deep chlorophyll maximum of warm open oceans. In these conditions, species are generally minute and do not form chains, but their diversity is poorly known; hence, it is here that molecular taxonomic surveys might uncover considerable new diversity. Diatoms do not contribute significantly to long-term carbon sequestration in any of these open oceanic communities.

6. LAST, BUT NOT LEAST RELEVANT: OTHER PHYTOPLANKTON TAXA

6.1. Stramenopiles Other Than Diatoms

Within the stramenopile lineage, a number of other phytoplankton groups exist besides diatoms (Fig. 1.1). Although less diverse, some of these groups have important ecological roles in marine ecosystems. They are usually flagellated cells with heterokont characteristics, that is, two unequal flagella, one being ornamented with hair-like structures called mastigonemes. They possess plastids acquired through secondary endosymbiosis, typically with chlorophylls *a* and *c*. Their main characteristics are summarized in Table 1.2.

Table 1.2 Main Characteristics of Stramenopiles Containing Phytoplanktonic Representatives Other Than Diatoms

Taxonomy	Main characteristics
Bolidophyceae	Exclusively marine picoplankton Fast swimming cells (Guillou *et al.*, 1999) Phylogenetically linked to Parmales (i.e. non-motile cells covered with silica plates that are anisometric with a radial organization similar to that of the valves of centric diatoms, (Ichinomiya *et al.*, 2011)
Dictyochophyceae (also called silicoflagellates by protistologists)	Characterized by tentacules or rhizopodia Presence of a siliceous skeleton in one phase of the life cycle in the order Dictyocales
Pelagophyceae	No real distinct morphological character but supported as a distinct lineage in all phylogenies The order Pelagomonadales contains mainly picoplanktonic representatives (e.g. *Pelagomonas*, *Aureococcus*), while the order Sarcinochrysidales contains both planktonic and benthic genera
Raphidophyceae	Characterised by the presence of ejectile bodies (trichocysts) Cells with two flagella, without cell wall Presence of mucilaginous bodies under plasmalemma in some genera Marine raphidophytes form a monophyletic clade and have a pigment composition distinct from their freshwater counterparts (Yamaguchi, Nakayama, Murakami, & Inouye, 2010)
Pinguiophyceae	No real distinct morphological character but supported as a distinct lineage in all phylogenies Characterised by the production of large amounts of omega-3 fatty acids (polyunsaturated fatty acids)

Table 1.2 Main Characteristics of Stramenopiles Containing Phytoplanktonic Representatives Other Than Diatoms—cont'd

Chrysophyceae/Synurophyceae	Taxa in both classes characterised by the ability to form silicified cysts or statospores Most species in these classes inhabit freshwater. Taxa covered by siliceous scales in both classes The Chrysophyceae have been re-defined recently (Andersen, 2004)
Eustigmatophyceae	Characterised by the absence of chlorophyll *c* (lost through evolution) Zoospores elongated with a basal swelling on the anterior flagellum. Stigma is in cytoplasm (and not in plastid)

Distinctive derived characters (synapomorphies) for each class are indicated when possible. However, it is sometimes difficult to define classes using morphological characters and a combination of ultrastructural and biochemical features is often needed, while some classes are only defined based on phylogenetic analyses.

The stramenopile group that is phylogenetically nearest to diatoms is the Bolidophyceae (Table 1.2). Bolidophyceae are regularly detected in molecular surveys and isolated in to culture from the marine environment, but very few quantitative studies have been performed and their ecological imprint is still unclear (Guillou, 2011). The dictyochophytes are frequently observed in the environment and are regularly detected in molecular surveys of planktonic communities (Massana & Pedros-Alio, 2008). Several species have been recently described (e.g. *Florenciella parvula*, Eikrem, Romari, Gall, Latasa, & Vaulot, 2004) or renamed and transferred from the raphidophyte genus *Chattonella* with which confusions were frequent (e.g. *Pseudochattonella* sp., Hosoi-Tanabe *et al.*, 2007). Some dictyochophyte species can produce icthyotoxins (Skjelbred, Horsberg, Tollefsen, Andersen, & Edvardsen, 2011). Pelagophytes are essentially known because of the ecosystem disruptive algal blooms of the brown tide species *Aureococcus anophagefferens* and *Aureoumbra lagunensis* (Gobler & Sunda, 2012). Other species such as *Pelagomonas* or *Pelagococcus* are frequently isolated from sea water, and their abundance, estimated through pigment signatures, indicate that they are probably non-negligible phytoplankton contributors in oligotrophic regions (Not *et al.*, 2008; Shi *et al.*, 2011). The Pinguiophyceae is a class of small-sized phytoplankton erected in 2002 (Kawachi *et al.*, 2002). One of their distinctive features is the production of large amounts of omega-3 fatty acids (polyunsaturated fatty acids), providing them with a yet unexploited potential for biotechnological applications (Kawachi *et al.*, 2002). They are rarely found in environmental surveys and do not seem be major contributors to phytoplanktonic communities (Fuller *et al.*, 2006).

The raphidophytes, chrysophytes, synurophytes and eustigmatophytes have marine representatives but are most abundant and diverse in freshwater. Yet, some species are of primary importance in marine ecosystems. Raphidophytes are found worldwide, mostly in coastal regions. The genera *Heterosigma* and *Chatonella* can cause important harmful effects in coastal waters and thus have a significant economic impact on aquaculture (Imai & Yamaguchi, 2012). Plastidial and mitochondrial genomes have recently been sequenced and analyzed (Masuda *et al.*, 2011). The most common marine eustigmatophyte genus is *Nannochloropsis*, a small-sized phytoplankton (2–4 μm) for which six species are currently described (Andersen, Brett, Potter, & Sexton, 1998). They have a characteristic pigment signature with violaxanthin and vaucheriaxanthin like as dominating carotenoids (Karlson, Potter, Kuylenstierna, & Andersen, 1996) and are essentially studied because they produce large quantities of fatty acids and in particular omega 3. Strains of *Nannochloropsis* are commonly used in aquaculture and because these microalgae are good candidates for green energy and blue biotechnologies (Cadoret, Garnier, & Saint-Jean, 2012), *Nannochloropsis* is one of the few phytoplankton taxa for which a full genome sequence is available (Kehou *et al.*, 2011; Oliver, Benemann, Niyogi, & Vick, 2011). Although regularly found in the environment, relatively few studies have been performed on the ecology of eustigmatophytes. Chrysophytes and synurophytes are closely related phylogenetically. Chrysophytes can be strictly phototrophs or heterotrophs or present a dual mixotrophic strategy (Preisig, Vørs, & Hällfors, 1991). A typical genus is *Ochromonas* comprising about 80 known species found in both freshwater and marine environments. Recent environmental molecular surveys of plankton diversity demonstrate the presence of novel clades of marine chrysophytes, in particular in the picoplankton size range, with no cultured representatives (Del Campo & Massana, 2011; Fuller *et al.*, 2006; Shi *et al.*, 2009).

6.2. The Cryptophytes

Cryptophytes are unicellular algae characterized by features including (1) asymmetrical cell shape, (2) presence of an invagination (either a tubular gullet, a furrow, a combination of furrow and gullet or a groove) lined with structures termed ejectosomes that discharge ribbon-like threads upon mechanical or chemical stress (Morrall & Greenwood, 1980), (3) a layered structure surrounding the cells that consists of proteinaceous inner and surface periplast components sandwiching the plasma membrane (Brett, Perasso, & Wetherbee, 1994), (4) a plastid surrounded by four membranes, with the periplastidial space containing a highly reduced remnant nucleus termed the nucleomorph (Gillott & Gibbs, 1980) and (5) a light-harvesting complex consisting of chlorophylls *a* and c_2, xanthophylls, and either red or blue phycobiliproteins (Hill & Rowan, 1989). Different combinations and/or concentrations of the pigments give cryptophytes brown, red or blue-green coloration. Cryptophyte cells have two unequal flagella, the longer one with two opposite rows of stiff flagellar hairs that provide reverse thrust, the shorter one with a single row of flagellar hairs (Hibberd, Greenwood, & Griffiths, 1971). The flagella are inserted subapically in the vestibule of an invagination often shifted to the cell's right, whereas the cell apex is shifted to the left.

Most cryptophytes are photosynthetic, but some have lost their photosynthetic pigments and returned to a heterotrophic mode of nutrition with retention of a leucoplast (remnant chloroplast) (Hoef-Emden, 2005). The phagotrophic *Goniomonas* is the only known cryptophyte genus without a plastid (Mcfadden, Gilson, & Hill, 1994) and also differs from plastid-containing cryptophytes in cell shape (flattened in lateral plane), cell invagination(s) and flagellar structure. Ultrastructural characters such as differences in periplast structure, type of invagination, flagellar apparatus and/or position of the nucleomorph have been used to define cryptophyte genera. In addition, each taxon produces only one of the seven known types of biliprotein. Cryptophytes thrive in all kinds of aqueous habitats, including marine, brackish and freshwater, but seem to be more diverse in marine than in freshwater habitats (Clay, Kugrens, & Lee, 1999; Novarino, 2003). Marine as well as freshwater cryptophytes are often abundant components of planktonic communities and may form blooms but are not known to have harmful impacts. Red-coloured genera like *Rhodomonas* and *Rhinomonas* are predominately marine, blue-green genera such as *Chroomonas* are present in both marine and freshwater habitats, while the brown to olive coloured genus *Cryptomonas* is restricted to freshwater habitats (Hoef-Emden & Melkonian, 2003). The origins and phylogenetic affiliations of cryptophytes are uncertain (see section 3.1).

6.3. The Chlorarachniophytes

Chorarachniophyta are a small group of algae, the first described members of which were amoeboid and associated with debris (Ishida, Yabuki, & Ota, 2007). They present a strong phylogenetic interest because they possess a four membrane plastid containing green algal pigments (chlorophyll *b* in particular) as well as a remnant of the endosymbiont nucleus called the nucleomorph, the genome of which has been sequenced for at least one species (Gilson *et al.*, 2006). At present, only 13 species and 8 genera are known. In recent years, planktonic chlororachniophytes have been described, one of picoplanktonic size (Moestrup & Sengco, 2001; Ota, Vaulot, Le Gall, Yabuki, & Ishida, 2009). Their ecological importance remains unknown, although 18S rRNA gene environmental sequences related to chlororachniophytes have been obtained from the Mediterranean Sea and the upwelling off Chile using specific primers (Ota, personal communication). Moreover, several new species have been isolated into culture from the Mediterranean Sea, hinting that this group could have a narrow biogeographical distribution (Ota & Vaulot, 2012).

6.4. The Euglenids

The Euglenozoa form a monophyletic lineage within the Excavata, a lineage that contains parasites, photo-autotrophs and predators. Typically, excavates have flagella inserted into a reservoir, paramylon (α-1,3 glucan) as the main carbohydrate reserve and a peculiar type of closed mitosis. The photosynthetic euglenids, the class Euglenophyceae, acquired a green plastid through secondary endosymbiosis (Marin, 2004). They mostly occur in freshwater habitats, but a few marine species belonging to the Eutreptiellales are known.

7. CONCLUDING REMARKS

Analysis of the diversity and ecology of phytoplankton has largely benefited from molecular data, and, as for many research fields in biology, phytoplankton research is entering a new era with the advent of high throughput sequencing technologies. Large-scale environmental meta-barcoding allows quasi-exhaustive analysis of the diversity of communities. Working at the community level allows integration of functional information, even for uncultured taxa (Toulza *et al.*, 2012, in this volume). In the near future, if carefully contextualized with environmental meta-data, the panel of '-omics' tools now available will facilitate eco-systemic approaches (Raes & Bork, 2008) to the analysis of phytoplankton communities and lead to a better understanding of phytoplankton ecology. It will also promote detailed experimental and physiological studies on particular taxa, ultimately fostering improvement of predictive models of phytoplankton distribution at global scales (Barton, Dutkiewicz, Flierl, Bragg, & Follows, 2010).

However, in this environmental genomic context, the improvement of knowledge on the diversity and ecology of phytoplankton strongly relies on high-quality reference databases and is tightly linked to our ability to clearly bridge molecular data and phenotypes. The potential to simultaneously and quantitatively assess the occurrence of phytoplankton taxa at relevant spatial and temporal resolution is also essential. These goals can only be achieved through multidisciplinary research involving strong taxonomic expertise and the development of high throughput microscopy and automatic molecular and imaging tools (Olson & Sosik, 2007; Preston *et al.*, 2011). While some phytoplankton species can be harmful for ecosystems and/or human activities, others provide great potential for providing natural products through blue biotechnology and alternative green energy (Larkum, Ross, Kruse, & Hankamer, 2011). Although little genomic data are currently available for phytoplankton (Rynearson & Palenik, 2011), genomics and postgenomics approaches will undoubtedly significantly contribute to unraveling the nature and implications of biological processes across ecological scales and will help addressing some of the current conceptual challenges in phytoplankton research and more generally in microbial ecology.

REFERENCES

Amato, A., Kooistra, W. H. C. F., Ghiron, J. H. L., Mann, D. G., Proschold, T., & Montresor, M. (2007). Reproductive isolation among sympatric cryptic species in marine diatoms. *Protist, 158*, 193–207.

Andersen, R. A. (2004). Biology and systematics of heterokont and haptophyte algae. *American Journal of Botany, 91*, 1508–1522.

Andersen, R. A., Brett, R. W., Potter, D., & Sexton, J. P. (1998). Phylogeny of the Eustigmatophyceae based upon 18S rDNA, with emphasis on *Nannochloropsis*. *Protist, 149*, 61–74.

Andersen, R. A., & Preisig, H. R. (2000). Class Pelagophyceae. In J. J. Lee, G. F. Leedale, & P. Bradbury (Eds.), *An illustrated guide to the Protozoa*. Lawrence, KS: Allen Press Inc.

Anderson, D. M., & Rengefors, K. (2006). Community assembly and seasonal succession of marine dinoflagellates in a temperate estuary: The importance of life cycle events. *Limnology and Oceanography, 51*, 860–873.

Archibald, J. (2012). The evolution of algae by secondary and tertiary endosymbiosis. *Advances in Botanical Research. 64*, 87–118.

Assmy, P., & Smetacek, V. (2009). Algal blooms. In M. Schaechter (Ed.), *Encyclopedia of microbiology*. Oxford: Elsevier.

Balzano, S., Marie, D., Gourvil, P., & Vaulot, D. (2012). Summer photosynthetic eukaryotic communities in the Beaufort Sea assessed by T-RFLP from flow cytometry sorted sample. *ISME Journal*.

Barton, A. D., Dutkiewicz, S., Flierl, G., Bragg, J., & Follows, M. J. (2010). Patterns of diversity in marine phytoplankton. *Science, 327*, 1509–1511.

Beaufort, L., Probert, I., De Garidel-Thoron, T., Bendif, E. M., Ruiz-Pino, D., Metzl, N., et al. (2011). Sensitivity of coccolithophores to carbonate chemistry and ocean acidification. *Nature, 476*, 80–83.

Bendif, E. M., Probert, I., Hervé, A., Billard, C., Goux, D., Lelong, C., et al. (2011). Integrative taxonomy of the Pavlovophyceae (Haptophyta): A reassessment. *Protist, 162*, 738–761.

Bergesch, M., Odebrecht, C., & Moestrup, O. (2008). Nanoflagellates from coastal waters of southern Brazil (32 degrees S). *Botanica Marina, 51*, 35–50.

Biddle, K. D., & Falkowski, P. G. (2004). Cell death in planktonic photosynthetic microorganisms. *Nature Reviews Microbiology, 2*, 644–655.

Bik, H. M., Porazinska, D. L., Creer, S., Caporaso, J. G., Knight, R., & Thomas, W. K. (2012). Sequencing our way towards understanding global eukaryotic biodiversity. *Trends in Ecology & Evolution, 27*, 233–243.

Bown, P. R. (1998). *Calcareous Nannofossil Biostratigraphy*. London: Chapman and Hall.

Boyd, P. W., Watson, A. J., Law, C. S., Abraham, E. R., Trull, T., Murdoch, R., et al. (2000). A mesoscale phytoplankton bloom in the polar southern ocean stimulated by iron fertilization. *Nature, 407*, 695–702.

Boyer, G. L., & Brand, L. E. (1998). Trace elements and harmful algal blooms. In D. A. Anderson, A. D. Cembella, & G. M. Hallegraeff (Eds.), *Physiological Ecology of Harmful Algae Blooms NATO ASI serie 41* (pp. 489–508). New York: Springer.

Brett, S. J., Perasso, L., & Wetherbee, R. (1994). Structure and development of the cryptomonad periplast: A review. *Protoplasma, 181*, 106–122.

Brussaard, C. P. D., Wilhelm, S. W., Thingstad, F., Weinbauer, M. G., Bratbak, G., Heldal, M., et al. (2008). Global-scale processes with a nanoscale drive: The role of marine viruses. *ISME Journal, 2*, 575–578.

Burkholder, J. M., Azanza, R. V., & Sako, Y. (2006). The ecology of harmful dinoflagellates. In E. Granéli, & J. T. Turner (Eds.), *Ecology of harmful algae*. Berlin, Heidelberg, New York: Springer.

Burki, F., Okamoto, N., & Keeling, P. J. (2011). Phylogenomics of Hacrobians, with focus on cryptophytes and katablepharids. *Journal of Phycology, 47*. S51–S51.

Burki, F., Shalchian-Tabrizi, K., Minge, M., Skjæveland, Å., Nikolaev, S. I., Jakobsen, K. S., et al. (2007). Phylogenomics reshuffles the eukaryotic supergroups. *PLoS ONE, 2*, e790.

Burki, F., Yuji, I., Jon, B., Archibald, J. M., Keeling, P. J., Cavalier-Smith, T., et al. (2009). Large-scale phylogenomic analyses reveal that two enigmatic protist lineages, Telonemia and Centroheliozoa, are related to photosynthetic Chromalveolates. *Genome Biology and Evolution, 1*, 231–238.

Butcher, R. W. (1952). Contribution to our knowledge of the smaller marine algae. *Journal of Marine Biology Association, United Kingdom, 31*, 175–191.

Cadoret, J., Garnier, M., & Saint-Jean, B. (2012). Microalgae, functional genomics and biotechnology. *Advances in Botanical Research*. 64, 285–341.

Casteleyn, G., Leliaert, F., Backeljau, T., Debeer, A.-E., Kotaki, Y., Rhodes, L., et al. (2010). Limits to gene flow in a cosmopolitan marine planktonic diatom. *Proceedings of the National Academy of Sciences of the United States of America, 107*, 12952–12957.

Cavalier-Smith, T. (1981). Eukaryotic kingdoms- 7 or 9. *Biosystems, 14*, 461–481.

Cavalier-Smith, T. (1999). Principles of protein and lipid targeting in secondary symbiogenesis: Euglenoid, Dinoflagellate, and Sporozoan plastid origins and the eucaryote family tree. *The Journal of Eukaryotic Microbiology, 46*, 347–366.

Cavalier-Smith, T. (2002). The phagotrophic origin of eukaryotes and phylogenetic classification of protozoa. *International Journal of Systematic and Evolutionary Microbiology, 52*, 297–354.

Cavalier-Smith, T., & Chao, E. E. (2004). Protalveolate phylogeny and systematics and the origins of Sporozoa and dinoflagellates (phylum Myzozoa nom. nov.). *European Journal of Protistology, 40*, 185–212.

Chambouvet, A., Morin, P., Marie, D., & Guillou, L. (2008). Control of toxic marine dinoflagellate blooms by serial parasitic killers. *Science, 322*, 1254–1257.

Chrétiennot-Dinet, M. J., Courties, C., Vaquer, A., Neveux, J., Claustre, H., Lautier, J., et al. (1995). A new marine picoeucaryote: *Ostreococcus tauri* gen et sp nov (Chlorophyta, Prasinophyceae). *Phycologia, 34*, 285–292.

Clay, B. L., Kugrens, P., & Lee, R. E. (1999). A revised classification of the Cryptophyta. *Botanical Journal of the Linnean Society, 131*, 131–151.

Collado-Fabri, S., Ulloa, O., & Vaulot, D. (2011). Structure and seasonal dynamics of the eukaryotic picophytoplankton community in a wind-driven coastal upwelling ecosystem. *Limnology and Oceanography, 56*, 2334–2346.

D'alelio, D., D'alcala, M. R., Dubroca, L., Sarno, D., Zingone, A., & Montresor, M. (2010). The time for sex: A biennial life cycle in a marine planktonic diatom. *Limnology and Oceanography, 55*, 106–114.

Daugbjerg, N., & Andersen, R. A. (1997). Phylogenetic analyses of the *rbcL* sequences from haptophytes and heterokont algae suggest their chloroplasts are unrelated. *Molecular Biology and Evolution, 14*, 1242–1251.

De Clerck, O., Bogaret, K., & Leliaert, F. (2012). Diversity and evolution of algae: Primary endosymbiosis. *Advances in Botanical Research*. 64, 55–86.

De Queiroz, K. (2007). Species concepts and species delimitation. *Systematic Biology, 56*, 879–886.

De Vargas, C., Aubry, M. P., Probert, I., & Young, J. R. (2007). Origin and evolution of coccolithophores: From coastal hunters to oceanic farmers. In P. G. Falkowski, & A. H. Knoll (Eds.), *Evolution of primary producers in the sea*. New York: Academic Press.

De Wit, R., & Bouvier, T. (2006). 'Everything is everywhere, but, the environment selects'; What did Baas Becking and Beijerinck really say? *Environmental Microbiology, 8*, 755–758.

Del Campo, J., & Massana, R. (2011). Emerging diversity within chrysophytes, choanoflagellates and bicosoecids based on molecular surveys. *Protist, 162*, 435–448.

Demir-Hilton, E., Sudek, S., Cuvelier, M. L., Gentemann, C. L., Zehr, J. P., & Worden, A. Z. (2011). Global distribution patterns of distinct clades of the photosynthetic picoeukaryote. *Ostreococcus*. *ISME Journal, 5*, 1095–1107.

Derelle, E., Ferraz, C., Rombauts, S., Rouze, P., Worden, A. Z., Robbens, S., et al. (2006). Genome analysis of the smallest free-living eukaryote *Ostreococcus tauri* unveils many unique features. *Proceedings of the National Academy of Sciences of the United States of America, 103*, 11647–11652.

Doblin, M. A., Blackburn, S. I., & Hallegraeff, G. M. (2000). Intraspecific variation in the selenium requirement of different geographic strains of the toxic dinoflagellate *Gymnodinium catenatum*. *Journal of Plankton Research, 22*, 421–432.

Edvardsen, B., Eikrem, W., Green, J. C., Andersen, R. A., Der Staay, S., & Medlin, L. K. (2000). Phylogenetic reconstructions of the Haptophyta inferred from 18S ribosomal DNA sequences and available morphological data. *Phycologia, 39*, 19–35.

Edvardsen, B., Eikrem, W., Throndsen, J., Saez, A. G., Probert, I., & Medlin, L. (2011). Ribosomal DNA phylogenies and a morphological revision provide the basis for a revised taxonomy of the Prymnesiales (Haptophyta). *European Journal of Phycology, 46*, 202–228.

Eikrem, W., Romari, K., Gall, F. L., Latasa, M., & Vaulot, D. (2004). *Florenciella parvula* gen. and sp. nov. (Dictyochophyceae, Heterokontophyta) a small flagellate isolated from the English Channel. *Phycologia, 43*, 658–668.

Eikrem, W., & Throndsen, J. (1990). The ultrastructure of *Bathycoccus* gen. nov. and *Bathycoccus prasinos* sp. nov., a non-motile picoplanktonic alga (Chlorophyta, Prasinophyceae) from the Mediterranean and Atlantic. *Phycologia, 29*, 344–350.

Falkowski, P. (2012). The power of plankton. *Nature, 483*, 17–20.

Falkowski, P. G., & Raven, J. A. (1997). *Aquatic photosynthesis*. Malden: Blackwell Scientific.

Fast, N. M., Kissinger, J. C., Roos, D. S., & Keeling, P. J. (2001). Nuclear-encoded, plastid-targeted genes suggest a single common origin for apicomplexan and dinoflagellate plastids. *Molecular Biology and Evolution, 18*, 418–426.

Fawley, M. W., Yun, Y., & Qin, M. (1999). The relationship between *Pseudoscourfieldia marina* and *Pycnococcus provasolii* (Prasinophycaea, Chlorophyta): Evidence from 18s rDNA sequence data. *Journal of Phycology, 35*, 838–843.

Field, C. B., Behrenfeld, M. J., Randerson, J. T., & Falkowski, P. (1998). Primary production of the biosphere: Integrating terrestrial and oceanic components. *Science, 281*, 237–240.

Finkel, Z. V., Beardall, J., Flynn, K. J., Quigg, A., Rees, T. A. V., & Raven, J. A. (2010). Phytoplankton in a changing world: Cell size and elemental stoichiometry. *Journal of Plankton Research, 32*, 119–137.

Finlay, B. J. (2002). Global dispersal of free-living microbial eukaryote species. *Science, 296*, 1061–1063.

Foster, R. A., Carpenter, E. J., & Bergman, B. (2006). Unicellular cyanobionts in open ocean dinoflagellates, radiolarians and tintinnids: Ultra-structural characterization and immuno-localization of phycoerythrin and nitrogenase. *Journal of Phycology, 42*, 453–463.

Foulon, E., Not, F., Jalabert, F., Cariou, T., Massana, R., & Simon, N. (2008). Ecological niche partitioning in the picoplanktonic green alga *Micromonas pusilla*: Evidence from environmental surveys using phylogenetic probes. *Environmental Microbiology, 10*, 2433–2443.

Freudenthal, H. D. (1962). *Symbiodinium gen.* nov. and *Symbiodinium microadriaticum* sp. nov ; A Zooxanthella: Taxonomy, life cycle and morphology. *Journal of Protozoology, 9*, 45–52.

Frias-Lopez, J., Thompson, A., Waldbauer, J., & Chisholm, S. W. (2009). Use of stable isotope-labelled cells to identify active grazers of picocyanobacteria in ocean surface waters. *Environmental Microbiology, 11*, 512–525.

Fuller, N. J., Tarran, G. A., Cummings, D. G., Woodward, M. S., Orcutt, K. M., Yallop, M., et al. (2006). Molecular analysis of photosynthetic picoeukaryotes community structure along an Arabian Sea transect. *Limnology and Oceanography, 51*, 2502–2514.

Garcia-Cuetos, L., Moestrup, O., Hansen, P. J., & Daugbjerg, N. (2010). The toxic dinoflagellate *Dinophysis acuminata* harbors permanent chloroplasts of cryptomonad origin, not kleptochloroplasts. *Harmful Algae, 9*, 25–38.

Gersonde, R., & Hardwood, D. M. (1990). Lower cretaceaous diatoms from ODP LEG 113 Site 693 (weddel sea). Part 1: Vegetative cells. *Proceedings of ODP Science Research, 113*, 365–402.

Gillott, M., & Gibbs, S. P. (1980). The cryptomonad nucleomorph: Its ultrastructure and evolutionary significance. *Journal of Phycology, 16*, 558–568.
Gilson, P. R., Su, V., Slamovits, C. H., Reith, M. E., Keeling, P. J., & Mcfadden, G. I. (2006). Complete nucleotide sequence of the chlorarachniophyte nucleomorph: Nature's smallest nucleus. *Proceedings of the National Academy of Sciences of the United States of America, 103*, 9566–9571.
Gobler, C. J., & Sunda, W. G. (2012). Ecosystem disruptive algal blooms of the brown tide species, *Aureococcus anophagefferens* and Aureoumbra lagunensis. *Harmful Algae, 14*, 36–45.
Gomez, F., Lopez-Garcia, P., & Moreira, D. (2011). Molecular phylogeny of dinophysoid dinoflagellates: The systematic position of *Oxyphysis Oxytoxoides* and the *Dinophysis hastata* group (Dinophysales, Dinophyceae). *Journal of Phycology, 47*, 393–406.
Gomez, F., Moreira, D., & Lopez-Garcia, P. (2010). Molecular phylogeny of noctilucoid dinoflagellates (noctilucales, dinophyceae). *Protist, 161*, 466–478.
Gordon, N., Angel, D. L., Neori, A., Kress, N., & Kimor, B. (1994). Heterotrophic dinoflagellates with symbiotic cyanobacteria and nitrogen limitation in the Gulf of Aqaba. *Marine Ecology Progress Series, 107*, 83–88.
Gould, S. B., Tham, W. H., Cowman, A. F., Mcfadden, G. I., & Waller, R. F. (2008). Alveolins, a new family of cortical proteins that define the protist infrakingdom Alveolata. *Molecular Biology and Evolution, 25*, 1219–1230.
Graham, L. E., & Wilcox, L. W. (2000). *Algae*. Upper Saddle River: Prentice-Hall.
Grimsley, N., Thomas, R., Kegel, J., Jacquet, S., Moreau, H., & Desdevises, Y. (2012). Genomics of Algal Host-Virus Interactions. *Advances in Botanical Research*. 64, 343–378.
Guillou, L. (2011). Characterization of the Parmales: Much more than the resolution of a taxonomic enigma. *Journal of Phycology, 47*, 2–4.
Guillou, L., Chrétiennot-Dinet, M. J., Medlin, L. K., Claustre, H., Loiseaux De Goer, S., & Vaulot, D. (1999). *Bolidomonas*: A new genus with two species belonging to a new algal class, the Bolidophyceae (Heterokonta). *Journal of Phycology, 35*, 368–381.
Guillou, L., Eikrem, W., Chretiennot-Dinet, M. J., Le Gall, F., Massana, R., Romari, K., et al. (2004). Diversity of picoplanktonic Prasinophytes assessed by direct nuclear SSU rDNA sequencing of environmental samples and novel isolates retrieved from oceanic and coastal marine ecosystems. *Protist, 155*, 193–214.
Guillou, L., Viprey, M., Chambouvet, A., Welsh, R. M., Kirkham, A. R., Massana, R., et al. (2008). Widespread occurrence and genetic diversity of marine parasitoids belonging to Syndiniales (Alveolata). *Environmental Microbiology, 10*, 3349–3365.
Guiry, M. D., & Guiry, G. M. (2012). *AlgaeBase*. Galway: National University of Ireland.
Gunderson, J. H., John, S. A., Boman, W. C., & Coats, D. W. (2002). Multiple strains of the parasitic dinoflagellate *Amoebophrya* exist in Chesapeake Bay. *Journal of Eukaryotic Microbiology, 49*, 469–474.
Hallegraeff, G. M. (2003). Harmful algal blooms: A global overview. In G. M. Hallegraeff, D. M. Anderson, & A. D. Cembella (Eds.), *Manual on harmful marine microalgae*. Paris: Intergovernmental Oceanographic Commission of UNESCO.
Hallegraeff, G. M. (2010). Ocean climate change, phytoplankton community responses, and Harmful Algal blooms: A formidable predictive challenge. *Journal of Phycology, 46*, 220–235.
Hanic, L. A., Sekimoto, S., & Bates, S. S. (2009). Oomycete and chytrid infections of the marine diatom *Pseudo-nitzschia pungens* (Bacillariophyceae) from Prince Edward Island, Canada. *Botany, 87*, 1096–1105.
Härnström, K., Ellegaard, M., Andersen, T. J., & Godhe, A. (2011). Hundred years of genetic structure in a sediment revived diatom population. *Proceedings of the National Academy of Sciences*.

Harper, J. T., & Keeling, P. J. (2003). Nucleus-encoded, plastid-targeted glyceraldehyde-3-phosphate dehydrogenase (GAPDH) indicates a single origin for chromalveolate plastids. *Molecular Biology and Evolution, 20*, 1730–1735.

Harwood, D. M., Chang, K. H., & Nikolaev, V. A. (2004). Late Jurassic to earliest Cretaceous diatoms from Jasong synthem, southern Korea: Evidence for a terrestrial origin. Abstracts, 18th International diatom symposium, *Miedzyzdroje,* Poland, 81.

Henley, W. J., Hironaka, J. L., Guillou, L., Buchheim, M. A., Buchheim, J. A., Fawley, M. W., et al. (2004). Phylogenetic analysis of the '*Nannochloris*-like' algae and diagnoses of *Picochlorum oklahomensis* gen. et sp. nov. (Trebouxiophyceae, Chlorophyta). *Phycologia, 43*, 641–652.

Hibberd, D. J., Greenwood, A. D., & Griffiths, H. B. (1971). Observations on the ultrastructure of the flagella and periplast in the Cryptophyceae. *British Phycology Journal, 6*, 61–72.

Hildebrand, M. (2008). Diatoms, biomineralization processes, and genomics. *Chemical reviews, 108*, 4855–4874.

Hill, D. R. A., & Rowan, K. S. (1989). The biliproteins of the Cryptophyceae. *Phycologia, 28*, 455–463.

Hinder, S. L., Hays, G. C., Brooks, C. J., Davies, A. P., Edwards, M., Walne, A. W., et al. (2011). Toxic marine microalgae and shellfish poisoning in the British isles: History, review of epidemiology, and future implications. *Environmental Health, 10*, 54.

Hoef-Emden, K. (2005). Multiple independent losses of photosynthesis in the genus *Cryptomonas* (Cryptophyceae)—Combined phylogenetic analyses of DNA sequences of the nuclear and the nucleomorph ribosomal operons. *Journal of Molecular Evolution, 60*, 183–195.

Hoef-Emden, K., & Melkonian, M. (2003). Revision of the genus *Cryptomonas* (Cryptophyceae): A combination of molecular phylogeny and morphology provides insights into a long-hidden dimorphism. *Protist, 154*, 371–409.

Hoppenrath, M., & Leander, B. S. (2010). Dinoflagellate phylogeny as inferred from heat shock protein 90 and ribosomal gene sequences. *Plos One, 5*.

Hori, T., & Green, J. C. (1994). Mitosis and cell division. In J. C. Green, & B. S. C. Leadbeater (Eds.), *The Haptophyte algae*. Oxford: Oxford University Press.

Hosoi-Tanabe, S., Honda, D., Fukaya, S., Otake, I., Inagaki, Y., & Sako, Y. (2007). Proposal of *Pseudochattonella verruculosa* gen. nov., comb. nov. (Dictyochophyceae) for a formar raphidophycean alga *Chattonella verruculosa*, based on 18S rDNA phylogeny and ultrastructural characteristics. *Phycological Research, 55*, 185–192.

Ichinomiya, M., Yoshikawa, S., Kamiya, M., Ohki, K., Takaichi, S., & Kuwata, A. (2011). Isolation and characterization of Parmales (Heterokonta/Heterokontophyta/Stramenopiles) from the Oyashio region, western north Pacific. *Journal of Phycology, 47*, 144–151.

Iglesias-Rodriguez, M. D., Halloran, P. R., Rickaby, R. E. M., Hall, I. R., Colmenero-Hidalgo, E., Gittins, J. R., et al. (2008). Phytoplankton calcification in a high-CO2 world. *Science, 320*, 336–340.

Imai, I., & Yamaguchi, M. (2012). Life cycle, physiology, ecology and red tide occurences of the fish-killing raphidophyte. *Chatonella. Harmful Algae, 14*, 46–70.

Ishida, K., Yabuki, A., & Ota, S. (2007). The chlorarachniophytes: Evolution and classification. In J. B., & J. L. (Eds.), *Unravelling the algae: The past, present, and future of algal systematics*. Taylor & Francis Group: CRC Press.

Ishii, K.-I., Iwataki, M., Matsuoka, K., & Imai, I. (2011). Proposal of identification criteria for resting spores of *Chaetoceros* species (Bacillariophyceae) from a temperate coastal sea. *Phycologia, 50*, 351–362.

Jeffrey, S. W. (1997). Application of pigment methods to oceanography. In S. W. Jeffrey, R. F. C. Mantoura, & S. W. Wright (Eds.), *Phytoplankton pigments in oceanography. A guide to advanced methods*. Paris: UNESCO publishing.

Jordan, R. W., Cros, L., & Young, J. R. (2004). A revised classification scheme for living haptophytes. *Micropaleontology, 50*, 55–79.

Karlson, B., Potter, D., Kuylenstierna, M., & Andersen, R. A. (1996). Ultrastructure, pigment composition, and 18S rRNA gene sequence for *Nannochloropsis granulata* sp. nov. (Monodopsidaceae, Eustigmatophyceae), a marine ultraplankter isolated from the Skagerrak, northeast Atlantic Ocean. *Phycologia, 35*, 253–260.

Kawachi, M., Inouye, I., Honda, D., O'kelly, C. J., Bailey, J. C., Bidigare, R. R., et al. (2002). The Pinguiophyceae *classis nova*, a new class of photosynthetic stramenopiles whose members produce large amount of omega-3 fatty acids. *Phycological Research, 50*, 31–47.

Kawachi, M., Inouye, I., Maeda, O., & Chihara, M. (1991). The haptonema as a food-capturing device—Observations on *Chrysochromulina hirta* (Prymnesiophyceae). *Phycologia, 30*, 563–573.

Kehou, P., Junjie, Q., Si, L., Dai, W., Zhu, B., Jin, Y., et al. (2011). Nuclear monoploidy and sexual propagation of *Nannochloropsis oceanica* (Eustigmatophyceae) as revealed by its genome sequence. *Journal of Phycology, 47*, 1425–1432.

Kooistra, W. H. C. F., Gersonde, R., Medlin, L. K., & Mann, D. G. (2007). The origin and evolution of the diatoms: Their adaptation to a planktonic existence. In P. G. Falkowski, & A. H. Knoll (Eds.), *Evolution of planktonic photoautotrophs* (pp. 207–249). Burlington (MA): Academic Press Inc.

Kooistra, W. H. C. F., Sarno, D., Hernández-Becerril, D. U., Assmy, P., Di Prisco, C., & Montresor, M. (2010). Comparative molecular and morphological phylogenetic analyses of taxa in the Chaetocerotaceae (Bacillariophyta). *Phycologia, 49*, 471–500.

Kroger, N., & Poulsen, N. (2008). Diatoms—From cell wall biogenesis to nanotechnology. *Annual Review Genetics, 42*, 83–107.

Labry, C., Denn, E. E. L., Chapelle, A., Fauchot, J., Youenou, A., Crassous, M. P., et al. (2008). Competition for phosphorus between two dinoflagellates: A toxic *Alexandrium minutum* and a non-toxic *Heterocapsa triquetra*. *Journal of Experimental Marine Biology and Ecology, 358*, 124–135.

Larkum, A. W. D., Ross, I. L., Kruse, O., & Hankamer, B. (2011). Selection, breeding and engineering of microalgae for bioenergy and biofuel production. *Trends in Biotechnology, 30*, 198–205.

Le Gall, F., Rigaut-Jalabert, F., Marie, D., Garczareck, L., Viprey, M., Godet, A., et al. (2008). Picoplankton diversity in the south-east Pacific Ocean from cultures. *Biogeosciences, 5*, 203–214.

Leander, B. S., & Keeling, P. J. (2003). Morphostasis in alveolate evolution. *Trends in Ecology & Evolution, 18*, 395–402.

Lee, R. E. (1999). *Phycology*. Cambridge: Cambridge University Press.

Leliaert, F., Verbruggen, H., & Zechman, F. W. (2011). Into the deep: New discoveries at the base of the green plant phylogeny. *BioEssays, 33*, 683–692.

Lewin, R. A., Krienitz, L., Goericke, R., Takeda, H., & Hepperle, D. (2000). *Picocystis salinarum* gen. et sp nov (Chlorophyta)—A new picoplanktonic green alga. *Phycologia, 39*, 560–565.

Liu, H., Aris-Brossou, S., Probert, I., & De Vargas, C. (2010). A time line of the environmental genetics of the haptophytes. *Molecular Biology and Evolution, 27*, 161–176.

Liu, H., Probert, I., Uitz, J., Claustre, H., Aris-Brossou, S., Frada, M., Not, F., & De Vargas, C. (2009). Extreme diversity in non-calcifying haptophytes explains a major pigment paradox in open oceans. *Proceedings of the National Academy of Sciences, 106*, 12803–12808.

Loebl, M., Cockshutt, A. M., Campbell, D. A., & Finkel, Z. V. (2010). Physiological basis for high resistance to photoinhibition under nitrogen depletion in *Emiliania huxleyi*. *Limnology and Oceanography, 55*, 2150–2160.

Lovejoy, C., Vincent, W. F., Bonilla, S., Roy, S., Martineau, M. J., Terrado, R., et al. (2007). Distribution, phylogeny, and growth of cold-adapted picoprasinophytes in arctic seas. *Journal of Phycology, 43*, 78–89.

Mann, D. G. (1999). The species concept in diatoms. *Phycologia, 38*, 437–495.

Mann, D. G., & Droop, S. J. M. (1996). Biodiversity, biogeography and conservation of diatoms. In J. Kristiansen (Ed.), *Biology of Freshwater Algae: Proceedings of the Workshop on Biogeography of Freshwater Algae, Developments in Hydrobiology* (pp. 118). Dordecht: Kluwer Academic Publishers.

Marañón, E., Holligan, P. M., Barciela, R., Gonzalez, N., Mourino, B., Pazo, M. J., et al. (2001). Patterns of phytoplankton size structure and productivity in contrasting open-ocean environments. *Marine Ecology—Progress Series, 216*, 43–56.

Marin, B. (2004). Origin and fate of chloroplasts in the Euglenoida. *Protist, 155*, 13–14.

Marin, B., & Melkonian, M. (2010). Molecular phylogeny and classification of the Mamiellophyceae class. nov. (Chlorophyta) based on sequence comparisons of the nuclear- and plastid-encoded rRNA operons. *Protist, 161*, 304–336.

Massana, R., & Pedros-Alio, C. (2008). Unveiling new microbial eukaryotes in the surface ocean. *Current Opinion in Microbiology, 11*, 213–218.

Massjuk, N. P. (2006). Chlorodendrophyceae class. nov. (Chlorophyta, Viridiplantae) in the Ukrainian flora: I. The volume, phylogenetic relations and taxonomical status. *Ukraine Botanical Journal, 63*, 601–614.

Masuda, I., Kamikawa, R., Ueda, M., Oyama, K., Sadaaki, Y., Inagaki, Y., et al. (2011). Mitochondrial genomes from two red tide forming raphidophycean algae *Heterosigma akashiwo* and *Chattonella marina* var. *marina*. *Harmful Algae, 10*, 130–137.

Mayr, E. (1969). The biological meaning of species. *Biological Journal of the Linnean Society* 311–320.

Mcfadden, G. I., Gilson, P. R., & Hill, D. R. A. (1994). *Goniomonas:* rRNA sequences indicate that this phagotrophic flagellate is a close relative of the host component of cryptomonads. *European Journal of Phycology, 29*, 29–32.

Mcfadden, G. I., Reith, M. E., Munholland, J., & Langunnasch, N. (1996). Plastid in human parasites. *Nature, 381*. 482–482.

Medlin, L., Saez, A. G., & Young, J. R. (2008). A molecular clock for coccolithophores and implications for selectivity of phytoplankton extinctions across the K/T boundary. *Marine Micropaleontology, 67*, 69–86.

Medlin, L. K., Metfies, K., Wiltshire, K., Mehl, H., & Valentin, K. (2006). Picoeukaryotic plankton diversity at the Helgoland time series site as assessed by three molecular methods. *Microbial Ecology, 52*, 53–71.

Mock, T., & Medlin, L. K. (2012). Genomics and genetics of diatoms. *Advances in Botanical Research*. 64, 245–284.

Moestrup, Ø., & Daugbjerg, N. (2007). On dinoflagellate phylogeny and classification. In J. Brodie, & J. Lewis (Eds.), *Unravelling the algae: The past, present, and future of algal systematics*. New York: CRC Press.

Moestrup, Ø., & Sengco, M. (2001). Ultrastructural studies on *Bigelowiella natans*, gen. et sp nov., a chlorarachniophyte flagellate. *Journal of Phycology, 37*, 624–646.

Mohammady, N. G. (2004). Total, free and conjugated sterolic forms in three micro-algae used in mariculture. *Zeitschrift Fur Naturforschung C a Journal of Biosciences, 59*, 619–624.

Monier, A., Welsh, R. M., Gentemann, C., Weinstock, G., Sodergren, E., Armburst, E. V., et al. (2012). Phosphate transporters in marine phytoplankton and their viruses: Cross-domain commonalities in viral-host gene exchanges. *Environmental Microbiology, 14*, 162–176.

Moon-Van Der Staay, S. Y., Van Der Staay, G. W. M., Guillou, L., Vaulot, D., Claustre, H., & Medlin, L. K. (2000). Abundance and diversity of Prymnesiophytes in the picoplankton community from the equatorial Pacific Ocean inferred from 18S rDNA sequences. *Limnology and Oceanography, 45*, 98–109.

Moreau, H., Verhelst, B., Couloux, A., Derelle, E., Rombauts, S., Grimsley, N., et al. (2012). Gene functionalities and genome structure in *Bathycoccus prasinos* reflect cellular specializations at the base of the green lineage. *Genome Biology, 13*(8), R74.

Morrall, S., & Greenwood, A. D. (1980). A comparison of the periodic substructures of the trichocysts of the Cryptophyceae and Prasinophyceae. *Biosystems, 12*, 71–83.

Murray, S., Ip, C. L. C., Moore, R., Nagahama, Y., & Fukuyo, Y. (2009). Are prorocentroid dinoflagellates monophyletic? A study of 25 species based on nuclear and mitochondrial genes. *Protist, 160*, 245–264.

Murray, S., Jorgensen, M. F., Ho, S. Y. M., Patterson, D. J., & Jermilin, L. S. (2005). Improving the analysis of dinoflagellate phylogeny based on rDNA. *Protist, 156*, 269–286.

Nagasaki, K., Tomaru, P. Y., Shirai, Y., Takao, Y., & Mizumoto, H. (2006). Dinoflagellate-infecting viruses. *Journal of Marine Biology Association United Kingdom, 86*, 469–474.

Noel, M. H., Kawachi, M., & Inouye, I. (2004). Induced dimorphic life cycle of a coccolithophorid *Calyptrosphaera sphaeroidea* (Prymnesiophyceae, Haptophyta). *Journal of Phycology, 40*, 112–129.

Not, F., Latasa, M., Marie, D., Cariou, T., Vaulot, D., & Simon, N. (2004). A single species, *Micromonas pusilla* (Prasinophyceae), dominates the eukaryotic picoplankton in the Western English Channel. *Applied and Environmental Microbiology, 70*, 4064–4072.

Not, F., Latasa, M., Scharek, R., Viprey, M., Karleskind, P., Balagué, V., et al. (2008). Phytoplankton diversity across the Indian Ocean: A focus on the picoplanktonic size fraction. *Deep—Sea Research Part I—Oceanographic Research Papers, 55*, 1456–1473.

Not, F., Valentin, K., Romari, K., Lovejoy, C., Massana, R., Töbe, K., et al. (2007). Picobiliphytes: A marine picoplanktonic algal group with unknown affinities to other eukaryotes. *Science, 315*, 252–254.

Novarino, G. (2003). A companion to the identification of cryptomonad flagellates (Cryptophyceae = Cryptomonadea). *Hydrobiologica, 502*, 225–270.

O'kelly, C. J., Sieracki, M. E., Thier, E. C., & Hobson, I. C. (2003). A transient bloom of Ostreococcus (Chlorophyta, Prasinophyceae) in West Neck Bay, Long Island, New York. *Journal of Phycology, 39*, 850–854.

O'malley, M. A. (2007). The nineteenth century roots of 'everything is everywhere'. *Nature Reviews Microbiology, 5*, 647–651.

Okamoto, N., Chantangsi, C., Horak, A., Leander, B. S., & Keeling, P. J. (2009). Molecular phylogeny and description of the novel katablepharid *Roombia truncata* gen. et sp nov., and establishment of the Hacrobia taxon nov. *PLoS ONE, 4*, e7080.

Oliver, K., Benemann, C. S. E., Niyogi, K. K., & Vick, B. (2011). High-efficiency homologous recombination in the oil-producing alga *Nannochloropsis* sp. *Proceedings of the National Academy of Sciences of the United States of America, 108*, 21265–21269.

Olson, R. J., & Sosik, H. M. (2007). A submersible imaging-in-flow instrument to analyze nano and microplankton: Imaging FlowCytobot. *Limnology and Oceanography: Methods, 5*, 195–203.

Ota, S., Kudo, A., & Ishida, K. (2011). *Gymnochlora dimorpha* sp. nov., a chlorarachniophyte with unique daughter cell behaviour. *Phycologia, 50*, 317–326.

Ota, S., & Vaulot, D. (2012). *Lotharella reticulosa* sp. nov.: A highly reticulated network forming Chlorarachniophyte from the Mediterranean Sea. *Protist, 163*, 91–104.

Ota, S., Vaulot, D., Le Gall, F., Yabuki, A., & Ishida, K. (2009). *Partenskyella glossopodia* gen. et sp. nov., the first report of chlorarachniophyte that lacks pyrenoid. *Protist, 160*, 137–150.

Paasche, E. (2002). A review of the coccolithophorid *Emiliania huxleyi* (Prymnesiophyceae), with particular reference to growth, coccolith formation, and calcification-photosynthesis interactions. *Phycologia, 40*, 503–529.

Park, M. G., Kim, S., Kim, H. S., Myung, G., Kang, Y. G., & Yih, W. (2006). First successful culture of the marine dinoflagellate *Dinophysis acuminata*. *Aquatic Microbial Ecology, 45*, 101–106.

Piganeau, G., Eyre-Walker, A., Grimsley, N., & Moreau, H. (2011a). How and Why DNA barcodes underestimate the diversity of microbial eukaryotes. *PLoS ONE, 6*, e16342.

Piganeau, G., Grimsley, N., & Moreau, H. (2011b). Genome diversity in the smallest marine photosynthetic eukaryotes. *Research in Microbiology, 162*, 570–577.

Pizay, M. D., Lemee, R., Simon, N., Cras, A. L., Laugier, J. P., & Dolan, J. R. (2009). Night and day morphologies in a planktonic dinoflagellate. *Protist, 160*, 565–575.

Potter, D., Lajeunesse, T. C., Saunders, G. W., & Anderson, R. A. (1997). Convergent evolution masks extensive biodiversity among marine coccoid picoplankton. *Biodiversity and Conservation, 6*, 99–107.

Preisig, H. R., Vørs, N., & Hällfors, G. (1991). Diversity of heterotrophic heterokont flagellates. In D. J. Patterson, & J. Larson (Eds.), *The biology of the free-living heterotrophic flagellates*. Oxford: Clarendon Press.

Preston, C. M., Harris, A., Ryan, J. P., Roman, B., Marin, R., Jensen, S., et al. (2011). Underwater application of quantitative PCR on an ocean mooring. *PLoS ONE, 6*, e22522.

Raes, J., & Bork, P. (2008). Molecular eco-systems biology: Towards an understanding of community function. *Nature Reviews Microbiology, 6*, 693–699.

Ragni, M., Airs, R. L., Leonardos, N., & Geider, R. J. (2008). Photoinhibition of PSII in Emiliania huxleyi (Haptophyta) under high light stress: The roles of photoacclimation, photoprotection, and photorepair. *Journal of Phycology, 44*, 670–683.

Ral, J. P., Derelle, E., Ferraz, C., Wattebled, F., Farinas, B., Corellou, F., et al. (2004). Starch division and partitioning. A mechanism for granule propagation and maintenance in the picophytoplanktonic green alga *Ostreococcus tauri*. *Plant Physiology, 136*, 3333–3340.

Reynolds, C. S. (2006). *The Ecology of Phytoplankton*. Cambridge: Cambridge University Press.

Rice, D. W., & Palmer, J. D. (2006). An exceptional horizontal gene transfer in plastids: Gene replacement by a distant bacterial paralog and evidence that haptophyte and cryptophyte plastids are sisters. *BMC Biology, 4*, 31.

Riebesell, U., Zondervan, I., Rost, B., Tortell, P. D., Zeebe, R. E., & Morel, F. M. M. (2000). Reduced calcification of marine plankton in response to increased atmospheric CO2. *Nature, 407*(6802), 364–367.

Rodriguez, F., Derelle, E., Guillou, L., Le Gall, F., Vaulot, D., & Moreau, H. (2005). Ecotype diversity in marine picoeukaryote *Ostreococcus* (Chlorophyta, Prasinophyceae). *Environmental Microbiology, 7*, 853–859.

Rodriguez, F., Varela, M., & Zapata, M. (2002). Phytoplankton assemblages in the Gerlache and Bransfield Straits (Antarctic Peninsula) determined by light microscopy and CHEMTAX analysis of HPLC pigment data. *Deep—Sea Research Part II - Oceanographic Research Papers, 49*, 723–747.

Romari, K., & Vaulot, D. (2004). Composition and temporal variability of picoeukaryote communities at a coastal site of the English Channel from 18S rDNA sequences. *Limnology and Oceanography, 49*, 784–798.

Rost, B., & Riebesell, U. (2004). Coccolithophores and the biological pump: Responses to environmental changes. In H. R. Thierstein, & J. R. Young, (Eds.), *Coccolithophores: From molecular processes to global impact*. Ascona, Switzerland.

Round, F. E., Crawford, R. M., & Mann, D. G. (Eds.). (1990). *The diatoms. Biology and morphology of the genera*. Cambridge University Press.

Roy, S., & Chattopadhyay, J. (2007). Towards a resolution of 'the paradox of the plankton': A brief overview of the proposed mechanisms. *Ecological complexity, 4*, 26–33.

Rynearson, T. A., & Palenik, B. (2011). Learning to read the oceans: Genomics of marine phytoplankton. *Advances in Marine Biology, 60*, 1–39.

Saez, A. G., Probert, I., Geisen, M., Quinn, P., Young, J. R., & Medlin, L. K. (2003). Pseudo-cryptic speciation in coccolithophores. *Proceedings of the National Academy of Sciences of the United States of America, 100*, 7163–7168.

Saldarriaga, J. F., Mcewan, M. L., Fast, N. M., Taylor, F. J. R., & Keeling, P. J. (2003). Multiple protein phylogenles show that *Oxyrrhis marina* and *Perkinsus marinus* are early branches of the dinoflagellate lineage. *International Journal of Systematic and Evolutionary Microbiology, 53*, 355–365.

Saldarriaga, J. F., Taylor, F. J. R., Cavalier-Smith, T., Menden-Deuer, S., & Keeling, P. J. (2004). Molecular data and the evolutionary history of dinoflagellates. *European Journal of Protistology, 40*, 85–111.

Saldarriaga, J. F., Taylor, F. J. R., Keeling, P. J., & Cavalier-Smith, T. (2001). Dinoflagellate nuclear SSU rRNA phylogeny suggests multiple plastid losses and replacement. *Journal of Molecular Evolution, 53*, 204–213.

Samadi, S., & Barberousse, A. (2006). The tree, the network, and the species. *Biological Journal of the Linnean Society, 89*, 509–521.

Schoemann, V., Becquevort, S., Stefels, J., Rousseau, W., & Lancelot, C. (2005). Phaeocystis blooms in the global ocean and their controlling mechanisms: A review. *Journal of Sea Research, 53*, 43–66.

Shalchian-Tabrizi, K., Reier-Roberg, K., Ree, D. K., Klaveness, D., & Brate, J. (2011). Marine-freshwater colonizations of haptophytes inferred from phylogeny of environmental 18S rDNA sequences. *Journal of Eukaryotic Microbiology, 58*, 315–318.

Shi, X. L., Lepère, C., Scanlan, D. J., & Vaulot, D. (2011). Plastid 16S rRNA gene diversity among eukaryotic picophytoplankton sorted by flow cytometry from the South Pacific Ocean. *PLoS ONE, 6*, e18979.

Shi, X. L., Marie, D., Jardillier, L., Scanlan, D. J., & Vaulot, D. (2009). Groups without cultured representatives dominate eukaryotic picophytoplankton in the oligotrophic South East Pacific Ocean. *PLoS ONE, 4*, e7657.

Siano, R., Alves-De-Souza, C., Foulon, E., Bendif, E. M., Simon, N., Guillou, L., et al. (2011). Distribution and host diversity of Amoebophryidae parasites across oligotrophic waters of the Mediterranean Sea. *Biogeosciences, 8*, 267–278.

Siano, R., Montresor, M., Probert, I., Not, F., & De Vargas, C. (2010). *Pelagodinium* gen. nov and *P-beii* comb. nov., a dinoflagellate symbiont of planktonic Foraminifera. *Protist, 161*, 385–399.

Sieburth, J. M., Keller, M. D., Johnson, P. W., & Myklestad, S. M. (1999). Widespread occurrence of the oceanic ultraplankter, *Prasinococcus capsulatus* (Prasinophyceae), the diagnostic "Golgi-decapore complex" and the newly described polysaccharide "capsulan". *Journal of Phycology, 35*, 1032–1043.

Silva, P. C. (2008). Historical review of attempts to decrease subjectivity in species identification, with particular regard to algae. *Protist, 159*, 153–161.

Simon, N., Cras, A.-L., Foulon, E., & Lemée, R. (2009). Diversity and evolution of marine phytoplankton. *Comptes Rendus De L'Academie Des Sciences Biologies, 332*, 159–170.

Sinninghe-Damsté, J. S., Muyzer, G., Abbas, B., Rampen, S. W., Masse, G., Allard, W. G., et al. (2004). The rise of the rhizosolenoid diatoms. *Science, 304*, 584–587.

Siokou-Frangou, I., Christaki, U., Mazzocchi, M. G., Montresor, M., D'alcala, M. R., Vaque, D., et al. (2010). Plankton in the open Mediterranean Sea: A review. *Biogeosciences, 7*, 1543–1586.

Six, C., Finkel, Z. V., Rodriguez, F., Marie, D., Partensky, F., & Campbell, D. A. (2008). Contrasting photoacclimation strategies in ecotypes of the eukayotic picoplankter. *Ostreococcus. Limnology and Oceanography, 53*, 255–265.

Skjelbred, B., Horsberg, T. E., Tollefsen, K. E., Andersen, T., & Edvardsen, B. (2011). Toxicity of the ichthyotoxic marine flagellate *Pseudochattonella* (Dictyochophyceae, Heterokonta) assessed by six bioassays. *Harmful Algae, 10*, 144–154.

Skovgaard, A., Massana, R., & Saiz, E. (2007). Parasitic species of the genus *Blastodinium* (Blastodiniphyceae) are peridinioid dinoflagellates. *Journal of Phycology, 43*, 553–560.

Slapeta, J., Lopez-Garcia, P., & Moreira, D. (2006). Global dispersal and ancient cryptic species in the smallest marine eukaryotes. *Molecular Biology and Evolution, 23*, 23–29.

Slapeta, J., Moreira, D., & Lopez-Garcia, P. (2005). The extent of protist diversity: Insights from molecular ecology of freshwater eukaryotes. *Proceedings of the Royal Society B—Biological Sciences, 272*, 2073–2081.

Smalley, G. W., & Coats, D. W. (2002). Ecology of red-tide dinoflagellate *Ceratium furca:* Distribution, mixotrophy, and grazing impact on ciliate populations of Chesapeake Bay. *The Journal of Eukaryotic Microbiology, 49*, 63–73.

Smayda, T. J. (1997). Harmful algal blooms: Their ecophysiology and general relevance to phytoplankton blooms in the sea. *Limnology and Oceanography, 42*, 1137–1153.

Smayda, T. J., & Reynolds, C. S. (2001). Community assembly in marine phytoplankton: Application of recent models to harmful dinoflagellate blooms. *Journal of Plankton Research, 23*, 447–461.

Smayda, T. J., & Reynolds, C. S. (2003). Strategies of marine dinoflagellate survival and some rules of assembly. *Journal of Sea Research, 49*, 95–106.

Smetacek, V. (2001). A watery arms race. *Nature, 411*. 745–745.

Sournia, A., Chretiennot-Dinet, M. J., & Ricard, M. (1991). Marine phytoplankton—How many species in the world ocean. *Journal of Plankton Research, 13*, 1093–1099.

Stoecker, D. K. (1999). Mixotrophy among dinoflagellates. *The Journal of Eukaryotic Microbiology, 46*, 397–401.

Stoecker, D. K., Johnson, M. D., De Vargas, C., & Not, F. (2009). Acquired phototrophy in aquatic protists. *Aquatic Microbial Ecology, 57*, 279–310.

Stoermer, E. F., & Smol, J. P. (1999). *The diatoms. Applications for the environmental and earth sciences*. Cambridge: Cambridge University Press.

Strom, S. L. (2008). Microbial ecology of ocean biogeochemistry: A community perspective. *Science, 320*, 1043–1045.

Suto, I. (2006). The explosive diversification of the diatom genus Chaetoceros across the Eocene/Oligocene and Oligocene/Miocene boundaries in the Norwegian Sea. *Marine Micropaleontology, 58*, 259–269.

Taylor, F. J. R. (1980). On dinoflagellate evolution. *Biosystems, 13*, 65–108.

Taylor, F. J. R., Hoppenrath, M., & Saldarriaga, J. F. (2008). Dinoflagellate diversity and distribution. *Biodiversity and Conservation, 17*(2), 407–418.

Taylor, F. J. R., & Pollingher, U. (1987). Ecology of dinoflagellates. In F. J. R. Taylor, (Ed.), *The biology of dinoflagellates*. London: Blackwell Scientific Publications.

Tett, P., & Barton, E. D. (1995). Why are there about 5000 species of phytoplankton in the sea. *Journal of Plankton Research, 17*, 1683–1704.

Thierstein, H. R., Geitzenauer, K. R., & Molfino, B. (1977). Global synchroneity of late quaternary coccolith datum levels—Validation by oxygen isotopes. *Geology, 5*, 400–404.

Tomas, C. R. (1997). *Identifying marine phytoplankton*. San Diego, CA: Academic Press.

Toulza, E., Blanc-Mathieu, R., Gourbiere, S., & Piganeau, G. (2012). Environmental genomics of microbial algae: Power and challenges of metagenomics. *Advances in Botanical Research*. 64, 379–423.

Treusch, A. H., Demir-Hilton, E., Vergin, K. L., Worden, A. Z., Carlson, C. A., Donatz, M. G., et al. (2011). Phytoplankton distribution patterns in the northwestern Sargasso Sea revealed by small subunit rRNA genes from plastids. *ISME Journal, 6*, 481–492.

Van Den Hoek, C., Mann, D. G., & Jahns, H. M. (1995). *Algae: An introduction to phycology*. Cambridge: Cambridge University Press.

Van Lenning, K., Latasa, M., Estrada, M., Saez, A. G., Medlin, L., Probert, I., et al. (2003). Pigment signatures and phylogenetic relationships of the pavlovophyceae (Haptophyta). *Journal of Phycology, 39*, 379–389.

Vaulot, D., Eikrem, W., Viprey, M., & Moreau, H. (2008). The diversity of small eukaryotic phytoplankton (< 3 μm) in marine ecosystems. *Fems Microbiology Reviews, 32*, 795–820.

Vaulot, D., Lepère, C., Toulza, E., De la Iglesia, R., Poulain, J., Gaboyer, F., et al. (2012). Metagenomes of the picoalga *Bathycoccus* from the Chile coastal upwelling. *PLoS ONE, 7*, e39648.

Véron, B., Dauguet, J. C., & Billard, C. (1996). Sterolic biomarkers in marine phytoplankton .1. Free and conjugated sterols of *Pavlova lutheri* (Haptophyta). *European Journal of Phycology, 31*, 211–215.

Viprey, M., Guillou, L., Férréol, M., & Vaulot, D. (2008). Wide genetic diversity of picoplanktonic green algae (Chloroplastida) in the Mediterranean Sea uncovered by a phylum-biased PCR approach. *Environmental Microbiology, 10*, 1804–1822.

Volkman, J. K., Farmer, C. L., Barrett, S. M., & Sikes, E. L. (1997). Unusual dihydroxysterols as chemotaxonomic markers for microalgae from the order Pavlovales (Haptophyceae). *Journal of Phycology, 33*, 1016–1023.

Weber, M., & Medina, M. (2012). The role of microalgal symbionts *(Symbiodinium)* in holobiont physiology. *Advances in Botanical Research*. 64, 119–140.

Wilson, W. H., Tarran, G. A., Schroeder, D. C., Cox, M., Oke, J., & Malin, G. (2002). Isolation of viruses responsible for the demise of an *Emiliania huxleyi* bloom in the English Channel. *Journal of the Marine Biological Association of the United Kingdom, 82*, 369–377.

Wisecaver, J. H., & Hackett, J. D. (2010). Transcriptome analysis reveals nuclear-encoded proteins for the maintenance of temporary plastids in the dinoflagellate *Dinophysis acuminata*. *BMC Genomics, 11*.

Worden, A. Z., Lee, J.-H., Mock, T., Rouzé, P., Simmons, M. P., Aerts, A. L., et al. (2009). Green evolution and dynamic adaptation revealed by genomes of the marine picoeukaryotes *Micromonas*. *Science, 324*, 268–272.

Yamaguchi, H., Nakayama, T., Murakami, A., & Inouye, I. (2010). Phylogeny and taxonomy of the Raphidophyceae (Heterokontophyta) and *Chlorinimonas sublosa* gen. et sp. nov., a new marine sand-dwelling raphidophyte. *Journal of Plant Research, 123*, 333–342.

Yoshida, M., Noel, M. H., Nakayama, T., Naganuma, T., & Inouye, I. (2006). A haptophyte bearing siliceous scales: Ultrastructure and phylogenetic position of *Hyalolithus neolepis* gen. et sp. nov. (Prymnesiophyceae, Haptophyta). *Protist, 157*, 213–234.

Young, J. R. (1994). Variation in *Emiliania huxleyi* coccolith morphology in samples from the Norwegian Ehux experiment, 1992. *Sarsia, 79*, 417–425.

Young, J. R., Didymus, J. M., Bown, P. R., Prins, B., & Mann, S. (1992). Crystal assembly and phylogenetic evolution of heterococcoliths. *Nature, 356*, 516–518.

Young, J. R., Geisen, M., Cros, L., Kleijne, A., Sprengel, C., & Probert, I. (2003). A guide to extant calcareous nannoplankton taxonomy. *Journal of Nannoplankton Research*, 1–125.

Zhang, H., Bhattacharya, D., & Lin, S. (2005). Phylogeny of dinoflagellates based on mitochondrial cytochrome B and nuclear small subunit rDNA sequence comparisons. *Journal of Phycology, 41*, 411–420.

Zingone, A., Sarno, D., Siano, R., & Marino, D. (2011). The importance and distinctiveness of small-sized phytoplankton in the Magellan Straits. *Polar Biology, 34*, 1269–1284.

CHAPTER TWO

Diversity and Evolution of Algae: Primary Endosymbiosis

Olivier De Clerck[1], Kenny A. Bogaert, Frederik Leliaert
Phycology Research Group, Biology Department, Ghent University, Krijgslaan 281 S8, 9000 Ghent, Belgium
[1]Corresponding author: E-mail: olivier.declerck@ugent.be

Contents

Abstract

Oxygenic photosynthesis, the chemical process whereby light energy powers the conversion of carbon dioxide into organic compounds and oxygen is released as a waste product, evolved in the anoxygenic ancestors of Cyanobacteria. Although there is still uncertainty about when precisely and how this came about, the gradual oxygenation of the Proterozoic oceans and atmosphere opened the path for aerobic organisms and ultimately eukaryotic cells to evolve. There is a general consensus that photosynthesis was acquired by eukaryotes through endosymbiosis, resulting in the enslavement of a cyanobacterium to become a plastid. Here, we give an update of the current understanding of the primary endosymbiotic event that gave rise to the Archaeplastida. In addition, we provide an overview of the diversity in the Rhodophyta, Glaucophyta and the Viridiplantae (excluding the Embryophyta) and highlight how genomic data are enabling us to understand the relationships and characteristics of algae emerging from this primary endosymbiotic event.

Advances in Botanical Research, Volume 64
ISSN 0065-2296,
http://dx.doi.org/10.1016/B978-0-12-391499-6.00002-5

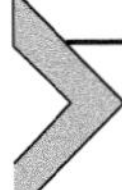

1. INTRODUCTION

1.1. Early Evolution of Oxygenic Photosynthesis

The origin of oxygenic photosynthesis has changed the face of our planet in all aspects. The first organisms that developed oxygenic photosynthesis are thought to have been the anoxygenic ancestors of Cyanobacteria (Allen & Martin, 2007), but when and how this came about remains a matter of debate (Farquhar, Zerkle, & Bekker, 2011; Hohmann-Marriott & Blankenship, 2011). Estimates based on geological and geochemical evidence and molecular phylogenetic analyses calibrated with the fossil record agree on a minimum age of 2.3 billion years ago (Tomitani, Knoll, Cavanaugh, & Ohno, 2006), but the origin of oxygenic photosynthesis may date back to 3.4 or even 3.8 billion years ago (Buick, 2008; Russell & Hall, 2006) (Fig. 2.1E). Because oxygenic photosynthesis involves the photolysis of water into electrons, protons and free oxygen, Cyanobacteria are singularly responsible for oxygenating the atmosphere and transforming a once reducing environment into an oxidising one (Holland, 2006).

With oxygen becoming gradually available as a very potent electron acceptor, the path lay open for aerobic organisms to evolve. Aerobes soon managed to maintain much more productive ecosystems as more energy per electron transfer could be harvested. Consequently, oceanic primary production increased an order of magnitude (Canfield, Rosing, & Bjerrum, 2006), permitting the evolution of more complex life forms (Catling, Glein, Zahnle, & McKay, 2005) and adapted or novel biochemical pathways (Falkowski, 2006; Raymond & Segré, 2006). The rising atmospheric oxygen is thought to have directly triggered cellular compartmentalization and eukaryogenesis (Fig. 2.1E). Atmospheric oxygen is thought to have constrained the topology of ancient transmembrane proteins by limiting the size and number of the external domains of transmembrane proteins (Acquisti, Kleffe, & Collins, 2007). When oxygen levels rose, the constraint likely decreased, permitting larger and more communication-related transmembrane proteins opening the door for subsequent compartmentalization. Alternatively, rising oxygen levels is speculated to have promoted cellular compartmentalization in order to protect the metabolic activities of the plasma membrane from rising levels of reactive oxygen species in the cellular environment (Gross & Bhattacharya, 2010). The fossil record (Javaux, 2011; Knoll, Javaux, Hewitt, & Cohen, 2006) and time-calibrated phylogenies (Hedges, Blair, Venturi, & Shoe, 2004; Parfrey, Lahr, Knoll, & Katz, 2011)

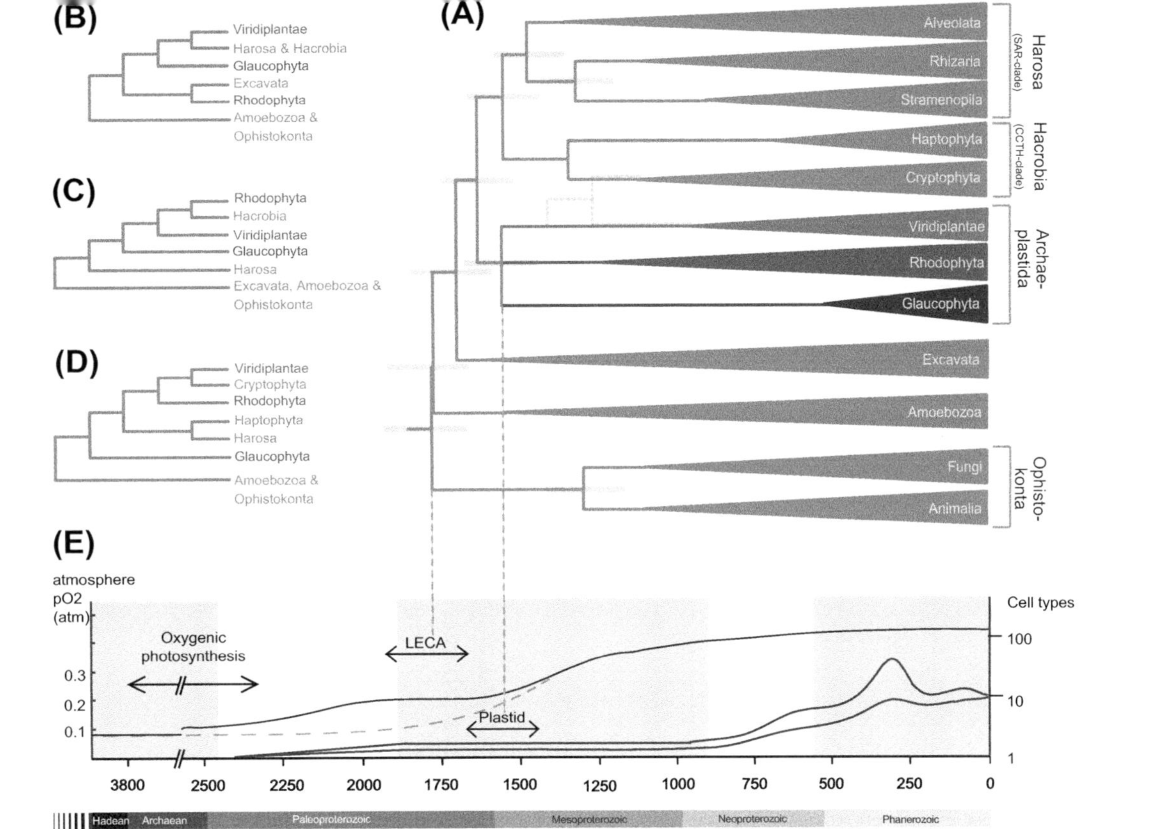

Figure 2.1 Relationships of Archaeplastida with main eukaryotic lineages and correlation between the rise in atmospheric oxygen and the evolution of organismal complexity. (A) Time-calibrated tree of extant eukaryotes (after Parfrey *et al.*, 2011). The tree topology is adjusted for the current uncertainty with respect to the branching order within the Archaeplastida. The dotted green line denotes the sister relationship between Viridiplantae and Cryptophyta in the analysis of Parfrey *et al.* (2011). Nodes are at mean divergence times and gray bars represent 95% highest probability density of node age. (B–D) Alternative topologies suggested by, respectively, Nozaki *et al.* (2009), Hampl *et al.* (2009) and Baurain *et al.* (2010). (E) Atmospheric partial oxygen pressure (blue lines) and cellular complexity (black line) (after Holland, 2006, and Hedges *et al.*, 2004). Blue lines denote maximum and minimum estimates of atmospheric O_2 partial pressure, respectively. Cellular complexity is defined as number of cell types. The black dashed line shows a more conservative interpretation of cellular complexity in the Proterozoic. The alternation of gray and white periods denotes the five different stages in oxygenation of the atmosphere according to Holland (2006). LECA: last eukaryotic common ancestor. See the colour plate.

suggest that the major eukaryotic lineages diverged already in the Paleoproterozoic era (2500–1600 Ma), but that diversity within major extant clades expanded later, beginning about 800 Ma, coinciding with the spread of oxygen through the Neoproterozoic oceans (Fig. 2.1A, E).

1.2. Origin of Plastids: Primary Endosymbiosis

Even though there is still considerable debate regarding the precise mechanisms and sequence of events that resulted in the first eukaryotic cell (de Duve, 2007; Embley & Martin, 2006; Martin & Muller, 1998; Poole & Neumann, 2011; Roger, 1999), there is a general consensus that photosynthetic eukaryotes emerged from a heterotrophic eukaryote which engulfed a cyanobacterium. The cyanobacterium was gradually enslaved and integrated into the cellular machinery as a new organelle: the plastid. This event has been termed primary endosymbiosis. The cyanobacterial origin of plastids is supported by overwhelming genetic evidence and ultrastructural similarities between plastids and their cyanobacterial relatives (Box 2.1). The original cyanobacterial genome underwent a drastic reduction with most genes either lost or transferred to the host nucleus,

BOX 2.1 Ultrastructural and Molecular Similarities Between Plastids and Cyanobacteria

Homology between (envelope) membranes	Presence of galactolipids, β-barrel proteins and occurrence of a peptidoglycan layer in Glaucophyta
Chloroplast DNA	Similarities in structure and gene content of the circular genome, organized into discrete nucleoids
Molecular phylogeny	Phylogenetic studies indicate that chloroplast gene sequences are nested within eubacterial homologs
Nuclear genes of cyanobacterial origin	Ample presence of genes of presumed cyanobacterial origin in the nuclear genome as a consequence of EGT
Ribosomes	Chloroplasts contain ribosomes that are 70S in size similar to prokaryotic ribosomes, as opposed to typical eukaryotic, cytosolic and endoplasmic-reticulum-associated, which are 80S in size; Inhibition by antibiotics (e.g. streptomycin, kanamycin) that affect ribosome function in free-living Eubacteria
Photosynthetic apparatus	The presence of two photosystems (PS I and PS II) in which a central chlorophyll *a* molecule is oxidized; electrolysis of H_2O as electron donor and release of O2; similarities in the electron transport chains; shared phycobilisomes between Cyanobacteria and Glaucophyta and Rhodophyta

termed endosymbiotic gene transfer (EGT). A fraction of the genome is retained within the primary plastid, minimally encoding its own protein synthesizing machinery and a number of genes involved in photosynthesis (Gould, Waller, & McFadden, 2008). Genes that have been transferred to the host nucleus are transcribed and translated in the host cytosol or endoplasmic reticulum and are targeted back to the chloroplast using a protein import system (Bhattacharya, Archibald, Weber, & Reyes-Prieto, 2007). In contrast to what might be intuitively expected, also gene products of host origin can be plastid-targeted and only a subset of cyanobacterial genes takes up a function in the organelle (Deusch *et al.*, 2008; Martin *et al.*, 2002). The overall emerging picture is one of large genomic impact of the symbiont on its host after primary endosymbiosis (Elias & Archibald, 2009), although the exact impact on the genomic content of Archaeplastida remains uncertain. Some phylogenomic analyses estimate the contributed genes to range around 20% of the total gene number when including a correction for the high rate of divergence (Deusch *et al.*, 2008; Martin *et al.*, 2002). Others calculate more modest percentages of chloroplast-derived genes, ranging around 5% while acknowledging these proportions are most likely underestimations due to high sequence divergence (Moustafa & Bhattacharya, 2008; Reyes-Prieto, Hackett, Soares, Bonaldo, & Bhattacharya, 2006). Next to sequence divergence, amelioration and modularity of transferred genes are thought to be additional complicating factors to detect horizontal gene transfer (Chan *et al.*, 2011). Remarkably, some phylogenomic analyses, with the exception of the glaucophyte study of Reyes-Prieto *et al.* (2006), indicate that more than 50% of the transferred genes have other functions, from metabolism to cell division, instead of being plastid targeted (Deusch *et al.*, 2008; Martin *et al.*, 2002).

Three extant groups of photosynthetic eukaryotes have primary plastids: the green plants, red algae and the glaucophytes. Together they make up the Archaeplastida. Even though the cyanobacterial origin of the plastids in these groups is beyond dispute, the number of endosymbiotic events and the relationships among the three lineages is more contentious (Delwiche, 1999, 2007). For a long time, variation in plastid structure and light-harvesting pigments has given credit to a polyphyletic origin of primary plastids, that is, the hypothesis that primary plastids resulted from multiple independent primary endosymbiotic events. Recent evidence points towards a single origin of primary plastids, which implies a single ancestor of the plastid as well as the monophyly of the three lineages that make up the Archaeplastida (Keeling, 2010; Rodriguez-Ezpeleta *et al.*, 2005). As pointed out by Larkum, Lockhart, and Howe (2007), however, support for a single origin of plastids should be treated with caution and several lines of evidence, which are predominantly based on phylogenetic tree methods, may not disprove all alternative scenarios of plastid acquisition. There is at least one exception to this rule: *Paulinella chromatophora*, a cercozoan amoeba with photosynthetic inclusion of cyanobacterial origin (Marin, Nowack, & Melkonian, 2005; Nowack *et al.*, 2011; Nowack & Grossman, 2012).

Even though several analyses provide moderate to strong support for a monophyletic Archaeplastida (Burki *et al.*, 2007, 2009; Hackett, Yoon, Li, Reyes-Prieto, Rummele, & Bhattacharya, 2007; Patron, Inagaki, & Keeling, 2007; Rodriguez-Ezpeleta *et al.*, 2005), other studies suggest that the Archaeplastida might be paraphyletic with respect to the Hacrobia (Burki, Okamoto, Pombert, & Keeling, 2012; Hampl *et al.*, 2009) or the entire Chromalveolata (Baurain *et al.*, 2010; Nozaki *et al.*, 2009) (Fig. 2.1A–D). The incongruence between analyses is likely caused by systematic biases including EGT as suggested by the high instability of resultant topologies of photosynthetic clades with varying levels of taxon sampling and missing data (Parfrey *et al.*, 2010). Indeed, gene sampling has been shown to account for at least some of the incongruence among the relationships of primary plastid lineages (Inagaki, Nakajima, Sato, Sakaguchi, & Hashimoto, 2009). The persistent incongruence of large concatenated data sets shows that a solution may not be found by increasing sequence length (Baurain *et al.*, 2010; Burki *et al.*, 2009; Hampl *et al.*, 2009). Instead when relaxing the assumption of vertical gene transfer by abandoning concatenation and choosing for a gene-by-gene approach, Chan *et al.* (2011) and Price *et al.* (2012) provide additional evidence for monophyly of red and green algae.

In the light of the persistent uncertainty on the monophyly of Archaeplastida, it may not come as a surprise that the relationships between green plants, red algae and glaucophytes are still unclear. Traditionally, glaucophytes are thought to have diverged before the red algae and green plants based on similarities of the plastid with cyanobacteria, such as the presence of a peptidoglycan layer surrounding the plastids (originally named 'cyanelles'). Phylogenetic gene analyses are unfortunately not conclusive on the relationships between the major clades of the Archaeplastida (Rodriguez-Ezpeleta *et al.*, 2005; Rodriguez-Ezpeleta, Philippe, Brinkmann, Becker, & Melkonian, 2007). Furthermore, several studies point towards an early diverging red algal lineage (Burki *et al.*, 2009; Hackett *et al.*, 2007; Patron *et al.*, 2007), although this result might be biased by the inclusion of clades with secondary plastids (Deschamps & Moreira, 2009). Therefore, analyses concentrating on EGT genes of cyanobacterial origin only might be more trustworthy. Even so, phylogenetic analyses are ambiguous either suggesting the glaucophytes (Reyes-Prieto & Bhattacharya, 2007) or green lineage (Deschamps & Moreira, 2009) as earliest diverging lineage within the Archaeplastida.

Under the assumption of a single origin of primary plastids, the question remains what kind of cyanobacterium participated in the origin of plastids. Unfortunately, due to the large divergence times and the considerable extent of horizontal gene transfer between cyanobacteria (Deusch *et al.* 2008), the phylogenetic signal of these relationships is seriously eroded. Some studies suggest a rather deep origin of plastids, predating diversification of most extant cyanobacterial lineages (Criscuolo & Gribaldo, 2011; Reyes-Prieto *et al.*, 2010; Rodriguez-Ezpeleta *et al.*, 2005; Sato, Wise, & Hoober, 2006), other studies suggest that plastids are more closely related to one of the contemporary clades such as N-fixing subsection I (Deschamps *et al.*, 2008; Falcon, Magallon, & Castillo, 2010) or filamentous heterocyst-forming subsection IV (Deusch *et al.*, 2008). In addition, it is difficult to determine when this primary endosymbiosis occurred. Estimates based on fossil evidence and biomarkers are widely divergent (Knoll, 1992). A recent calibrated phylogeny of Parfrey *et al.* (2011) corroborates earlier studies with an estimated ages of the clade containing Viridiplantae, red algae and glaucophytes around 1.5–1.6 billion years ago (Hedges *et al.*, 2004; Yoon, Hackett, Ciniglia, Pinto, & Bhattacharya, 2004) (Fig. 2.1E). Following the origin of Archaeplastida, photosynthesis spread widely among diverse eukaryotic groups via secondary and tertiary endosymbiotic events (Archibald, 2009; Gould *et al.*, 2008; Keeling, 2010). Overviews of the intricate histories of plastid acquisition are provided in the next chapter of this volume (Archibald, 2012).

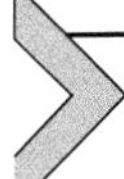

2. RED ALGAE

2.1. Red Algae Defined

The red algae or Rhodophyta are a distinct lineage of eukaryotic algae, containing about 5000–6000 species of mostly multicellular marine algae. The red algae are distinguishable among eukaryotic lineages by a combination of biochemical and ultrastructural features, some of which they share with Glaucophyta and Cyanobacteria. First, red algal plastids lack chlorophyll accessory pigments. Instead light energy is directed to the reaction centre by phycobiliproteins (phycocyanin, allophycocyanin and phycoerythrin). Light-harvesting antennae pigments are grouped in hemispherical protein complexes, phycobilisomes, anchored to the thylakoids. These are not stacked in grana like in the Viridiplantae but lie singly and more or less equidistant in the plastid stroma. One of the most distinctive characters of the red algae is the complete absence of flagella and centrioles in all life stages, which affects mitosis and, at least in some groups, their life cycles (Graham, Graham, & Wilcox, 2009; Maggs, Verbruggen, & De Clerck, 2007; Saunders & Hommersand, 2004; van den Hoek, Mann, & Jahns, 1995; Yoon, Müller, Sheath, Ott, & Bhattacharya, 2006b; Yoon, Zuccarello, & Bhattacharya, 2010).

From the early twentieth century until very recently, red algae were classified in two distinct groups, most commonly treated as classes, Bangiophyceae and Florideophyceae, within a single phylum, Rhodophyta. This dichotomy in the classification is reflected in the morphological complexity that characterizes the red algae, with the Bangiophyceae uniting the morphologically simple forms (unicells or undifferentiated filaments and blades) and the Florideophyceae containing the more complex growth forms. Growth in the Florideophyceae is essentially filamentous, but individual filaments may aggregate to form a pseudoparenchymatous tissue. Growth forms include filaments, blades, elaborately branched thalli as well as calcified crusts (coralline algae). A wealth of molecular and ultrastructural data, however, made clear that this traditional classification did not reflect the antiquity and diversity of the rhodophytes (Müller, Lynch, & Sheath, 2010; Müller *et al.*, 2001; Oliveira & Bhattacharya, 2000). The structurally simple Bangiophyceae is composed of a series of radiations that define the ancestral lineages of the red algae, and part of the traditionally circumscribed Bangiophyceae is more closely related to the Florideophyceae. Hence, a new classification was originally proposed by Saunders and Hommersand (2004) and subsequently refined by Yoon *et al.* (2006b). The phylum Rhodophyta is now subdivided into two subphyla, Cyanidiophytina and Rhodophytina, and seven classes, Cyanidiophyceae, Bangiophyceae, Compsopogonophyceae, Florideophyceae, Porphyridiophyceae, Rhodellophyceae and Stylonematophyceae (Fig. 2.2). The diversity contained in the Compsopogonophyceae, Porphyridiophyceae, Rhodellophyceae and Stylonematophyceae is still ill-defined as can be witnessed by the many unnamed lineages that typically adorn phylogenetic trees (Scott *et al.*, 2008; West, Zuccarello, Scott, West, & Karsten, 2007; Yang *et al.*, 2010; Yokoyama *et al.*, 2009; Zuccarello, Kikuchi, & West,

2010; Zuccarello, Oellermann, West, & De Clerck, 2009; Zuccarello, West, & Kikuchi, 2008; Zuccarello *et al.*, 2011).

Red algae are an ancient lineage (Xiao, Zhang, & Knoll, 1998; Yoon *et al.*, 2004). A 1.2-billion-year-old fossil, *Bangiomorpha pubescens* (Butterfield, 2000), which bears a lot of resemblance to extant *Bangia* species, is regarded as the oldest taxonomically resolved eukaryotic fossil. The taxonomic affinity of *Bangiomorpha* was long contested, with some authors (e.g. Cavalier-Smith, 2006) advocating that *Bangiomorpha* is a blue-green alga or a mixture of at least two species of blue-green algae, possibly related to the Stigonematales. *Bangiomorpha* being a eukaryotic fossil would indicate a Mesoproterozoic origin of red algae and by extension all major lineages of the eukaryotes, which contradicts the hypothesis of Cavalier-Smith that places eukaryogenesis at 850 Ma (Cavalier-Smith, 2010). A red algal nature of *Bangiomorpha*, however, is not in conflict with the most recent timing of eukaryotic diversification using calibrated phylogenies (Parfrey *et al.*, 2011) or the interpretation of several microfossils that place the origin of the eukaryotes at around 1800 Ma (Knoll *et al.*, 2006). In addition to the fossils indicative of a Mesoproterozoic origin of red algae, the remarkably well-preserved multicellular red algae from the Doushantuo Formation in southern China (ca. 600 Ma) are clear proof that red algae had already radiated prior to the Precambrian radiation.

2.2. Cyanidiophytes

The updated classification of the red algae also better reflects the ultrastructural and ecological diversity of the group. Of special interest are the Cyanidiophyceae, a group of unicellular and presumably asexual algae, which live in thermoacidophilic conditions that are detrimental to most eukaryotic live on earth (Barbier *et al.*, 2005; Ciniglia, Yoon, Pollio, Pinto, & Bhattacharya, 2004; Matsuzaki *et al.*, 2004; Yoon *et al.*, 2006a). Yoon *et al.* (2004) suggested that the Cyanidiophyceae diverged from the remaining red algae prior to the secondary endosymbiotic event that gave rise to the Chromalveolata. Even though a reinterpretation of this event is necessary now that the chromalveolate hypothesis is being increasingly challenged (Archibald, 2012; Baurain *et al.*, 2010), it does vouch for the antiquity of the divergences that separate the deep red algal lineages. Because the Cyanidiophyceae are one of few eukaryotic groups that thrive in environments that are otherwise dominated by Archaea and Bacteria, their enzymes are of special interest to the biotechnology and pharmaceutical industry. It is, therefore, not surprising that *Cyanidioschyzon merolae* was the first eukaryotic alga for which a full genome sequence was available (Matsuzaki *et al.*, 2004; Misumi *et al.*, 2005) and the genome of the related *Galdieria sulphuraria* is currently being sequenced (Barbier *et al.*, 2005). The small and compact genome of *C. merolae*, combined with an extremely reduced cell architecture (one nucleus, mitochondrion and chloroplast and the absence of a cell wall), makes it an ideal organisms to address the synchronization and mechanisms of the organellar division during cytokinesis (Kuroiwa, 1998). Only seven Cyanidiophyceae species are currently described, but this number severely underestimates the diversity. Applying environmental sequencing from a single locality in Italy, Ciniglia *et al.* (2004) unveiled considerable cryptic diversity and within-lineage (often 'intraspecific') sequence divergence that is comparable, for example, to between-order divergences in the non-Cyanidiales red algae.

2.3. Of Nori and Red Seaweed

The great majority of red algae are multicellular, marine seaweeds, with an enormous range of morphologies and complex haplodiploid life histories, which involve additional zygote amplification stages resulting in large numbers of spores from a single fertilization (Verbruggen *et al.*, 2010). Red seaweeds belong nearly exclusively to two classes, Bangiophyceae (in the narrow sense of the new classification) and Florideophyceae (Fig. 2.2). Traditionally, two genera belonging to a single family, Bangiaceae, have been recognized in the Bangiophyceae. Unbranched uniseriate to multiseriate filaments have been placed in the genus *Bangia* and blades in the genus *Porphyra* (Sutherland *et al.*, 2011). The latter genus is commonly known as 'nori', which is cultivated as one of the most profitable mariculture crops in the north-western Pacific (Niwa *et al.*, 2009). Molecular phylogenetic analyses resulted in the recognition of 15 genera, 7 filamentous and 8 foliose, which are only separable on molecular rather than morphological grounds. Regrettably, this taxonomic vigour also resulted in the fact that the *Porphyra*, which has wrapped sushi for decades, has now become a *Pyropia* (Zuccarello, 2011).

The near-morphological stasis that characterizes the Bangiophyceae, contrasts sharply with the wealth of growth forms that is encountered in its sister taxon, the Florideophyceae. DNA sequence data have progressively refined an ordinal classification (e.g. Choi, Kraft, Lee, & Saunders, 2002; Harper & Saunders, 2001; Le Gall & Saunders, 2007; Saunders, Chiovitti, & Kraft, 2004), which was originally entirely based on morphological

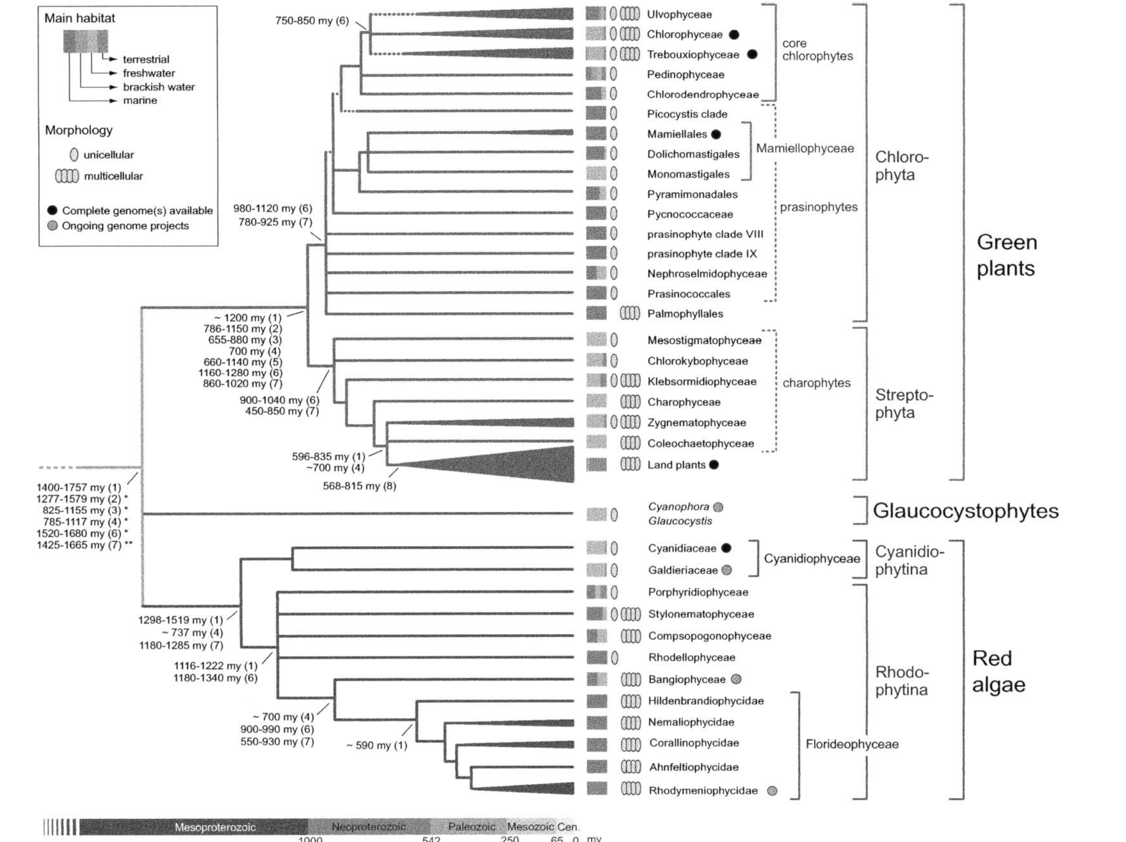

Figure 2.2 Relationships among major groups of Archaeplastida based on published molecular phylogenetic data (Leliaert *et al.*, 2012; Yoon *et al.*, 2010) with indication of main habitats and morphology. Divergence times are based on molecular clock studies calibrated with the fossil record: (1) Yoon *et al.* (2004), (2) Hedges *et al.* (2004), (3) Douzery, Snell, Bapteste, Delsuc, and Philippe (2004), (4) Berney and Pawlowski (2006), (5) Roger and Hug (2006) (r8s-PL method), (6) Herron *et al.* (2009) and (7) Parfrey *et al.* (2011). *excluding glaucophytes; **including cryptomonads. See the colour plate.

and reproductive features (Maggs *et al.*, 2007). The Florideophyceae now includes some 25 orders grouped into five subclasses, Hildenbrandiophycidae, Nemaliophycidae, Ahnfeltiophycidae, Rhodymeniophycidae and Corallinophycidae (Le Gall & Saunders, 2007; Maggs *et al.*, 2007; Saunders & Hommersand, 2004), which are reasonably well supported by ultrastructural characters. Ordinal relationships, however, remain at least partly unresolved (Verbruggen *et al.*, 2010) and form the motivation of a Red Tree of Life project (http://dblab.rutgers.edu/redtol/).

Genomic data of red seaweeds are currently limited to a number of organelle genomes and expressed sequence tag (EST) libraries of commercially important species such as *Chondrus*, *Gracilaria* and *Porphyra* (Asamizu *et al.*, 2003; Collen *et al.*, 2006; Hagopian, Reis, Kitajima, Bhattacharya, & De Oliveira, 2004; Nikaido *et al.*, 2000) and the coralline red alga *Calliarthron tuberculosum* (Chan *et al.*, 2011), but this may change soon with whole genome project of the carragenophyte *Chondrus crispus* and *Porphyra umbilicalis* in progress.

3. GREEN PLANTS (VIRIDIPLANTAE)

3.1. Green Plants Defined

The green plant clade (Viridiplantae) includes green algae and embryophytic land plants and is one of the main groups of photosynthetic eukaryotes. Green plants are diverse in terms of species number, morphology, biochemistry and ecology. Monophyly of the group is well established based on ultrastructural, biochemical and molecular data (Leliaert *et al.*, 2012; Lewis & McCourt, 2004).

Green plants share a number of unique characteristics. The chloroplasts are surrounded by a double membrane, have thylakoids grouped in lamellae and contain chlorophyll *a* and *b* along with some accessory pigments including carotenoids and xanthophylls. Pyrenoids (when present) are embedded within the chloroplast and are surrounded by starch, which is the main reserve polysaccharide. Cell walls (when present) are generally composed of cellulose. Many green algae are flagellates or have flagellate cells in some stage of the life cycle. The flagella (generally two or four on a cell) are isokont, which means that they are similar in structure, although they may differ in length. The region between the flagellar axoneme and the basal body is characterized by a stellate structure (Graham *et al.*, 2009; van den Hoek *et al.*, 1995).

Apart from these unifying ultrastructural and biochemical features, green plants are extremely diverse morphologically. They range from unicells with sizes comparable to bacteria to large and complex multicellular or siphonal life forms. Although the described species diversity of land plants (including over 250,000 species) exceeds that of green algae (about 15,000 named species), green algae encompass a greater cytomorphological, biochemical and reproductive diversity, which reflect their old evolutionary age (Leliaert *et al.*, 2012). The progenitor of green plants was likely a unicellular flagellate or at least had flagellate stages in its life cycle. Colonial and multicellular forms have evolved multiple times in several lineages, including the Streptophyta, Ulvophyceae, Chlorophyceae, Trebouxiophyceae and Palmophyllales (Fig. 2.2).

Green plants are also ecologically very diverse. They are especially abundant in freshwater (most charophytes, Chlorophyceae and Trebouxiophyceae) and marine environments (Ulvophyceae and prasinophytes; Not *et al.*, 2012), but some have adapted to specific habitats, such as dry land (Lewis & Lewis, 2005; López-Bautista, Rindi, & Guiry, 2006), arctic (De Wever *et al.*, 2009) and marine deep water environments (Zechman *et al.*, 2010). Several members of the core chlorophytes live in symbiosis with a diverse array of eukaryotes (Friedl & Bhattacharya, 2002; Kerney *et al.*, 2011; Lewis & Muller-Parker, 2004) or have adopted a heterotrophic life style as parasites (Joubert & Rijkenberg, 1971; Sudman, 1974). Embryophytes have dominated terrestrial habitats for millions of years; some land plants have adapted secondarily to freshwater or marine environments.

3.2. Evolutionary History of Green Plants

Green plants have played a significant ecological role for millions of years (Leliaert, Verbruggen, & Zechman, 2011; O'Kelly, 2007). The ecological importance of green algae has been mainly in marine and freshwater environments. The origin of land plants from a green algal ancestor was a key event in the evolution of life on earth. This event initiated the development of the entire terrestrial ecosystem and has led to environmental changes on a global scale (Kenrick & Crane, 1997). Time-calibrated phylogenies, calibrated with the scarce fossil record, have estimated the origin of the green plant lineage somewhere between 700 and 1500 Ma (Berney & Pawlowski, 2006; Hedges *et al.*, 2004; Herron, Hackett, Aylward, & Michod, 2009; Yoon *et al.*, 2004). An early split in the evolution of green plants gave rise to two main clades: the Chlorophyta and Streptophyta (Leliaert *et al.*, 2012; Lemieux, Otis, & Turmel, 2007; Rodriguez-Ezpeleta *et al.*, 2007) (Fig. 2.2).

The Chlorophyta probably diversified as unicellular algae in the Meso- and Neoproterozoic. These green algae were dominant in the oceanic phytoplankton of the Paleozoic as evidenced by fossil deposits of resistant outer walls of prasinophytic cysts, known as phycomata (Colbath, 1983; Knoll, 1992; O'Kelly, 2007; Tappan, 1980). This early radiation of Chlorophyta was important to the eukaryotic greening that shaped the geochemistry of our planet (Worden *et al.*, 2009). During the Mesozoic, the dominance of marine green algae in the phytoplankton gradually decreased as they were largely displaced by the red-plastid-containing dinoflagellates, coccolithophores and diatoms (Falkowski *et al.*, 2004; Leliaert *et al.*, 2011; O'Kelly, 2007; Simon, Cras, Foulon, & Lemee, 2009). These ancestral green unicells gave rise to modern prasinophytes and the core Chlorophyta that diversified as unicellular and multicellular organism in marine, freshwater and terrestrial habitats.

The Streptophyta probably originated in the Neoproterozoic and diversified as unicellular algae in freshwater environments (Becker & Marin, 2009). Two important groups of multicellular charophytes diversified during the Paleozoic: the conjugating green algae (Zygnematophyceae) and stoneworts (Charophyceae) (Becker & Marin, 2009). Similar to the situation in marine environments, red-plastid-containing dinoflagellates, diatoms and chrysophytes gradually took over the green dominance in Early Cretaceous and Cenozoic freshwater ecosystems (Becker & Marin, 2009). Ancestral charophytes invaded the land during the mid-Ordovician and early Silurian (480–430 million years ago), giving rise to the land plants (Delaux, Nanda, Mathé, Sejalon-Delmas, & Dunand, 2012; McCourt, Delwiche, & Karol, 2004).

Molecular phylogenetic studies have drastically reshaped our views of green plant evolution and continue to do so (Leliaert *et al.*, 2012; Marin, 2012; Timme, Bachvaroff, & Delwiche, 2012). However, many uncertainties remain, especially about the deepest branches of the green plants. One of the main goals of the Green Algal Tree of Life Project (http://alleyn.eeb.uconn.edu/gratol/) is to resolve relationships among the main green algal lineages. Phylogenetic hypotheses are critical in providing an evolutionary framework for comparative genomic studies. In the following section, we give a brief overview of the major green plant lineages and their relationships.

3.3. Chlorophyta

The Chlorophyta form a large and morphologically diverse clade of marine, freshwater and terrestrial green algae. The flagellar apparatus in this clade is characterized by a symmetrical cruciate root system wherein rootlets of variable (*X*) numbers of microtubules alternate with rootlets composed of two microtubules to form a '*X*-2-*X*-2' arrangement. The orientation of this flagellar root system has been an important character for defining the main groups of Chlorophyta. Molecular data have revealed several major chlorophytan clades. Several early diverging clades of unicellular algae, collectively termed the prasinophytes, form a paraphyletic assemblage at the base of the chlorophytan tree. These clades are relatively species poor compared to the three principal clades of the core Chlorophyta: Ulvophyceae, Trebouxiophyceae and Chlorophyceae (Fig. 2.2).

Prasinophytes form a heterogeneous assemblage of mostly unicellular algae with diverse cell shapes that are naked, covered by walls or organic body scales; flagella are present or absent (Leliaert *et al.*, 2011; Melkonian, 1990; Sym & Pienaar, 1993). Mitotic processes, biochemical features and photosynthetic pigments are equally diverse, reflecting the paraphyletic nature of the group (Fawley, Yun, & Qin, 2000; Guillou *et al.*, 2004; Latasa, Scharek, Le Gall, & Guillou, 2004; Nakayama *et al.*, 1998; Zingone *et al.*, 2002). Prasinophytes are predominantly found in marine environments, although several species also occur in freshwater. About 10 distinct prasinophyte lineages have been identified, but their phylogenetic affinities remain largely unresolved (Leliaert *et al.*, 2011, 2012; Marin & Melkonian, 2010; Turmel, Gagnon, O'Kelly, Otis, & Lemieux, 2009) (Fig. 2.2).

The Nephroselmidophyceae includes flagellates with complex scale covering and is possibly one of the earliest diverging chlorophytan lineages (Turmel *et al.*, 2009). Although *Nephroselmis* is one of the few prasinophytes where sexual reproduction has been observed (Suda, Watanabe, & Inouye, 2004), genomic evidence, such as the identification of meiosis-related genes, indicates that sexual reproduction is probably more widespread among prasinophytes (Derelle *et al.*, 2006; Worden *et al.*, 2009). Future population genomic studies may enable us to estimate the prevalence of sexual recombination in algae (Toulza, Knoll, Cavanaugh, & Ohno, 2012).

The Mamiellophyceae includes the morphologically and ecologically diverse Mamiellales and two smaller clades, the Monomastigales and Dolichomastigales (Marin & Melkonian, 2010; Nakayama, Kawachi, & Inouye, 2000; Sym & Pienaar, 1993; Turmel *et al.*, 2009; Zingone *et al.*, 2002). The Mamiellales include marine and freshwater flagellates and coccoid forms. Species of *Ostreococcus* and *Micromonas* are among the smallest eukaryotes known, with cell sizes of 0.5–2 μm (Derelle *et al.*, 2006; Palenik *et al.*, 2007; Worden *et al.*, 2009) and are important components of marine picoeukaryotic communities (Leliaert *et al.*, 2012; Not *et al.*, 2004;

O'Kelly, Sieracki, Thier, & Hobson, 2003; Vaulot, Eikrem, Viprey, & Moreau, 2008). We refer to Chapter 10 (Toulza *et al.*, 2012) for a review on environmental genomics in the Mamiellales and other microalgae.

The Pyramimonadales includes large flagellates covered by complex body scales found in marine and freshwater environments. Some species are unique among green plants in possessing a food uptake apparatus (Moestrup, Inouye, & Hori, 2003), which has been interpreted as a character that might have been inherited from a phagotrophic ancestor of the green plants (O'Kelly, 2007).

The *Picocystis* clade includes the coccoid *Picocystis* from saline lakes. Together with some undescribed coccoids ('CCMP1205 clade'), these prasinophytes might form the closest sister lineages of the core chlorophytes, although strong support is lacking (Marin & Melkonian, 2010).

Several other prasinophytic groups have uncertain phylogenetic affinities. These include the Pycnococcaceae, a clade of marine flagellate and coccoid species (Nakayama, Suda, Kawachi, & Inouye, 2007; Turmel *et al.*, 2009); the Prasinococcales, a clade of marine coccoids (Hasegawa *et al.*, 1996; Sieburth, Keller, Johnson, & Myklestad, 1999) and two clades ('clades VIII and IX') that are known from environmental sequencing only (Lepère, Vaulot, & Scanlan, 2009; Shi, Marie, Jardillier, Scanlan, & Vaulot, 2009; Viprey, Guillou, Ferréol, & Vaulot, 2008).

The Palmophyllales includes green algae from dimly lit benthic marine habitats. These algae feature a unique type of multicellularity, forming well-defined macroscopic bodies composed of small spherical cells embedded in a firm gelatinous matrix. Phylogenetic analysis either places the Palmophyllales as the sister clade to all other Chlorophyta or allies it with the Prasinococcales (Leliaert *et al.*, 2011; Zechman *et al.*, 2010).

The core Chlorophyta evolved from one of the ancestral prasinophytic lineages probably somewhere in the Neoproterozoic (Herron *et al.*, 2009). The core Chlorophyta includes the species-poor and early-diverging Pedinophyceae (marine and freshwater uniflagellates) and Chlorodendrophyceae (marine and freshwater quadriflagellates), and the large and diverse clades, Trebouxiophyceae, Ulvophyceae and Chlorophyceae (TUC) (Leliaert *et al.*, 2012; Marin, 2012). The TUC clades include a wide variety of morphological forms and eco-physiological features. Unlike the prasinophytes, where sexual reproduction has rarely been observed, the core chlorophytes encompass a large diversity of life cycle strategies, many of which involve sexual reproduction. Marine members of the Ulvophyceae generally have life cycles involving an alternation between two free-living multicellular phases (a haploid gametophyte and diploid sporophyte). Many freshwater Chlorophyceae and Trebouxiophyceae have a haploid vegetative phase and a single-celled, often dormant zygote as the diploid phase. Conversely, terrestrial members of the core chlorophytes are mainly asexual (Rindi, 2011). We refer to Chapter 6 (Umen and Olson, 2012) for a review on the evolution of sex in the chlorophycean green algae *Chlamydomonas* and *Volvox*.

A new mode of cell division likely evolved in the clade uniting the Chlorodendrophyceae and the TUC clade and was subsequently lost in the Ulvophyceae (Leliaert *et al.*, 2012). This type of cell division is mediated by a phycoplast, which is an array of microtubules oriented parallel to the plane of cell division, determining the formation of a new cell wall (Graham *et al.*, 2009; van den Hoek *et al.*, 1995). Morphological and eco-physiological adaptations probably allowed successful radiation of the Trebouxiophyceae and Chlorophyceae in freshwater and terrestrial habitats and diversification of the Ulvophyceae along marine shorelines (Becker & Marin, 2009; Cocquyt, Verbruggen, Leliaert, & De Clerck, 2010; Leliaert *et al.*, 2012).

The relationships among the core chlorophytan lineages are difficult to resolve, probably as a result of their antiquity and the short time span of diversification (Cocquyt *et al.*, 2010; O'Kelly, 2007). Furthermore, some phylogenetic studies showed that at least the Trebouxiophyceae and Ulvophyceae might not be monophyletic (e.g. Lü *et al.*, 2011; Turmel, Otis, & Lemieux, 2009; Zuccarello *et al.*, 2009) (but see Marin, 2012; Cocquyt *et al.*, 2010).

The Trebouxiophyceae includes flagellates, coccoids, colonies and multicellular filaments and blades. The group is predominantly freshwater or terrestrial; some members occur in brackish or marine habitats. Many species are photosynthetic symbionts with lichen fungi, various protists, invertebrates and plants; others have evolved a free-living or parasitic heterotrophic life style (Friedl & Rybalka, 2011; Leliaert *et al.*, 2012). Analysis of the complete genome of *Chlorella variabilis* NC64A (an endosymbiont of the ciliate *Paramecium bursaria*) has provided insights into the genetic facilitation of an endosymbiotic lifestyle (Blanc *et al.*, 2010). In particular, expansion of protein families containing protein–protein interaction domains and adhesion domains could have been involved in adaptation to symbiosis. Although *Chlorella* (and many other members of Trebouxiophyceae) has been assumed to be asexual and non-motile, meiosis- and flagella-specific proteins have been found in its genome, suggestive of cryptic sex and involvement of a flagella-derived structure in sexual reproduction (Blanc *et al.*, 2010).

The Chlorophyceae includes flagellates, coccoids and various colonial and multicellular forms. The group occurs mainly in freshwater and to a lesser extent in terrestrial habitats; some are marine (Klochkova *et al.*, 2008). Five main lineages have been recognized: the speciose and diverse Sphaeropleales and Chlamydomonadales including some of the most common freshwater phytoplankters, and the smaller clades, Chaetophorales, Oedogoniales and Chaetopeltidales (Leliaert *et al.*, 2012). The unicellular flagellate *Chlamydomonas* has been extensively studied as a model for photosynthesis, chloroplast biogenesis, flagellar assembly and function, cell–cell recognition, circadian rhythm and cell cycle control (Grossman *et al.*, 2003). The colonial *Volvox* has served as a model for the evolution of multicellularity, cell differentiation and colony motility (Herron & Michod, 2008; Kirk, 2003). Analysis of the complete genomes of *Chlamydomonas reinhardtii* and *Volvox carteri* has provided important genetic insights into the evolution of multicellularity and sex (Merchant *et al.*, 2007; Prochnik *et al.*, 2010, Umen and Olson, 2012). The Ulvophyceae includes unicells and multicellular algae as well as giant-celled forms with unique cellular characteristics (Cocquyt *et al.*, 2010; Leliaert *et al.*, 2012). Ulvophytes are generally known as macroalgae growing along marine coasts (green seaweeds). Species in the Ulvales, Bryopsidales and Cladophorales frequently dominate rocky shores, tropical lagoons and reefs. Some species of *Ulva* can form extensive, free-floating blooms, known as green tides (Ye *et al.*, 2011). *Caulerpa* and *Codium* species are notorious for their invasive nature (Williams & Smith, 2007). Several ulvophytes (e.g. *Ulva* and *Cladophora*) have secondarily adapted to freshwater environments. The Trentepohliales is atypical with respect to both morphology and ecology, occurring exclusively in terrestrial habitats (López-Bautista & Chapman, 2003). Some early diverging lineages (Oltmannsiellopsidales and *Ignatius*) include microscopic organisms occurring in freshwater or terrestrial habitats, indicating that the ancestral ulvophytes may have been freshwater or terrestrial unicells (Cocquyt *et al.*, 2010).

3.4. Streptophyta and the Origin of Land Plants

The Streptophyta include a paraphyletic assemblage of green algae (charophytes) and the land plants. Charophytes range in morphology from unicellular to complex multicellular organisms and occur in freshwater or moist terrestrial habitats. Streptophyta share a number of unique traits, including motile cells (when present) with two subapically inserted flagella and an asymmetrical flagellar apparatus that contains a distinctive multilayered structure and parallel basal bodies; open mitosis with a persistent mitotic spindle and several unique enzymes (Leliaert *et al.*, 2012). There are six main lineages of charophytes: Mesostigmatophyceae, Chlorokybophyceae, Klebsormidiophyceae, Zygnematophyceae, Charophyceae and Coleochaetophyceae (McCourt *et al.*, 2004) (Fig. 2.2). Many phylogenetic studies have aimed to resolve the relationship among these lineages and in particular to determine the origins of land plants (Karol, McCourt, Cimino, & Delwiche, 2001; Lemieux *et al.*, 2007; Rodriguez-Ezpeleta *et al.*, 2007; Timme *et al.*, 2012; Wodniok *et al.*, 2011).

Mesostigma (Mesostigmatophyceae) and *Chlorokybus* (Chlorokybophyceae) form the earliest-diverging streptophytic lineages (Timme *et al.*, 2012) (Fig. 2.2). *Mesostigma* is a flagellate covered with diverse organic scales and is found in freshwater habitats. *Chlorokybus* forms packets of a few non-motile cells and grows in moist terrestrial environments (McCourt *et al.*, 2004). The freshwater or terrestrial filamentous Klebsormidiophyceae diverged after the Mesostigmatophyceae and Chlorokybophyceae.

In contrast to these three early-diverging lineages that undergo cell division by furrowing, the remaining lineages (Charophyceae, Zygnematophyceae, Coleochaetophyceae and the land plants) evolved a new mechanism of cell division involving a phragmoplast, which consists of an array of microtubules oriented perpendicularly to the plane of cell division, determining the formation of the cell plate and new cell wall. Most of these later-diverging streptophytes also have cell walls with plasmodesmata, facilitating cytoplasmic communication between cells and development of complex tissues (Graham, Cook, & Busse, 2000). Species in the three early-diverging lineages have never been observed to reproduce sexually in contrasts to the remaining streptophytes where sex is widespread (McCourt *et al.*, 2004).

The Zygnematophyceae (conjugating green algae) is a species-rich and morphologically diverse clade, including non-motile unicells, filaments and small colonial forms. Sexual reproduction occurs by a unique process of conjugation, involving fusion of non-motile gametes. Flagellate stages are completely absent. The Charophyceae (stoneworts) include freshwater algae with complex macroscopic bodies composed of a main axis with whorled branches. Growth is by an apical meristematic cell. Sexual reproduction is oogamous with oogonia and antheridia surrounded by sterile cells. Charophyceae are well represented in the fossil record, which is a large diversity extending back to the Silurian (McCourt *et al.*, 2004).

The Coleochaetophyceae is a small clade of branched filaments that sometimes form discoid parenchymatous thalli (Graham, 1984). Based on morphological similarities with embryophytes, *Coleochaete* has traditionally been put forward as the closest relative of land plants (Graham, 1984; Graham *et al.*, 2000). For example, some species of *Coleochaete* have corticated zygotes that are retained on the mother plant from which they receive nourishment via placental transfer cells with wall ingrowths. Also, cytokinesis and phragmoplast formation are similar to land plants (Graham *et al.*, 2000).

Identifying the closest living relative of land plants has proven to be a difficult task (Cocquyt *et al.*, 2010; Karol *et al.*, 2001; Lemieux *et al.*, 2007; Rodriguez-Ezpeleta *et al.*, 2007; Turmel *et al.*, 2009; Turmel, Otis, & Lemieux, 2006, 2007). Recent studies based on broad phylogenomic sampling have suggested the Zygnematophyceae, or a clade uniting the Zygnematophyceae and Coleochaetophyceae, as sister lineage of the land plants (Timme *et al.*, 2012; Wodniok *et al.*, 2011; Laurin-Lemay, Brinkmann, & Philippe, 2012). The inferred relationship between Coleochaetophyceae and land plants is in line with earlier morphology-based hypotheses (Graham, 1984; Graham *et al.*, 2000), and the relationship between Zygnematophyceae and land plants is supported by some cellular and molecular features, including similarities in auxin signalling (De Smet *et al.*, 2011) and chloroplast movement (Wada, Kagawa, & Sato, 2003).

4. GLAUCOPHYTES

The Glaucophyta (also known as Glaucocystophyta) is a small and inconspicuous group of unicellular algae found in freshwater and terrestrial environments (Baldauf, 2008; Bhattacharya & Schmidt, 1997; Kies & Kremer, 1990; Schenk, 2001). The importance of the group lies mainly in its critical phylogenetic position, branching deeply within the Archaeplastida (Fig. 2.1) (Bhattacharya, Helmchen, Bibeau, & Melkonian, 1995; Moreira, Le Guyader, & Philippe, 2000; Nozaki *et al.*, 2009; Price *et al.*, 2012; Reyes-Prieto & Bhattacharya, 2007; Rodriguez-Ezpeleta *et al.*, 2005).

Glaucophytes are unique among photosynthetic eukaryotes in that they contain unusual plastids (originally named 'cyanelles') with several characteristics reminiscent of Cyanobacteria: plastids are surrounded by a prokaryote-type peptidoglycan wall (except for *Glaucosphaera vacuolata*) and contain only chlorophyll *a* and phycobilins (Steiner & Löffelhardt, 2002; Steiner, Yusa, Pompe, & Löffelhardt, 2005). Similar to red algae, plastids have unstacked thylakoids and light-harvesting proteins organized into phycobilisomes. About 13 species have been described in five genera: *Glaucocystis*, including coccoid cells with a cellulosic wall and two short rudimentary flagella; *Cyanophora* and *Peliaina* are wall-less flagellates with two heterokont flagella and *Gloeochaete* and *Cyanoptyche*, including non-motile cells in a gelatinous matrix, reproducing by motile or non-motile spores (Schenk, 2001).

To date, the genome of a single glaucophyte has been sequenced (Price *et al.*, 2012; Stirewalt, Michalowski, Löffelhardt, Bohnert, & Bryant, 1995). Genomic data and a better understanding of the phylogenetic position of glaucophytes will provide valuable insights into the endosymbiotic origin and evolution of plastids in eukaryotes.

5. ARCHAEPLASTIDA GENOME STUDIES

Genomic data are rapidly accumulating. To date, about 10 complete genomes have been sequenced, but several other genome projects are ongoing (Tirichine & Bowler, 2011). Whole-genome data provide a great resource for analysis of eukaryotic genome evolution and user friendly online platforms for exploring this genome information is becoming increasingly available (e.g. Pico-Plaza, http://bioinformatics.psb.ugent.be/pico-plaza/; Van Bel *et al.*, 2012).

BOX 2.2 Glossary

Archaeplastida	A group of photosynthetic eukaryotes that are hypothesized to have evolved from a common ancestor with a primary plastid comprising the green plants (Viridiplantae), red algae (Rhodophyta) and the glaucophytes (Glaucophyta)
Biomarker	Organic molecules derived from distinctive cellular components in geological deposits indicative for the existence of certain organisms in specific time periods (e.g. 2-methyl hopanoids as evidence for cyanobacteria or okenone for purple bacteria)
Calibrated phylogeny	Phylogeny in which the branch lengths are proportional to time using geo-paleontological date (e.g. fossils, biomarkers and geological events) or substitution rates as scaling parameters
Chromalveolata	Group consisting of chlorophyll *c* containing phototrophs and some clades that secondarily lost their plastids, hypothesized to have evolved from a common ancestor with a secondary red algal derived plastid, including Alveolata, Stramenopila, Haptophyta and Cryptophyta
Concatenation	Data set in which sequence information of multiple genes is combined and analyzed together assuming all genes share a common history

Continued

BOX 2.2 Glossary—cont'd

Endosymbiosis	The term may refer either to one organism living inside another organism or to the process in which an organism is engulfed and subsequently enslaved by a host and thereby progressively integrated in the cellular machinery of the host as an organelle. The latter is often referred to as symbiogenesis
Endosymbiotic gene transfer	The process by which genes are transferred from the endosymbiont into the host genome during and following endosymbiosis
Gene-by-gene approach	Phylogenetic approach in which the evolutionary history of each gene is analyzed separately as opposed to concatenating all genes in a single matrix under the assumption of a single evolutionary history
Hacrobia	Group comprising Haptophyta and Cryptophyta, also known as the CCTH clade
Harosa	Group comprising of Stramenopila, Alveolata and Rhizaria, also known as the SAR clade
Horizontal gene transfer	Inheritance of genes from one organism or species to another that is not its descendent or ancestor
Oxygenic photosynthesis	The chemical process whereby light energy powers the conversion of carbon dioxide into organic compounds and oxygen is released as a waste product
Peptidoglycan layer	Biopolymer, also known as murein, consisting of a sugar of alternating residues of β-(1,4) linked N-acetylglucosamine and N-acetylmuramic acid and crosslinking peptides outside the (inner) plasma membrane of Eubacteria and plastids of Glaucophyta
Plastid	An endosymbiont derived, membrane-bound organelle that carries out photosynthesis in eukaryotes
Primary/secondary/ tertiary plastid	Plastid that is the result of, respectively one, two or three endosymbiotic events
Vertical gene transfer	Inheritance of genes from one organism to its offspring.

ACKNOWLEDGEMENTS

Funding was provided by the Research Foundation—Flanders (research grant G.0142.05, a doctoral fellowship to K.A.B. and a postdoctoral fellowships to F. L.).

REFERENCES

Acquisti, C., Kleffe, J., & Collins, S. (2007). Oxygen content of transmembrane proteins over macroevolutionary time scales. *Nature, 445*, 47–52.

Allen, J. F., & Martin, W. (2007). Evolutionary biology—Out of thin air. *Nature, 445*, 610–612.

Archibald, J. M. (2009). The puzzle of plastid evolution. *Current Biology, 19*, R81–R88.

Archibald, J. (2012). The evolution of algae by secondary and tertiary endosymbiosis. *Advances in Botanical Research, 64*, 87–118.

Asamizu, E., Nakajima, M., Kitade, Y., Saga, N., Nakamura, Y., & Tabata, S. (2003). Comparison of RNA expression profiles between the two generations of *Porphyra yezoensis* (Rhodophyta), based on expressed sequence tag frequency analysis. *Journal of Phycology, 39*, 923–930.

Baldauf, S. L. (2008). An overview of the phylogeny and diversity of eukaryotes. *Journal of Systematics and Evolution, 46*, 263–273.

Barbier, G., Oesterhelt, C., Larson, M. D., Halgren, R. G., Wilkerson, C., Garavito, R. M., et al. (2005). Comparative genomics of two closely related unicellular thermo-acidophilic red algae, *Galdieria sulphuraria* and *Cyanidioschyzon merolae*, reveals the molecular basis of the metabolic flexibility of *Galdieria sulphuraria* and significant differences in carbohydrate metabolism of both algae. *Plant Physiology, 137*, 460–474.

Baurain, D., Brinkmann, H., Petersen, J., Rodriguez-Ezpeleta, N., Stechmann, A., Demoulin, V., et al. (2010). Phylogenomic evidence for separate acquisition of plastids in cryptophytes, haptophytes, and stramenopiles. *Molecular Biology and Evolution, 27*, 1698–1709.

Becker, B., & Marin, B. (2009). Streptophyte algae and the origin of embryophytes. *Annals of Botany, 103*, 999–1004.

Berney, C., & Pawlowski, J. (2006). A molecular time-scale for eukaryote evolution recalibrated with the continuous microfossil record. *Proceedings of the Royal Society B-Biological Sciences, 273*, 1867–1872.

Bhattacharya, D., Archibald, J. M., Weber, A. P. M., & Reyes-Prieto, A. (2007). How do endosymbionts become organelles? Understanding early events in plastid evolution. *BioEssays, 29*, 1239–1246.

Bhattacharya, D., Helmchen, T., Bibeau, C., & Melkonian, M. (1995). Comparisons of nuclear-encoded small-subunit ribosomal RNAs reveal the evolutionary position of the Glaucocystophyta. *Molecular Biology and Evolution, 12*, 415–420.

Bhattacharya, D., & Schmidt, H. A. (1997). Division Glaucocystophyta. *Plant Systematics and Evolution Supplement, 11*, 139–148.

Blanc, G., Duncan, G., Agarkova, I., Borodovsky, M., Gurnon, J., Kuo, A., et al. (2010). The *Chlorella variabilis* NC64A genome reveals adaptation to photosymbiosis, coevolution with viruses, and cryptic sex. *Plant Cell, 22*, 2943–2955.

Buick, R. (2008). When did oxygenic photosynthesis evolve? *Philosophical Transactions of the Royal Society B-Biological Sciences, 363*, 2731–2743.

Burki, F., Inagaki, Y., Brate, J., Archibald, J. M., Keeling, P. J., Cavalier-Smith, T., et al. (2009). Large-scale phylogenomic analyses reveal that two enigmatic protist lineages, Telonemia and Centroheliozoa, are related to photosynthetic Chromalveolates. *Genome Biology and Evolution, 1*, 231–238.

Burki, F., Okamoto, N., Pombert, J.-F. O., & Keeling, P. J. (2012). The evolutionary history of haptophytes and cryptophytes: Phylogenomic evidence for separate origins. *Proceedings of the Royal Society B: Biological Sciences, 279*, 2246–2254.

Burki, F., Shalchian-Tabrizi, K., Minge, M., Skjaeveland, A., Nikolaev, S. I., Jakobsen, K. S., et al. (2007). Phylogenomics reshuffles the eukaryotic supergroups. *Plos One, 2*.

Butterfield, N. J. (2000). *Bangiomorpha pubescens* n. gen., n. sp.: implications for the evolution of sex, multicellularity, and the Mesoproterozoic/Neoproterozoic radiation of eukaryotes. *Paleobiology, 26*, 386–404.

Canfield, D. E., Rosing, M. T., & Bjerrum, C. (2006). Early anaerobic metabolisms. *Philosophical Transactions of the Royal Society B-Biological Sciences, 361*, 1819–1834.

Catling, D. C., Glein, C. R., Zahnle, K. J., & McKay, C. P. (2005). Why O(2) is required by complex life on habitable planets and the concept of planetary "oxygenation time. *Astrobiology, 5*, 415–438.

Cavalier-Smith, T. (2006). Phylogeny and megasystematics of phagotrophic heterokonts (kingdom Chromista). *Journal of Molecular Evolution, 62*, 388–420.

Cavalier-Smith, T. (2010). Deep phylogeny, ancestral groups and the four ages of life. *Philosophical Transactions of the Royal Society B-Biological Sciences, 365*, 111–132.

Chan, C. X., Yang, E. C., Banerjee, T., Yoon, H. S., Martone, P. T., Estevez, J. M., et al. (2011). Red and green algal monophyly and extensive gene sharing found in a rich repertoire of red algal genes. *Current Biology, 21*, 328–333.

Choi, H. G., Kraft, G. T., Lee, I. K., & Saunders, G. W. (2002). Phylogenetic analyses of anatomical and nuclear SSU rDNA sequence data indicate that the Dasyaceae and Delesseriaceae (Ceramiales, Rhodophyta) are polyphyletic. *European Journal of Phycology, 37*, 551–569.

Ciniglia, C., Yoon, H. S., Pollio, A., Pinto, G., & Bhattacharya, D. (2004). Hidden biodiversity of the extremophilic Cyanidiales red algae. *Molecular Ecology, 13*, 1827–1838.

Cocquyt, E., Verbruggen, H., Leliaert, F., & De Clerck, O. (2010). Evolution and cytological diversification of the green seaweeds (Ulvophyceae). *Molecular Biology and Evolution, 27*, 2052–2061.

Colbath, G. K. (1983). Fossil prasinophycean phycomata (Chlorophyta) from the Silurian Bainbridge formation, Missouri, USA. *Phycologia, 22*, 249–265.

Collen, J., Roeder, V., Rousvoal, S., Collin, O., Kloareg, B., & Boyen, C. (2006). An expressed sequence tag analysis of thallus and regenerating protoplasts of *Chondrus crispus* (Gigartinales, Rhodophyceae). *Journal of Phycology, 42*, 104–112.

Criscuolo, A., & Gribaldo, S. (2011). Large-scale phylogenomic analyses indicate a deep origin of primary plastids within cyanobacteria. *Molecular Biology and Evolution, 28*, 3019–3032.

de Duve, C. (2007). Essay—The origin of eukaryotes: A reappraisal. *Nature Reviews Genetics, 8*, 395–403.

De Smet, I., Voss, U., Lau, S., Wilson, M., Shao, N., Timme, R. E., et al. (2011). Unraveling the evolution of auxin signaling. *Plant Physiology, 155*, 209–221.

De Wever, A., Leliaert, F., Verleyen, E., Vanormelingen, P., Van der Gucht, K., Hodgson, D. A., et al. (2009). Hidden levels of phylodiversity in Antarctic green algae: Further evidence for the existence of glacial refugia. *Proceedings of the Royal Society B-Biological Sciences, 276*, 3591–3599.

Delaux, P.-M., Nanda, A. K., Mathé, C., Sejalon-Delmas, N., & Dunand, C. (2012). Molecular and biochemical aspects of plant terrestrialization. *Perspectives in Plant Ecology, Evolution and Systematics, 14*, 49–59.

Delwiche, C. F. (1999). Tracing the thread of plastid diversity through the tapestry of life. *American Naturalist, 154*, S164–S177.

Delwiche, C. F. (2007). Algae in the warp and weave of life: Bound by plastids. In J. Brodie, & J. Lewis (Eds.), *Unravelling the algae: The past, present, and future of algal systematics* (pp. 7–20). London: Taylor and Francis.

Derelle, E., Ferraz, C., Rombauts, S., Rouze, P., Worden, A. Z., Robbens, S., et al. (2006). Genome analysis of the smallest free-living eukaryote *Ostreococcus tauri* unveils many unique features. *Proceedings of the National Academy of Sciences U S A, 103*, 11647–11652.

Deschamps, P., Colleoni, C., Nakamura, Y., Suzuki, E., Putaux, J. L., Buleon, A., et al. (2008). Metabolic symbiosis and the birth of the plant kingdom. *Molecular Biology and Evolution, 25*, 536–548.

Deschamps, P., & Moreira, D. (2009). Signal conflicts in the phylogeny of the primary photosynthetic eukaryotes. *Molecular Biology and Evolution, 26*, 2745–2753.

Deusch, O., Landan, G., Roettger, M., Gruenheit, N., Kowallik, K. V., Allen, J. F., et al. (2008). Genes of cyanobacterial origin in plant nuclear genomes point to a heterocyst-forming plastid ancestor. *Molecular Biology and Evolution, 25*, 748–761.

Douzery, E. J. P., Snell, E. A., Bapteste, E., Delsuc, F., & Philippe, H. (2004). The timing of eukaryotic evolution: Does a relaxed molecular clock reconcile proteins and fossils? *Proceedings of the National Academy of Sciences, 101*, 15386–15391.

Elias, M., & Archibald, J. M. (2009). Sizing up the genomic footprint of endosymbiosis. *BioEssays, 31*, 1273–1279.

Embley, T. M., & Martin, W. (2006). Eukaryotic evolution, changes and challenges. *Nature, 440*, 623–630.

Falcon, L. I., Magallon, S., & Castillo, A. (2010). Dating the cyanobacterial ancestor of the chloroplast. *Isme Journal, 4*, 777–783.

Falkowski, P. G. (2006). Evolution—Tracing oxygen's imprint on Earth's metabolic evolution. *Science, 311*, 1724–1725.

Falkowski, P. G., Katz, M. E., Knoll, A. H., Quigg, A., Raven, J. A., Schofield, O., et al. (2004). The evolution of modern eukaryotic phytoplankton. *Science, 305*, 354–360.

Farquhar, J., Zerkle, A., & Bekker, A. (2011). Geological constraints on the origin of oxygenic photosynthesis. *Photosynthesis Research, 107*, 11–36.

Fawley, M. W., Yun, Y., & Qin, M. (2000). Phylogenetic analyses of 18S rDNA sequences reveal a new coccoid lineage of the Prasinophyceae (Chlorophyta). *Journal of Phycology, 36*, 387–393.

Friedl, T., & Bhattacharya, D. (2002). In J. Seckbach (Ed.), *Symbiosis* (pp. 341–357). Dordrecht: Kluwer Academic Publishers.

Friedl, T., & Rybalka, N. (2011). Systematics of the green algae: A brief introduction to the current status. *Progress in Botany, 73*, 259–280.

Gould, S. B., Waller, R. R., & McFadden, G. I. (2008). Plastid evolution. *Annual Review of Plant Biology, 59*, 491–517.

Graham, L. E. (1984). *Coleochaete* and the origin of land plants. *American Journal of Botany, 71*, 603–608.

Graham, L. E., Cook, M. E., & Busse, J. S. (2000). The origin of plants: Body plan changes contributing to a major evolutionary radiation. *Proceedings of the National Academy of Sciences, 97*, 4535–4540.

Graham, L. E., Graham, J. M., & Wilcox, L. W. (2009). *Algae* (2nd ed.). San Francisco: Pearson Education.

Gross, J., & Bhattacharya, D. (2010). Uniting sex and eukaryote origins in an emerging oxygenic world. *Biology Direct, 5*, 53.

Grossman, A. R., Harris, E. E., Hauser, C., Lefebvre, P. A., Martinez, D., Rokhsar, D., et al. (2003). *Chlamydomonas reinhardtii* at the crossroads of genomics. *Eukaryot Cell, 2*, 1137–1150.

Guillou, L., Eikrem, W., Chretiennot-Dinet, M. J., Le Gall, F., Massana, R., Romari, K., et al. (2004). Diversity of picoplanktonic prasinophytes assessed by direct nuclear SSU rDNA sequencing of environmental samples and novel isolates retrieved from oceanic and coastal marine ecosystems. *Protist, 155*, 193–214.

Hackett, J. D., Yoon, H. S., Li, S., Reyes-Prieto, A., Rummele, S. E., & Bhattacharya, D. (2007). Phylogenomic analysis supports the monophyly of cryptophytes and haptophytes and the association of Rhizaria with Chromalveolates. *Molecular Biology and Evolution, 24*, 1702–1713.

Hagopian, J. C., Reis, M., Kitajima, J. P., Bhattacharya, D., & De Oliveira, M. C. (2004). Comparative analysis of the complete plastid genome sequence of the red alga *Gracilaria tenuistipitata* var. liui provides insights into the evolution of rhodoplasts and their relationship to other plastids. *Journal of Molecular Evolution, 59*, 464–477.

Hampl, V., Hug, L., Leigh, J. W., Dacks, J. B., Lang, B. F., Simpson, A. G. B., et al. (2009). Phylogenomic analyses support the monophyly of Excavata and resolve relationships among eukaryotic "supergroups. *Proceedings of the National Academy of Sciences ,Current Biology Current Biology Current Biology U S A, 106*, 3859–3864.

Harper, J. T., & Saunders, G. W. (2001). Molecular systematics of the Florideophyceae (Rhodophyta) using nuclear large and small subunit rDNA sequence data. *Journal of Phycology, 37*, 1073–1082.

Hasegawa, T., Miyashita, H., Kawachi, M., Ikemoto, H., Kurano, N., Miyachi, S., et al. (1996). *Prasinoderma coloniale* gen. et sp. nov., a new pelagic coccoid prasinophyte from the western Pacific ocean. *Phycologia, 35*, 170–176.

Hedges, S. B., Blair, J. E., Venturi, M. L., & Shoe, J. L. (2004). A molecular timescale of eukaryote evolution and the rise of complex multicellular life. *BMC Evolutionary Biology, 4*:2.

Herron, M. D., Hackett, J. D., Aylward, F. O., & Michod, R. E. (2009). Triassic origin and early radiation of multicellular volvocine algae. *Proceedings of the National Academy of Sciences U S A, 106*, 3254–3258.

Herron, M. D., & Michod, R. E. (2008). Evolution of complexity in the volvocine algae: Transitions in individuality through Darwin's eye. *Evolution, 62*, 436–451.

Hohmann-Marriott, M. F., & Blankenship, R. E. (2011). Evolution of photosynthesis. In S. S. Merchant, W. R. Briggs, & D. Ort (Eds.), *Annual review of plant biology*, Vol. 62 (pp. 515–548).

Holland, H. D. (2006). The oxygenation of the atmosphere and oceans. *Philosophical Transactions of the Royal Society B: Biological Sciences, 361*, 903–915.

Inagaki, Y., Nakajima, Y., Sato, M., Sakaguchi, M., & Hashimoto, T. (2009). Gene sampling can bias multi-gene phylogenetic inferences: the relationship between red algae and green plants as a case study. *Molecular Biology and Evolution, 26*, 1171–1178.

Javaux, E. (2011). Early eukaryotes in Precambrian oceans. In M. Gargaud, P. Lopez-Gracia, & H. Martin (Eds.), *Origins and evolution of life: An astrobiological perspective* (pp. 414–449). Cambridge: University Press.

Joubert, J. J., & Rijkenberg, F. H. J. (1971). Parasitic green algae. *Annual Review of Phytopathology, 9*, 45–64.

Karol, K. G., McCourt, R. M., Cimino, M. T., & Delwiche, C. F. (2001). The closest living relatives of land plants. *Science, 294*, 2351–2353.

Keeling, P. J. (2010). The endosymbiotic origin, diversification and fate of plastids. *Philosophical Transactions of the Royal Society B-Biological Sciences, 365*, 729–748.

Kenrick, P., & Crane, P. R. (1997). The origin and early evolution of plants on land. *Nature, 389*, 33–39.

Kerney, R., Kim, E., Hangarter, R. P., Heiss, A. A., Bishop, C. D., & Hall, B. K. (2011). Intracellular invasion of green algae in a salamander host. *Proceedings of the National Academy of Sciences U S A, 108*, 6497–6502.

Kies, L., & Kremer, B. P. (1990). Phylum Glaucocsytophyta. In L. Margulis, J. O. Corliss, M. Melkonian, & D. J. Chapman (Eds.), *Handbook of Protoctista. The structure, cultivation, habitats and life histories of the eukaryotic microorganisms and their descendants exclusive of animals, plants and fungi* (pp. 152–166). Boston: Jones and Bartlett Publishers.

Kirk, D. L. (2003). Seeking the ultimate and proximate causes of *Volvox* multicellularity and cellular differentiation. *Integrative and Comparative Biology, 43*, 247–253.

Klochkova, T. A., Cho, G. Y., Boo, S. M., Chung, K. W., Kim, S. J., & Kim, G. H. (2008). Interactions between marine facultative epiphyte *Chlamydomonas* sp (Chlamydomonadales, Chlorophyta) and ceramiaceaen algae (Rhodophyta). *Journal of Environmental Biology, 29*, 427–435.

Knoll, A. H. (1992). The early evolution of eukaryotes—A geological perspective. *Science, 256*, 622–627.

Knoll, A. H., Javaux, E. J., Hewitt, D., & Cohen, P. (2006). Eukaryotic organisms in Proterozoic oceans. *Philosophical Transactions of the Royal Society B-Biological Sciences, 361*, 1023–1038.

Kuroiwa, T. (1998). The primitive red algae *Cyanidium caldarium* and *Cyanidioschyzon merolae* as model system for investigating the dividing apparatus of mitochondria and plastids. *BioEssays, 20*, 344–354.

Larkum, A. W. D., Lockhart, P. J., & Howe, C. J. (2007). Shopping for plastids. *Trends in Plant Science, 12*, 189–195.

Latasa, M., Scharek, R., Le Gall, F., & Guillou, L. (2004). Pigment suites and taxonomic groups in Prasinophyceae. *Journal of Phycology, 40*, 1149–1155.
Laurin-Lemay, S., Brinkmann, H., & Philippe, H. (2012). Origin of land plants revisited in the light of sequence contamination and missing data. *Current Biology, 22*, R593–R594.
Le Gall, L., & Saunders, G. W. (2007). A nuclear phylogeny of the Florideophyceae (Rhodophyta) inferred from combined EF2, small subunit and large subunit ribosomal DNA: Establishing the new red algal subclass Corallinophycidae. *Molecular Phylogenetics and Evolution, 43*, 1118–1130.
Leliaert, F., Smith, D. R., Moreau, H., Herron, M. D., Verbruggen, H., Delwiche, C. F., et al. (2012). Phylogeny and molecular evolution of the green algae. *Critical Reviews in Plant, 31*, 1–46.
Leliaert, F., Verbruggen, H., & Zechman, F. W. (2011). Into the deep: New discoveries at the base of the green plant phylogeny. *BioEssays, 33*, 683–692.
Lemieux, C., Otis, C., & Turmel, M. (2007). A clade uniting the green algae *Mesostigma viride* and Chlorokybus atmophyticus represents the deepest branch of the Streptophyta in chloroplast genome-based phylogenies. *BMC Biology, 5*, 2.
Lepère, C., Vaulot, D., & Scanlan, D. J. (2009). Photosynthetic picoeukaryote community structure in the South East Pacific Ocean encompassing the most oligotrophic waters on Earth. *Environmental Microbiology, 11*, 3105–3117.
Lewis, L. A., & Lewis, P. O. (2005). Unearthing the molecular phylodiversity of desert soil green algae (Chlorophyta). *Systematic Biology, 54*, 936–947.
Lewis, L. A., & McCourt, R. M. (2004). Green algae and the origin of land plants. *American Journal of Botany, 91*, 1535–1556.
Lewis, L. A., & Muller-Parker, G. (2004). Phylogenetic placement of "Zoochlorellae" (Chlorophyta), algal symbiont of the temperate sea anemone Anthopleura elegantissima. *Biological Bulletin, 207*, 87–92.
López-Bautista, J. M., & Chapman, R. L. (2003). Phylogenetic affinities of the Trentepohliales inferred from small-subunit rDNA. *Journal of Systematic and Evolutionary Microbiology, 53*, 2099–2106.
López-Bautista, J. M., Rindi, F., & Guiry, M. D. (2006). Molecular systematics of the subaerial green algal order Trentepohliales: An assessment based on morphological and molecular data. *Journal of Systematic and Evolutionary Microbiology, 56*, 1709–1715.
Lü, F., Xü, W., Tian, C., Wang, G., Niu, J., Pan, G., & Hu, S. (2011). The *Bryopsis hypnoides* plastid genome: multimeric forms and complete nucleotide sequence. *PLoS One, 6*, e14663.
Maggs, C. A., Verbruggen, H., & De Clerck, O. (2007). Molecular systematics of red algae: Building future structures on firm foundations. In J. Brodie, & J. Lewis (Eds.), *Unravelling the algae: The past, present, and future of algal systematics* (pp. 103–121). Taylor and Francis.
Marin, B. (2012). Nested in the Chlorellales or independent class? Phylogeny and classification of the Pedinophyceae (Viridiplantae) revealed by molecular phylogenetic analyses of complete nuclear and plastid-encoded rRNA operons. *Protist, 163*, 778–805.
Marin, B., & Melkonian, M. (2010). Molecular phylogeny and classification of the Mamiellophyceae class. nov. (Chlorophyta) based on sequence comparisons of the nuclear- and plastid-encoded rRNA operons. *Protist, 161*, 304–336.
Marin, B., Nowack, E. C. M., & Melkonian, M. (2005). A plastid in the making: Evidence for a second primary endosymbiosis. *Protist, 156*, 425–432.
Martin, W., & Muller, M. (1998). The hydrogen hypothesis for the first eukaryote. *Nature, 392*, 37–41.
Martin, W., Rujan, T., Richly, E., Hansen, A., Cornelsen, S., Lins, T., et al. (2002). Evolutionary analysis of Arabidopsis, cyanobacterial, and chloroplast genomes reveals

plastid phylogeny and thousands of cyanobacterial genes in the nucleus. *Proceedings of the National Academy of Sciences U S A, 99*, 12246–12251.

Matsuzaki, M., Misumi, O., Shin-i, T., Maruyama, S., Takahara, M., Miyagishima, S.-y., et al. (2004). Genome sequence of the ultrasmall unicellular red alga *Cyanidioschyzon merolae* 10D. *Nature, 428*, 653–657.

McCourt, R. M., Delwiche, C. F., & Karol, K. G. (2004). Charophyte algae and land plant origins. *Trends in Ecology & Evolution, 19*, 661–666.

Melkonian, M. (1990). Phylum chlorophyta. Class prasinophyceae. In L. Margulis, J. O. Corliss, M. Melkonian, & D. J. Chapman (Eds.), *Handbook of Protoctista. The structure, cultivation, habitats and life histories of the eukaryotic microorganisms and their descendants exclusive of animals, plants and fungi* (pp. 600–607). Boston: Jones and Bartlett Publishers.

Merchant, S. S., Prochnik, S. E., Vallon, O., Harris, E. H., Karpowicz, S. J., Witman, G. B., et al. (2007). The *Chlamydomonas* genome reveals the evolution of key animal and plant functions. *Science, 318*, 245–251.

Misumi, O., Matsuzaki, M., Nozaki, H., Miyagishima, S., Mori, T., Nishida, K., et al. (2005). *Cyanidioschyzon merolae* genome. A tool for facilitating comparable studies on organelle biogenesis in photosynthetic eukaryotes. *Plant Physiology, 137*, 567–585.

Moestrup, Ø, Inouye, I., & Hori, T. (2003). Ultrastructural studies on *Cymbomonas tetramitiformis* (Prasinophyceae). I. General structure, scale microstructure, and ontogeny. *Canadian Journal of Botany, 81*, 657–671.

Moreira, D., Le Guyader, H., & Philippe, H. (2000). The origin of red algae and the evolution of chloroplasts. *Nature, 405*, 69–72.

Moustafa, A., & Bhattacharya, D. (2008). PhyloSort: A user-friendly phylogenetic sorting tool and its application to estimating the cyanobacterial contribution to the nuclear genome of Chlamydomonas. *BMC Evolutionary Biology, 8*.

Müller, K. M., Lynch, M. D. J., & Sheath, R. G. (2010). Bangiophytes: From one class to six; Where do we go from here? In J. Seckbach, & D. J. Chapman (Eds.), *Red Algae in the Genomic Age, vol. 13. Cellular Origin, Life in Extreme Habitats and Astrobiology* (pp. 241–259) The Netherlands: Springer.

Müller, K. M., Oliveira, M. C., Sheath, R. G., & Bhattacharya, D. (2001). Ribosomal DNA phylogeny of the Bangiophycidae (Rhodophyta) and the origin of secondary plastids. *American Journal of Botany, 88*, 1390–1400.

Nakayama, T., Kawachi, M., & Inouye, I. (2000). Taxonomy and the phylogenetic position of a new prasinophycean alga, Crustomastix didyma gen. & sp. nov. (Chlorophyta). *Phycologia, 39*, 337–348.

Nakayama, T., Marin, B., Kranz, H. D., Surek, B., Huss, V. A. R., Inouye, I., et al. (1998). The basal position of scaly green flagellates among the green algae (Chlorophyta) is revealed by analyses of nuclear-encoded SSU rRNA sequences. *Protist, 149*, 367–380.

Nakayama, T., Suda, S., Kawachi, M., & Inouye, I. (2007). Phylogeny and ultrastructure of *Nephroselmis* and *Pseudoscourfieldia* (Chlorophyta), including the description of *Nephroselmis anterostigmatica* sp. nov. and a proposal for the Nephroselmidales ord. nov. *Phycologia, 46*, 680–697.

Nikaido, I., Asamizu, E., Nakajima, M., Nakamura, Y., Saga, N., & Tabata, S. (2000). Generation of 10,154 expressed sequence tags from a leafy gametophyte of a marine red alga, *Porphyra yezoensis. DNA Research, 7*, 223–227.

Niwa, K., Iida, S., Kato, A., Kawai, H., Kikuchi, N., Kobiyama, A., et al. (2009). Genetic diversity and introgression in two cultivated species (*Porphyra yezoensis* and *Porphyra tenera*) and closely related wild species of *Porphyra* (Bangiales, Rhodophyta). *Journal of Phycology, 45*, 493–502.

Not, F., Latasa, M., Marie, D., Cariou, T., Vaulot, D., & Simon, N. (2004). A single species, *Micromonas pusilla* (Prasinophyceae), dominates the eukaryotic picoplankton in the western English channel. *Applied and Environmental Microbiology, 70*, 4064–4072.

Not, F., Siano, R., Kooistra, W. H. C. F., Simon, N., Vaulot, D., & Probert, I. (2012). Diversity and ecology of eukaryotic marine phytoplankton. *Advances in Botanical Research, 64*, 1–53.

Nowack, E. C. M., & Grossman, A. R. (2012). Trafficking of protein into the recently established photosynthetic organelles of *Paulinella chromatophora*. *Proceedings of the National Academy of Sciences, 109*, 5340–5345.

Nowack, E. C. M., Vogel, H., Groth, M., Grossman, A. R., Melkonian, M., & Glockner, G. (2011). Endosymbiotic gene transfer and transcriptional regulation of transferred genes in *Paulinella chromatophora*. *Molecular Biology and Evolution, 28*, 407–422.

Nozaki, H., Maruyama, S., Matsuzaki, M., Nakada, T., Kato, S., & Misawa, K. (2009). Phylogenetic positions of Glaucophyta, green plants (Archaeplastida) and Haptophyta (Chromalveolata) as deduced from slowly evolving nuclear genes. *Molecular Phylogenetics and Evolution, 53*, 872–880.

O'Kelly, C. J. (2007). The origin and early evolution of green plants. In P. G. Falkowski, & A. H. Knoll (Eds.), *Evolution of primary producers in the sea* (pp. 287–309). Burlington, MA: Elsevier Academic Press.

O'Kelly, C. J., Sieracki, M. E., Thier, E. C., & Hobson, I. C. (2003). A transient bloom of *Ostreococcus* (Chlorophyta, Prasinophyceae) in West Neck Bay, Long Island, New York. *Journal of Phycology, 39*, 850–854.

Oliveira, M. C., & Bhattacharya, D. (2000). Phylogeny of the Bangiophycidae (Rhodophyta) and the secondary endosymbiotic origin of algal plastids. *American Journal of Botany, 87*, 482–492.

Palenik, B., Grimwood, J., Aerts, A., Rouze, P., Salamov, A., Putnam, N., et al. (2007). The tiny eukaryote *Ostreococcus* provides genomic insights into the paradox of plankton speciation. *Proceedings of the National Academy of Sciences U S A, 104*, 7705–7710.

Parfrey, L. W., Grant, J., Tekle, Y. I., Lasek-Nesselquist, E., Morrison, H. G., Sogin, M. L., et al. (2010). Broadly sampled multigene analyses yield a well-resolved eukaryotic tree of life. *Systematic Biology, 59*, 518–533.

Parfrey, L. W., Lahr, D. J. G., Knoll, A. H., & Katz, L. A. (2011). Estimating the timing of early eukaryotic diversification with multigene molecular clocks. *Proceedings of the National Academy of Sciences, 108*, 13624–13629.

Patron, N. J., Inagaki, Y., & Keeling, P. J. (2007). Multiple gene phylogenies support the monophyly of cryptomonad and haptophyte host lineages. *Current Biology, 17*, 887–891.

Poole, A. M., & Neumann, N. (2011). Reconciling an archaeal origin of eukaryotes with engulfment: A biologically plausible update of the Eocyte hypothesis. *Research in Microbiology, 162*, 71–76.

Price, D. C., Chan, C. X., Yoon, H. S., Yang, E. C., Qiu, H., Weber, A. P., et al. (2012). *Cyanophora paradoxa* genome elucidates origin of photosynthesis in algae and plants. *Science, 335*, 843–847.

Prochnik, S. E., Umen, J., Nedelcu, A. M., Hallmann, A., Miller, S. M., Nishii, I., et al. (2010). Genomic analysis of organismal complexity in the multicellular green alga *Volvox carteri*. *Science, 329*, 223–226.

Raymond, J., & Segré, D. (2006). The effect of oxygen on biochemical networks and the evolution of complex life. *Science, 311*, 1764–1767.

Reyes-Prieto, A., & Bhattacharya, D. (2007). Phylogeny of nuclear-encoded plastid-targeted proteins supports an early divergence of glaucophytes within plantae. *Molecular Biology and Evolution, 24*, 2358–2361.

Reyes-Prieto, A., Hackett, J. D., Soares, M. B., Bonaldo, M. F., & Bhattacharya, D. (2006). Cyanobacterial contribution to algal nuclear genomes a primarily limited to plastid functions. *Current Biology, 16*, 2320–2325.

Reyes-Prieto, A., Yoon, H. S., Moustafa, A., Yang, E. C., Andersen, R. A., Boo, S. M., et al. (2010). Differential gene retention in plastids of common recent origin. *Molecular Biology and Evolution, 27*, 1530–1537.

Rindi, F. (2011). Terrestrial green algae: Systematics, biogeography and expected responses to climate change. In T. Hodkinson, S. Jones, S. Waldren, & J. Parnell (Eds.), *Climate change, ecology and systematics* (pp. 201–227). Cambridge: Cambridge University Press.

Rodriguez-Ezpeleta, N., Brinkmann, H., Burey, S. C., Roure, B., Burger, G., Löffelhardt, W., et al. (2005). Monophyly of primary photosynthetic eukaryotes: Green plants, red algae, and glaucophytes. *Current Biology, 15,* 1325–1330.

Rodriguez-Ezpeleta, N., Philippe, H., Brinkmann, H., Becker, B., & Melkonian, M. (2007). Phylogenetic analyses of nuclear, mitochondrial, and plastid multigene data sets support the placement of Mesostigma in the Streptophyta. *Molecular Biology and Evolution, 24,* 723–731.

Roger, A. J. (1999). Reconstructing early events in eukaryotic evolution. *American Naturalist, 154,* S146–S163.

Roger, A. J., & Hug, L. A. (2006). The origin and diversification of eukaryotes: problems with molecular phylogenetics and molecular clock estimation. *Philosophical Transactions of the Royal Society B-Biological Sciences, 361,* 1039–1054.

Russell, M. J., & Hall, A. J. (2006). The onset and early evolution of life. In S. E. Kesler, & H. Ohmoto (Eds.), *Evolution of early earth's atmosphere, hydrosphere and biosphere—Constraints from ore depositis* (pp. 1–32). Boulder, Colorado: Geological Society of America.

Sato, N., Wise, R. R., & Hoober, J. K. (2006). *Origin and evolution of plastids: genomic view on the unification and diversity of plastids. The Structure and Function of Plastids, vol. 23. Advances in Photosynthesis and Respiration.* The Netherlands: Springer. (pp. 75–102).

Saunders, G. W., Chiovitti, A., & Kraft, G. T. (2004). Small-subunit rDNA sequences from representatives of selected families of the Gigartinales and Rhodymeniales (Rhodophyta). 3. Delineating the Gigartinales sensu stricto. *Canadian Journal of Botany, 82,* 43–74.

Saunders, G. W., & Hommersand, M. H. (2004). Assessing Red algal supraordinal diversity and taxonomy in the context of contemporary systematic data. *American Journal of Botany, 91,* 1494–1507.

Schenk, H. E. A. (2001). Glaucocystophytes. In *eLS.* John Wiley & Sons, Ltd.

Scott, J., Yokoyama, A., Billard, C., Fresnel, J., Haras, Y., West, K. A., et al. (2008). *Neorhodella cyanea,* a new genus in the Rhodellophyceae (Rhodophyta). *Phycologia, 47,* 560–572.

Shi, X. L., Marie, D., Jardillier, L., Scanlan, D. J., & Vaulot, D. (2009). Groups without cultured representatives dominate eukaryotic picophytoplankton in the oligotrophic South East Pacific Ocean. *PLoS One, 4,* e7657.

Sieburth, J. M., Keller, M. D., Johnson, P. W., & Myklestad, S. M. (1999). Widespread occurrence of the oceanic ultraplankter, Prasinococcus capsulatus (Prasinophyceae), the diagnostic "Golgi-decapore complex" and the newly described polysaccharide "capsulan. *Journal of Phycology, 35,* 1032–1043.

Simon, N., Cras, A. L., Foulon, E., & Lemee, R. (2009). Diversity and evolution of marine phytoplankton. *Comptes Rendus Biologies, 332,* 159–170.

Steiner, J. M., & Löffelhardt, W. (2002). Protein import into cyanelles. *Trends in Plant Science, 7,* 72–77.

Steiner, J. M., Yusa, F., Pompe, J. A., & Löffelhardt, W. (2005). Homologous protein import machineries in chloroplasts and cyanelles. *Plant Journal, 44,* 646–652.

Stirewalt, V., Michalowski, C., Löffelhardt, W., Bohnert, H., & Bryant, D. (1995). Nucleotide sequence of the cyanelle genome from Cyanophora paradoxa. *Plant Molecular Biology Reporter, 13,* 327–332.

Suda, S., Watanabe, M. M., & Inouye, I. (2004). Electron microscopy of sexual reproduction in Nephroselmis olivacea (Prasinophyceae, Chlorophyta). *Phycological Research, 52,* 273–283.

Sudman, M. S. (1974). Protothecosis: Critical review. *American Journal of Clinical Pathology, 61,* 10–19.

Sutherland, J. E., Lindstrom, S. C., Nelson, W. A., Brodie, J., Lynch, M. D. J., Hwang, M. S., et al. (2011). A new look at an ancient order: generic revision of the Bangiales (Rhodophyta). *Journal of Phycology, 47,* 1131–1151.

Sym, S. D., & Pienaar, R. N. (1993). The class Prasinophyceae. In F. E. Round, & D. J. Chapman (Eds.), *Progress in phycological research* (pp. 281–376). Bristol: Biopress Ltd.
Tappan, H. (1980). *Palaeobiology of plant protists*. San Francisco: Freeman.
Timme, R. E., Bachvaroff, T. R., & Delwiche, C. F. (2012). Broad phylogenomic sampling and the sister lineage of land plants. *PLoS One, 7*, e29696.
Tirichine, L., & Bowler, C. (2011). Decoding algal genomes: Tracing back the history of photosynthetic life on Earth. *The Plant Journal, 66*, 45–57.
Tomitani, A., Knoll, A. H., Cavanaugh, C. M., & Ohno, T. (2006). The evolutionary diversification of cyanobacteria: Molecular-phylogenetic and paleontological perspectives. *Proceedings of the National Academy of Sciences U S A, 103*, 5442–5447.
Toulza, E., Blanc-Mathieu, R., Gourbiere, S., & Piganeau, G. (2012). Environmental genomics of microbial algae: Power and challenges of metagenomics. *Advances in Botanical Research, 64*, 379–423.
Turmel, M., Gagnon, M.-C., O'Kelly, C. J., Otis, C., & Lemieux, C. (2009). The chloroplast genomes of the green algae *Pyramimonas, Monomastix*, and *Pycnococcus* shed new light on the evolutionary history of prasinophytes and the origin of the secondary chloroplasts of euglenids. *Molecular Biology and Evolution, 26*, 631–648.
Turmel, M., Otis, C., & Lemieux, C. (2006). The chloroplast genome sequence of *Chara vulgaris* sheds new light into the closest green algal relatives of land plants. *Molecular Biology and Evolution, 23*, 1324–1338.
Turmel, M., Otis, C., & Lemieux, C. (2007). An unexpectedly large and loosely packed mitochondrial genome in the charophycean green alga Chlorokybus atmophyticus. *BMC Genomics, 8*, 137.
Turmel, M., Otis, C., & Lemieux, C. (2009). The chloroplast genomes of the green algae *Pedinomonas minor, Parachlorella kessleri*, and *Oocystis solitatia* reveal a shared ancestry between the Pedinomonadales and Chlorellales. *Molecular Biology and Evolution, 26*, 2317–2331.
Umen, J., & Olson, B. (2012). Genomics of Volvocine Algae. *Advances in Botanical Research, 64*, 185–243.
Van Bel, M., Proost, S., Wischnitzki, E., Movahedi, S., Scheerlinck, C., Van de Peer, Y., et al. (2012). Dissecting plant genomes with the PLAZA comparative genomics platform. *Plant Physiology, 158*, 590–600.
van den Hoek, C., Mann, D. G., & Jahns, H. M. (1995). *Algae: an introduction to phycology*. Cambridge University Press.
Vaulot, D., Eikrem, W., Viprey, M., & Moreau, H. (2008). The diversity of small eukaryotic phytoplankton (≤ 3 μm) in marine ecosystems. *FEMS Microbiology Reviews, 32*, 795–820.
Verbruggen, H., Maggs, C., Saunders, G., Le Gall, L., Yoon, H. S., & De Clerck, O. (2010). Data mining approach identifies research priorities and data requirements for resolving the red algal tree of life. *BMC Evolutionary Biology, 10*, 16.
Viprey, M., Guillou, L., Ferréol, M., & Vaulot, D. (2008). Wide genetic diversity of picoplanktonic green algae (Chloroplastida) in the Mediterranean Sea uncovered by a phylum-biased PCR approach. *Environmental Microbiology, 10*, 1804–1822.
Wada, M., Kagawa, T., & Sato, Y. (2003). Chloroplast movement. *Ann Rev Plant Biol, 54*, 455–468.
West, J. A., Zuccarello, G. C., Scott, J. L., West, K. A., & Karsten, U. (2007). *Rhodaphanes brevistipitata* gen. et sp nov., a new member of the Stylonematophyceae (Rhodophyta). *Phycologia, 46*, 440–449.
Williams, S. L., & Smith, J. E. (2007). A global review of the distribution, taxonomy, and impacts of introduced seaweeds. *Annual Review of Plant Biology, 38*, 327–359.
Wodniok, S., Brinkmann, H., Glockner, G., Heidel, A., Philippe, H., Melkonian, M., et al. (2011). Origin of land plants: Do conjugating green algae hold the key? *BMC Evolutionary Biology, 11*, 104.

Worden, A. Z., Lee, J. H., Mock, T., Rouze, P., Simmons, M. P., Aerts, A. L., et al. (2009). Green evolution and dynamic adaptations revealed by genomes of the marine picoeukaryotes *Micromonas*. *Science, 324*, 268–272.

Xiao, S. H., Zhang, Y., & Knoll, A. H. (1998). Three-dimensional preservation of algae and animal embryos in a Neoproterozoic phosphorite. *Nature, 391*, 553–558.

Yang, E. C., Scott, J., West, J. A., Orlova, E., Gauthier, D., Kupper, F. C., et al. (2010). New taxa of the Porphyridiophyceae (Rhodophyta): *Timspurckia oligopyrenoides* gen. et sp. nov. and *Erythrolobus madagascarensis* sp. nov. *Phycologia, 49*, 604–616.

Ye, N. H., Zhang, X. W., Mao, Y. Z., Liang, C. W., Xu, D., Zou, J., et al. (2011). 'Green tides' are overwhelming the coastline of our blue planet: Taking the world's largest example. *Ecological Research, 26*, 477–485.

Yokoyama, A., Scott, J. L., Zuccarello, G. C., Kajikawa, M., Hara, Y., & West, J. A. (2009). *Corynoplastis japonica* gen. et sp nov and Dixoniellales ord. nov (Rhodellophyceae, Rhodophyta) based on morphological and molecular evidence. *Phycological Research, 57*, 278–289.

Yoon, H. S., Ciniglia, C., Wu, M., Comeron, J. M., Pinto, G., Pollio, A., et al. (2006a). Establishment of endolithic populations of extremophilic Cyanidiales (Rhodophyta). *BMC Evolutionary Biology, 6*.

Yoon, H. S., Hackett, J. D., Ciniglia, C., Pinto, G., & Bhattacharya, D. (2004). A Molecular timeline for the origin of photosynthetic eukaryotes. *Molecular Biology and Evolution, 21*, 809–818.

Yoon, H. S., Müller, K. M., Sheath, R. G., Ott, F. D., & Bhattacharya, D. (2006b). Defining the major lineages of red algae (Rhodophyta). *Journal of Phycology, 42*, 482–492.

Yoon, H. S., Zuccarello, G. C., & Bhattacharya, D. (2010). Evolutionary history and taxonomy of red algae. In J. Seckbach, & D. J. Chapman (Eds.), *Red algae in the genomic age, vol. 13. Cellular origin, life in extreme habitats and astrobiology* (pp. 25–42). Netherlands: Springer.

Zechman, F. W., Verbruggen, H., Leliaert, F., Ashworth, M., Buchheim, M. A., Fawley, M. W., et al. (2010). An unrecognized ancient lineage of green plants persists in deep marine waters. *Journal of Phycology, 46*, 1288–1295.

Zingone, A., Borra, M., Brunet, C., Forlani, G., Kooistra, W. H. C. F., & Procaccini, G. (2002). Phylogenetic position of Crustomastix stigmatica sp. nov. and *Dolichomastix tenuilepis* in relation to the Mamiellales (Prasinophyceae, Chlorophyta). *Journal of Phycology, 38*, 1024–1039.

Zuccarello, G. C., Kikuchi, N., & West, J. A. (2010). Molecular phylogeny of the crustose Erythropeltidales (Compsopogonophyceae, Rhodophyta): new genera *Pseudoerythrocladia* and *Madagascaria* and the evolution of the upright habit. *Journal of Phycology, 46*, 363–373.

Zuccarello, G. C., Oellermann, M., West, J. A., & De Clerck, O. (2009). Complex patterns of actin molecular evolution in the red alga *Stylonema alsidii* (Stylonematophyceae, Rhodophyta). *Phycological Research, 57*, 59–65.

Zuccarello, G. C., West, J. A., & Kikuchi, N. (2008). Phylogenetic relationships within the Stylonematales (Stylonematophyceae, Rhodophyta): Biogeographic patterns do not apply to *Stylonema alsidii*. *Journal of Phycology, 44*, 384–393.

Zuccarello, G. C., Yoon, H. S., Kim, H., Sun, L., de Goer, S. L., & West, J. A. (2011). Molecular phylogeny of the upright Erythropeltidales (Compsopogonophyceae, Rhodophyta): multiple cryptic lineages of *Erythrotrichia carnea*. *J Phycol, 47*, 627–637.

Zuccarello, J. (2011). What are you eating? It may be nori, but it is probably not *Porphyra* anymore. *Journal of Phycology, 47*, 967–968.

CHAPTER THREE

The Evolution of Algae by Secondary and Tertiary Endosymbiosis

John M. Archibald[1]

Canadian Institute for Advanced Research, Program in Integrated Microbial Biodiversity, Department of Biochemistry and Molecular Biology, Dalhousie University, Halifax NS B3H 4R2, Canada

[1]Corresponding author: E-mail: jmarchib@dal.ca

Contents

Abstract

The endosymbiotic origin of plastids (chloroplasts) from cyanobacteria was a pivotal event in eukaryotic evolution. By giving certain eukaryotes the ability to carry out photosynthesis it contributed to the oxygenation of the biosphere and provided the impetus for subsequent endosymbiotic events. On multiple occasions, so-called 'primary' plastids have been passed from one eukaryotic lineage to another by the process of secondary endosymbiosis, that is the assimilation of a primary plastid-bearing alga by a eukaryotic heterotroph. Despite the prominence of primary plastid-bearing organisms such as green algae and their land plant descendents, the bulk of the algal biodiversity present on Earth today is known to have acquired photosynthesis via eukaryote–eukaryote endosymbioses. Such organisms include bloom-forming haptophytes, diatoms and giant kelp. In this chapter, I summarize the known diversity of secondary plastid-bearing algae and provide an overview of the current state of knowledge of how their plastids have evolved. Despite numerous advances made possible by the comparative genomics revolution, many fundamental questions about the tempo and mode of secondary plastid diversification remain. This includes the

Advances in Botanical Research, Volume 64
ISSN 0065-2296,
http://dx.doi.org/10.1016/B978-0-12-391499-6.00003-7

extent to which cryptic tertiary plastid acquisitions have contributed to the spread of plastids.

1. INTRODUCTION

Next to the origin of the eukaryotic cell itself, it is difficult to imagine a more important event in the evolution of complex life than the advent of photosynthesis in eukaryotes. Oxygenic photosynthesis, a process that uses sunlight to produce organic compounds and liberates oxygen as a bi-product, first evolved in the ancestors of modern-day cyanobacteria more than two billion years ago (Tomitani, Knoll, Cavanaugh & Ohno, 2006). Some time after the eukaryotic cell evolved (Embley & Martin, 2006; Koonin, 2010) but long before the evolution of multicellular life (Canfield, Poulton, & Narbonne, 2007), an ancestral heterotrophic eukaryote forged an intimate symbiotic relationship with cyanobacteria. These cyanobacteria were presumably initially ingested and digested as food. Yet, their persistence within the cytoplasm of their host would undoubtedly have been advantageous due to the photosynthetic products they generated. Over time, these once transient cyanobacteria became permanent intracellular residents and evolved into plastids, or chloroplasts, the light-harvesting organelles so well understood by plant biologists today. The oxygen released by these early eukaryotic phototrophs further contributed to the transformation of the atmosphere (Katz, Finkel, Grzebyk, Knoll, & Falkowski, 2004), a transformation thought to have played a role in the diversification of complex animals approximately 500 million years ago (see Canfield *et al.*, 2007, and references therein for detailed review).

How did plastids first evolve? Not surprisingly, most of what we know about the transition from endosymbiont to organelle comes from comparing the biology of present-day cyanobacteria and plastids. Cyanobacterial genomes are typically between 2 and 8 Mbp in size and possess ~2000–8000 genes (Beck, Knoop, Axmann, & Steuer, 2012). In stark contrast, plastid genomes are invariably less than 0.5 Mbp in size and have at most ~200 protein genes (Kim & Archibald, 2009; Martin & Herrmann, 1998), despite the fact that 1000 or more proteins are needed to maintain proper organelle function (e.g. Terashima, Specht, & Hippler, 2011). Where do the 'missing' proteins come from? The nuclear genomes of photosynthetic eukaryotes harbour many genes of cyanobacterial provenance whose protein products are post-translationally targeted back to the organelle (Jarvis & Soll, 2001;

Mcfadden, 1999). As we shall see, the movement of DNA from endosymbiont/organelle to host, a process referred to as endosymbiotic gene transfer, or EGT (Martin, Brinkmann, Savonna, & Cerff, 1993; Timmis, Ayliffe, Huang, & Martin, 2004) (Fig. 3.1A), is a recurring theme in the evolution and diversification of plastids in eukaryotes.

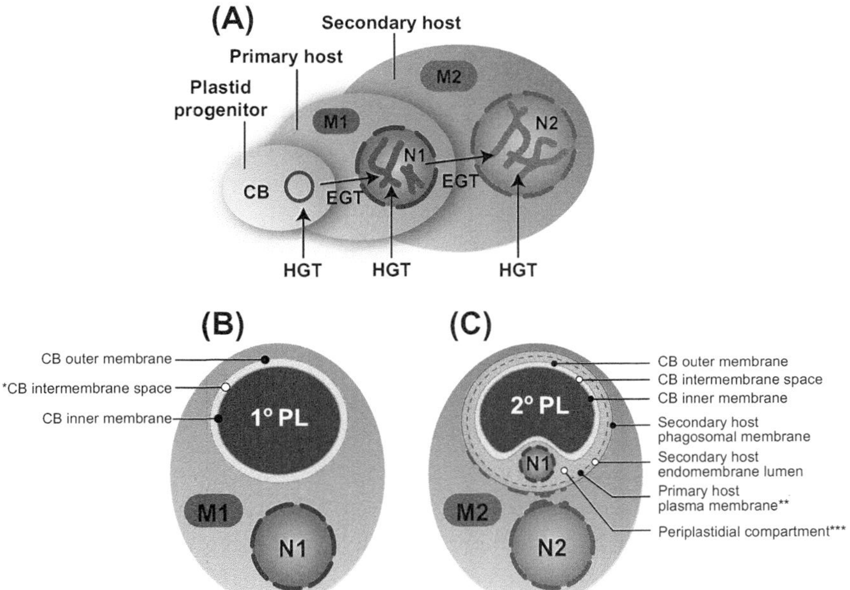

Figure 3.1 Plastid evolution by primary and secondary endosymbiosis. (A) Diagram showing gene flow in photosynthetic eukaryotes, beginning with endosymbiotic gene transfer (EGT) from the cyanobacterial (CB) progenitor of the plastid to the primary host nucleus (N1). In secondary endosymbiosis, another round of EGT occurs, in this case from the primary host nucleus to that of the secondary host (N2). Horizontal gene transfer (HGT) can also impact genome evolution at any stage. Gene transfers involving the mitochondria (M) of the primary and secondary hosts are omitted for simplicity. (B) Generic primary plastid-bearing alga. The primary plastids (PL) of red, green and glaucophyte algae are surrounded by two membranes. In the case of glaucophytes, the intermembrane space (*) contains a layer of peptidoglycan, presumably inherited from the cyanobacterial plastid progenitor. (C) A generic secondary plastid-bearing alga. Depending on the organism, these plastids are surrounded by three or four membranes. In the case of the three-membrane plastids of euglenids and peridinin-containing dinoflagellates, the primary host plasma membrane (**) is believed to have been lost. The remnant cytosol of the secondary endosymbiont is referred to as the periplastidial compartment (***). In cryptophytes and chlorarachniophytes, the primary host nucleus (N1) persists in a remnant form referred to as a nucleomorph. Nucleomorphs have been lost from the periplastidial compartments of all other secondary plastid-bearing algae. Refer to text for further discussion. See the colour plate.

The biochemistry, molecular biology and genetics of plastids speak strongly to their cyanobacterial ancestry. This includes consideration of their pigmentation, light-harvesting mechanisms and protein import apparatus as well as the gene content and structure of their genomes (Aronsson & Jarvis, 2009; Durnford *et al.*, 1999; Gould, Waller, & Mcfadden, 2008; Kim & Archibald, 2009; Mcfadden, 1999; Weihe, Liere, & Borner, 2012). Current evidence suggests—but does not prove—that plastids evolved directly from cyanobacteria on a single occasion (Palmer, 2003; Rodriguez-Ezpeleta *et al.*, 2005), perhaps as early as 1.5 billion years ago (Parfrey, Lahr, Knoll, & Katz, 2011; Yoon, Hackett, Ciniglia, Pinto, & Bhattacharya, 2004). To which group of extant cyanobacteria plastids are most closely related is not known.

Three eukaryotic lineages possess so-called 'primary' plastids: red algae, glaucophyte (or glaucocystophyte) algae and green algae, the latter group being the algal stock from which land plants evolved (Fig. 3.2) (Delwiche, Andersen, Bhattacharya, Mishler, & Mccourt, 2004; Gould *et al.*, 2008; Lewis & McCourt, 2004; Reyes-Prieto, Weber, & Bhattacharya, 2007). Red, green and glaucophyte plastids are surrounded by two membranes, which are thought to correspond to the inner and outer membrane envelopes of their cyanobacterial progenitors (Kim & Archibald, 2009) (Fig. 3.1B). Interested readers are referred to the previous chapter for discussion of the origin of primary plastids and the taxonomic composition of the organisms that harbour them (De Clerck, Bogaret, & Leliaert, 2012). In this chapter, I focus on the complex ways in which photosynthesis has spread indirectly across the eukaryotic tree by secondary and tertiary endosymbiosis, that is eukaryote–eukaryote endosymbiotic mergers. These processes have given rise to some of the most abundant and ecologically significant eukaryotic phototrophs on the planet (Fig. 3.2). They have also proven surprisingly difficult to understand despite more than a decade of genomics-enabled research.

2. SECONDARY, SERIAL SECONDARY AND TERTIARY ENDOSYMBIOSIS

Biologists have long had difficulty reconciling the tremendous diversity of photosynthetic eukaryotes with a simple model of vertical plastid inheritance. Taylor (1974) and Gibbs (1978) were among the first to entertain the possibility of horizontal plastid transmission, that is the idea that plastids could spread by endosymbiosis between evolutionarily distinct

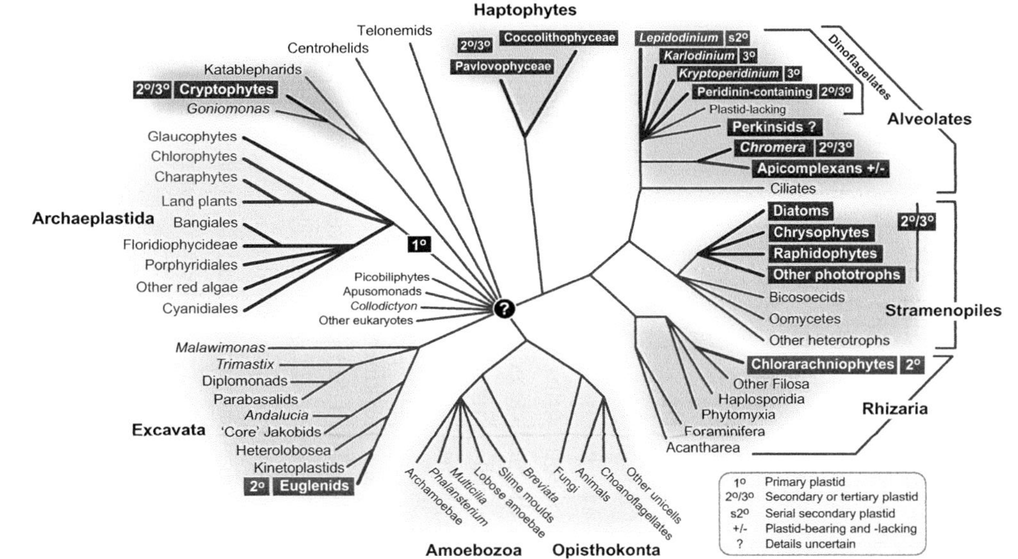

Figure 3.2 Distribution of plastids across the eukaryotic tree of life. The tree topology represents a synthesis of published data as of March 2012. The tree is unrooted and branch lengths are not to scale. Primary plastids are believed to have evolved from cyanobacteria in a common ancestor shared by red algae, green algae and glaucophyte algae. Subsequent secondary endosymbiotic events resulted in the spread of both red and green algal plastids to other eukaryotic lineages, including chlorarachniophytes and euglenids. Based on our current understanding of the higher order relationships between the major eukaryotic groups, secondary plastids of red algal ancestry appear to have spread by tertiary endosymbiosis. The donor and recipient lineages in such endosymbioses are not known. Green and red boxes indicate lineages harbouring green- and red-algal-derived plastids, respectively. Some apicomplexans have non-photosynthetic plastids and perkinsids appear to as well. Refer to text for discussion and references. See the colour plate.

eukaryotes. In pondering the model lab alga *Euglena gracilis*, for example, Gibbs noted that while its plastid is clearly green algal in nature—it possesses the signature chlorophyll $a + b$ pigmentation—*E. gracilis* cells do not *look* like green algae. Gibbs posited that a heterotrophic eukaryote had ingested a green alga and retained only its 'useful part', the plastid (Gibbs, 1978, 2006). Consistent with this hypothesis was the fact that the *E. gracilis* plastid has three membranes, not two as in green algae. The presence of additional membranes is now seen as a hallmark of plastids derived from eukaryote–eukaryote endosymbioses. In the sections, that follow I summarize the known diversity of so-called 'secondary' and 'tertiary' plastid bearing organisms. Of the three primary algal lineages described above, both green and red algae are known to have 'donated' their plastids to other eukaryotes in this fashion.

2.1. Green-Algal-Derived Plastids

Euglena gracilis is a member of the euglenids, an abundant and well-studied lineage of marine and freshwater protists characterized by the presence of a pellicle, a series of proteinaceous strips beneath the outer membrane. Together with their flagella, the pellicle contributes to the locomotion of euglenid cells and can give the cell a striped appearance under the scanning electron microscope (Leander, Witek, & Farmer, 2001b). Not all euglenids are photosynthetic; the more basal lineages are heterotrophs (Leander, Triemer, & Farmer, 2001a), as are the kinetoplastids and diplonemids, to which euglenids are most closely related (Simpson, Stevens, & Lukes, 2006). Plastid-bearing 'euglenophytes' such as *E. graciis* are characterized by the presence of a three membrane-bound photosynthetic organelle whose genome possesses strong green algal signatures (Hallick *et al.*, 1993), even though the nucleocytoplasmic component of these organisms shows no affinity for green algae (Fig. 3.2).

The chlorarachniophytes are a somewhat less well known but equally important algal lineage whose members also harbour chlorophyll $a + b$-pigmented plastids of green algal provenance. They are amoeboflagellate algae that appear to be limited to marine habitats. Relatively few chlorarachniophyte genera and species have been described (Hibberd & Norris, 1984; Ishida, Green, & Cavalier-Smith, 1999; Moestrup & Sengco, 2001; Ota, Silver, Archibald, & Ishida, 2009a; Ota, Ueda, & Ishida, 2005, 2007; Ota, Vaulot, Le Gall, Yabuki, & Ishida, 2009b). From the perspective of plastid evolution, chlorarachiophytes are of particular interest by virtue of the fact that the nucleus of the algal endosymbiont—the 'nucleomorph'—persists in close association with the plastid (Hibberd & Norris, 1984; Moestrup & Sengco, 2001). Chlorarachniophyte plastids have four bounding membranes; the nucleomorph resides within the periplastial compartment, that is the residual green algal cytosol sandwiched between the inner and outer membrane pairs (Fig. 3.1C). Characterization of the nucleomorph and its genome has provided the definitive proof that secondary endosymbiosis has occurred (Archibald, 2007; Douglas, Murphy, Spencer, & Gray, 1991; Mcfadden, Gilson, Hofmann, Adcock, & Maier, 1994b; Moore & Archibald, 2009). Unlike euglenophytes, whose green algal endosymbiont nucleus has disappeared, chlorarachniophytes have essentially been 'caught in the act'.

A variety of molecular phylogenetic and comparative genomic data have been brought to bear on the origin of the chlorarachniophyte plastid and nucleomorph. Early investigations were limited to single-gene analyses of plastid small subunit ribosomal DNA (SSU rDNA) and various proteins (e.g. Ishida, Cao, Hasegawa, Okada, & Hara, 1997; Ishida *et al.*, 1999) and have now been complemented with complete genome sequence data. The plastid genome of the model chlorarachniophyte *Bigelowiella natans* has been sequenced and is demonstrably green algal (Rogers, Gilson, Su, Mcfadden, & Keeling, 2007), and the *B. natans* nucleomorph genome is, in essence, a green algal nuclear genome in miniature (Gilson *et al.*, 2006).

Despite the general similarities of their plastids, available evidence suggests that the host components of chlorarachniophyte and euglenophyte algae are unrelated. While the euglenophytes belong to the 'supergroup' Excavata (Hampl *et al.*, 2009), the chlorarachniophytes are the sole plastid-bearing group within the Rhizaria, a diverse protist lineage that includes foraminiferans, test-forming heterotrophic protists famous for their prominence in the fossil record (Nikolaev *et al.*, 2004; Pawlowski *et al.*, 2003) (Fig. 3.2). Advances in our understanding of global eukaryotic phylogeny will be elaborated upon below. Here, it is important to note that not only are the host components of chlorarachniophytes and euglenophytes unrelated, their plastids are not specifically related to one another in the context of green algal phylogeny. Whereas the chlorarachniophyte plastid shows specific ties to the ulvophyte–trebouxiophyte–chlorophyte subgroup of green algae, the euglenophyte plastid appears as a distinct branch on the green plastid line (Rogers *et al.*, 2007; Turmel, Gagnon, O'Kelly, Otis, & Lemieux, 2009). Analyses of nucleus-encoded, plastid-targeted proteins such as psbO produce a similar result (e.g. Takahashi *et al.*, 2007). Collectively, these data suggest that the chlorarachniophyte and euglenophyte plastids are the product of independent secondary endosymbioses involving different hosts and different green algal endosymbionts (Archibald, 2009; Keeling, 2004, 2009; Reyes-Prieto *et al.*, 2007).

2.2. Red-Algal-Derived Plastids in Haptophytes, Stramenopiles and Cryptophytes

The diversity of algae with green-algal-derived plastids pales in comparison to those with plastids of red algal ancestry. The most abundant, species-rich and ecologically significant of such organisms are the haptophytes, stramenopiles and dinoflagellates (Not *et al.*, 2012). Each of these lineages has its peculiarities, and from the perspective of plastid evolution, an important factor is that the latter two groups are by no means exclusively photosynthetic. As a result, it has proven much more difficult to infer how, and from where, the plastids in these organisms have evolved.

Haptophytes are unicellular biflagellate marine and freshwater phototrophs typified by the presence of a 'haptonema', a flagellum-like protrusion of ambiguous function (Andersen, 2004). Haptophytes are generally divided into two groups, the Pavlovophyceae and Coccolithophyceae (Fig. 3.2). The best known members reside within the latter group and are characterized by the presence of coccoliths, prominent and often elaborate calcareous extracellular plates. The bloom-forming haptophyte *Emiliania huxleyi* is a model haptophyte species. Stramenopiles (heterokonts) are an exceptionally diverse assemblage of plastid-containing and plastid-lacking organisms. By virtue of their ornate silica shells and significant contributions to marine primary productivity (Nelson, Tréguer, Brzezinski, Leynaert, & Queguiner, 1995), unicellular diatoms are one of the best known stramenopile algal lineages (see review on diatoms in Chapter VII of this volume, Mock & Medlin, 2012). At the opposite end of the size spectrum is giant kelp, stramenopiles that can grow to more than 40 m in length (a genomic perspective on the brown alga *Ectocarpus* appears in Chapter V of this volume, The *Ectocarpus* Genome Consortium, 2012). Non-photosynthetic and plastid-lacking stramenopile lineages include oomycetes such as the causative agent of potato blight, *Phytophthora*, and the human intestinal parasite, *Blastocystis*. As we shall see, whether some or all these heterotrophic stramenopiles descend from photosynthetic ancestors is a subject of much debate.

The plastids of both haptophytes and photosynthetic stramenopiles have chlorophyll $a + c$ pigmentation and are surrounded by four membranes (Andersen, 2004). The outermost membrane is continuous with the nuclear envelope and is studded with ribosomes; the plastids in these organisms thus reside within the lumen of the endoplasmic reticulum (ER), an arrangement historically referred to as the 'chloroplast ER' (Gibbs, 1981). ER-localized plastids are also found in a third algal group, the cryptophytes. Cryptophytes are bi-flagellate freshwater and marine algae with chlorophyll $a + c$ pigmented plastids. As in stramenopiles, cryptophytes are closely allied with plastid-lacking protists, in this case, members of the genus *Goniomonas* (Mcfadden, Gilson, & Hill, 1994a; Shalchian-Tabrizi *et al.*, 2008) (Fig. 3.2). Together these two groups are generally referred to as the cryptomonads. Complete plastid genome sequence analyses have bolstered the results of early single-gene analyses (e.g. Daugbjerg & Andersen, 1997; Müller, Oliveira, Sheath, & Bhattacharya, 2001; Oliveira & Bhattacharya, 2000) in demonstrating a red algal origin for the plastids of stramenopiles (Oudot-Le Secq *et al.*, 2007), haptophytes (Sanchez-Puerta, Bachvaroff, & Delwiche, 2005) and cryptophytes (Douglas & Penny, 1999; Khan *et al.*, 2007).

Unlike stramenopiles and haptophytes, the cryptophyte periplastidial compartment retains a nucleomorph (Greenwood, 1974; Greenwood, Griffiths, & Santore, 1977) (Fig. 3.1C), whose genome provides strong support for a red algal ancestry for the cryptophyte secondary endosymbiont (Douglas *et al.*, 2001; Hoef-Emden, Mann, & Melkonian, 2002; Lane *et al.*, 2007; Moore & Archibald, 2009; Tanifuji *et al.*, 2011). Although clearly the product of separate endosymbioses involving red and green algae, the nucleomorph genomes in cryptophytes and chlorarachniophytes have converged upon similar sizes and structures. These similarities include (i) highly reduced genomes between ~0.3 and 1 Mbp in size, (ii) a three-chromosome architecture and (iii) rDNA operons located on the sub-telomeric region of each chromosome end (see Archibald, 2007; Archibald & Lane, 2009; Moore & Archibald, 2009, and references therein for detailed review). Nucleomorph genomes are dominated by 'housekeeping' proteins involved in transcription, translation and protein folding/degradation essential for the expression of a small set of plastid-targeted proteins whose genes have not (yet) been transferred to the host cell nuclear genome. The question of whether the cryptophyte and/or chlorarachniophyte nucleomorph genomes are 'frozen' in their current state or might eventually be eliminated, as has occurred in other algae that acquired photosynthesis in this manner, is an active area of research. What are needed are more nucleomorph genome sequences from diverse species within both cryptophytes and chlorarachniophytes in order to better understand the process of genome reduction. Information gleaned from host nuclear genomes will also be important. The Joint Genome Institute's Community Sequencing Program has sequenced the nuclear genomes of the cryptophyte *Guillardia theta* and the chlorarachniophyte *Bigelowiella natans* (http://www.jgi.doe.gov/sequencing/why/50026.html), analyses of which will provide an estimate of the frequency of nucleomorph-to-host-nucleus gene transfer in the two lineages.

2.3. Alveolate Plastids: Red and Green

The alveolates are a large and complex protist lineage comprised five sub-lineages: apicomplexans, dinoflagellates, perkinsids, ciliates and the recently discovered chromerids. The name 'alveolate' refers to the prominent cortical alveoli (sacs or cavities) that lay beneath their cell surface. These four groups form a robust monophyletic assemblage in molecular phylogenies (e.g. Bachvaroff, Handy, Place, & Delwiche, 2011; Van De Peer, Van Der Auwera, & De Wachter, 1996); plastids are patchily distributed both within and between the groups (Fig. 3.2). Consequently, alveolates feature prominently in discussions of plastid evolution.

Apicomplexans are unicellular spore-forming parasites of animals. The causative agent of malaria, *Plasmodium falciparum*, is one of the best known members of the group, and the discovery that it and other apicomplexans such as *Toxoplasma gondii* possessed a remnant plastid—now called an apicoplast—came as a surprise (Köhler *et al.*, 1997; Mcfadden, Reith, Munholland, & Lang-Unnasch, 1996; Wilson *et al.*, 1996) (the organelle had in fact been known for decades but was not recognized as such; see Mcfadden & Waller, 1997, and references therein). As unlikely as it seems given their current lifestyle, apicomplexans evolved from algal ancestors. While they no longer use their apicoplast for harnessing the energy of sunlight, genomic and biochemical data indicate that the apicoplast is the site of various essential cellular processes, including isoprenoid and haeme biosynthesis (Mazumdar, Wilson, Masek, & Hunter, 2006; Seeber & Soldati-Favre, 2010; Waller & Mcfadden, 2005; Waller *et al.*, 1998). Although clearly plastid like, the apicoplast genomes of apicomplexans such as *Plasmodium* (Wilson *et al.*, 1996) and *Eimeria* (Cai, Fuller, Mcdougald, & Zhu, 2003) are highly reduced (~35 kb in size) and difficult to interpret from an evolutionary perspective. The apicoplast has four membranes (although there is some controversy on this point; Hopkins *et al.*, 1999; Kohler, 2005; Köhler *et al.*, 1997; Tomova, Geerts, Muller-Reichert, Entzeroth, & Humbel, 2006) and so is an obvious candidate for being of secondary endosymbiotic origin. Initially proposed to be green algal in origin (Köhler *et al.*, 1997), available data best support the hypothesis that the apicoplast is in fact of red algal ancestry (see Keeling, 2009, 2010, and references therein for review).

Not all apicomplexans possess apicoplasts. For example, *Cryptosporidium* is a well-studied human gastrointestinal parasite that appears to lack such an organelle, as do gregarines, which inhabit marine invertebrates (Toso & Omoto, 2007). The genomes of two *Cryptosporidium* species, *C. hominis* and *C. parvum*, have been completely sequenced and no evidence for the existence of genes for plastid-targeted proteins was found (Abrahamsen *et al.*, 2004; Xu *et al.*, 2004; Zhu, Marchewka, & Keithly, 2000). Nevertheless, the presence of several algal/cyanobacterial-like genes in their nuclear genomes (e.g. genes for the enzymes glucose-6-phosphate, uridine kinase/uracil phosphoribosyltransferase and leucine aminopeptidase) has been interpreted to mean that the ancestors of *Cryptosporidium* had such an organelle in the past (Huang *et al.*, 2004). This seems reasonable enough given the close evolutionary relationship between *Cryptosporidium* and apicoplast-harbouring apicomplexans (Bachvaroff *et al.*, 2011), although the extent to which algal/cyanobacterial genes in the genomes of plastid-lacking protists truly speak to a photosynthetic ancestry is a controversial topic. We shall revisit this issue below.

Dinoflagellates are far and away the most complex of all eukaryotic phototrophs in terms of their plastid diversity. The most common type of plastid in the group is surrounded by three (or occasionally two) membranes and contains chlorophyll $a + c$ pigmentation as well as peridinin (Kim & Archibald, 2009; Schnepf & Elbrächter, 1999). These are the so-called peridinin-containing dinoflagellates and are generally thought to have plastids of secondary endosymbiotic origin. Other dinoflagellates have plastids taken by tertiary endosymbiosis from some of the red algal plastid-bearing organisms described above. For example, *Karinia* and *Karlodinium* have haptophyte-derived plastids (Tengs *et al.*, 2000) and *Kryptoperidinium* has a diatom-derived plastid (Dodge, 1969; Inagaki, Dacks, Doolittle, Watanabe, & Ohama, 2000) that was acquired very 'recently'; an apparently unreduced diatom nucleus and mitochondria reside next to the plastid (Chesnick, Hooistra, Wellbrock, & Medlin, 1997; Imanian,

Carpenter, & Keeling, 2007; Imanian & Keeling, 2007; McEwan & Keeling, 2004). *Dinophysis acuminata* is a particularly interesting dinoflagellate with a plastid derived from a cryptophyte (Hackett, Maranda, Yoon, & Bhattacharya, 2003; Schnepf & Elbrächter, 1988). Remarkably, the cryptophyte 'organelle' is first acquired by the ciliate *Myrionecta rubra*, which is in turn ingested by *D. acuminata* (Park, Park, Kim, & Yih, 2008). Genomic data suggest that the *D. acuminata* cryptophyte-derived plastid is a transient entity, which must regularly be re-acquired by phagotrophy (Kim & Archibald, 2010; Wisecaver & Hackett, 2010).

The dinoflagellate *Lepidodinium* harbours a four membrane-bound plastid derived from a green alga of the 'core chlorophyte' line (Hansen, Botes, & De Salas, 2007; Matsumoto *et al.*, 2011; Watanabe, Suda, Inouye, Sawaguchi, & Chihara, 1990) (Fig. 3.2). This constitutes an example of what is referred to as 'serial secondary' endosymbiosis. Recent investigation of the transcriptome of *L. chlorophorum* revealed the presence of genes for plastid-targeted proteins from a wide variety of sources, not only green algae but also peridinin-containing dinoflagellates (Minge *et al.*, 2010), suggesting that *Lepidodinium* once had a 'typical' dinoflagellate plastid but replaced it with one from a green alga. All in all, dinoflagellates exhibit remarkable plasticity in their photosynthetic apparatus, appearing to have 'swapped out' their original plastids for new ones on multiple occasions (Hackett, Anderson, Erdner, & Bhattacharya, 2004; Shalchian-Tabrizi *et al.*, 2006). The tendency for dinoflagellates to do this is reflected in the mosaic nature of their genomes (Nosenko *et al.*, 2006; Patron, Waller, Archibald, & Keeling, 2005; Shalchian-Tabrizi *et al.*, 2006; Takishita *et al.*, 2008) and is probably related to their mixotrophic lifestyle, that is they are capable of both phototrophic and heterotrophic growth (Stoecker, 1999).

As impressive as this diversity is, it is important to recognize that only approximately 50% of dinoflagellate species are actually photosynthetic and/or possess a plastid (Taylor, 1987). So did the ancestor of present-day dinoflagellates have a plastid? It is not yet known with certainty, but insights gleaned from several additional members of the alveolates have brought us closer to an answer. The first is the Perkinsida, a group of parasitic protists that are considered by some to be basal dinoflagellates and others to be an independent alveolate lineage. Despite being heterotrophic, a handful of genes of probable algal origin have been found in the nuclear genome of the oyster parasite *Perkinsus marinus*, including the same plastid-type isoprenoid biosynthetic genes found in apicomplexans (Fernandez Robledo *et al.*, 2011; Matsuzaki, Kita, & Nozaki, 2006; Matsuzaki, Kuroiwa, Kuroiwa, Kita, &

Nozaki, 2008; Stelter, El-Sayed, & Seeber, 2007). Significantly, these genes encode proteins with amino terminal extensions reminiscent of the bipartite leader sequences known to target proteins to plastids of secondary endosymbiotic origin (Gould *et al.*, 2006; Gould, *et al.*, 2008), and ultrastructural and immunolocalization studies have identified a tiny membranous sub-cellular compartment that could be a remnant plastid (Matsuzaki *et al.*, 2008; Teles-Grilo *et al.*, 2007). Genes for putative plastid-targeted proteins have been found in the nuclear genomes of several other heterotrophic and once-presumed-to-be plastid-lacking dinoflagellates, including *Oxyrrhis marina* (Slamovits & Keeling, 2008) and *Crypthecodinium cohnii* (Sanchez-Puerta, Lippmeier, Apt, & Delwiche, 2007b), suggesting that plastids may exist in many and perhaps even all dinoflagellates, regardless of their lifestyle.

One of the most important recent advances in our understanding of alveolate plastid evolution has come from the discovery of the chromerids. *Chromera velia* is a unicellular alga found living among the coral in Sydney Harbor, Australia (Moore *et al.*, 2008). *C. velia* has a four membrane-bound plastid with chlorophyll *a* pigmentation and is clearly an alveolate: it has cortical alveoli and preliminary molecular phylogenies suggested that it is the closest free-living relative of apicomplexan parasites (Moore *et al.*, 2008; Obornik, Janouskovec, Chrudimsky, & Lukes, 2008) (Fig. 3.2). The plastid genomes of *C. velia* and a related un-named organism, strain CCMP1355, have recently been sequenced (Janouškovec, Horak, Obornik, Lukes, & Keeling, 2010) and lend further support to this hypothesis. Apicomplexan plastid genomes are highly divergent and (obviously) lack genes associated with photosynthesis, and dinoflagellate plastids have ultra-fast-evolving genes that reside on single-gene 'minicircles' (Nisbet, Koumandou, Barbrook, & Howe, 2004; Nisbet *et al.*, 2008; Sanchez-Puerta, Bachvaroff, & Delwiche, 2007a; Zhang, Green, Cavalier-Smith, 1999, 2000). In contrast, chromerid plastid genomes have comparatively slowly evolving sequences and are conservative in structure; they are packed with useful information. Having considered the chromerid genomes and associated plastid data, Janouškovec *et al.* (2010) concluded that "... the extant plastids of apicomplexans and dinoflagellates were inherited by linear descent from a common red algal endosymbiont." In short, chromerids appear to be a 'missing link' between the dinoflagellates and apicomplexans (Keeling, 2008).

So if the common ancestor of dinoflagellates and apicomplexans had a plastid, how much further back in time might this organelle have been acquired? This question brings us to the 'chromalveolate' hypothesis of Cavalier-Smith (1999). The underlying principle is one of parsimony: evolving a secondary plastid is 'difficult' because with each secondary endosymbiotic event hundreds of nuclear genes for plastid proteins must move from the primary host nucleus to the secondary host nucleus (Fig. 3.1A) and the machinery for importing these proteins into the nascent plastid must be established anew (Cavalier-Smith, 1999; Gould *et al.*, 2008). Therefore, all things being equal one should invoke the minimum number of endosymbioses necessary to explain the observed distribution of plastids. It is far from certain, however, whether an emphasis on plastid gain or plastid loss brings us closer to an accurate explanation of the distribution of red-algal-derived plastids. As of this writing the answer seems to be a bit of both.

3. TESTING EVOLUTIONARY HYPOTHESES

Determining whether the plastids of any two algal lineages are or are not likely to be the product of the same endosymbiotic event is a complex problem. Not only must one show that the ultrastructure, biochemistry and molecular phylogeny of their plastids are consistent with a single origin, the relationship between their host components must also be considered (Larkum, Lockhart, & Howe, 2007). This has proven to be straightforward in some instances but not in others.

3.1. Early Steps Towards a Realistic Picture of Plastid Evolution

The test of plastid and host congruence/incongruence was considered above when discussing the secondary plastids of euglenophytes and chlorarachniophytes. In this case, not only do their plastids appear to trace back to distinct green algal lineages (Rogers *et al.*, 2007), their hosts belong to different eukaryotic lineages (Hampl *et al.*, 2009; Nikolaev *et al.*, 2004; Pawlowski *et al.*, 2003) (Fig. 3.2). Therefore, despite an initial proposal for a single origin (Cavalier-Smith, 1999), most workers in the field are firm in their belief that the euglenophyte and chlorarachniophyte plastids are the product of two separate secondary endosymbioses (Archibald, 2009; Keeling, 2010; Reyes-Prieto *et al.*, 2007).

In considering the red-algal-derived plastids of stramenopiles, haptophytes and cryptophytes, Cavalier-Smith proposed that their photosynthetic organelles stem from a single red algal secondary endosymbiotic event in their common ancestor: these three lineages were referred to as 'chromists' (Cavalier-Smith, 1986). He subsequently proposed that the chromist plastid was even more ancient, having evolved by secondary endosymbiosis in a common ancestor shared between chromists and alveolates: chromalveolates (Cavalier-Smith, 1999). Such a scenario is parsimonious from the perspective of plastid gain (i.e. a single event), but it hinges on whether or not chromists and alveolates are actually related to one another. Since its inception, enthusiasm for the chromalveolate hypothesis has ebbed and flowed in response to the discovery of new lineages (e.g. chromerids) and a tidal wave of genomic data.

Early analyses of plastid genes such as ribulose-1,5-bisphosphate carboxylase/oxygenase (RuBisCO) and SSU rDNA suggested that photosynthetic stramenopiles, haptophytes and cryptophytes had acquired their plastids from different red algal endosymbionts (Daugbjerg & Andersen, 1997; Müller *et al.*, 2001; Oliveira & Bhattacharya, 2000). However, with improved taxon sampling, larger data sets, and more sophisticated phylogenetic methods, plastid sequences from these three lineages do in fact branch together in the context of red algal phylogeny, albeit with varied support (Bachvaroff, Sanchez-Puerta, & Delwiche, 2005; Iida, Takishita, Ohshima, & Inagaki, 2007; Khan *et al.*, 2007; Yoon, Hackett, Pinto, & Bhattacharya, 2002). This is consistent with, but not conclusive evidence for, the hypothesis that their plastids are the product of a single secondary endosymbiotic event. Unfortunately, apicomplexans must be excluded from such analyses because their apicoplast genomes lack many of the genes useful for plastid phylogeny, and the divergent nature of dinoflagellate plastid sequences makes them frustratingly prone to generating methodological artefacts (Bachvaroff *et al.*, 2005; Sanchez-Puerta *et al.*, 2007a).

Analyses of nuclear genes for plastid-targeted proteins allow a much broader spectrum of 'chromalveolate' taxa to be considered and have produced intriguing results. For example, endosymbiotic gene replacements in glyceraldehyde-3-phosphate dehydrogenase (Fast, Kissinger, Roos, & Keeling, 2001; Harper & Keeling, 2003) and fructose-1,6-bisphosphate (Patron, Rogers, & Keeling, 2004) were found to be shared between chromists and some alveolates. The simplest interpretation is that these replacements, which are generally believed to be 'rare' events, took place in their common ancestor, supporting a shared ancestry of these organisms and their plastids. However, it has been argued that such genes are not reliable markers of vertical (i.e. organismal) inheritance: they have been shown to have evolved by horizontal (or lateral) gene transfer (HGT) on at least several occasions (e.g. Qian & Keeling, 2001; Takishita, Ishida, & Maruyama, 2003; Takishita, Yamaguchi, Maruyama, & Inagaki, 2009; Takishita *et al.*, 2008) and their shared presence in chromalveolate taxa can equally be explained by nucleus-to-nucleus gene transfers in the context of tertiary endosymbiosis (Archibald, 2009; Bodył, 2005, 2006; Bodył, Stiller, & Mackiewicz, 2009; Sanchez-Puerta & Delwiche, 2008). Hypotheses about the evolution of plastids based on single-gene analyses have therefore become increasingly difficult to defend.

3.2. A Phylogenomic Framework for Eukaryotes and Plastids

A fruitful line of inquiry has been the application of 'phylogenomics' to the question of deep eukaryotic phylogeny. This approach, which involves the analysis of dozens to hundreds of proteins linked together as a single concatenate (Delsuc, Brinkmann, & Philippe, 2005), has led to the establishment of five or six eukaryotic 'supergroups' (Keeling *et al.*, 2005; Roger & Simpson, 2009; Simpson & Roger, 2004). These large assemblages may or may not be of equivalent taxonomic rank, but together they provide a useful framework against which plastid gain / loss scenarios can be considered. In some cases, the new supergroups, and the sub-groups within them, neatly correspond to previously known or suspected monophyletic assemblages, for example the primary plastid-bearing Archaeplastida (or Plantae) (Cavalier-Smith, 1998; Chan *et al.*, 2011; Moreira, Le Guyader, & Phillipp, 2000) and the Excavata (Hampl *et al.*, 2009; Simpson, 2003). Other supergroups have emerged solely on the basis of molecular phylogeny and were entirely unexpected.

The schematic tree shown in Fig. 3.2 represents a synthesis of the results of various multi-gene studies published in the last decade or so (most notably (Burki, Shalchian-Tabrizi, & Pawlowski, 2008; Burki *et al.*, 2007, 2009; Hackett *et al.*, 2007; Moreira *et al.*, 2000; Parfrey *et al.*, 2010; Patron, Inagaki, & Keeling, 2007), including a recent 258-protein analysis by Burki, Okamoto, Pombert, & Keeling (2012). From the perspective of plastid evolution, it is significant in showing that the chromalveolate lineages—that is chromists (stramenopiles + haptophytes + cryptophytes) and alveolates—do *not* constitute a monophyletic clade. The Rhizaria, to which the chlorarachniophytes belong, form a robust sister relationship with stramenopiles and alveolates, an unexpected trio dubbed the 'SAR' clade (stramenopiles + alveolates + Rhizaria) (Burki *et al.*, 2008; Hackett *et al.*, 2007). A sister relationship between the host components of cryptophytes and haptophytes has been observed in several analyses (e.g. Burki *et al.*, 2008, 2009;

Hampl *et al.*, 2009; Patron *et al.*, 2007), a result that is particularly interesting in light of the fact that their plastid genomes share an HGT involving the ribosomal protein gene *rpL36* (Rice & Palmer, 2006). However, the latest Burki *et al.* topology places haptophytes sister to the SAR clade, with the cryptophytes showing weak affiliations to the Archaeplastida (Burki *et al.*, 2012) (Fig. 3.2).

A complementary approach to testing the chromalveolate hypothesis was recently published by Baurain *et al.* (2010). These authors assessed the phylogenetic signal contained in the nuclear, mitochondrial and plastid genomes of the so-called CASH taxa (cryptophytes, alveolates, stramenopiles and haptophytes), reasoning that if chlorophyll-*c*-containing algae shared a common plastid-bearing ancestor, then their monophyly should be recoverable from similar amounts of sequence data taken from these three genomes. Baurain *et al.* reject the chromalveolate hypothesis, finding that the plastid genomes of CASH lineages appear to have diverged from one another more recently than their mitochondrial and nuclear genomes did (Baurain *et al.*, 2010).

What are we to conclude from the apparent falsification of the chromalveolate hypothesis? First and foremost, it means that additional endosymbioses must be invoked to account for the current distribution of red-algal-derived plastids across the eukaryotic tree of life. This is not a new idea. Bodył and colleagues (Bodył, 2005, 2006; Bodył *et al.*, 2009) have questioned the *a priori* assumption that endosymbiosis is 'difficult' and should be invoked sparingly, arguing that "... an alternative model of serial tertiary endosymbioses is more consistent with available data, and should be taken into account in phylogenomic investigations of eukaryotic diversity" (Bodył *et al.*, 2009). At the present time, it is far from clear who the donor and recipient lineages involved in such tertiary endosymbioses were. Most authors favour a single secondary endosymbiosis of a red alga, perhaps in an ancestral cryptophyte, followed by tertiary plastid movement to or from haptophytes, plastid-bearing stramenopiles and plastid-bearing alveolates (see Archibald, 2009; Bodył *et al.*, 2009; Sanchez-Puerta & Delwiche, 2008; Teich, Zauner, Baurain, Brinkmann, & Petersen, 2007, and references therein for comprehensive review).

3.3. The Genomic 'Footprint' of Endosymbiosis: Do We Know it to See it?

The nuclear genomes of red, green and glaucophyte algae harbour hundreds of genes of cyanobacterial origin (Matsuzaki *et al.*, 2004; Merchant *et al.*,

2007; Price *et al.*, 2012; Terashima *et al.*, 2011). The textbook explanation for the origin and persistence of these genes is that (i) they moved to the nucleus from the cyanobacterial progenitor of the plastid by EGT during the course of evolution and (ii) their protein products are targeted back to the plastid post-translationally. While adequate, this explanation fails to capture the true complexity of the situation. In a landmark 2002 study, Martin and colleagues showed that <50% of the ~4500 genes of cyanobacterial origin in the *Arabidopsis* genome actually encode proteins that appear to be targeted to the plastid, with the majority predicted to participate in a wide range of plastid-independent functions (Martin *et al.*, 2002). Different studies have produced somewhat different estimates (e.g. Bhattacharya, Archibald, Weber, & Reyes-Prieto, 2007; Deusch *et al.*, 2008; Moustafa & Bhattacharya, 2008; Reyes-Prieto, Hackett, Soares, Bonaldo, & Bhattacharya, 2006; Sato, Ishikawa, Fujiwara, & Sonoike, 2005, and references therein), but the take-home message is the same: "... there is no evolutionary 'homing device' that automatically directs the protein product of a transferred gene back to the organelle of its provenance. Instead, the products of genes acquired from endosymbionts can explore all targeting possibilities within the cell" (Martin, 2010).

In the context of secondary endosymbiosis, a second round of EGT moves these cyanobacterial genes from the primary host nucleus to that of the secondary host (Fig. 3.1A). Genes of eukaryotic ancestry are presumably transferred between the two nuclei as well (Lane & Archibald, 2008). Combined with the potential for DNA acquisition by HGT (Bowler *et al.*, 2008; Keeling & Palmer, 2008) (Fig. 3.1A), the emerging picture is one of extensive genome mosaicism in photosynthetic eukaryotes, with the proteomes of the various subcellular compartments being of mixed heritage. For example, genes of both red *and* green algal ancestry have been found in the nuclear genomes of the green algal plastid-bearing chlorarachniophyte *Bigelowiella natans* (Archibald, Rogers, Toop, Ishida, & Keelimg, 2003), and the red algal plastid-harbouring diatoms (Moustafa *et al.*, 2009) and chromerids (Woehle, Dagan, Martin, & Gould, 2011). Interpretations of these data range from proposals for ancient cryptic endosymbioses to an emphasis on HGT (see Elias & Archibald, 2009; Woehle *et al.*, 2011, for recent discussion). One thing most researchers would agree on is that it is difficult to be confident of the precise evolutionary history of any given gene.

All this matters when it comes to searching for genomic 'footprints' of endosymbiosis in non-photosynthetic and presumed-to-be plastid-lacking eukaryotes. As noted above, genes for putative plastid-targeted proteins have provided indirect evidence for the existence of cryptic plastids in various heterotrophic dinoflagellates (Slamovits & Keeling, 2008, Sanchez-Puerta *et al.*, 2007b, Fernandez Robledo *et al.*, 2011). This furthers strengthens the photosynthetic link between dinoflagellates and apicomplexans, a link that seems increasingly tenable with the discovery of chromerids (Keeling, 2008; Moore *et al.*, 2008). However, for other organisms that have featured in the chromalveolate debate, such as heterotrophic stramenopiles and the ciliates (Fig. 3.2), the evidence is much more controversial. The genome of the oomycete stramenopile *Phytophthora* was originally purported to harbour many hundreds of genes of red algal/cyanobacterial ancestry (Tyler *et al.*, 2006) and was regularly cited in support of the idea of secondary plastid loss in this organism (e.g. Keeling, 2009; Lane & Archibald, 2008; Slamovits & Keeling, 2008). However, a reanalysis by Stiller, Huang, Ding, Tian, & Goodwillie (2009) suggests that the number of algal-like genes in *Phytophthora* is no higher than 'background', which in this particularly study was defined as the number of algal/cyanobacterial genes found in the genomes of amoebozoans, organisms that are not expected to have endosymbiotically derived genes (Fig. 3.2).

In the case of ciliates, a vast exclusively heterotrophic alveolate lineage (Fig. 3.2), 16 genes with algal/cyanobacterial signatures were discovered in the genome of *Tetrahymena* and were taken as evidence for a photosynthetic ancestry (Reyes-Prieto, Moustafa, & Bhattacharya, 2008). However, the authors of the original *Tetrahymena* genome sequence did not arrive at this conclusion (Eisen *et al.*, 2006) and given that ciliates have a penchant for harbouring algal symbionts (e.g. Summerer, Sonntag, & Sommaruga, 2007), it seems equally likely that this handful of algal/cyanobacterial genes are the product of HGT, not ancient EGT (Elias & Archibald, 2009). Interestingly, a recent comprehensive survey of the transcriptome of the katablepharid *Rhoombia* failed to reveal a single convincing case of a gene of apparent algal/plastid origin (Burki *et al.*, 2012). This is significant given the sister relationship of katablepharids to *Goniomonas* and the plastid-bearing cryptophytes (Okamoto, Chantangsi, Horak, Leander, & Keeling, 2009; Okamoto & Inouye, 2005) (Fig. 3.2). What can be said at the present time is that the only reasonably strong case of outright plastid loss in eukaryotes is in the apicomplexan *Cryptosporidium*, and even here the potential cyanobacterial/algal footprint in the genome is not very large and is complicated by HGT (Huang *et al.*, 2004). Taking stock of the situation, Stiller (2011) has argued for improved experimental design and increased statistical rigor when screening genomes for endosymbiotically derived genes. Whether or not an endosymbiotic footprint is observed in a given genome is presently still very much in the eye of the beholder.

4. CONCLUSIONS AND PROSPECTUS

Comparative genomics paints a remarkably complex picture of the evolution of photosynthesis in eukaryotes. It is clear that no simple endosymbiotic scenario will suffice and we are many years away from a complete understanding of the tempo and mode of plastid diversification. There are several reasons for this. Despite its great promise, the burgeoning field of phylogenomics is still very much in its infancy and the tree of eukaryotes is still a work in progress. The position of the root is still far from clear (Roger & Simpson, 2009) and there are many lineages that remain 'in limbo' to varying degrees, telonemids, centrohelids, apusomonads and *Collodictyon* to name a few (Burki *et al.*, 2009, 2012; Kim, Simpson, & Graham, 2006; Okamoto *et al.*, 2009; Parfrey *et al.*, 2010; Zhao *et al.*, 2012) (Fig. 3.2). As well, entirely new groups of organisms occupying potentially key phylogenetic positions continue to be discovered, most notably the (pico)biliphytes and rappemonads, both of which have been proposed to have phylogenetic connections to haptophytes and cryptophytes (Cuvelier *et al.*, 2008; , Kim *et al.*, 2011; Not *et al.*, 2007; Yoon *et al.*, 2011). Neither group has been brought into culture and determining where they and other 'orphan' lineages fit on the eukaryotic tree will be a difficult but important task.

Focusing just on the 'well-understood' algal lineages, why is it that different phylogenomic analyses with different taxonomic sampling can produce different results, even when 100+ proteins are analyzed? The popular but now uncertain monophyly of haptophytes and cryptophytes is a prominent example (e.g. Burki *et al.*, 2008, 2012; Patron *et al.*, 2007). Even the monophyly of primary plastid-bearing eukaryotes, often taken as 'a given', is not always supported in large multi-gene phylogenies and has in fact long been questioned (Nozaki *et al.*, 2007, 2009; Palmer, 2003; Stiller, 2007; Stiller & Hall, 1997). Understanding the biological and/or methodological reasons for these discrepancies is one of the most fundamental challenges facing researchers in the area of eukaryotic systematics today. Given the critical importance of a robust phylogenomic framework for testing evolutionary hypotheses, future investigations must strive for analytical consistency and rigor and to identify and account for incongruent signals in multi-gene data sets. Methods for assessing phylogenetic congruence have been developed (e.g. de Vienne, Ollier, & Aguileta, 2012; Leigh, Lapointe, Lopez, & Bapteste, 2011a; Leigh, Susko, Baumgartner, & Roger, 2008; Leigh, Schliep, Lopez, & Bapteste, 2011b), but they are often computationally intensive and not commonly applied. Continued refinement of such methods will hopefully lead to their increased use and to the establishment of 'standard' multi-gene data sets and bioinformatics pipelines for deep eukaryotic phylogenomics.

Despite these uncertainties, in light of the realization that tertiary endosymbiosis appears to have played a significant role in the spread of plastids (beyond the clear-cut 'recent' examples within dinoflagellates), it would seem time to reconsider the similarities in the cellular and molecular biology of secondary/tertiary plastid-bearing algae, in particular the chloroplast ER of haptophytes, cryptophytes and photosynthetic stramenopiles (Cavalier-Smith, 1999, 2000). Rather than representing a shared derived character in the three groups, the host-derived signal peptide secretion system appears to have been co-opted to traffic proteins into secondary/tertiary plastids on numerous occasions. The 'recycling' of pre-existing protein trafficking machinery presumably represents the 'easiest' solution to the problem of EGT and protein import into a newly acquired plastid. As complex and haphazard as it appears to be, there may be universal principles underlying the evolution of a complex photosynthetic organelle.

ACKNOWLEDGEMENTS

I thank the Canadian Institute for Advanced Research, Integrated Microbial Biodiversity Program, the Canadian Institutes of Health Research (CIHR), the Natural Sciences and Engineering Research Council of Canada and the Tula Foundation for research support. The CIHR New Investigator Program is also acknowledged for salary support.

REFERENCES

Abrahamsen, M. S., Templeton, T. J., Enomoto, S., Abrahante, J. E., Zhu, G., Lancto, C. A., et al. (2004). Complete genome sequence of the apicomplexan, *Cryptosporidium parvum*. *Science, 304*, 441–445.

Andersen, R. A. (2004). Biology and systematics of heterokont and haptophyte algae. *American Journal of Botany, 91*, 1508–1522.

Archibald, J. M. (2007). Nucleomorph genomes: Structure, function, origin and evolution. *Bioessays, 29*, 392–402.

Archibald, J. M. (2009). The puzzle of plastid evolution. *Current Biology, 19*, R81–R88.

Archibald, J. M., & Lane, C. E. (2009). Going, going, not quite gone: Nucleomorphs as a case study in nuclear genome reduction. *Journal of Heredity, 100*, 582–590.

Archibald, J. M., Rogers, M. B., Toop, M., Ishida, K., & Keeling, P. J. (2003). Lateral gene transfer and the evolution of plastid-targeted proteins in the secondary plastid-containing alga *Bigelowiella natans*. *Proceedings of the National Academy of Sciences of the United States of America, 100*, 7678–7683.

Aronsson, H., & Jarvis, P. (2009). The chloroplast protein import apparatus, its components, and their roles. In H. Aronsson, & A. S. Sandelius (Eds.), *The chloroplast-interactions with the environment*. Berlin: Springer-Verlag.

Bachvaroff, T. R., Handy, S. M., Place, A. R., & Delwiche, C. F. (2011). Alveolate phylogeny inferred using concatenated ribosomal proteins. *Journal of Eukaryotic Microbiology, 58*, 223–233.

Bachvaroff, T. R., Sanchez-Puerta, M. V., & Delwiche, C. F. (2005). Chlorophyll c-containing plastid relationships based on analyses of a multigene data set with all four chromalveolate lineages. *Molecular Biology and Evolution, 22*, 1772–1782.

Baurain, D., Brinkmann, H., Petersen, J., Rodriguez-Ezpeleta, N., Stechmann, A., Demoulin, V., et al. (2010). Phylogenomic evidence for separate acquisition of plastids in cryptophytes, haptophytes, and stramenopiles. *Molecular Biology and Evolution, 27*, 1698–1709.

Beck, C., Knoop, H., Axmann, I. M., & Steuer, R. (2012). The diversity of cyanobacterial metabolism: Genome analysis of multiple phototrophic microorganisms. *BMC Genomics, 13*, 56.

Bhattacharya, D., Archibald, J. M., Weber, A. P., & Reyes-Prieto, A. (2007). How do endosymbionts become organelles? Understanding early events in plastid evolution. *Bioessays, 29*, 1239–1246.

Bodył, A. (2005). Do plastid-related characters support the chromalveolate hypothesis? *Journal of Phycology, 41*, 712–719.

Bodył, A. (2006). Did the peridinin plastid evolve through tertiary endosymbiosis? A hypothesis. *European Journal of Phycology, 41*, 435–448.

Bodył, A., Stiller, J. W., & Mackiewicz, P. (2009). Chromalveolate plastids: Direct descent or multiple endosymbioses? *Trends in Ecology and Evolution, 24*, 119–121.

Bowler, C., Allen, A. E., Badger, J. H., Grimwood, J., Jabbari, K., Kuo, A., et al. (2008). The *Phaeodactylum* genome reveals the evolutionary history of diatom genomes. *Nature, 456*, 239–244.

Burki, F., Inagaki, Y., Brate, J., Archibald, J. M., Keeling, P. J., Cavalier-Smith, T., et al. (2009). Large-scale phylogenomic analyses reveal that two enigmatic protist lineages, Telonemia and Centroheliozoa, are related to photosynthetic chromalveolates. *Genome Biology and Evolution, 1*, 231–238.

Burki, F., Okamoto, N., Pombert, J. F., & Keeling, P. J. (2012). The evolutionary history of haptophytes and cryptophytes: Phylogenomic evidence for separate origins. *Proceedings of the Royal Society of London B, 279*, 2246–2254.

Burki, F., Shalchian-Tabrizi, K., Minge, M., Skjaeveland, Å, Nikolaev, S. I., Jakobsen, K. S., et al. (2007). Phylogenomics reshuffles the eukaryotic supergroups. *PLoS One, 8*, e790.

Burki, F., Shalchian-Tabrizi, K., & Pawlowski, J. (2008). Phylogenomics reveals a new 'megagroup' including most photosynthetic eukaryotes. *Biology Letters, 4*, 366–369.

Cai, X., Fuller, A. L., Mcdougald, L. R., & Zhu, G. (2003). Apicoplast genome of the coccidian *Eimeria tenella*. *Gene, 321*, 39–46.

Canfield, D. E., Poulton, S. W., & Narbonne, G. M. (2007). Late-Neoproterozoic deep-ocean oxygenation and the rise of animal life. *Science, 315*, 92–95.

Cavalier-Smith, T. (1986). The kingdom Chromista: Origin and Systematics. *Progress in Phycological Researcg, 4*, 309–347.

Cavalier-Smith, T. (1998). A revised six-kingdom system of life. *Biological Reviews of the Cambridge Philosophical Society, 73*, 203–266.

Cavalier-Smith, T. (1999). Principles of protein and lipid targeting in secondary symbiogenesis: Euglenoid, dinoflagellate, and sporozoan plastid origins and the eukaryote family tree. *Journal of Eukaryotic Microbiology, 46*, 347–366.

Cavalier-Smith, T. (2000). Membrane heredity and early chloroplast evolution. *Trends in Plant Science, 5*, 174–182.

Chan, C. X., Yang, E. C., Banerjee, T., Yoon, H. S., Martone, P. T., Estevez, J. M., et al. (2011). Red and green algal monophyly and extensive gene sharing found in a rich repertoire of red algal genes. *Current Biology, 21*, 328–333.

Chesnick, J. M., Hooistra, W. H., Wellbrock, U., & Medlin, L. K. (1997). Ribosomal RNA analysis indicates a benthic pennate diatom ancestry for the endosymbionts of the dinoflagellates *Peridinium foliaceum* and *Peridinium balticum* (Pyrrhophyta). *Journal of Eukaryotic Microbiology, 44*, 314–320.

Cuvelier, M. L., Ortiz, A., Kim, E., Moehlig, H., Richardson, D. E., Heidelberg, J. F., et al. (2008). Widespread distribution of a unique marine protistan lineage. *Environmental Microbiology, 10*, 1621–1634.

Daugbjerg, N., & Andersen, R. A. (1997). Phylogenetic analyses of the *rbcL* sequences from haptophytes and heterokont algae suggest their chloroplasts are unrelated. *Molecular Biology and Evolution, 14*, 1242–1251.

De Clerck, O., Bogaret, K., & Leliaert, F. (2012). Diversity and evolution of algae: Primary endosymbiosis. *Advances in Botanical Research, 64*, 55–86.

de Vienne, D., Ollier, S., & Aguileta, G. (2012). Phylo-MCOA: A fast and efficient method to detect outlier genes and species in phylogenomics using multiple co-inertia analysis. *Molecular Biology and Evolution, 29*, 1587–1598.

Delsuc, F., Brinkmann, H., & Philippe, H. (2005). Phylogenomics and the reconstruction of the tree of life. *Nature Reviews Genetics, 6*, 361–375.

Delwiche, C. F., Andersen, R. A., Bhattacharya, D., Mishler, B. D., & Mccourt, R. M. (2004). Algal evolution and the early radiation of green plants. In J. Cracraft, & M. J. Donoghue (Eds.), *Assembling the tree of life* (pp. 121–137). NewYork: Oxford University Press, Inc.

Deusch, O., Landan, G., Roettger, M., Gruenheit, N., Kowallik, K. V., Allen, J. F., et al. (2008). Genes of cyanobacterial origin in plant nuclear genomes point to a heterocyst-forming plastid ancestor. *Molecular Biology and Evolution, 25*, 748–761.

Dodge, J. D. (1969). Observations on the fine structure of the eyespot and associated organelles in the dinoflagellate *Glenodinium foliaceum*. *Journal of Cell Science, 5*, 479–493.

Douglas, S. E., Murphy, C. A., Spencer, D. F., & Gray, M. W. (1991). Cryptomonad algae are evolutionary chimaeras of two phylogenetically distinct unicellular eukaryotes. *Nature, 350*, 148–151.

Douglas, S. E., & Penny, S. L. (1999). The plastid genome of the cryptophyte alga, *Guillardia theta*: Complete sequence and conserved synteny groups confirm its common ancestry with red algae. *Journal of Molecular Evolution, 48*, 236–244.

Douglas, S. E., Zauner, S., Fraunholz, M., Beaton, M., Penny, S., Deng, L., et al. (2001). The highly reduced genome of an enslaved algal nucleus. *Nature, 410*, 1091–1096.

Durnford, D. G., Deane, J. A., Tan, S., Mcfadden, G. I., Gantt, E., & Green, B. R. (1999). A phylogenetic assessment of the eukaryotic light-harvesting antenna proteins, with implications for plastid evolution. *Journal of Molecular Evolution, 48*, 59–68.

Eisen, J. A., Coyne, R. S., Wu, M., Wu, D., Thiagarajan, M., Wortman, J. R., et al. (2006). Macronuclear genome sequence of the ciliate Tetrahymena thermophila, a model eukaryote. *PLoS Biology, 4*, e286.

Elias, M., & Archibald, J. M. (2009). Sizing up the genomic footprint of endosymbiosis. *Bioessays, 31*, 1273–1279.

Embley, T. M., & Martin, W. (2006). Eukaryotic evolution, changes and challenges. *Nature, 440*, 623–630.

Fast, N. M., Kissinger, J. C., Roos, D. S., & Keeling, P. J. (2001). Nuclear-encoded, plastid-targeted genes suggest a single common origin for apicomplexan and dinoflagellate plastids. *Molecular Biology and Evolution, 18*, 418–426.

Fernandez Robledo, J. A., Caler, E., Matsuzaki, M., Keeling, P. J., Shanmugam, D., Roos, D. S., et al. (2011). The search for the missing link: A relic plastid in *Perkinsus*? *International Journal of Parasitology, 41*, 1217–1229.

Gibbs, S. P. (1978). The chloroplasts of *Euglena* may have evolved from symbiotic green algae. *Canadian Journal of Botany, 56*, 2883–2889.

Gibbs, S. P. (1981). The chloroplast endoplasmic reticulum: Structure, function, and evolutionary significance. *International Review of Cytology, 72*, 49–99.

Gibbs, S. P. (2006). Looking at life: From binoculars to the electron microscope. *Annual Review of Plant Biology, 57*, 1–17.

Gilson, P. R., Su, V., Slamovits, C. H., Reith, M. E., Keeling, P. J., & Mcfadden, G. I. (2006). Complete nucleotide sequence of the chlorarachniophyte nucleomorph: Nature's smallest nucleus. *Proceedings of the National Academy of Sciences of the United States of America, 103*, 9566–9571.

Gould, S. B., Sommer, M. S., Kroth, P. G., Gile, G. H., Keeling, P. J., & Maier, U. G. (2006). Nucleus-to-nucleus gene transfer and protein retargeting into a remnant cytoplasm of cryptophytes and diatoms. *Molecular Biology and Evolution, 23*, 2413–2422.

Gould, S. B., Waller, R. F., & Mcfadden, G. I. (2008). Plastid evolution. *Annual Review of Plant Biology, 59*, 491–517.

Greenwood, A. D. (1974). The Cryptophyta in relation to phylogeny and photosynthesis. *Proceedings of the 8th International Congress Electron Microscopy, 2*, 566–567.

Greenwood, A. D., Griffiths, H. B., & Santore, U. J. (1977). Chloroplasts and cell compartments in Cryptophyceae. *British Phycological Journal, 12*, 119.

Hackett, J. D., Anderson, D. M., Erdner, D. L., & Bhattacharya, D. (2004). Dinoflagellates: A remarkable evolutionary experiment. *American Journal of Botany, 91*, 1523–1534.

Hackett, J. D., Maranda, L., Yoon, H. S., & Bhattacharya, D. (2003). Phylogenetic evidence for the cryptophyte origin of the plastid of *Dinophysis* (Dinophysiales, Dinophyceae). *Journal of Phycology, 39*, 440–448.

Hackett, J. D., Yoon, H. S., Li, S., Reyes-Prieto, A., Rummele, S. E., & Bhattacharya, D. (2007). Phylogenomic analysis supports the monophyly of cryptophytes and haptophytes and the association of rhizaria with chromalveolates. *Molecular Biology and Evolution, 24*, 1702–1713.

Hallick, R. B., Hong, L., Drager, R. G., Favreau, M. R., Monfort, A., Orsat, B., et al. (1993). Complete sequence of *Euglena gracilis* chloroplast DNA. *Nucleic Acids Research, 21*, 3537–3544.

Hampl, V., Hug, L., Leigh, J. W., Dacks, J. B., Lang, B. F., Simpson, A. G., et al. (2009). Phylogenomic analyses support the monophyly of Excavata and resolve relationships among eukaryotic "supergroups". *Proceedings of the National Academy of Sciences of the United States of America, 106*, 3859–3864.

Hansen, G., Botes, L., & De Salas, M. (2007). Ultrastructure and large subunit rDNA sequence of *Lepidodinium viride* reveal a close relationship to *Lepidodinium chlorophorum* comb. nov. (= *Gymnodiuium chlorophorum. Phycological Research, 55*, 25–41.

Harper, J. T., & Keeling, P. J. (2003). Nucleus-encoded, plastid-targeted glyceraldehyde-3-phosphate dehydrogenase (GAPDH) indicates a single origin for chromalveolate plastids. *Molecular Biology and Evolution, 20*, 1730–1735.

Hibberd, D. J., & Norris, R. E. (1984). Cytology and ultrastructure of *Chlorarachnion reptans* (Chlorarachniophyta divisio nova, Chlorarachniophyceae classis nova). *Journal of Phycology, 20*, 310–330.

Hoef-Emden, K., Marin, B., & Melkonian, M. (2002). Nuclear and nucleomorph SSU rDNA phylogeny in the Cryptophyta and the evolution of cryptophyte diversity. *Journal of Molecular Evolution, 55*, 161–179.
Hopkins, J., Fowler, R., Krishna, S., Wilson, I., Mitchell, G., & Bannister, L. (1999). The plastid in *Plasmodium falciparum* asexual blood stages: A three-dimensional ultrastructural analysis. *Protist, 150*, 283–295.
Huang, J., Mullapudi, N., Lancto, C. A., Scott, M., Abrahamsen, M. S., & Kissinger, J. C. (2004). Phylogenomic evidence supports past endosymbiosis, intracellular and horizontal gene transfer in Cryptosporidium parvum. *Genome Biology, 5*, R88.
Iida, K., Takishita, K., Ohshima, K., & Inagaki, Y. (2007). Assessing the monophyly of chlorophyll-c containing plastids by multi-gene phylogenies under the unlinked model conditions. *Molecular Phylogenetics and Evolution, 45*, 227–238.
Imanian, B., Carpenter, K. J., & Keeling, P. J. (2007). Mitochondrial genome of a tertiary endosymbiont retains genes for electron transport proteins. *Journal of Eukaryotic Microbiology, 54*, 146–153.
Imanian, B., & Keeling, P. J. (2007). The dinoflagellates *Durinskia baltica* and *Kryptoperidinium foliaceum* retain functionally overlapping mitochondria from two evolutionarily distinct lineages. *BMC Evolutionary Biology, 7*, 172.
Inagaki, Y., Dacks, J. B., Doolittle, W. F., Watanabe, K. I., & Ohama, T. (2000). Evolutionary relationship between dinoflagellates bearing obligate diatom endosymbionts: Insight into tertiary endosymbiosis. *International Journal of Systematic and Evolutionary Microbiology, 50*, 2075–2081.
Ishida, K., Cao, Y., Hasegawa, M., Okada, N., & Hara, Y. (1997). The origin of chlorarachniophyte plastids, as inferred from phylogenetic comparisons of amino acid sequences of EF-Tu. *Journal of Molecular Evolution, 45*, 682–687.
Ishida, K., Green, B. R., & Cavalier-Smith, T. (1999). Diversification of a chimaeric algal group, the chlorarachniophytes: Phylogeny of nuclear and nucleomorph small-subunit rRNA genes. *Molecular Biology and Evolution, 16*, 321–331.
Janouškovec, J., Horak, A., Obornik, M., Lukes, J., & Keeling, P. J. (2010). A common red algal origin of the apicomplexan, dinoflagellate, and heterokont plastids. *Proceedings of the National Academy of Sciences of the United States of America, 107*, 10949–10954.
Jarvis, P., & Soll, J. (2001). Toc, Tic, and chloroplast protein import. *Biochimica et Biophysica Acta, 1541*, 64–79.
Katz, M. E., Finkel, Z. V., Grzebyk, D., Knoll, A. H., & Falkowski, P. G. (2004). Evolutionary trajectories and biogeochemical impacts of marine eukaryotic phytoplankton. *Annual Review of Ecology Evolution and Systematics, 35*, 523–556.
Keeling, P. (2004). A brief history of plastids and their hosts. *Protist, 155*, 3–7.
Keeling, P. J. (2008). Evolutionary biology: Bridge over troublesome plastids. *Nature, 451*, 896–897.
Keeling, P. J. (2009). Chromalveolates and the evolution of plastids by secondary endosymbiosis. *Journal of Eukaryotic Microbiology, 56*, 1–8.
Keeling, P. J. (2010). The endosymbiotic origin, diversification and fate of plastids. *Philosophical Transactions of the Royal Society of London B Biological Sciences, 365*, 729–748.
Keeling, P. J., Burger, G., Durnford, D. G., Lang, B. F., Lee, R. W., Pearlman, R. E., et al. (2005). The tree of eukaryotes. *Trends in Ecology and Evolution, 20*, 670–676.
Keeling, P. J., & Palmer, J. D. (2008). Horizontal gene transfer in eukaryotic evolution. *Nature Reviews Genetics, 9*, 605–618.
Khan, H., Parks, N., Kozera, C., Curtis, B. A., Parsons, B. J., Bowman, S., et al. (2007). Plastid genome sequence of the cryptophyte alga *Rhodomonas salina* CCMP1319: Lateral transfer of putative DNA replication machinery and a test of chromist plastid phylogeny. *Molecular Biology and Evolution, 24*, 1832–1842.

Kim, E., & Archibald, J. M. (2009). Diversity and evolution of plastids and their genomes. In H. Aronsson, & A. S. Sandelius (Eds.), *The chloroplast-interactions with the environment* (pp. 1–39). Berlin: Springer-Verlag.

Kim, E., & Archibald, J. M. (2010). Plastid evolution: Gene transfer and the maintenance of 'stolen' organelles. *BMC Biology, 8*, 73.

Kim, E., Harrison, J. W., Sudek, S., Jones, M. D., Wilcox, H. M., Richards, T. A., et al. (2011). Newly identified and diverse plastid-bearing branch on the eukaryotic tree of life. *Proceedings of the National Academy of Sciences of the United States of America, 108*, 1496–1500.

Kim, E., Simpson, A. G., & Graham, L. E. (2006). Evolutionary relationships of apusomonads inferred from taxon-rich analyses of 6 nuclear encoded genes. *Molecular Biology and Evolution, 23*, 2455–2466.

Kohler, S. (2005). Multi-membrane-bound structures of Apicomplexa: I. the architecture of the *Toxoplasma gondii* apicoplast. *Parasitology Research, 96*, 258–272.

Köhler, S., Delwiche, C. F., Denny, P. W., Tilney, L. G., Webster, P., Wilson, R. J. M., et al. (1997). A plastid of probable green algal origin in apicomplexan parasites. *Science, 275*, 1485–1489.

Koonin, E. V. (2010). The origin and early evolution of eukaryotes in the light of phylogenomics. *Genome Biology, 11*, 209.

Lane, C. E., & Archibald, J. M. (2008). The eukaryotic tree of life: Endosymbiosis takes its TOL. *Trends in Ecology and Evolution, 23*, 268–275.

Lane, C. E., Van Den Heuvel, K., Kozera, C., Curtis, B. A., Parsons, B., Bowman, S., et al. (2007). Nucleomorph genome of *Hemiselmis andersenii* reveals complete intron loss and compaction as a driver of protein structure and function. *Proceedings of the National Academy of Sciences of the United States of America, 104*, 19908–19913.

Larkum, A. W., Lockhart, P. J., & Howe, C. J. (2007). Shopping for plastids. *Trends in Plant Science, 12*, 189–195.

Leander, B. S., Triemer, R. E., & Farmer, M. A. (2001a). Character evolution in heterotrophic euglenids. *European Journal of Protistology, 37*, 337–356.

Leander, B. S., Witek, R. P., & Farmer, M. A. (2001b). Trends in the evolution of the euglenid pellicle. *Evolution, 55*, 2215–2235.

Leigh, J. W., Lapointe, F.-J., Lopez, P., & Bapteste, E. (2011a). Evaluating phylogenetic congruence in the post-genomic era. *Genome Biology and Evolution, 3*, 571–587.

Leigh, J. W., Schliep, K., Lopez, P., & Bapteste, E. (2011b). Let them fall where they may: Congruence analysis in massive phylogenetically messy datasets. *Molecular Biology and Evolution, 28*, 2773–2785.

Leigh, J. W., Susko, E., Baumgartner, M., & Roger, A. J. (2008). Testing congruence in phylogenomic analysis. *Systematic Biology, 57*, 104–115.

Lewis, L. A., & Mccourt, R. M. (2004). Green algae and the origin of land plants. *American Journal of Botony, 91*, 1535–1556.

Martin, W. (2010). Evolutionary origins of metabolic compartmentalization in eukaryotes. *Philosophical Transactions of the Royal Society of London B Biological Sciences, 365*, 847–855.

Martin, W., Brinkmann, H., Savonna, C., & Cerff, R. (1993). Evidence for a chimeric nature of nuclear genomes: Eubacterial origin of eukaryotic glyceraldehyde-3-phosphate dehydrogenase genes. *Proceedings of the National Academy of Sciences of the United States of America, 90*, 8692–8696.

Martin, W., & Herrmann, R. G. (1998). Gene transfer from organelles to the nucleus: How much, what happens, and Why? *Plant Physiology, 118*, 9–17.

Martin, W., Rujan, T., Richly, E., Hansen, A., Cornelsen, S., Lins, T., et al. (2002). Evolutionary analysis of *Arabidopsis*, cyanobacterial, and chloroplast genomes reveals plastid phylogeny and thousands of cyanobacterial genes in the nucleus. *Proceedings of the National Academy of Sciences of the United States of America, 99*, 12246–12251.

Matsumoto, T., Shinozaki, F., Chikuni, T., Yabuki, A., Takishita, K., Kawachi, M., et al. (2011). Green-colored plastids in the dinoflagellate genus *Lepidodinium* are of core chlorophyte origin. *Protist, 162*, 268–276.

Matsuzaki, M., Kita, K., & Nozaki, H. (2006). Orthologs of plastid isoprenoids biosynthesis pathway genes from *Perkinsus*. *Journal of Plant Research, 119*, 122.

Matsuzaki, M., Kuroiwa, H., Kuroiwa, T., Kita, K., & Nozaki, H. (2008). A cryptic algal group unveiled: A plastid biosynthesis pathway in the oyster parasite *Perkinsus marinus*. *Molecular Biology and Evolution, 25*, 1167–1179.

Matsuzaki, M., Misumi, O., Shin, I. T., Maruyama, S., Takahara, M., Miyagishima, S. Y., et al. (2004). Genome sequence of the ultrasmall unicellular red alga *Cyanidioschyzon merolae* 10D. *Nature, 428*, 653–657.

Mazumdar, J., Wilson, E. H., Masek, K., Hunter, C. A., & Striepen, B. (2006). Apicoplast fatty acid synthesis is essential for organelle biogenesis and parasite survival in *Toxoplasma gondii*. *Proceedings of the National Academy of Sciences of the United States of America, 103*, 13192–13197.

McEwan, M. L., & Keeling, P. J. (2004). HSP90, tubulin and actin are retained in the tertiary endosymbiont genome of Kryptoperidinium foliaceum. *Journal of Eukaryotic Microbiology, 51*, 651–659.

Mcfadden, G. I. (1999). Plastids and protein targeting. *Journal of Eukaryotic Microbiology, 46*, 339–346.

Mcfadden, G. I., Gilson, P. R., & Hill, D. R. A. (1994a). *Goniomonas*: rRNA sequences indicate that this phagotrophic flagellate is a close relative of the host component of cryptomonads. *European Journal of Phycology, 29*, 29–32.

Mcfadden, G. I., Gilson, P. R., Hofmann, C. J., Adcock, G. J., & Maier, U. G. (1994b). Evidence that an amoeba acquired a chloroplast by retaining part of an engulfed eukaryotic alga. *Proceedings of the National Academy of Sciences of the United States of America, 91*, 3690–3694.

Mcfadden, G. I., Reith, M., Munholland, J., & Lang-Unnasch, N. (1996). Plastid in human parasites. *Nature, 381*, 482.

Mcfadden, G. I., & Waller, R. F. (1997). Plastids in parasites of humans. *Bioessays, 19*, 1033–1040.

Merchant, S. S., Prochnik, S. E., Vallon, O., Harris, E. H., Karpowicz, S. J., Witman, G. B., et al. (2007). The *Chlamydomonas* genome reveals the evolution of key animal and plant functions. *Science, 318*, 245–250.

Minge, M. A., Shalchian-Tabrizi, K., Torresen, O. K., Takishita, K., Probert, I., Inagaki, Y., et al. (2010). A phylogenetic mosaic plastid proteome and unusual plastid-targeting signals in the green-colored dinoflagellate *Lepidodinium chlorophorum*. *BMC Evolutionary Biology, 10*, 191.

Mock, T., & Medlin, L. K. (2012). Genomics and Genetics of Diatoms. *Advances in Botanical Research, 64*, 245–284.

Moestrup, O., & Sengco, M. (2001). Ultrastructural studies on *Bigelowiella natans*, gen. et sp. nov., a chlorarachniophyte flagellate. *Journal of Phycology, 37*, 624–646.

Moore, C. E., & Archibald, J. M. (2009). Nucleomorph genomes. *Annual Review of Genetics, 43*, 251–264.

Moore, R. B., Obornik, M., Janouskovec, J., Chrudimsky, T., Vancova, M., Green, D. H., et al. (2008). A photosynthetic alveolate closely related to apicomplexan parasites. *Nature, 452*, 900.

Moreira, D., Le Guyader, H., & Phillippe, H. (2000). The origin of red algae and the evolution of chloroplasts. *Nature, 405*, 69–72.

Moustafa, A., Beszteri, B., Maier, U. G., Bowler, C., Valentin, K., & Bhattacharya, D. (2009). Genomic footprints of a cryptic plastid endosymbiosis in diatoms. *Science, 324*, 1724–1726.

Moustafa, A., & Bhattacharya, D. (2008). PhyloSort: A user-friendly phylogenetic sorting tool and its application to estimating the cyanobacterial contribution to the nuclear genome of *Chlamydomonas*. *BMC Evolutionary Biology, 8*, 6.

Müller, K. M., Oliveira, M. C., Sheath, R. G., & Bhattacharya, D. (2001). Ribosomal DNA phylogeny of the Bangiophycidae (Rhodophyta) and the origin of secondary plastids. *American Journal of Botany, 88*, 1390–1400.

Nelson, D. M., Tréguer, P., Brzezinski, M. A., Leynaert, A., & Queguiner, B. (1995). Production and dissolution of biogenic silica in the ocean: Revised global estimates, comparison with regional data and relationship to biogenic sedimentation. *Global Biogeochemical Cycles, 9*, 359–372.

Nikolaev, S. I., Berney, C., Fahrni, J. F., Bolivar, I., Polet, S., Mylnikov, A. P., et al. (2004). The twilight of Heliozoa and rise of Rhizaria, an emerging supergroup of amoeboid eukaryotes. *Proceedings of the National Academy of Sciences of the United States of America, 101*, 8066–8071.

Nisbet, R. E., Hiller, R. G., Barry, E. R., Skene, P., Barbrook, A. C., & Howe, C. J. (2008). Transcript analysis of dinoflagellate plastid gene minicircles. *Protist, 159*, 31–39.

Nisbet, R. E., Koumandou, L. V., Barbrook, A. C., & Howe, C. J. (2004). Novel plastid gene minicircles in the dinoflagellate *Amphidinium operculatum*. *Gene, 331*, 141–147.

Nosenko, T., Lidie, K. L., Van Dolah, F. M., Lindquist, E., Cheng, J. F., & Bhattacharya, D. (2006). Chimeric plastid proteome in the Florida "red tide" dinoflagellate Karenia brevis. *Molecular Biology and Evolution, 23*, 2026–2038.

Not, F., Siano, R., Kooistra, W. H. C. F., Simon, N., Vaulot, D., & Probert, I. (2012). Diversity and ecology of eukaryotic marine phytoplankton. *Advances in Botanical Research, 64*, 1–53.

Not, F., Valentin, K., Romari, K., Lovejoy, C., Massana, R., Tobe, K., et al. (2007). Picobiliphytes: A marine picoplanktonic algal group with unknown affinities to other eukaryotes. *Science, 315*, 253–255.

Nozaki, H., Iseki, M., Hasegawa, M., Misawa, K., Nakada, T., Sasaki, N., et al. (2007). Phylogeny of primary photosynthetic eukaryotes as deduced from slowly evolving nuclear genes. *Molecular Biology and Evolution, 24*, 1592–1595.

Nozaki, H., Maruyama, S., Matsuzaki, M., Nakada, T., Kato, S., & Misawa, K. (2009). Phylogenetic positions of Glaucophyta, green plants (Archaeplastida) and Haptophyta (Chromalveolata) as deduced from slowly evolving nuclear genes. *Molecular Phylogenetics and Evolution, 53*, 872–880.

Obornik, M., Janouskovec, J., Chrudimsky, T., & Lukes, J. (2008). Evolution of the apicoplast and its host: To autotrophy and back again. *International Journal of Parasitology, 39*, 1–12.

Okamoto, N., Chantangsi, C., Horak, A., Leander, B. S., & Keeling, P. J. (2009). Molecular phylogeny and description of the novel katablepharid *Roombia truncata* gen. et sp. nov., and establishment of the Hacrobia taxon nov. *PLoS One, 4*, e7080.

Okamoto, N., & Inouye, I. (2005). The katablepharids are a distant sister group of the Cryptophyta: A proposal for Katablepharidophyta Divisio Nova/Katablepharida Phylum Novum based on SSU rDNA and beta-tubulin phylogeny. *Protist, 156*, 163–179.

Oliveira, M. C., & Bhattacharya, D. (2000). Phylogeny of the Bangiophycidae (Rhodophyta) and the secondary endosymbiotic origin of algal plastids. *American Journal of Botany, 87*, 482–492.

Ota, S., Silver, T. D., Archibald, J. M., & Ishida, K.-I. (2009a). *Lotharella oceanica* sp. nov.—a new planktonic chlorarachniophyte studied by light and electron microscopy. *Phycologia, 48*, 315–323.

Ota, S., Ueda, K., & Ishida, K.-I. (2005). *Lotharella vacuolata* sp. nov., a new species of chlorarachniophyte algae, and time-laps video observations on its unique post-cell division behavior. *Phycological Research, 53*, 275–286.

Ota, S., Ueda, K., & Ishida, K.-I. (2007). *Norrisiella sphaerica* gen. et sp. nov., a new coccoid chlorarachniophyte from Baja California, Mexico. *Journal of Plant Research, 120*, 661–670.

Ota, S., Vaulot, D., Le Gall, F., Yabuki, A., & Ishida, K.-I. (2009b). *Partenskyella glossopodia* gen. et sp. nov., the first report of a chlorarachniophyte that lacks a pyrenoid. *Protist, 160*, 137–150.

Oudot-Le Secq, M. P., Grimwood, J., Shapiro, H., Armbrust, E. V., Bowler, C., & Green, B. R. (2007). Chloroplast genomes of the diatoms *Phaeodactylum tricornutum* and *Thalassiosira pseudonana*: Comparison with other plastid genomes of the red lineage. *Molecular Genetics and Genomics, 277*, 427–439.

Palmer, J. D. (2003). The symbiotic birth and spread of plastids: How many times and whodunnit? *Journal of Phycology, 39*, 4–11.

Parfrey, L. W., Grant, J., Tekle, Y. I., Lasek-Nesselquist, E., Morrison, H. G., Sogin, M. L., et al. (2010). Broadly sampled multigene analyses yield a well-resolved eukaryotic tree of life. *Systematic Biology, 59*, 518–533.

Parfrey, L. W., Lahr, D. J., Knoll, A. H., & Katz, L. A. (2011). Estimating the timing of early eukaryotic diversification with multigene molecular clocks. *Proceedings of the National Academy of Sciences of the United States of America, 108*, 13624–13629.

Park, M. G., Park, J. S., Kim, M., & Yih, W. (2008). Plastid dynamics during survival of *Dinophysis caudata* without its ciliate prey. *Journal of Phycology, 44*, 1154–1163.

Patron, N. J., Inagaki, Y., & Keeling, P. J. (2007). Multiple gene phylogenies support the monophyly of cryptomonad and haptophyte host lineages. *Current Biology, 17*, 887–891.

Patron, N. J., Rogers, M. B., & Keeling, P. J. (2004). Gene replacement of fructose-1,6-bisphosphate aldolase supports the hypothesis of a single photosynthetic ancestor of chromalveolates. *Eukaryotic Cell, 3*, 1169–1175.

Patron, N. J., Waller, R. F., Archibald, J. M., & Keeling, P. J. (2005). Complex protein targeting to dinoflagellate plastids. *Journal of Molecular Biology, 348*, 1015–1024.

Pawlowski, J., Holzmann, M., Berney, C., Fahrni, J., Gooday, A. J., Cedhagen, T., et al. (2003). The evolution of early Foraminifera. *Proceedings of the National Academy of Sciences of the United States of America, 100*, 11494–11498.

Price, D. C., Chan, C. X., Yoon, H. S., Yang, E. C., Qiu, H., Weber, A. P., et al. (2012). *Cyanophora paradoxa* genome elucidates origin of photosynthesis in algae and plants. *Science, 335*, 843–847.

Qian, Q., & Keeling, P. J. (2001). Diplonemid glyceraldehyde-3phosphate dehydrogenase (GAPDH) and prokaryote-to-eukaryote lateral gene transfer. *Protist, 152*, 193–201.

Reyes-Prieto, A., Hackett, J. D., Soares, M. B., Bonaldo, M. F., & Bhattacharya, D. (2006). Cyanobacterial contribution to algal nuclear genomes is primarily limited to plastid functions. *Current Biology, 16*, 2320–2325.

Reyes-Prieto, A., Moustafa, A., & Bhattacharya, D. (2008). Multiple genes of apparent algal origin suggest ciliates may once have been photosynthetic. *Current Biology, 18*, 956–962.

Reyes-Prieto, A., Weber, A. P., & Bhattacharya, D. (2007). The origin and establishment of the plastid in algae and plants. *Annual Review of Genetics, 41*, 147–168.

Rice, D. W., & Palmer, J. D. (2006). An exceptional horizontal gene transfer in plastids: Gene replacement by a distant bacterial paralog and evidence that haptophyte and cryptophyte plastids are sisters. *BMC Biology, 4*, 31.

Rodriguez-Ezpeleta, N., Brinkmann, H., Burey, S. C., Roure, B., Burger, G., Loffelhardt, W., et al. (2005). Monophyly of primary photosynthetic eukaryotes: Green plants, red algae, and glaucophytes. *Current Biology, 15*, 1325–1330.

Roger, A. J., & Simpson, A. G. (2009). Evolution: Revisiting the root of the eukaryote tree. *Current Biology, 19*, R165–R167.

Rogers, M. B., Gilson, P. R., Su, V., Mcfadden, G. I., & Keeling, P. J. (2007). The complete chloroplast genome of the chlorarachniophyte *Bigelowiella natans*: Evidence for independent origins of chlorarachniophyte and euglenid secondary endosymbionts. *Molecular Biology and Evolution, 24*, 54–62.

Sanchez-Puerta, M. V., Bachvaroff, T. R., & Delwiche, C. F. (2005). The complete plastid genome sequence of the haptophyte *Emiliania huxleyi*: A comparison to other plastid genomes. *DNA Research, 12*, 151–156.

Sanchez-Puerta, M. V., Bachvaroff, T. R., & Delwiche, C. F. (2007a). Sorting wheat from chaff in multi-gene analyses of chlorophyll c-containing plastids. *Molecular Phylogenetics and Evolution, 44*, 885–897.

Sanchez-Puerta, M. V., & Delwiche, C. F. (2008). A hypothesis for plastid evolution in chromalveolates. *Journal of Phycology, 44*, 1097–1107.

Sanchez-Puerta, M. V., Lippmeier, J. C., Apt, K. E., & Delwiche, C. F. (2007b). Plastid genes in a non-photosynthetic dinoflagellate. *Protist, 158*, 105–117.

Sato, N., Ishikawa, M., Fujiwara, M., & Sonoike, K. (2005). Mass identification of chloroplast proteins of endosymbiont origin by phylogenetic profiling based on organism-optimized homologous protein groups. *Genome Informatics, 16*, 56–68.

Schnepf, E., & Elbrächter, M. (1988). Cryptophycean-like double membrane-bound plastid chloroplast in the dinoflagellate, *Dinophysis* Ehrenb.: Evolutionary, phylogenetic and toxicological implications. *Botanica Acta, 101*, 196–203.

Schnepf, E., & Elbrächter, M. (1999). Dinophyte chloroplasts and phylogeny—a review. *Grana, 38*, 81–97.

Seeber, F., & Soldati-Favre, D. (2010). Metabolic pathways in the apicoplast of apicomplexa. *International Review of Cell and Molecular Biology, 281*, 161–228.

Shalchian-Tabrizi, K., Brate, J., Logares, R., Klaveness, D., Berney, C., & Jakobsen, K. S. (2008). Diversification of unicellular eukaryotes: Cryptomonad colonizations of marine and fresh waters inferred from revised 18S rRNA phylogeny. *Environmental Microbiology, 10*, 2635–2644.

Shalchian-Tabrizi, K., Minge, M. A., Cavalier-Smith, T., Nedreklepp, J. M., Klaveness, D., & Jakobsen, K. S. (2006). Combined heat shock protein 90 and ribosomal RNA sequence phylogeny supports multiple replacements of dinoflagellate plastids. *Journal of Eukaryotic Microbiology, 53*, 217–224.

Simpson, A. G. (2003). Cytoskeletal organization, phylogenetic affinities and systematics in the contentious taxon Excavata (Eukaryota). *International Journal of Systematic and Evolutionary Microbiology, 53*, 1759–1777.

Simpson, A. G., Stevens, J. R., & Lukes, J. (2006). The evolution and diversity of kinetoplastid flagellates. *Trends in Parasitology, 22*, 168–174.

Simpson, A. G. B., & Roger, A. J. (2004). The real 'kingdoms' of eukaryotes. *Current Biology, 14*, R693–R696.

Slamovits, C. H., & Keeling, P. J. (2008). Plastid-Derived Genes in the Non-Photosynthetic Alveolate *Oxyrrhis marina*. *Molecular Biology and Evolution, 25*, 1297–1306.

Stelter, K., El-Sayed, N. M., & Seeber, F. (2007). The expression of a plant-type ferredoxin redox system provides molecular evidence for a plastid in the early dinoflagellate *Perkinsus marinus*. *Protist, 158*, 119–130.

Stiller, J. W. (2007). Plastid endosymbiosis, genome evolution and the origin of green plants. *Trends in Plant Science, 12*, 391–396.

Stiller, J. W. (2011). Experimental design and statistical rigor in phylogenomics of horizontal and endosymbiotic gene transfer. *BMC Evolutionary Biology, 11*, 259.

Stiller, J. W., & Hall, B. D. (1997). The origin of red algae: Implications for plastid evolution. *Proceedings of the National Academy of Sciences of the United States of America, 94*, 4520–4525.

Stiller, J. W., Huang, J., Ding, Q., Tian, J., & Goodwillie, C. (2009). Are algal genes in nonphotosynthetic protists evidence of historical plastid endosymbioses? *BMC Genomics, 10*, 484.

Stoecker, D. K. (1999). Mixotrophy among dinoflagellates. *Journal of Eukaryotic Microbiology, 46*, 397–401.

Summerer, M., Sonntag, B., & Sommaruga, R. (2007). An experimental test of the symbiosis specificity between the ciliate *Paramecium bursaria* and strains of the unicellular green alga *Chlorella*. *Environmental Microbiology, 9*, 2117–2122.

Takahashi, F., Okabe, Y., Nakada, T., Sekimoto, H., Ito, M., Kataoka, H., et al. (2007). Origins of the secondary plastids of Euglenophyta and Chlorarachniophyta as revealed by an analysis of the plastid-targeting, nuclear-encoded gene *psbO*. *Journal of Phycology, 43*, 1302–1309.

Takishita, K., Ishida, K., & Maruyama, T. (2003). An enigmatic GAPDH gene in the symbiotic dinoflagellate genus *Symbiodinium* and its related species (the order Suessiales): Possible lateral gene transfer between two eukaryotic algae, dinoflagellate and euglenophyte. *Protist, 154*, 443–454.

Takishita, K., Kawachi, M., Noel, M. H., Matsumoto, T., Kakizoe, N., Watanabe, M. M., et al. (2008). Origins of plastids and glyceraldehyde-3-phosphate dehydrogenase genes in the green-colored dinoflagellate *Lepidodinium chlorophorum*. *Gene, 410*, 26–36.

Takishita, K., Yamaguchi, H., Maruyama, T., & Inagaki, Y. (2009). A hypothesis for the evolution of nuclear-encoded, plastid-targeted glyceraldehyde-3-phosphate dehydrogenase genes in "chromalveolate" members. *PLoS One, 4*, e4737.

Tanifuji, G., Onodera, N. T., Wheeler, T. J., Dlutek, M., Donaher, N., & Archibald, J. M. (2011). Complete nucleomorph genome sequence of the nonphotosynthetic alga *Cryptomonas paramecium* reveals a core nucleomorph gene set. *Genome Biology and Evolution, 3*, 44–54.

Taylor, F. J. R. (1974). Implications and extensions of the serial endosymbiosis theory of the origin of eukaryotes. *Taxon, 23*, 229–258.

Taylor, F. J. R. E. (1987). *The biology of dinoflagellates*. Oxford: Blackwell Scientific Publications.

Teich, R., Zauner, S., Baurain, D., Brinkmann, H., & Petersen, J. (2007). Origin and distribution of Calvin Cycle Fructose and Sedoheptulose Bisphosphatases in Plantae and complex algae: A single secondary origin of complex red plastids and subsequent propagation via tertiary endosymbioses. *Protist, 158*, 263–276.

Teles-Grilo, M. L., Tato-Costa, J., Duarte, S. M., Maia, A., Casal, G., & Azevedo, C. (2007). Is there a plastid in *Perkinsus atlanticus* (Phylum Perkinsozoa)? *European Journal of Protistology, 43*, 163–167.

Tengs, T., Dahlberg, O. J., Shalchian-Tabrizi, K., Klaveness, D., Rudi, K., Delwiche, C. F., et al. (2000). Phylogenetic analyses indicate that the 19'hexanoyloxy-fucoxanthin-containing dinoflagellates have tertiary plastids of haptophyte origin. *Molecular Biology and Evolution, 17*, 718–729.

Terashima, M., Specht, M., & Hippler, M. (2011). The chloroplast proteome: A survey from the *Chlamydomonas reinhardtii* perspective with a focus on distinctive features. *Current Genetics, 57*, 151–168.

The Ectocarpus Genome Consortium. (2012). The *Ectocarpus* genome and brown algal genomics. *Advances in Botanical Research, 64*, 141–184.

Timmis, J. N., Ayliffe, M. A., Huang, C. Y., & Martin, W. (2004). Endosymbiotic gene transfer: Organelle genomes forge eukaryotic chromosomes. *Nature Reviews Genetics, 5*, 123–135.

Tomitani, A., Knoll, A. H., Cavanaugh, C. M., & Ohno, T. (2006). The evolutionary diversification of cyanobacteria: Molecular-phylogenetic and paleontological perspectives. *Proceedings of the National Academy of Sciences of the United States of America, 103*, 5442–5447.

Tomova, C., Geerts, W. J. C., Muller-Reichert, T., Entzeroth, R., & Humbel, B. M. (2006). New comprehension of the apicoplast of *Sarcocystis* by transmission electron tomography. *Biology of the Cell, 98*, 535–545.

Toso, M. A., & Omoto, C. K. (2007). *Gregarina niphandrodes* may lack both a plastid genome and organelle. *Journal of Eukaryotic Microbiology, 54*, 66–72.

Turmel, M., Gagnon, M. C., O'Kelly, C. J., Otis, C., & Lemieux, C. (2009). The chloroplast genomes of the green algae *Pyramimonas, Monomastix*, and *Pycnococcus* shed new light on the evolutionary history of prasinophytes and the origin of the secondary chloroplasts of euglenids. *Molecular Biology and Evolution, 26*, 631–648.

Tyler, B. M., Tripathy, S., Zhang, X., Dehal, P., Jiang, R. H., Aerts, A., et al. (2006). *Phytophthora* genome sequences uncover evolutionary origins and mechanisms of pathogenesis. *Science, 313*, 1261–1266.

Van De Peer, Y., Van Der Auwera, G., & De Wachter, R. (1996). The evolution of stramenopiles and alveolates as derived by "substitution rate calibration" of small ribosomal subunit RNA. *Journal of Molecular Evolution, 42*, 201–210.

Waller, R. F., Keeling, P. J., Donald, R. G., Striepen, B., Handman, E., Lang-Unnasch, N., et al. (1998). Nuclear-encoded proteins target to the plastid in *Toxoplasma gondii* and *Plasmodium falciparum. Proceedings of the National Academy of Sciences of the United States of America, 95*, 12352–12357.

Waller, R. F., & Mcfadden, G. I. (2005). The apicoplast: A review of the derived plastid of apicomplexan parasites. *Current Issues in Molecular Biology, 7*, 57–79.

Watanabe, M. M., Suda, S., Inouye, I., Sawaguchi, I., & Chihara, M. (1990). *Lepidodinium viride* gen et sp. nov. (Gymnodiniales, Dinophyta), a green dinoflagellate with a chlorophyll *a*- and *b*-containing endosymbiont. *Journal of Phycology, 26*, 741–751.

Weihe, A., Liere, K., & Borner, T. (2012). Transcription and transcription regulation in chloroplasts and mitochondria of higher plants. In C. E. Bullerwell (Ed.), *Organelle genetics: Evolution of organelle genomes and gene expression* (pp. 297–325). Springer-Verlag.

Wilson, R. J. M. I., Denny, P. W., Preiser, D. J., Rangachari, K., Roberts, K., Roy, A., et al. (1996). Complete gene map of the plastid-like DNA of the malaria parasite *Plasmodium falciparum. Journal of Molecular Biology, 261*, 155–172.

Woehle, C., Dagan, T., Martin, W. F., & Gould, S. B. (2011). Red and problematic green phylogenetic signals among thousands of nuclear genes from the photosynthetic and apicomplexa-related *Chromera velia. Genome Biology and Evolution, 3*, 1220–1230.

Wisecaver, J. H., & Hackett, J. D. (2010). Transcriptome analysis reveals nuclear-encoded proteins for the maintenance of temporary plastids in the dinoflagellate *Dinophysis acuminata. BMC Genomics, 11*, 366.

Xu, P., Widmer, G., Wang, Y., Ozaki, L. S., Alves, J. M., Serrano, M. G., et al. (2004). The genome of *Cryptosporidium hominis. Nature, 431*, 1107–1112.

Yoon, H. S., Hackett, J. D., Ciniglia, C., Pinto, G., & Bhattacharya, D. (2004). A molecular timeline for the origin of photosynthetic eukaryotes. *Molecular Biology and Evolution, 21*, 809–818.

Yoon, H. S., Hackett, J. D., Pinto, G., & Bhattacharya, D. (2002). The single, ancient origin of chromist plastids. *Proceedings of the National Academy of Sciences of the United States of America, 99*, 15507–15512.

Yoon, H. S., Price, D. C., Stepanauskas, R., Rajah, V. D., Sieracki, M. E., Wilson, W. H., et al. (2011). Single-cell genomics reveals organismal interactions in uncultivated marine protists. *Science, 332*, 714–717.

Zhang, Z., Green, B. R., & Cavalier-Smith, T. (1999). Single gene circles in dinoflagellate chloroplast genomes. *Nature, 400*, 155–159.

Zhang, Z., Green, B. R., & Cavalier-Smith, T. (2000). Phylogeny of ultra-rapidly evolving dinoflagellate chloroplast genes: A possible common origin for sporozoan and dinoflagellate plastids. *Journal of Molecular Evolution, 51*, 26–40.

Zhao, S., Burki, F., Brate, J., Keeling, P. J., Klaveness, D., & Shalchian-Tabrizi, K. (2012). *Collodictyon*–An Ancient Lineage in the Tree of Eukaryotes. *Molecular Biology and Evolution, 29*, 1557–1568.

Zhu, G., Marchewka, M. J., & Keithly, J. S. (2000). *Cryptosporidium parvum* appears to lack a plastid genome. *Microbiology, 146*, 315–321.

CHAPTER FOUR

The Role of Microalgal Symbionts (*Symbiodinium*) in Holobiont Physiology

Michele X. Weber[1] and Mónica Medina
Molecular and Cell Biology, University of California, Merced, Merced, California 95343, USA
[1]Corresponding author: E-mail: mxweber@gmail.com

Contents

Abstract

Many marine organisms host microalgal symbionts. Complex holobionts require adaptations to accommodate closely associated lifestyles when algae live within host tissues. Six categories of physiological adaptation in *Symbiodinium*-holobiont systems include pathways related to nutrient transfer, carbon concentration, nitrogen recycling, calcification, oxidative stress and cell–cell communication. Combining traditional physiological measurements with genomic and transcriptomic analysis informs hypotheses about underpinning mechanisms. As methods advance, large, next generation sequence data sets make broad comparative investigations possible. Integrating data from multiple partners provides insight on physiological solutions to cooperative living.

1. INTRODUCTION

Single-celled symbionts play an integral role in the evolution and function of many organisms. Most plants cooperate with mycorrhizae fungi as well as diverse microbes living within their roots and leaves. Lichens are

Advances in Botanical Research, Volume 64
ISSN 0065-2296,
http://dx.doi.org/10.1016/B978-0-12-391499-6.00004-9

complex symbioses between fungi and either green algae or cyanobacteria. *Paramecium* host *Chlorella* but the relationship has evolved to be facultative, meaning both partners can live independently. Some well-characterized obligative marine examples include the microbial symbionts in the squid bioluminescent organ and the coral microbial community. The natural world includes many such complex holobionts, where a multicellular host and its diverse associated microbial assemblage live intimately together and reciprocally affect the evolution of genes and lineages. It is becoming increasingly clear that most organisms associate with diverse microbes throughout their ontogeny. In endosymbiotic systems, where the symbiont lives inside the cells of its host, processes become integrated and evolution and development are tightly coupled.

What is a *holobiont*? Margulis (1993) defined the term as, 'The holobiont is the product, temporary or permanent, of the association between its constituent bionts (partners)'. Iglesias-Prieto and Trench (1997) used *holosymbiont* to describe highly integrated systems where host and symbiont evolve together. Holobiont was adopted by Rowan (1998) to describe corals together with their microalgal symbionts. Rohwer, Seguritan, Azam, and Knowlton (2002) used holobiont to refer to the entire microbiome associated with corals. This latter use stresses that a diverse microbial community is also associated with the coral colony and that it may include fungi, bacteria, archea, other unicellular eukaryotes and viruses. The holobiont has become an accepted concept in coral biology where the host organism is a colony composed of thousands of individual polyps and distinct microalgal and microbial assemblages are associated with the animal tissue and the mucus and layered within the skeleton. In this review, we will focus on dinoflagellate components of the holobiont, but it is important to consider them in the perspective of a diverse complex system.

Scleractinian corals host dinoflagellates from the genus *Symbiodinium* (Freudenthal, 1962). Dinoflagellates are a diverse group of algae that have acquired their photosynthetic activity through a secondary endosymbiosis. We refer to Chapter I of this volume for a review of dinoflagellate diversity and ecology (Not *et al.*, 2012) and to Chapter III which covers the evolution of this group of algae (Archibald, 2012). Although the diverse members of the *Symbiodinium* lineage have been historically called zooxanthellae (from zoo = animal and xanth = yellow), this term is categorical and includes multiple groups of algal symbionts. Here we refer to a specific group of microalgal symbionts by their Latin generic name, *Symbiodinium*; however, we recognize that this may not be an accurate representation of the generic rank. With increasing attention and advancing sequencing methods, more diversity has been discovered within this genus than within many other orders of dinoflagellates (Pochon & Gates, 2010; Rowan & Powers, 1992). Nine lineages within *Symbiodinium* are composed of diverse phylotypes, many of which appear to be functionally unique and occupy defined ecological niches and roles within and across holobionts (Lajeunesse *et al.*, 2010; Pochon & Gates, 2010; Stat, Morris, & Gates, 2008). Closely related lineages of *Symbiodinium* are hosted by diverse eukaryotic organisms (Fig. 4.1). Single celled hosts include foraminifera, radiolarians and ciliates. Metazoan hosts include porifera, acoelomorpha, cnidaria and mollusca. Associations between these hosts and *Symbiodinium* form the basis of the food web in coral reef ecosystems. These associations form the basis of the food web in coral reef ecosystems. However the stability of these symbioses is threatened by climate change and as global temperatures rise, the symbiotic relationships become unstable and break down in a phenomenon known as coral bleaching.

Here, we review the functional role of *Symbiodinium* in marine invertebrates focusing on corals and anemones. For these systems, we summarize physiological characters that may be under selection pressure in symbiosis systems. We will highlight algal contributions to holobiont physiology, including the novel insights contributed from genomics, transcriptomics and metabolomics. Finally, we will discuss the evolutionary trajectory of holobionts; asking if physiological processes and functionality may be affected by changing conditions and what we can learn about it using modern methods.

Physiological processes evolved to accommodate algal cells living inside host cells (Fig. 4.2). Adaptations that facilitate symbiosis in a variety of marine invertebrates and single-celled eukaryotes are associated with six major categories of physiological processes: (A) Transport pathways shuttle products of photosynthesis from alga to host. (B) Light enhanced calcification, a phenomenon in which corals and other hosts with calcified skeletons exploit photosynthesis to precipitate calcium carbonate structures. (C) Carbon concentration mechanisms in which carbon from seawater is concentrated within animal cells to provide algal symbionts with raw materials for photosynthesis. (D) Nitrogen recycling provides additional raw materials for photosynthesis in nutrient limited seawater. (E) Reduction of oxidative stress induced by photosynthesis and high light conditions. (F) Cell–cell communication and molecular recognition pathways as part of host immune systems, which allow entry and residence to potential symbionts. Each of these functional categories is related to symbiosis (see citations in following section

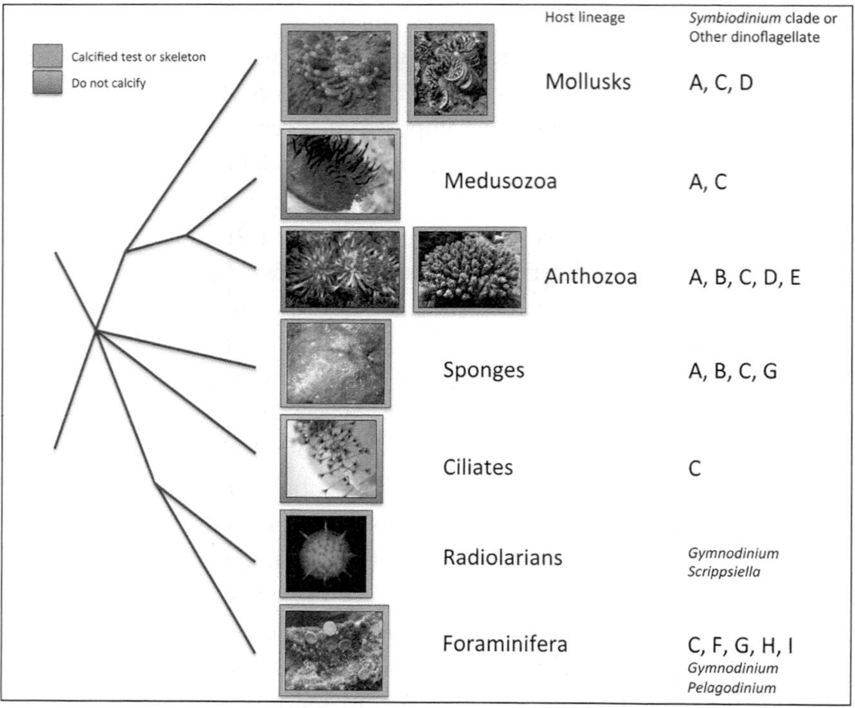

Figure 4.1 Diversity of *Symbiodinium* and hosts. *Symbiodinium* and other closely related lineages of dinoflagellates are hosted by diverse organisms across the tree of life. Mollusk host lineages include nudibranchs and giant clams. Cnidarian host lineages include jellyfish, soft corals, hard corals and anemones. Several sponge genera, including Cliona, host *Symbiodinum*. Single-celled eukaryote hosts include ciliates, radiolarians and foraminifera. Symbiont-produced photosynthate is implicated in the calcification of some of these organisms: giant clams, scleractinian corals, sponges, radiolarians, foraminifera (orange boxes). *(Photo credits:* Phyllodesmium *and* Tridacna, *Michele Weber;* Cassiopeia, *Erika Díaz-Almeyda;* Xenia *and* Acropora, *Michele Weber;* Cliona, *Deborah Gochfeld,* Maristentor, *Chris Lobban; acantharian, Scott Fay; soritid foraminifera, Michele Weber).* For colour version of this figure, the reader is referred to the online version of this book.

for relevant work in each category) and allows the holobiont to function as a complex and efficient photosynthetic meta-organism.

Genomic data for both host and algal symbiont will help clarify the selection pressures imposed by symbiotic lifestyles on the physiology of each partner. Although the field is moving rapidly, we do not yet know as much about the algal genomes compared to the developing body of research on cnidarian genomes. We have the complete genome for one coral: *Acropora digitifera* (Shinzato *et al.*, 2011); and four more in progress including *Acropora millepora, Stylophora pistilata, Montastraea faveolata, Acropora palmata* (C. Voolstra,

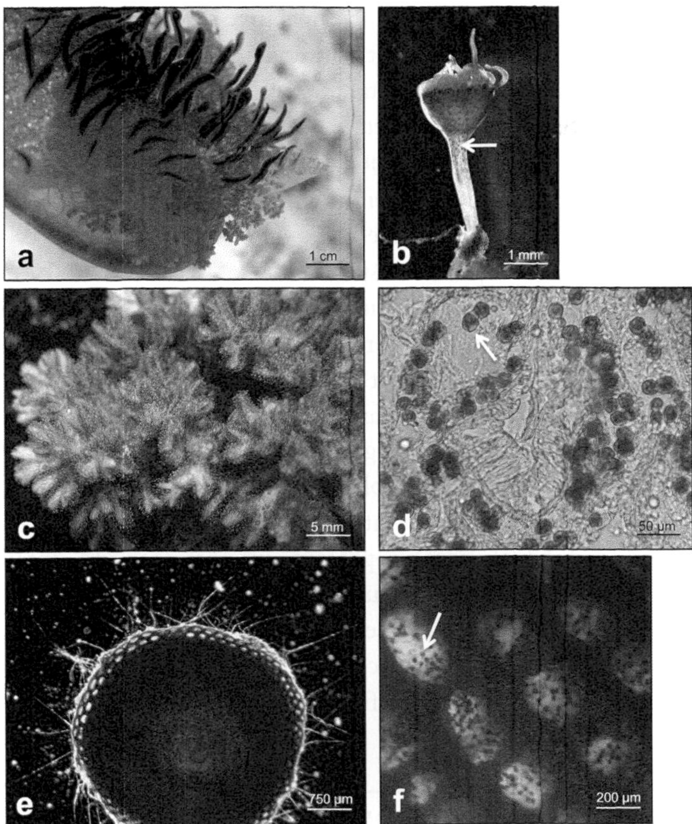

Figure 4.2 ***Symbiodinium* cells.** In most organisms, *Symbiodinium* are housed inside host cells. Images show symbionts living inside the tissues of jellyfish (a and b), coral (c and d) and soritid foraminifera (e and f). White arrows indicate individual dinoflagellates. *(Photo credits: a and b, Erika Díaz-Almeyda; c, Michele Weber; d, Nancy Cabanillas-Terán; e and f, Scott Fay).* For colour version of this figure, the reader is referred to the online version of this book.

I. Baums, personal communication). The complete genome for the aposymbiotic anemone, *Nematastella vectensis*, is available (Putnam et al. 2007) and the hydra, *Hydra magnipapillata*, was also recently published (Chapman *et al.*, 2010), and the genome of the symbiotic anemone, *Aiptasia pallida,* is in progress (J. Schwartz, personal communication). Although ESTs have been generated for several other orders of dinoflagellates, only a few studies have been published on *Symbiodinium* (Bayer *et al.*, 2012; Leggat, Hoegh-Guldberg, Dove, & Yellowlees, 2007; Leggat, Yellowlees, & Medina, 2011; Voolstra *et al.*, 2009b). With more coral genomes becoming available and genomes for several *Symbiodinium* lineages in progress (N. Satoh, C. Voolstra, personal communication), powerful comparative studies using symbiotic and aposymbiotic hosts and the relevant symbionts, will become possible.

Genomics, transcriptomics, metabolomics and interactomics are 'omics' fields that will contribute perspectives on the evolution of physiological processes. Growing collections of next generation sequence data for various partners will allow comparative genomic studies to better understand how holobiont physiologies have adapted to exploit cooperative lifestyles across diverse lineages. Transcriptome response and EST libraries can be used to characterize expressed genes and understand what parts of the genomes are involved in which processes. Experiments monitoring gene expression can suggest proteins and pathways that manage interactions between host and microalgal symbiont. When transcriptome data generate functional hypotheses, they can be tested using RNA interference (RNAi) methods to knock down those genes. Metabolomics will clarify how metabolites are produced and shared between intimately associated partners. Increasing attention to individual protein–protein interactions and interactomes (Medina & Sachs, 2010) will allow us to propose models and mechanisms for the evolving physiological processes. Physiological genomics takes advantage of these system wide approaches to consider physiological processes within an integrated evolutionary framework.

2. HOLOBIONT PHYSIOLOGY

2.1. Nutrient Transfer

Ancestral cnidarians were filter feeders but many derived lineages supplement heterotrophy with the products of photosynthesis provided by microalgal endosymbionts. It was first shown in 1958 and then observed further by Muscatine, Trench and their colleagues that nutrients of algal origin were incorporated into host tissues (Muscatine, 1967; Muscatine & Hand, 1958; Muscatine, Mccloskey, & Marian, 1981; Trench, 1971a, 1971b, 1971c). Subsequent studies inferred host pathways that function to move metabolites from symbiont to host. Algal-derived nutrients supplement host heterotrophy when they are transported through the symbiosome into gastrodermal cells. Experimental evidence showed that the net sugars produced by photosynthesis were greater than the products of algal metabolism. These data suggested that excess algal-produced sugars were transported out of the algal cell to meet host metabolism requirements (Allemand, Furla, & Benazet-Tambutté, 1998). Using bleaching experiments to 'knockout' potential transport pathways, Luo *et al.* (2009) demonstrated that certain lipid bodies are responsible for moving algal-synthesized lipids to the host. These bodies were not present in aposymbiotic corals. Proteins involved in lipid transfer were differentially expressed between symbiotic and aposymbiotic anemone transcriptomes (Ganot *et al.*, 2011). The *Acropora digitifera* genome revealed that this coral does not have genes for an enzyme required for cysteine biosynthesis. This result suggested that cysteine may be synthesized by the algae and then transported back to the host along with photosynthate (Shinzato *et al.*, 2011).

Little is known about how the algae transfer nutrients to their hosts. However, an early-controlled experiment on cultured *Symbiodinium* showed that they released more sugar in the presence of host tissue (Trench, 1971c). A study of *Symbiodinium* ESTs from cultured algae suggested that symbiosis may strengthen selection pressures and described 'symbiosis-specific' sequences as generally 'evolving quickly' in those dinoflagellate lineages (Voolstra *et al.*, 2009b). ESTs generated from *Symbiodinium* collected from coral hosts revealed over 7% of sequences involved in carbohydrate transport metabolism and over 3% of sequences involved in amino acid transport and metabolism: both categories that potentially include translocation of these products (Leggat *et al.*, 2007). In addition, a membrane spanning sugar phosphate permease was identified as part of this EST library and may function to conduct carbohydrates from alga to coral (Leggat *et al.*, 2007). Transcripts with membrane-transport-related functionality were obtained just under 10% of sequences for two lineages of cultured *Symbiodinium* (Bayer *et al.*, 2012).

2.2. Calcification

Biomineralization is common in marine environments and scleractinian corals build massive calcium carbonate structures that form the multi-dimensional reef framework. Some mollusks and various single-celled eukaryote groups also produce calcium carbonate shells and tests (Fig. 4.1). A large number of these skeleton building organisms host algal symbionts. Hosts utilize a portion of algal fixed carbon to fuel metabolism. But some of the energy derived from photosynthesis is used to build an organic matrix and subsequent biomineralization of the skeleton or test. These activities appear to increase in the presence of light; however, calcification happens distant from the site of photosynthesis and it is unclear if algal cells play a more direct role in calcification processes.

Several models have been proposed to explain how animals utilize algal photosynthesis to precipitate calcium carbonate (Allemand *et al.*, 2004; Goreau, 1959; Muscatine, 1990). Many authors and multiple lines of evidence suggest that calcification is light enhanced, meaning that in the presence of light, deposition of aragonite calcium carbonate increases (Goreau, 1959). An experiment conducted in aquaria showed that providing glycerol to bleached corals increased their rate of calcification. Bleached corals lack symbionts and therefore have fewer resources available to devote to calcification processes. Exogenous resources provided to starving hosts increased calcification rates because they simulated the resources that would be reallocated from symbiont to host under normal conditions. Although glycerol, a primary product of photosynthesis, did not accelerate calcification rates in healthy corals, these results suggested that biomineralization pathways are photosynthesis driven (Colombo-Pallotta, Rodríguez-Román, & Iglesias-Prieto, 2010). Exogenous inhibitors were used to show that calcification could be artificially decelerated in the light but not in the dark (Al-Horani, Al-Rousan, Manasrah, & Rasheed, 2005). Under thermal stress or elevated CO_2 conditions, three coral species exhibited differential expression of calcification-associated genes (Desalvo *et al.*, 2008; Vidal-Dupiol *et al.*, 2009; Moya *et al.*, 2012). These lines of evidence supported the hypothesis that calcification is a photosynthesis-driven process because thermal stress induces bleaching and loss of symbionts so that hosts subsequently lack the extra resources as products of photosynthesis.

Carbonic anhydrase helps deliver carbon to the symbiont for photosynthesis and it may also function to transport bicarbonate to the site of calcium carbonate deposition in the host. Inhibition of carbonic anhydrase dramatically decreased calcification rates (Al-Horani *et al.*, 2005). A unique form of carbonic anhydrase was localized at the site of precipitation of calcium carbonate in the coral *Stylophora pistillata*. These data suggested a role in a carbon-concentrating mechanism; the enzyme may convert metabolic carbon into bicarbonate, which can then be precipitated as aragonite skeleton (Bertucci, Tambutté, Supuran, Allemand, & Zoccola, 2011; Furla, Allemand, & Orsenigo, 2000a; Moya *et al.*, 2008).

Analysis of an EST library for *Symbiodinium in hospite* (living inside coral tissues) identified transport enzymes that may move glycerol generated by photosynthesis to the site of calcification where it could supply energy required for calcification (Colombo-Pallotta *et al.*, 2010; Leggat *et al.*, 2007). Analyzing data from a complete *Symbiodinium* genome will generate additional candidate genes coding for proteins that transport products of photosynthesis to the site of calcification. Comparing genomes from symbiotic and free living lineages will further clarify the role of algal symbionts in biomineralization.

2.3. Carbon Concentration

Metabolic production of CO_2 from both host and symbiont are not enough to supply carbon for algal photosynthesis. Dissolved inorganic carbon (DIC) is a principle raw ingredient of photosynthesis but it is limited within host tissues. Seawater is the primary source of carbon but algal symbionts do not directly contact seawater once they are sequestered within host tissues. Therefore, ancestral organisms must have evolved the ability to concentrate carbon from the surrounding seawater (Raven, 1991) and deliver it directly to *Symbiodinium* cells.

Membrane-bound carbonic anhydrase appears to play an important role in moving carbon from the seawater environment, across multiple layers of membrane, into the host cell and then into the algal cell (Furla, Galgani, Durand, & Allemand, 2000b; Furla *et al.*, 2000a, 2005; Leggat *et al.*, 2002; Weis, 1993; Weis, Smith, & Muscatine, 1989). Carbonic anhydrase is present in host transcriptomes. The Pacific anemone, *Anthopleura elegantissima*, is a facultative host and transcription of carbonic anhydrase was upregulated in the presence of *Symbiodinium* (Weis & Reynolds, 1999). In a Mediterranean anemone, *Anemonia viridis*, transcription of multiple different carbonic anhydrases increases in symbiotic polyps (Ganot *et al.*, 2011). Carbon concentration and the presence of DIC in these symbiotic non-calcifying anthozoans provide materials for photosynthesis; however, some inorganic carbon is likely used to produce calcium carbonate skeletons in scleractinian corals and other biomineralizing organisms such as giant clams and foraminifera.

It is not clear how much the algal symbionts participate in this process or whether they just passively receive carbon resources provided by their hosts. A membrane-associated ATPase protein expressed by *Symbiodinium in hospite* (but not in cultured algae) may function to adjust pH so that host-initiated pathways can more easily deliver carbon to the chloroplasts (Bertucci, Tambutté, Tambutté, Allemand, & Zoccola, 2010). Transcriptome data from *Symbiodinium* revealed several sequences potentially related to carbon transport but it was not possible to determine which membrane they spanned (Bayer *et al.*, 2012). Therefore, it is not clear if these proteins receive DIC from the host or whether they move it around once it is already inside the algal cell (Bayer *et al.*, 2012). Although algal EST data do not show direct evidence for carbon-concentrating mechanisms, identifying enzymes in the *Symbiodinium* genome that adjust pH would suggest the presence of a mechanism that may allow DIC to diffuse into localized regions where it is required.

2.4. Nitrogen Recycling

Nitrogen is a critical nutrient that is growth limiting for many organisms, particularly organisms in tropical oceans (Muscatine & Porter, 1977). Symbiotic algae use nitrogen as a raw ingredient for photosynthesis and acquire it by several mechanisms. Holobionts can absorb inorganic nitrogen from the water column. Although it is not known whether the host or algal symbiont is primarily responsible, this capability is not physiologically possible in most aposymbiotic organisms. Alternatively, heterotrophic animal hosts can capture and ingest zooplankton. When metabolism releases ammonium, it can be transferred to *Symbiodinium*. Algae then return usable nitrogen to the host in the form of amino acids (Wang & Douglas, 1999; Yellowlees, Rees, & Leggat, 2008). This active recycling of nitrogen increases the supply of useful products for both symbiont and host that might otherwise require additional energetically costly filtering processes. However, the host must also limit the amount of nitrogen accessible to the algal symbionts in order to prevent overgrowth (Wang & Douglas, 1998; Yellowlees *et al.*, 2008).

Symbiodinium ESTs revealed that over 3% of the sequences were similar to prokaryote transporters with functions attributed to ammonium transport and amino acid transport and metabolism. Some fraction of these are likely involved in moving nitrogen acquired by the algae back to the host in the form of amino acids. Although these sequences are not present in other dinoflagellate ESTs to date, they are present in other eukaryotes (Leggat *et al.*, 2007). Genomic resources may further identify proteins and pathways that have evolved to regulate transport and recycling processes. An interesting start might experiment with cultured versus algae living *in hospite*. If exogenous nitrogen is supplied, are there differences in algal gene expression?

2.5. Oxidative Stress

Life in reef environments is stressful because the water is shallow and clear, with limited capacity to filter out damaging ultra violet radiation (UVR).

Symbiodinium require light energy in order to photosynthesize. They force their hosts to adapt to these high-risk habitats in the photic zone where free radicals and other oxidants increase mutation rates and detrimentally alter DNA. Although cellular mechanisms can repair this damage, after a certain threshold, permanent damage from oxidative stress results in death (Tchernov *et al.*, 2011). Host organisms have evolved certain morphologies and specialized structures that increase horizontal surface area and maximize the number of photons reaching chlorophyll pigments in their algal symbionts. However, these morphological adaptations increase the risk of oxidative stress and must be balanced by other adaptations that evolve to protect host tissues.

One defence against oxidative stress is increased expression of enzymes that either prevent or repair cell damage. Enzymes that quench free radicals are most common in corals collected from shallow habitats (Shick *et al.*, 1995). For example, ferritin is a protein that sequesters metals that would otherwise catalyze free radicals. Microarrays were used to measure differentially expressed genes in *Anthopleura elegantissima* individuals that were exposed to increased UVR and/or increased heat. One of the most dramatic upregulated sequences was ferritin (Richier, Rodriguez-Lanetty, Schnitzler, & Weis, 2008). These proteins were also highly expressed in two species of *Acropora* (Schwarz *et al.*, 2008). Microarray analysis compared symbiont-hosting *Anemonia viridis* to aposymbiotic anemones and revealed differential expression of genes with known antioxidant activity (Ganot *et al.*, 2011; Moya, Ganot, Furla, & Sabourault, 2012). Evaluating expression patterns for corals under thermal stress also showed up-regulation of antioxidants (Desalvo, Sunagawa, Voolstra, & Medina, 2010b; Desalvo *et al.*, 2008). High irradiance levels intensified cellular response to high temperature (Tchernov *et al.*, 2011). Since high temperature conditions are often associated with high light, holobiont response mechanisms to oxidative stress initiated by both stresses may have evolved together.

An EST library constructed from symbionts isolated from *Acropora aspera* indicated that members of a family of peroxidases are expressed in *Symbiodinium*. These genes had not been previously recognized in dinoflagellates but they are known to have a function related to oxidative stress in prokaryotes (Leggat *et al.*, 2007). Sequences encoding proteins that may have antioxidant properties were identified in the transcriptome from two different lineages of *Symbiodinium*. These proteins have been well studied in plants where oxidative stress is associated with photosynthesis. According to this study, many of the sequences with oxidative stress response functions were more highly expressed in *Symbiodinium* than in other photosynthesizing lineages (Bayer *et al.*, 2012).

Another strategy to protect tissues in the high light conditions is to produce sunscreens that prevent additional oxidative stress. Microsporine-like amino acids (MAAs) are secondary metabolites, which act as sunscreens by absorbing excess UVR and quenching free radicals (Garcia-Pichel, Wingard, & Castenholz, 1993; Shick *et al.*, 1995). Experimental and field-collected data demonstrated that MAA concentration in coral holobionts varied with depth and time of day, both of which affect UVR levels (Gleason, 1993; Shick, 2004; Shick *et al.*, 1995; Yakovleva & Hidaka, 2004).

Whether these compounds are produced by host or by dinoflagellate symbionts is unknown. Basal lineages of *Symbiodinium* produced MAAs in culture but more derived lineages did not (Banaszak, Lajeunesse, & Trench, 2000). MAAs were more diverse when *Symbiodinium* lived in corals but the coral host may have altered precursor molecules produced by the algae (Shick, 2004). The *A. digitifera* genome revealed evidence that at least some MAAs are produced by corals themselves (Shinzato *et al.*, 2011). None of the available *Symbiodinium* EST data sets provided specific evidence for the production of MAAs; however, all agreed that many sequences were functionally unidentified and that these may have important symbiosis specific roles (Bayer *et al.*, 2012; Leggat *et al.*, 2007; Voolstra *et al.*, 2009b). A symbiotic lifestyle exerts selection pressures that modify physiological components for both partners. And products that would benefit the holobiont by protecting both the symbionts and the host in high light conditions are just one group of candidate functions for these unidentified algal sequences.

2.6. Cell–Cell Communication and Recognition

Finally, a broad but critical component of holobiont physiology that evolved to accommodate symbiosis includes cellular recognition and immune-system-related pathways. Within this category are two critical types of communication: (i) the partners must recognize each other to establish a partnership, (ii) the host immune system must allow the alga access to host tissues and accommodate the symbiont living in its tissues. Many groups of *Symbiodinium* are not obligate and have a free-living phase of their life cycle. Other lineages do not survive in culture, suggesting that they are more adapted to a symbiotic lifestyle inside the host (Stat *et al.*, 2008). However, many hosts will not survive if they do not find the appropriate *Symbiodinium* cells during a brief phase in their larval development. Once acquired, the host immune system must recognize the symbiont as an essential component of the holobiont. Both these processes involve mechanisms of intercellular communication between single-celled algae and metazoans (or other eukaryotic hosts).

Symbiodinium is a diverse lineage but most hosts maintain some degree of specificity (Lajeunesse *et al.*, 2010). Many hosts reacquire *Symbiodinium* at every generation, but a minority of taxa transfer symbiont cells from mother to daughter organisms via vertical transmission. In the case of foraminifera (granuloureticulose shelled unicellular eukaryotes), the mother cell can divide and each daughter cell hosts a fraction of the original *Symbiodinium* population. In some corals, egg cells are packaged with a complement of the maternal *Symbiodinium* cells. However, for most scleractinian corals as well as sponges and mollusks, gametes are released into the water column where they meet and fuse (Fadlallah, 1983; Harrison *et al.*, 1984). Therefore, it is necessary for the developing planktonic larva to procure a population of symbionts from free-living algal populations (Harii, Yasuda, Rodriguez-Lanetty, Irie, & Hidaka, 2009).

Metazoans have evolved pathways to identify foreign cells that invade their tissues triggering inflammatory response and phagocytosis. Compared to the genome of aposymbiotic anemone *Nematostella*, immunity-associated components were much more complex in the *Symbiodinium*-hosting coral *Acropora digitifera* (Shinzato *et al.*, 2011). Comparing ESTs from two different species of *Acropora* showed that genes associated with the immune system were under positive selection (Iguchi, Shinzato, Foret, & Miller, 2011; Voolstra *et al.*, 2011). During symbiosis breakdown, the coral, *Pocillopora damicornus*, exhibited down regulation of lectin proteins associated with the onset of symbiosis (Vidal-Dupiol *et al.*, 2009; Wood-Charlson, Hollingsworth, Krupp, & Weis, 2006). Several molecules involved in immunological response, which evolved deep in cnidarian history, were identified in *Acropora millepora*. Expression was localized to the gastroderm, where symbionts are maintained (Kvennefors *et al.*, 2010). These diverse proteins likely evolved to play a role in both recognition and maintenance of the symbiosis.

Perhaps homologous sequences associated with recognition evolved in a more ancestral cnidarian lineage. In a study of aposymbiotic and symbiotic anemones, *A. viridis* expressed multiple isoforms of a homolog to vertebrate recognition proteins. One form was strongly suppressed in symbiotic anemones compared to aposymbiotic anemones (Ganot *et al.*, 2011). This suggested that symbionts play a role in suppressing host immune systems to grant themselves 'safe passage'. Also identified in anemone hosts, Sym32 was first labelled a symbiosis gene because it was highly expressed in the symbiotic state (Reynolds, Schwarz, & Weis, 2000). It was localized to the membranes that contained the algal cells within the host suggesting that it played a role in communication between the partners (Schwarz & Weis, 2003). In *A. viridis*, the ortholog to this sequence was highly upregulated in the symbiotic state, as was Calumenin, which may help regulate Sym32 (Ganot *et al.*, 2011).

In their analysis of ESTs from cultured *Symbiodinium,* Voolstra *et al.* (2009b) found sequences with functions related to cell–cell communication. They proposed that free-living dinoflagellates sense their surrounding environment when they are not protected by host tissue (Voolstra *et al.*, 2009b). However, it is also possible that these same sequences produced proteins that were utilized in recognition between free-living algae and competent host larvae. Onset of symbiosis does not induce a strong transcriptomic signature in corals (Schnitzler & Weis, 2010; Voolstra *et al.*, 2009a). However, under experimental conditions, there was a strong host transcriptional response when coral larvae rejected an incompatible lineage of *Symbiodinium*. These results also suggested that successful symbionts evolved strategies to avoid host immune response (Voolstra *et al.*, 2009a).

It is possible that the immune systems of both host and symbiont are becoming more tightly associated as they evolve together as a holobiont. The host must 'learn' to recognize the symbiont and facilitate its entrance into host tissues. *Symbiodinium* likely evolved from an infectious ancestor. As is seen in pathogenic prokaryotes, certain products expressed by the microorganisms can interrupt the host immune system or even mask the presence of infecting organisms. A precursor of this type would likely evolve in a symbiotic system to signal the arrival of a compatible partner and promote the establishment of a cooperative relationship.

3. PHYSIOLOGICAL GENOMICS

Increasingly, genomics and other 'omics' inspire new frontiers of research targeting physiological adaptations to symbiosis. Experiments that address host and algal symbiont responses to specific environmental conditions offer new testable hypotheses. Gene expression analysis can identify candidate proteins that function in interactions and dialogues between hosts and symbionts. New molecular methods are being used to test the function of candidate proteins involved in these physiological pathways.

Experiments that analyze *Symbiodinium* transcriptome response to conditions that disrupt symbiosis provide initial clues into understanding how dynamic differential gene expression affects physiological processes. For example, applying thermal stress to cultured *Symbiodinium* revealed physiological responses related to reductions in key photosynthesis proteins. This effect was partially due to decreased transcription and there was no measurable difference in protein degradation. These data suggested that the remainder of the decrease in photosynthesis proteins in bleached corals resulted from differentially translated transcripts (Takahashi, Whitney, Itoh, Maruyama, & Badger, 2008).

Another experiment used microarrays to test the effects of thermal stress on cultured *Symbiodinium*. The data showed differential expression in 4% of the genes and suggested that *Symbiodinium* grown at high temperatures might experience cell damage. Expression patterns revealed increased damage to protein and DNA as well as decreased protein production in thermally stressed *Symbiodinium*. However, physiological data indicated acclimation to high temperature because the electron transport rate in photosystem II showed no difference between treatments. This experiment showed that physiological response and gene expression are not perfectly correlated (E. Díaz-Almeyda, C. Voolstra, M.K. DeSalvo, N. Masuoka and M. Medina, unpublished observations).

Other stresses also affect algal transcription. Osmotic stress induced cultured algae to upregulate transcription of enzymes related to glycerol production and release. These data suggested that host signalling may actively increase algal transcription related to photosynthesis. By limiting the available carbon, they may also modulate algal population densities (L.P. Suescún-Bolívar, P.E. Thomé and R. Iglesias-Prieto, unpublished observations). Investigations are in progress to test the effects of irradiance and variable pH on both algal transcriptome and algal physiological processes. These experiments underline the advantages of combining 'omics' methods with traditional measurements of physiological proxies to understand the evolution of holobiont physiological solutions under changing environmental conditions. Scaling up and using RNA-seq to evaluate algal transcriptome responses under variable experimental conditions will likely advance existing mechanistic hypotheses. And comparative analysis of large-scale transcriptomes from various *Symbiodinium* cultures and populations *in hospite* will contribute to hypotheses regarding physiological diversification within this important lineage of microalgal symbionts.

Algal diversity affects host transcriptome response to symbiosis breakdown. Some lineages of *Symbiodinium* are more physiologically sensitive and alter their transcription of important photosynthetic machinery when thermally stressed (Takahashi *et al.*, 2008). Microarray data showed that coral gene expression under conditions of thermal stress was more dependent on algal genotype than degree of stress, emphasizing the degree to which algal physiology impacts host transcriptome (Desalvo *et al.*, 2010a). As the evolution of diversity within *Symbiodinium* is further explored using more powerful sequencing strategies, we will learn how physiological characters diverged and adapted in host specific versus generalist lineages of algae.

Several additional studies documented host transcriptome response to stress and subsequent symbiosis breakdown. In order to better understand mechanisms of bleaching, corals were thermally stressed to test the ability of specific host-produced enzymes to mitigate the effects of reactive oxygen species produced by algae (Tchernov *et al.*, 2011). Certain genes related to oxidative stress and cell–cell communication were differentially expressed in coral hosts during naturally changing temperature regimes in the field (Seneca *et al.*, 2010). It therefore seems reasonable to expect, and worthwhile to explore, how different host taxa might also drive specific transcriptome responses in particular *Symbiodinium* lineages. These differentially expressed genes evident in host transcriptomes during bleaching may be part of symbiosis breakdown, and we would predict a similar response from functionally equivalent genes in the algal partners. Reciprocal experiments to evaluate the role of thermal stress in algae are now feasible and in progress thanks to the new genomic tools being released by the *Symbiodinium* research community. Corresponding communication and interaction with the host, viewed from the algal perspective, will be a critical next step in understanding how algal symbionts react to changing conditions in the short term and adapt to changing environments in the long term.

Increasingly intimate contact and eventual incorporation of infectious symbionts into host larvae influenced the evolution of holobiont physiology. Therefore, the onset of symbiosis is a critical life history stage. Comparing the genomic response of cultured *Symbiodinium* to dinoflagellates living *in hospite* during the onset of symbiosis will provide insight regarding the integration of physiological components. Evaluating the transcriptome of coral larvae at the onset of symbiosis has shown that although recognition is critical, transcriptomic response is limited in successful infections (Schnitzler & Weis, 2010; Voolstra *et al.*, 2009a). Even microarrays with limited features showed that for incompatible partners, a large portion of the host transcriptome is differentially expressed as the coral larva rejects that symbiont (Voolstra *et al.*, 2009a). Genes associated with immune systems and cell–cell communication, presumably including communication at onset of symbiosis, were under positive selection and were evolving quickly in the *Acropora* genome (Voolstra *et al.*, 2011). As mentioned above, further evaluating the onset of symbiosis using next generation sequencing will add an algal perspective to these data collected from metazoan hosts and inform hypotheses about the subsequent evolution of the maintenance of symbiosis.

Following up on observed differences in expression, we need to experimentally validate proposed physiological solutions starting with protein function. Several methods have been developed to test function in recombinant protein constructs. RNAi as well as more traditional methods for knocking down gene function have been employed in *Symbiodinium* (M.F. Ortiz-Matamoros, T. Islas-Flores, B. Voigt, D. Volkmann, D. Menzel, F. Baluska and M.A. Villanueva, unpublished observations). These methods will facilitate testing of mechanistic hypotheses for individual genes that are expressed differently under different conditions. Using yeast two-hybrid screens to test protein–protein interaction hypotheses, could also be adapted to the *Symbiodinium*-host system (Medina & Sachs, 2010). A more holistic view of the interactome will become possible as we optimize molecular methods to validate functionality of interacting protein components associated with both partners.

Large amounts of next generation sequence data make it possible to further investigate the evolution of physiological processes in diverse hosts and symbionts. The *Symbiodinium* ESTs that were recently published are the largest public collection for any dinoflagellate (Bayer *et al.*, 2012). Several *Symbiodinium* genomes from known cultures within clades B and C are in progress (N. Satoh, C. Voolstra, personal communication). Comparative genomics across divergent *Symbiodinium* genomes will provide insight into gene family evolution, genome rearrangements, copy number variation, conserved regulatory regions, etc., associated with candidate genes identified in physiological experiments. Downstream editing and post-transcriptional modification can also affect physiological processes and interactions between partners (e.g. Takahashi *et al.*, 2008). Inheritance of non-transcriptional processes also affect phenotypes (Jablonka & Raz, 2009).

As methods to explore variation in holobiont epigenomes and the effect of epigenetics on physiological processes are developed, it will help explain phenotypes and patterns that do not correspond with observed transcriptional responses.

Physiological solutions evolved to allow organisms to survive and succeed via cooperative interactions with phylogenetically divergent partners. The mutualistic associations that sustain reef communities evolved at several different hierarchical levels within the genome. Our capability to integrate 'omics' insights from each partner is improving as technology advances. Single gene and single protein hypotheses are now testable with modern molecular methods; however, large-scale inferences about the evolution of interkingdom interactions are within sight. This is an exciting opportunity and paves the way for new interpretations about the evolution of integrated physiological processes and meta-organism interactomes.

REFERENCES

Al-Horani, F. A., Al-Rousan, S. A., Manasrah, R. S., & Rasheed, M. Y. (2005). Coral calcification: Use of radioactive isotopes and metabolic inhibitors to study the interactions with photosynthesis and respiration. *Chemistry and Ecology, 21*, 325–335.

Allemand, D., Ferrier-Pagès, C., Furla, P., Houlbrèque, F., Puverel, S., Reynaud, S., et al. (2004). Biomineralisation in reef-building corals: From molecular mechanisms to environmental control. *Comptes Rendus Palevol, 3*, 453–467.

Allemand, D., Furla, P., & Benazet-Tambutté, S. (1998). Mechanisms of carbon acquisition for endosymbiont photosynthesis in Anthozoa. *Canadian Journal of Botany-Revue Canadienne De Botanique, 76*, 925–941.

Archibald, J. (2012). The evolution of algae by secondary and tertiary endosymbiosis. *Advances in Botanical Research, 63*, 87–118.

Banaszak, A. T., Lajeunesse, T. C., & Trench, R. K. (2000). The synthesis of mycosporine-like amino acids (MAAs) by cultured, symbiotic dinoflagellates. *Journal of Experimental Marine Biology and Ecology, 249*, 219–233.

Bayer, T., Aranda, M., Sunagawa, S., Yum, L. K., Desalvo, M. K., Lindquist, E., Coffroth, M. A., Voolstra, C. R., & Medina, M. (2012). *Symbiodinium* transcriptomes: genome insights into the dinoflagellate symbionts of reef-building corals. *PLoS ONE, 7*, e35269.

Bertucci, A., Tambutté, E., Tambutté, S., Allemand, D., & Zoccola, D. (2010). Symbiosis-dependent gene expression in coral-dinoflagellate association: cloning and characterization of a P-type H+-ATPase gene. *Proceedings of the Royal Society Biological Sciences Series B, 277*, 87–95.

Bertucci, A., Tambutté, S., Supuran, C. T., Allemand, D., & Zoccola, D. (2011). A new coral carbonic anhydrase in *Stylophora pistillata. Marine Biotechnology (New York), 13*, 992–1002.

Chapman, J. A., Kirkness, E. F., Simakov, O., Hampson, S. E., Mitros, T., Weinmaier, T., et al. (2010). The dynamic genome of Hydra. *Nature (London), 464*, 592–596.

Colombo-Pallotta, M. F., Rodríguez-Román, A., & Iglesias-Prieto, R. (2010). Calcification in bleached and unbleached *Montastraea faveolata*: Evaluating the role of oxygen and glycerol. *Coral Reefs, 29*, 899–907.

Desalvo, M. K., Sunagawa, S., Fisher, P. L., Voolstra, C. R., Iglesias-Prieto, R., & Medina, M. (2010a). Coral host transcriptomic states are correlated with *Symbiodinium* genotypes. *Molecular Ecology, 19*, 1174–1186.

Desalvo, M. K., Sunagawa, S., Voolstra, C. R., & Medina, M. (2010b). Transcriptomic responses to heat stress and bleaching in the elkhorn coral *Acropora palmata. Marine Ecology Progress Series, 402*, 97–113.

Fadlallah, Y. H. (1983). Sexual reproduction, development and larval biology in scleractinian corals. *Coral Reefs, 2*, 129–150.

Freudenthal, H. D. (1962). Sym-biodinium gen. nov. and *Symbiodinium microadriaticum* sp. nov., a zooxanthella: Taxonomy, life cycle, and morphology. *Jour Protozool, 9*, 45–52.

Furla, P., Allemand, D., & Orsenigo, M.-N. (2000a). Involvement of H+-ATPase and carbonic anhydrase in inorganic carbon uptake for endosymbiont photosynthesis. *American Journal of Physiology, 278*, R870–R881.

Furla, P., Allemand, D., Shick, J. M., Ferrier-Pagès, C., Richier, S., Plantivaux, A., et al. (2005). The symbiotic anthozoan: A physiological chimera between alga and animal. *Integrative and Comparative Biology, 45*, 595–604.

Furla, P., Galgani, I., Durand, I., & Allemand, D. (2000b). Sources and mechanisms of inorganic carbon transport for coral calcification and photosynthesis. *Journal of Experimental Biology, 203*, 3445–3457.

Ganot, P., Moya, A., Magnone, V., Allemand, D., Furla, P., & Sabourault, C. (2011). Adaptations to endosymbiosis in a cnidarian-dinoflagellate association: Differential gene expression and specific gene duplications. *Plos Genetics, 7*, 17.

Garcia-Pichel, F., Wingard, C. E., & Castenholz, R. W. (1993). Evidence regarding the UV sunscreen role of a mycosporine-like compound in the cyanobacterium *Gloeocapsa* sp. *Applied and Environmental Microbiology, 59*, 170–176.

Gleason, D. F. (1993). Differential effects of ultraviolet radiation on green and brown morphs of the Caribbean coral Porites astreoides. *Limnology and Oceanography, 38*, 1452–1463.

Goreau, T. F. (1959). The physiology of skeleton formation in corals. I. A method for measuring the rate of calcium deposition by corals under different conditions. *Biological Bulletin, 116*, 59–75.

Harii, S., Yasuda, N., Rodriguez-Lanetty, M., Irie, T., & Hidaka, M. (2009). Onset of symbiosis and distribution patterns of symbiotic dinoflagellates in the larvae of scleractinian corals. *Marine Biology, 156*, 1203–1212.

Harrison, P. L., Babcock, R. C., Bull, G. D., Oliver, J. K., Wallace, C. C., & Willis, B. L. (1984). Mass spawning in tropical reef corals. *Science, 223*, 1186–1189.

Iglesias-Prieto, R., & Trench, R. K. (1997). Photoadaptation, photoacclimation and niche diversification in invertebrate-dinoflagellate symbioses. *Proceedings of the 8th International Coral Reef Symposium.*

Iguchi, A., Shinzato, C., Foret, S., & Miller, D. J. (2011). Identification of fast-evolving genes in the scleractinian coral *Acropora* using comparative EST analysis. *PLoS One, 6*, e20140.

Jablonka, E., & Raz, G. (2009). Transgenerational epigenetic inheritance: Prevalence, mechanisms, and implications for the study of heredity and evolution. *Quarterly Review of Biology, 84*, 131–176.

Kvennefors, E. C. E., Leggat, W., Kerr, C. C., Ainsworth, T. D., Hoegh-Guldberg, O., & Barnes, A. C. (2010). Analysis of evolutionarily conserved innate immune components in coral links immunity and symbiosis. *Developmental & Comparative Immunology, 34*, 1219–1229.

Lajeunesse, T. C., Pettay, D. T., Sampayo, E. M., Phongsuwan, N., Brown, B., Obura, D. O., et al. (2010). Long-standing environmental conditions, geographic isolation and host-symbiont specificity influence the relative ecological dominance and

genetic diversification of coral endosymbionts in the genus Symbiodinium. *Journal of Biogeography, 37*, 785–800.
Leggat, W., Hoegh-Guldberg, O., Dove, S., & Yellowlees, D. (2007). Analysis of an EST library from the dinoflagellate (*Symbiodinium* sp.) symbiont of reef-building corals. *Journal of Phycology, 43*, 1010–1021.
Leggat, W., Marendy, E. M., Baillie, B., Whitney, S. M., Ludwig, M., Badger, M. R., et al. (2002). Dinoflagellate symbioses: Strategies and adaptations for the acquisition and fixation of inorganic carbon. *Functional Plant Biology, 29*, 309–322.
Leggat, W., Yellowlees, D., & Medina, M. (2011). Recent progress in *Symbiodinium* transcriptomics. *Journal of Experimental Marine Biology and Ecology, 408*, 120–125.
Luo, Y. J., Wang, L. H., Chen, W. N. U., Peng, S. E., Tzen, J. T. C., Hsiao, Y. Y., et al. (2009). Ratiometric imaging of gastrodermal lipid bodies in coral-dinoflagellate endosymbiosis. *Coral Reefs, 28*, 289–301.
Margulis, L. (1993). *Symbiosis in cell evolution*. New York: W.H. Freeman.
Medina, M., & Sachs, J. L. (2010). Symbiont genomics, our new tangled bank. *Genomics, 95*, 129–137.
Moya, A., Tambutté, S., Bertucci, A., Tambutté, É, Lotto, S., Vullo, D., et al. (2008). Carbonic anhydrase in the scleractinian coral *Stylophora pistillata*—Characterization, localization, and role in biomineralization. *Journal of Biological Chemistry, 283*, 25475–25484.
Moya, A., Huisman, L., Ball, E. E., Hayward, D. C., Grasso, L. C., Chua, C. M., et al. (2012). Whole Transcriptome Analysis of the Coral Acropora millepora Reveals Complex Responses to CO2-driven Acidification during the Initiation of Calcification. *Molecular Ecology, 21*, 2440–2454.
Moya, A., Ganot, P., Furla, P., & Sabourault, C. (2012). The transcriptomic response to thermal stress is immediate, transient and potentiated by ultraviolet radiation in the sea anemone Anemonia viridis. *Molecular Ecology, 21*, 1158–1174.
Muscatine, L. (1967). Glycerol excretion by symbiotic algae from corals and tridacna and its control by the host. *Science, 156*, 516–519.
Muscatine, L. (1990). The role of symbiotic algae in carbon and energy flux in reef corals. *Coral Reefs Ecosystems of the World, 25*, 75–87.
Muscatine, L., & Hand, C. (1958). Direct evidence for the transfer of materials from symbiotic algae to the tissues of a coelenterate. *Proceedings of the National Academic Science, 44*, 1259–1263.
Muscatine, L., Mccloskey, L. R., & Marian, R. E. (1981). Estimating the daily contribution of carbon from zooxanthellae to coral animal respiration. *Limnology and Oceanography, 26*, 601–611.
Muscatine, L., & Porter, J. W. (1977). Reef corals mutualistic symbioses adapted to nutrient-poor environments. *Bioscience, 27*, 454–460.
Not, F., Siano, R., Kooistra, W. H. C. F., Simon, N., Vaulot, D., & Probert, I. (2012). Diversity and ecology of eukaryotic marine phytoplankton. *Advances in Botanical Research, 63*, 1–53.
Pochon, X., & Gates, R. D. (2010). A new *Symbiodinium* clade (Dinophyceae) from soritid foraminifera in Hawai'i. *Molecular Phylogenetics and Evolution, 56*, 492–497.
Putnam, N. H., Srivastava, M., Hellsten, U., Dirks, B., Chapman, J., Salamov, A., et al. (2007). Sea anemone genome reveals ancestral eumetazoan gene repertoire and genomic organization. *Science, 317*, 86–94.
Raven, J. A. (1991). Implications of inorganic carbon utilization: Ecology, evolution, and geochemistry. *Canadian Journal of Botany, 69*, 908–924.
Reynolds, W. S., Schwarz, J. A., & Weis, V. M. (2000). Symbiosis-enhanced gene expression in cnidarian-algal associations: Cloning and characterization of a cDNA, sym32, encoding a possible cell adhesion protein. *Comparative Biochemistry and Physiology Part A Molecular and Integrative Physiology, 126A*, 33–44.

Richier, S., Rodriguez-Lanetty, M., Schnitzler, C. E., & Weis, V. M. (2008). Response of the symbiotic cnidarian Anthopleura elegantissima transcriptome to temperature and UV increase. *Comparative Biochemistry and Physiology Part D Genomics & Proteomics, 3*, 283–289.

Rohwer, F., Seguritan, V., Azam, F., & Knowlton, N. (2002). Diversity and distribution of coral-associated bacteria. *Marine Ecology Progress Series, 243*, 1–10.

Rowan, R. (1998). Diversity and ecology of zooxanthellae on coral reefs. *Journal of Phycology, 34*, 407–417.

Rowan, R., & Powers, D. A. (1992). Ribosomal RNA sequences and the diversity of symbiotic dinoflagellates zooxanthellae. *Proceedings of the National Academy of Sciences of the United States of America, 89*, 3639–3643.

Schnitzler, C. E., & Weis, V. M. (2010). Coral larvae exhibit few measurable transcriptional changes during the onset of coral-dinoflagellate endosymbiosis. *Marine Genomics, 3*, 107–116.

Schwarz, J. A., Brokstein, P. B., Voolstra, C., Terry, A. Y., Miller, D. J., Szmant, A. M., et al. (2008). Coral life history and symbiosis: Functional genomic resources for two reef building Caribbean corals, Acropora palmata and Montastraea faveolata. *BMC Genomics, 9*, 97.

Schwarz, J. A., & Weis, V. M. (2003). Localization of a symbiosis-related protein, sym32, in the Anthopleura elegantissima-Symbiodinium muscatinei association. *Biology Bulletin, 205*, 339–350.

Seneca, F. O., Foret, S., Ball, E. E., Smith-Keune, C., Miller, D. J., & Van Oppen, M. J. H. (2010). Patterns of gene expression in a scleractinian coral undergoing natural bleaching. *Marine Biotechnology (New York), 12*, 594–604.

Shick, J. M. (2004). The continuity and intensity of ultraviolet irradiation affect the kinetics of biosynthesis, accumulation, and conversion of mycosporine-like amino acids (MAAS) in the coral *Stylophora pistillata*. *Limnology and Oceanography, 49*, 442–458.

Shick, J. M., Lesser, M. P., Dunlap, W. C., Stochaj, W. R., Chalker, B. E., & Won, J. W. (1995). Depth-dependent responses to solar ultraviolet radiation and oxidative stress in the zooxanthellate coral *Acropora microphthalma*. *Marine Biology, 122*, 41–51.

Shinzato, C., Shoguchi, E., Kawashima, T., Hamada, M., Hisata, K., Tanaka, M., et al. (2011). Using the *Acropora digitifera* genome to understand coral responses to environmental change. *Nature (London), 476*, 320.

Stat, M., Morris, E., & Gates, R. D. (2008). Functional diversity in coral-dinoflagellate symbiosis. *Proceedings of the National Academy of Sciences of the United States of America, 105*, 9256–9261.

Suescún-Bolivar Luis Parmenio; Iglesias-Prieto, Roberto and Thome, Patricia Elena. in press. Induction of Glycerol Synthesis and Release in Cultured Symbiodinium. PLoS ONE in press.

Takahashi, S., Whitney, S., Itoh, S., Maruyama, T., & Badger, M. (2008). Heat stress causes inhibition of the de novo synthesis of antenna proteins and photobleaching in cultured Symbiodinium. *Proceedings of the National Academy of Sciences of the United States of America, 105*, 4203–4208.

Tchernov, D., Kvitt, H., Haramaty, L., Bibby, T. S., Gorbunov, M. Y., Rosenfeld, H., et al. (2011). Apoptosis and the selective survival of host animals following thermal bleaching in zooxanthellate corals. *Proceedings of the National Academy of Sciences of the United States of America, 108*, 9905–9909.

Trench, R. K. (1971a). The physiology and biochemistry of zooxanthellae symbiotic with marine coelenterates part 1 the assimilation of photosynthetic products of zooxanthellae by 2 marine coelenterates. *Proceedings of the Royal Society of London Series B Biological Sciences, 177*, 225–235.

Trench, R. K. (1971b). The physiology and biochemistry of zooxanthellae symbiotic with marine coelenterates part 2 liberation of fixed carbon-14 by zooxanthellae in-vitro. *Proceedings of the Royal Society of London Series B Biological Sciences, 177*, 237–250.
Trench, R. K. (1971c). The physiology and biochemistry of zooxanthellae symbiotic with marine coelenterates part 3 the effect of homogenates of host tissues on the excretion of photosynthetic products in-vitro by zooxanthellae from 2 marine coelenterates. *Proceedings of the Royal Society of London Series B Biological Sciences, 177*, 251–264.
Vidal-Dupiol, J., Adjeroud, M., Roger, E., Foure, L., Duval, D., Mone, Y., et al. (2009). Coral bleaching under thermal stress: putative involvement of host/symbiont recognition mechanisms. *BMC Physiology, 9*, 14.
Voolstra, C. R., Schwarz, J. A., Schnetzer, J., Sunagawa, S., Desalvo, M. K., Szmant, A. M., et al. (2009a). The host transcriptome remains unaltered during the establishment of coral-algal symbioses. *Molecular Ecology, 18*, 1823–1833.
Voolstra, C. R., Sunagawa, S., Matz, M. V., Bayer, T., Aranda, M., Buschiazzo, E., et al. (2011). Rapid evolution of coral proteins responsible for interaction with the environment. *PLoS One, 6*, e20392.
Voolstra, C. R., Sunagawa, S., Schwarz, J. A., Coffroth, M. A., Yellowlees, D., Leggat, W., et al. (2009b). Evolutionary analysis of orthologous cDNA sequences from cultured and symbiotic dinoflagellate symbionts of reef-building corals (Dinophyceae: Symbiodinium). Comparative Biochemistry and Physiology Part D. *Genomics & Proteomics, 4*, 67–74.
Wang, J., & Douglas, A. E. (1998). Nitrogen recycling or nitrogen conservation in an alga-invertebrate symbiosis? *Journal of Experimental Biology, 201*, 2445–2453.
Wang, J. T., & Douglas, A. E. (1999). Essential amino acid synthesis and nitrogen recycling in an alga-invertebrate symbiosis. *Marine Biology, 135*, 219–222.
Weis, V. M. (1993). Effect of dissolved inorganic carbon concentration on the photosynthesis of the symbiotic sea anemone Aiptasia pulchella Carlgren: Role of carbonic anhydrase. *J Exp Mar Biol Ecol, 174*, 209–225.
Weis, V. M., & Reynolds, W. S. (1999). Carbonic anhydrase expression and synthesis in the sea anemone Anthopleura elegantissima are enhanced by the presence of dinoflagellate symbionts. *Physiological and Biochemical Zoology, 72*, 307–316.
Weis, V. M., Smith, G. J., & Muscatine, L. (1989). A "CO2 supply" mechanism in zooxanthellate cnidarians: role of carbonic anhydrase. *Mar Biol, 100*, 195–202.
Wood-Charlson, E. M., Hollingsworth, L. L., Krupp, D. A., & Weis, V. M. (2006). Lectin/glycan interactions play a role in recognition in a coral/dinoflagellate symbiosis. *Cellular Microbiology, 8*, 1985–1993.
Yakovleva, I., & Hidaka, M. (2004). Diel fluctuations of mycosporine-like amino acids in shallow-water scleractinian corals. *Marine Biology, 145*, 863–873.
Yellowlees, D., Rees, T. A. V., & Leggat, W. (2008). Metabolic interactions between algal symbionts and invertebrate hosts. *Plant Cell and Environment, 31*, 679–694.
Desalvo, M. K., Voolstra, C. R., Sunagawa, S., Schwarz, J. A., Stillman, J. H., Coffroth, M. A., et al. (2008). Differential gene expression during thermal stress and bleaching in the Caribbean coral *Montastraea faveolata*. *Molecular Ecology, 17*, 3952–3971.

CHAPTER FIVE

The *Ectocarpus* Genome and Brown Algal Genomics

The Ectocarpus Genome Consortium

J. Mark Cock[*,†,1], Lieven Sterck[‡,§], Sophia Ahmed[*,†], Andrew E. Allen[¶], Grigoris Amoutzias[‡,§,2], Veronique Anthouard[‖], François Artiguenave[‖], Alok Arun[*,†], Jean-Marc Aury[‖], Jonathan H. Badger[¶], Bank Beszteri[#,3], Kenny Billiau[‡,§], Eric Bonnet[‡,§,4], John H. Bothwell[**,††,‡‡], Chris Bowler[§§,¶¶], Catherine Boyen[*,†], Colin Brownlee[‡‡], Carl J. Carrano[‖‖], Bénédicte Charrier[*,†], Ga Youn Cho[*,†], Susana M. Coelho[*,†], Jonas Collén[*,†], Gildas Le Corguillé[##], Erwan Corre[##], Laurence Dartevelle[*,†], Corinne Da Silva[‖], Ludovic Delage[*,†], Nicolas Delaroque[***], Simon M. Dittami[*,†], Sylvie Doulbeau[†††], Marek Elias[‡‡‡], Garry Farnham[‡‡], Claire M.M. Gachon[§§§], Olivier Godfroy[*,†], Bernhard Gschloessl[*,†], Svenja Heesch[*,†], Kamel Jabbari[‖], Claire Jubin[‖], Hiroshi Kawai[¶¶¶], Kei Kimura[‖‖‖], Bernard Kloareg[*,†], Frithjof C. Küpper[§§§,5], Daniel Lang[###], Aude Le Bail[*,†], Rémy Luthringer[*,†], Catherine Leblanc[*,†], Patrice Lerouge[****], Martin Lohr[††††], Pascal J. Lopez[§§,6], Nicolas Macaisne[*,†], Cindy Martens[‡,§], Florian Maumus[§§], Gurvan Michel[*,†], Diego Miranda-Saavedra[‡‡‡‡], Julia Morales[§§§§,¶¶¶¶], Hervé Moreau[‖‖‖‖], Taizo Motomura[‖‖‖], Chikako Nagasato[‖‖‖], Carolyn A. Napoli[####], David R. Nelson[*****], Pi Nyvall-Collén[*,†], Akira F. Peters[*,†,7], Cyril Pommier[†††††], Philippe Potin[*,†], Julie Poulain[‖], Hadi Quesneville[†††††], Betsy Read[‡‡‡‡‡], Stefan A. Rensing[###], Andrés Ritter[*,†,§§§§§], Sylvie Rousvoal[*,†], Manoj Samanta[¶¶¶¶¶], Gaelle Samson[‖], Declan C. Schroeder[‡‡], Delphine Scornet[*,†], Béatrice Ségurens[‖], Martina Strittmatter[§§§], Thierry Tonon[*,†], James W. Tregear[†††], Klaus Valentin[#], Peter Von Dassow[‖‖‖‖‖], Takahiro Yamagishi[¶¶¶], Pierre Rouzé[‡,§], Yves Van de Peer[‡,§], Patrick Wincker[‖]

[*]UPMC Université Paris 6, The Marine Plants and Biomolecules Laboratory, UMR 7139, Station Biologique de Roscoff, Place Georges Teissier, BP74, 29682 Roscoff Cedex, France

ISSN 0065-2296,
http://dx.doi.org/10.1016/B978-0-12-391499-6.00005-0

[†]CNRS, UMR 7139, Laboratoire International Associé Dispersal and Adaptation in Marine Species, Station Biologique de Roscoff, Place Georges Teissier, BP74, 29682 Roscoff Cedex, France
[‡]Department of Plant Systems Biology, VIB, 9052 Ghent, Belgium
[§]Department of Plant Biotechnology and Bioinformatics, Ghent University, 9052 Ghent, Belgium
[¶]J. Craig Venter Institute, San Diego, California 92121, USA
[‖]CEA, DSV, Institut de Génomique, Génoscope, 2 rue Gaston Crémieux, CP5706, 91057 Evry, France
[#]Alfred Wegener Institute for Polar and Marine Research, Am Handelshafen 12, 27570 Bremerhaven, Germany
[**]School of Biological Sciences, Queen's University Belfast, Belfast BT9 7BL, United Kingdom
[††]Queen's University Marine Laboratory, Portaferry, Co. Down BT22 1PF, United Kingdom
[‡‡]Marine Biological Association of the United Kingdom, The Laboratory, Citadel Hill, Plymouth PL1 2PB, United Kingdom
[§§]Institut de Biologie de l'Ecole Normale Supérieure (IBENS), Centre National de la Recherche Scientifique UMR8197, Ecole Normale Supérieure, 75005 Paris, France
[¶¶]Stazione Zoologica, Villa Comunale, I 80121 Naples, Italy
[‖‖]San Diego State University, San Diego, California 92182-1030, USA
[##]Computer and Genomics Resource Centre, FR 2424, Station Biologique de Roscoff, Place Georges Teissier, BP74, 29682 Roscoff Cedex, France
[***]Fraunhofer Institute for Cell Therapy and Immunology IZI, Perlickstrasse 1, 04103 Leipzig, Germany
[†††]IRD, IRD/CIRAD Palm Developmental Biology Group, UMR 1097 DIAPC, 34394 Montpellier, France
[‡‡‡]Department of Biology and Ecology, Faculty of Science, University of Ostrava, Chittussiho 10, 710 00 Ostrava, Czech Republic
[§§§]Scottish Association for Marine Science, Culture Collection for Algae and Protozoa, Scottish Marine Institute, Oban PA37 1QA, United Kingdom
[¶¶¶]Kobe University Research Center for Inland Seas, 1-1, Rokkodai, Nadaku, Kobe 657-8501, Japan
[‖‖‖]Muroran Marine Station, Field Science Center for Northern Biosphere, Hokkaido University, Muroran 051-0003, Hokkaido, Japan
[###]Faculty of Biology, University of Freiburg, Hauptstr. 1, 79104 Freiburg, Germany
[****]Laboratoire Glyco-MEV EA 4358, IFRMP 23, Université de Rouen, 76821 Mont-Saint-Aignan, France
[††††]Institut für Allgemeine Botanik, Johannes Gutenberg-Universität Mainz, 55099 Mainz, Germany
[‡‡‡‡]World Premier International (WPI) Immunology Frontier Research Center (IFReC), Osaka University, 3-1 Yamadaoka, Suita, 565-0871 Osaka, Japan
[§§§§]UPMC Université Paris 6, UMR 7150 Mer & Santé, Equipe Traduction Cycle Cellulaire et Développement, Station Biologique de Roscoff, 29680 Roscoff, France
[¶¶¶¶]CNRS, UMR 7150 Mer & Santé, Station Biologique de Roscoff, 29680 Roscoff, France
[‖‖‖‖]Laboratoire ARAGO, BP44, 66651 Banyuls-sur-mer, France
[####]Bio5 Institute and Department of Plant Sciences, University of Arizona, Tucson, Arizona 85719, USA
[*****]Department of Molecular Sciences, University of Tennessee Health Science Center, Suite G01, Memphis, Tennessee 38163, USA
[†††††]Unité de Recherches en Génomique-Info (UR INRA 1164), INRA, Centre de recherche de Versailles, bat.18, RD10, Route de Saint Cyr, 78026 Versailles Cedex, France
[‡‡‡‡‡]Biological Sciences, California State University, San Marcos, California 92096-0001, USA
[§§§§§]Departamento de Ecología, Center for Advanced Studies in Ecology & Biodiversity, Facultad de Ciencias Biológicas, Pontificia Universidad Católica de Chile, Santiago, Chile
[¶¶¶¶¶]Systemix Institute, Redmond, Washington 98053, USA
[‖‖‖‖‖]CNRS, UMR 7144, Evolution du Plancton et PaleOceans, Station Biologique de Roscoff, Place Georges Teissier, BP74, 29682 Roscoff Cedex, France
[1]Corresponding author: Email: cock@sb-roscoff.fr
[2]Department of Biochemistry and Biotechnology, University of Thessaly, Larisa, 41221, Greece.
[3]Present addresses: Department of Microbiology, Oregon State University, Corvallis, Oregon 97331, USA.
[4]Present addresses: Institut Curie, 26 rue d'Ulm, Paris, F-75248 France; INSERM, U900, Paris, F-75248 France; and Mines ParisTech, Fontainebleau, F-77300 France.
[5]Present addresses: Oceanlab, University of Aberdeen, Main Street, Newburgh AB41 6 AA, Scotland, UK.
[6]Present addresses: CNRS UMR 7208 BOREA, Muséum National d'Histoire Naturelle, 75005, Paris.
[7]Present addresses: Bezhin Rosko, 40 Rue des pêcheurs, 29250 Santec, France.

Contents

Abstract

Brown algae are important organisms both because of their key ecological roles in coastal ecosystems and because of the remarkable biological features that they have acquired during their unusual evolutionary history. The recent sequencing of the complete genome of the filamentous brown alga *Ectocarpus* has provided unprecedented access to the molecular processes that underlie brown algal biology. Analysis of the genome sequence, which exhibits several unusual structural features, identified genes that are predicted to play key roles in several aspects of brown algal metabolism, in the construction of the multicellular bodyplan and in resistance to biotic and abiotic stresses. Information from the genome sequence is currently being used in combination with other genomic, genetic and biochemical tools to further investigate these and other aspects of brown algal biology at the molecular level. Here, we review some of the major discoveries that emerged from the analysis of the *Ectocarpus* genome sequence, with a particular focus on the unusual genome structure, inferences about brown algal evolution and novel aspects of brown algal metabolism.

1. INTRODUCTION

1.1. The Brown Algae

Brown algae (or Phaeophyceae) are a group of multicellular algae that belong to the stramenopile lineage (also known as heterokonts). They occur almost exclusively in marine environments, particularly rocky coastlines in temperate regions of the globe. Brown algae are often the main primary producers of such ecosystems and therefore play an important ecological role, creating habitats for a broad range of other marine organisms. As a consequence, there has been considerable interest in understanding the biology and ecology of the brown algae. These organisms have also attracted interest for a number of other reasons. The stramenopiles are very distantly related to well-studied groups such as the opisthokonts (animals and fungi) and the green lineage (which includes land plants); the common ancestor of these major lineages dating back to the crown radiation of the eukaryotes more than a billion years ago (Yoon, Hackett, Ciniglia, Pinto & Bhattacharya, 2004). During this long period of evolutionary time, the brown algae have evolved many unusual characteristics that are not found in the other groups, including a number of features that have exquisitely adapted these organisms for the harsh environment of the intertidal and subtidal zones. Brown algae exhibit novel features even at the basic level of their cell biology. For example, they acquired their plastid via a process of secondary endosymbiosis involving the capture of a red alga (Archibald, 2012, in this volume; Keeling, 2004), an event that had a major consequence both on the ultrastructure of the cell and on the composition of the nuclear genome (as a result of gene transfers from the endosymbiont). Brown algae are also remarkable in that they are one of only a small number of eukaryotic groups to have evolved complex multicellularity (Cock *et al.*, 2010).

1.2. *Ectocarpus*, a Model Organism For The Brown Algae

Over the past two decades, the adoption of genomic approaches such as genome sequencing and efficient methods to analyse gene function has allowed remarkable progress in our understanding of the biology of selected model organisms in the animal, plant and fungal lineages. During this time, it became clear that it would be necessary to select an analogous model organism for the brown algae if similar approaches were to be applied to this group. In 2004, several potential brown algal model species were compared, leading to a proposition to develop genomic and genetic tools and techniques for the filamentous brown alga *Ectocarpus* (Fig. 5.1; Peters, Marie, Scornet, Kloareg & Cock, 2004). *Ectocarpus* was selected because mature thalli are small, highly fertile and progress rapidly through the life cycle (Müller, Kapp & Knippers, 1998), characteristics that are essential for the application of genetic approaches. It had also been shown that basic genetic

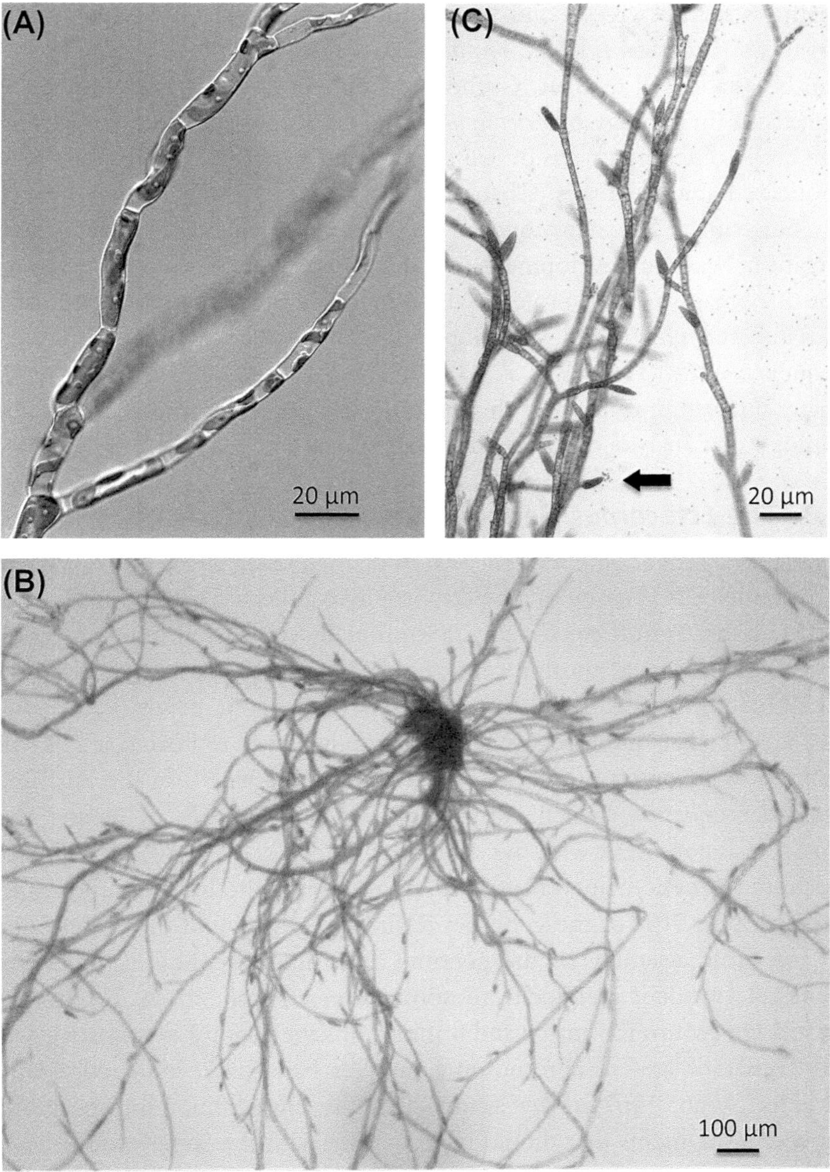

Figure 5.1 The filamentous brown alga *Ectocarpus*. Photographs are of the strain Ec32, which was used to obtain the complete genome sequence. (A) partheno-sporophyte filaments, (B) gametophyte filaments bearing plurilocular gametangia, (C) mature gametophyte filaments releasing gametes from plurilocular gametangia (arrow). See the colour plate.

methods such as crosses and segregation analysis could be used with this organism (Bräutigam, Klein, Knippers & Müller, 1995) and axenic cultures can be established (Müller, Gachon & Küpper, 2008). In addition, several aspects of the biology of *Ectocarpus* had been studied, including taxonomy, life cycle, different aspects of cell biology and metabolism and responses to biotic and abiotic stresses (Charrier *et al.*, 2008).

One of the key steps towards the emergence of *Ectocarpus* as a model organism was the development of a complete genome sequencing project for this organism. This project, which involved more than 30 laboratories, was initiated in 2006 and was completed with the publication of the genome sequence in 2010 (Cock *et al.*, 2010). The following sections describe the many interesting features of this genome and the insights into brown algal biology that analysis of the genome has afforded.

1.3. The *Ectocarpus* Genome Project

The *Ectocarpus* genome sequence was obtained using strain Ec32, which is a male meiotic offspring of a field sporophyte collected in 1988 in San Juan de Marcona, Peru (Peters *et al.*, 2008). Initially considered to belong to the species *E. siliculosus*, more recent phylogenetic analyses suggest that strain Ec32 belongs to a so far unnamed species of the same genus (Peters *et al.*, 2010a). The size of the genome in this strain had been estimated at 214 Mbp using flow cytometry (Peters *et al.*, 2004) and the length of the assembled genome sequence, which comprised 1561 supercontigs of greater than 2 kbp, was consistent with this estimation (Cock *et al.*, 2010). A sequence-anchored genetic map was used to assign 325 of the longest supercontigs (137 Mbp or 70% of the genome) to linkage groups and thereby produce a large-scale assembly of the genome sequence by concatenating supercontigs to produce pseudochromosomes (Heesch *et al.*, 2010). The number of linkage groups (26 major and 8 minor linkage groups) is consistent with previous estimates of chromosome number based on cytogenetic studies (Müller, 1966, 1967) if we assume that the eight minor linkage groups represent fragments that should be associated with the larger groups.

2. STRUCTURE OF THE *ECTOCARPUS* GENOME

2.1. Large-scale Structure

Analysis of gene and transposon density along the *Ectocarpus* pseudochromosome sequences did not reveal any obvious large-scale structures that could have represented centromeres or heterochromatic knobs, although it is possible that such regions were lost during the assembly stage if they are very rich in repeated sequences. Linkage group 30 was particularly rich in transposons and poor in genes compared to the other major linkage groups. No evidence was found for large-scale duplication events such as genome duplications. This is unusual for an organism that belongs to a group that has evolved complex multicellularity.

2.2. Gene Structure and Gene Organization

Ectocarpus genes contain many long introns (seven per gene on average, with an average size of 704 bp) and, in consequence, introns make up an exceptionally large percentage of the genome (40.4%). Only 5.3% of the predicted genes lack introns completely; this is the smallest fraction for any eukaryotic genome reported to date. The 3' untranslated regions are also very long for a genome of this size (average size: 855 bp). Mouse genes have a comparable mean 3'UTR length, despite the fact that the mouse genome is more than 13 times larger. It is possible that the long *Ectocarpus* 3'UTRs contain regulatory elements in which case messenger RNA (mRNA)s bearing different 3' regulatory signals could be generated by the use of alternative polyadenylation sites. However, the frequency of the use of alternative polyadenylation sites did not appear to be particularly elevated in *Ectocarpus* (Cock *et al.*, 2010).

The features described in the previous paragraph are typical of large expanded genomes, but the *Ectocarpus* genome also exhibits a number of features more typical of small compact genomes. For example, a significant proportion (61.5%) of the 16,256 predicted protein-coding genes is arranged in an alternating manner along the chromosome, so that adjacent genes are on opposite strands (Fig. 5.2; Cock *et al.*, 2010). The proportion

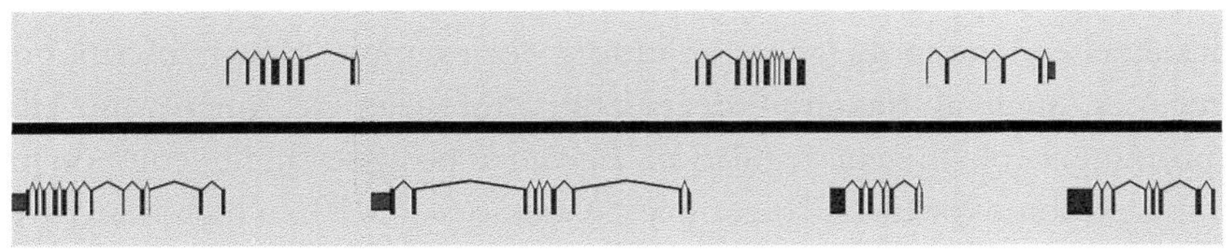

Figure 5.2 Diagram showing the alternating pattern of gene organization typical of many regions of the *Ectocarpus* genome. Coding exons are shown as dark blue bars and untranslated regions as light blue bars. Lines joining exons represent introns. Genes above the black line are transcribed from left to right and genes below the black line are transcribed from right to left. See the colour plate.

of alternating genes is comparable with that found in highly compact genomes such as those of *Ostreococcus tauri* (12.6 Mbp, 61.5% alternating genes) and *Phaeodactylum tricornutum* (27.4 Mbp, 64.2% alternating genes), while the percentage of alternating genes in larger genomes normally tends to decrease as a function of genome size (Cock *et al.*, 2010; note that 50% alternating genes is expected if the genes are randomly organized in the genome). One factor that may contribute to the high number of alternating genes is the relative rarity of tandem duplications in *Ectocarpus* because tandem duplications will tend to disorganize arrays of alternating genes. Only 823 of the 16,256 *Ectocarpus* genes are part of tandem duplications. It is not clear whether the organization of many *Ectocarpus* genes into alternating arrays is important for gene function. In yeast, adjacent genes tend to be co-regulated (Kruglyak & Tang, 2000) and this effect is more marked for divergently expressed genes than for either convergently expressed or same-strand gene pairs (Kensche, Oti, Dutilh, & Huynen, 2008; Trinklein *et al.*, 2004). An analysis of stress-response data indicated that adjacent genes also showed a greater degree of co-regulation than non-adjacent genes in *Ectocarpus*, but this effect was not dependent on the relative orientation of the genes (Cock *et al.*, 2010). This is consistent with the fact that there is almost exactly the same number of divergently and convergently transcribed gene pairs in the genome (4078 and 4076 pairs, respectively).

Another feature typical of compact genomes is that the intergenic regions between the 5' ends of divergently transcribed genes (i.e. genes on opposite strands with adjacent 5' ends) are often very short in *Ectocarpus* (29% are less than 400 bp long). This feature may help stabilize arrays of alternating genes by grouping the genes closely together (Hurst, Williams, & Pál, 2002). Another consequence of short intergenic regions is that either promoter regions must be very compact or regulatory elements must be located elsewhere in the genes, in introns, for example. Post-transcriptional processes, such as transcript degradation, for example, might also play a prominent role in gene regulation. *Ectocarpus* possesses all the components of the exosome (or PM/Scl) complex (Cock *et al.*, 2010), which is involved in the degradation not only of mRNA but also ribosomal RNA (rRNA) and many species of small RNA (Belostotsky, 2009). Gene expression may also be regulated at other steps, such as intron splicing (this may be particularly important given the intron-rich nature of the *Ectocarpus* genes) and mRNA translation. Detailed annotations of the genes involved in these processes in *Ectocarpus* have been carried out (Cock *et al.*, 2010).

The large number of introns in the *Ectocarpus* genome could potentially allow the production of multiple transcripts from individual genes. A genome-wide analysis of alternative transcripts based on 91,041 expressed sequence tag (EST) sequences indicated that only a small proportion of genes (less than 3%) produced alternative transcripts, but it is likely that this EST data set was insufficient to exhaustively describe this phenomenon in *Ectocarpus*, and deeper sequencing of the trancriptome is expected to allow the identification of additional alternative transcripts in the future.

2.3. Non-coding RNAs

In addition to mRNA transcripts, *Ectocarpus* cells produce many endogenous small RNAs. An analysis of more than seven million small RNA sequences from sporophyte and gametophyte tissues identified 24,132 unique small RNA sequences, which mapped to 1,031,522 loci in the *Ectocarpus* genome (Cock *et al.*, 2010). For small RNAs that mapped to intergenic regions, transposons, introns or exons, the largest size group in each case was 21 nucleotides, indicating that a proportion of these molecules may have specific functions that impose a size constraint. Small RNAs originating from rRNA and transfer RNA (tRNA), on the other hand, did not exhibit this size bias, suggesting that they may simply be degradation products of longer transcripts.

Micro RNAs (miRNAs) are small (21–25 nucleotide) RNAs generated by Dicer enzymes (ribonucleases of the RNAseIII family) by processing imperfect stem-loop structures in longer RNA transcripts. miRNAs are incorporated into silencing complexes, which include Argonaute proteins, allowing these complexes to target-specific nucleic acid sequences in the cell. miRNAs have been shown to have important regulatory roles in many cellular and developmental pathways in both green plants and animals (Carthew & Sontheimer, 2009; Voinnet, 2009). *Ectocarpus* also appears to employ this system. Using a set of stringent rules (Meyers *et al.*, 2008), 26 miRNA sequences were identified in the *Ectocarpus* genome (Cock *et al.*, 2010). In addition, *Ectocarpus* possesses both a Dicer and an Argonaute gene, and there are two RNA-dependent RNA polymerase (RdRP) homologues, which may have a role in the amplification of double-stranded RNA molecules. The small RNA machinery in *Ectocarpus* seems to be rather simple compared to other eukaryotes; many organisms possess multiple copies of Dicer, Argonaute and RdRP proteins (Carthew & Sontheimer, 2009; Voinnet, 2009). This diversification has been shown to be linked to functional specialization in many cases, for example, between small interfering RNA and miRNA silencing. miRNAs regulate their targets in the cell by binding to regions of partial or complete base complementarity. A total of 71 potential target sequences were identified for 12 of the 26 *Ectocarpus* miRNAs. Surprisingly, three quarters of these targets contain leucine-rich repeat (LRR) domains, including 28 members of a large Ras of complex proteins (ROCO) GTPase family and 5 tetratricopeptide repeat (TPR) -containing proteins (see Table 5.1 for an updated list of predicted *Ectocarpus* miRNA targets). LRR proteins are involved in recognition and transduction events linked to immunity in both plants and animals (Kumar, Kawai, & Akira, 2011; Meyers, Kaushik, & Nandety, 2005) and the *Ectocarpus* LRR ROCO genes may be involved in similar processes, with exon shuffling allowing an adaptive immune response (Zambounis, Elias, Sterck, Maumus & Gachon, 2012). miRNAs may regulate this important class of molecule, which would be consistent with recent results showing that miRNAs act as master regulators of NB-LRR genes in land plants (Zhai *et al.*, 2011). The discovery of miRNAs in *Ectocarpus* taken together with the miRNAs previously described in animals, green plants and slime moulds (Hinas *et al.*, 2007), indicated an ancient origin for these important regulatory molecules. More recently, miRNAs have also been described in the diatom *P. tricornutum* (Huang, He & Wang, 2011).

A whole genome tiling array approach identified 8,741 expressed regions longer than 200 nucleotides located outside the 16,256 predicted genes. Many of these regions represent potential candidates for genes encoding non-coding RNAs (Cock *et al.*, 2010). A large proportion of the expressed regions correspond to repeated elements and the vast majority (8706) are not conserved in the *Thalassiosira pseudonana* genome, suggesting that they have originated since the divergence from diatoms.

2.4. Repeated Sequences

Repeated sequences make up a significant proportion of the *Ectocarpus* genome (22.7%), dominated by transposable elements (TEs) and unclassified repeats that represent about 12.5 and 9.9% of the genome, respectively (Table 5.2). The TEs are mainly retrotransposons, including LTR retrotransposons (such as Ty1/copia, Ty3/gypsy and DIRS/Ngaro-like elements), TRIM/LARD-like elements and non-LTR retrotransposons, as well as DNA transposons of both subclass I (such as Harbinger, JERKY and POGO-like elements) and subclass II (Helitrons) (Cock *et al.*, 2010). The

Table 5.1 *Ectocrpus* Genes Predicted to be MicroRNA Targets

miRBase miRNA ID Number*	Target Gene Locus ID	Target Gene Domains	Target Gene Description
3463	Esi0041_0132	Kinesin	AGAP010519-PA
3453	Esi0010_0186	WSC	Conserved hypothetical protein
3453	Esi0057_0062		Conserved hypothetical protein
3457	Esi0012_0076	WSC	Conserved hypothetical protein
3468	Esi0380_0014		Conserved hypothetical protein
3454a, 3454b, 3454c, 3454e	Esi0062_0049		Conserved hypothetical protein
3454a, 3454b, 3454c	Esi0033_0024		Conserved hypothetical protein
3454d, 3454f	Esi0046_0125		Hypothetical aspartate carbamoyltransferase
3452	Esi0269_0025	LRR	Hypothetical LRR protein
3454a, 3454b, 3454c	Esi0106_0041	LRR	Hypothetical LRR protein
3454a, 3454b, 3454c	Esi0106_0045	LRR	Hypothetical LRR protein
3454a, 3454b, 3454c	Esi0106_0087	LRR	Hypothetical LRR protein
3454a, 3454b, 3454c	Esi0165_0070	LRR	Hypothetical LRR Protein
3454a, 3454b, 3454c	Esi0191_0069	LRR	Hypothetical LRR protein
3454a, 3454b, 3454c	Esi0450_0002	LRR	Hypothetical LRR protein
3454a, 3454b, 3454c, 3454d, 3454e	Esi0106_0084	LRR	Hypothetical LRR protein
3454a, 3454b, 3454c, 3454d, 3454e, 3454f	Esi0015_0089	LRR	Hypothetical LRR protein
3454a, 3454b, 3454c, 3454d, 3454e, 3454f	Esi0015_0097	LRR	Hypothetical LRR protein
3454a, 3454b, 3454c, 3454d, 3454f	Esi0015_0090	LRR	Hypothetical LRR protein
3454a, 3454b, 3454c, 3454d, 3454f	Esi0191_0070	LRR	Hypothetical LRR protein
3454a, 3454b, 3454c, 3454e	Esi0055_0135	LRR	Hypothetical LRR protein
3454a, 3454b, 3454c, 3454e	Esi0191_0006	LRR	Hypothetical LRR protein
3454a, 3454b, 3454c, 3454e	Esi0191_0017	LRR	Hypothetical LRR protein
3454a, 3454b, 3454c, 3454e	Esi0236_0019	LRR	Hypothetical LRR protein

(Continued)

Table 5.1 *Ectocrpus* Genes Predicted to be MicroRNA Targets—cont'd

miRBase miRNA ID Number*	Target Gene Locus ID	Target Gene Domains	Target Gene Description
3454d, 3454f	Esi0328_0036	LRR	Hypothetical LRR protein
3454e	Esi0055_0077	LRR	Hypothetical LRR protein
3454e	Esi0106_0039	LRR	Hypothetical LRR Protein
3454e	Esi0200_0010	LRR	Hypothetical LRR protein
3454g	Esi0029_0096	LRR	Hypothetical LRR protein
3454e	Esi0085_0019	LRR	Hypothetical LRR protein
3454a, 3454b, 3454c, 3454e	Esi0144_0013		Hypothetical potassium transporter
3457	Esi0198_0016		Hypothetical protein
3452	Esi0269_0037		Likely pseudogene
3452	Esi0269_0028	LRR	LRR protein
3454d, 3454f	Esi0088_0028	LRR	LRR protein
3452	Esi0269_0017	LRR	LRR protein
3454a, 3454b, 3454c, 3454d, 3454e, 3454f	Esi0011_0207	LRR	LRR-GTPase of the ROCO family
3454a, 3454b, 3454c, 3454d, 3454e, 3454f	Esi0138_0012	LRR	LRR-GTPase of the ROCO family
3454a, 3454b, 3454c, 3454d, 3454e, 3454f	Esi0141_0028	LRR	LRR-GTPase of the ROCO family
3454a, 3454b, 3454c, 3454d, 3454f	Esi0031_0015	LRR	LRR-GTPase of the ROCO family
3454a, 3454b, 3454c, 3454d, 3454f	Esi0032_0115	LRR	LRR-GTPase of the ROCO family
3454a, 3454b, 3454c, 3454e	Esi0164_0034	LRR	LRR-GTPase of the ROCO family
3454a, 3454b, 3454c, 3454e	Esi0264_0029	LRR	LRR-GTPase of the ROCO family
3454a, 3454b, 3454c, 3454e	Esi0265_0008	LRR	LRR-GTPase of the ROCO family
3454d, 3454f	Esi0032_0138	LRR	LRR-GTPase of the ROCO family
3454a, 3454b, 3454c	Esi0026_0057	LRR	LRR-GTPase of the ROCO family, putative pseudogene

3454a, 3454b, 3454c	Esi0041_0085	LRR	LRR-GTPase of the ROCO family, putative pseudogene
3454a, 3454b, 3454c	Esi0054_0138	LRR	LRR-GTPase of the ROCO family, putative pseudogene
3454a, 3454b, 3454c	Esi0138_0081	LRR	LRR-GTPase of the ROCO family, putative pseudogene
3454a, 3454b, 3454c, 3454d, 3454e, 3454f	Esi0014_0084	LRR	LRR-GTPase of the ROCO family, putative pseudogene
3454a, 3454b, 3454c, 3454e	Esi0416_0018	LRR	LRR-GTPase of the ROCO family, putative pseudogene
3454d, 3454f	Esi0027_0032	LRR	LRR-GTPase of the ROCO family, putative pseudogene
3454d, 3454f	Esi0112_0069	LRR	LRR-GTPase of the ROCO family, putative pseudogene
3454e	Esi0562_0010	LRR	LRR-GTPase of the ROCO family, putative pseudogene
3457	Esi0008_0050	TPR	NB-ARC and TPR repeat-containig protein
3457	Esi0008_0175	TPR	NB-ARC and TPR repeat-containing protein—likely pseudogene
3457	Esi0380_0021	TPR	NB-ARC and TPR repeat-containing protein—likely pseudogene
3453	Esi0007_0219		Ran-GTPase activating protein
3454a, 3454b, 3454c	Esi0141_0085	LRR	ROCO gene-associated coding fragment
3454a, 3454b, 3454c, 3454d, 3454e, 3454f	Esi0032_0177	LRR	ROCO gene-associated coding fragment
3454a, 3454b, 3454c, 3454d, 3454e, 3454f	Esi0047_0155	LRR	ROCO gene-associated coding fragment
3454a, 3454b, 3454c, 3454d, 3454e, 3454f	Esi0264_0036	LRR	ROCO gene-associated coding fragment
3454a, 3454b, 3454c, 3454d, 3454f	Esi0011_0242	LRR	ROCO gene-associated coding fragment

(Continued)

Table 5.1 *Ectocrpus* Genes Predicted to be MicroRNA Targets—cont'd

miRBase miRNA ID Number*	Target Gene Locus ID	Target Gene Domains	Target Gene Description
3454a, 3454b, 3454c, 3454d, 3454f	Esi0138_0092	LRR	ROCO gene-associated coding fragment
3454a, 3454b, 3454c, 3454e	Esi0032_0139	LRR	ROCO gene-associated coding fragment
3454a, 3454b, 3454c, 3454e	Esi0281_0054	LRR	ROCO gene-associated coding fragment
3454d, 3454f	Esi0014_0085	LRR	ROCO gene-associated coding fragment
3454e	Esi0041_0084	LRR	ROCO gene-associated coding fragment
3463	Esi0036_0127		Similar to CG3714-PA, isoform A
3457	Esi0126_0047	TPR	TPR repeat-containing protein
3457	Esi0274_0033	TPR	TPR repeat-containing protein, putative pseudogene

LRR, leucine-rich repeat; TPR, tetratricopeptide repeat; ROCO, Ras of complex proteins; WSC, Wall and stress response component.
* Add esi-MIR in front of the four digit number to construct the complete miRBase ID number.

Table 5.2 Abundance of Different Classes of Repeated Element in the *Ectocarpus* Genome

Class	Subclass	Category	Coverage (bp)	Genome Coverage (%)
Class 1	LTR-retrotransposon	Ty1/copia	4817138	2.40
		Ty3/gypsy	4348141	2.17
		DIRS/Ngaro	2918206	1.45
		TRIM/LARD	2600089	1.30
	Non LTR-retrotransposon	LINE	4074762	2.03
Class 2	Subclass 1	TIR	2042086	1.02
		TIR putative	228020	0.11
		non-autonomous TIR	2445103	1.22
	Subclass 2	Helitron	891778	0.44
Tandem repeat			721010	0.36
Unclassified repeat			19817249	9.88

LTR; long terminal repeats; TRIM, terminal-repeat retrotransposon in miniature; LARD; large retrotransposon derivative.

most abundant unclassified repeat in the *Ectocarpus* genome is a 676 nucleotide element dubbed Sower that accounts for about 1.5% of the genome.

Analysis of EST data indicated that several TE sequences (particularly Ty1/copia elements) were expressed at unexpectedly high levels in unstressed tissues grown under laboratory conditions (Cock *et al.*, 2010). In green plants and animals, TE silencing is mediated by mechanisms related to RNA interference, which usually involve methylation of DNA in the silenced regions of the genome (Malone & Hannon, 2009). High-performance liquid chromatography (HPLC) analysis of *Ectocarpus* genomic DNA did not detect any methylcytosine (mC) and several TE families were shown to be resistant to digestion with the mC-specific endonuclease McrBC, again indicating that these elements are not methylated (Cock *et al.*, 2010). It is therefore possible that the observed expression of TE sequences in *Ectocarpus* is a result of these elements being only weakly silenced in the absence of a DNA methylation system. Interestingly, genome-wide analysis of the small RNA sequences described above indicated that these sequences were derived preferentially from TE-rich regions of the genome, suggesting that these molecules play a role in TE silencing despite the lack of DNA methylation. In other organisms that lack DNA methylation, such as *Caenorhabditis elegans*, silencing is thought to be mediated by alternative chromatin signals such as histone modifications (Vastenhouw & Plasterk, 2004). It is possible that similar mechanisms operate in *Ectocarpus*.

Recently, an increase in the abundance of transcripts corresponding to several different TEs (especially long interspersed element (LINEs)) was detected during infection by the pathogen *Eurychasma dicksonii*, hinting at mechanisms that regulate TE expression (Grenville-Briggs *et al.*, 2011). This observation suggests that TE-mediated generation of genetic diversity may occur in response to biotic stress. It also confirms that most TEs are repressed under normal laboratory conditions.

2.5. An Integrated Viral Genome

One particularly interesting feature of the *Ectocarpus* nuclear genome sequence was the presence of a single copy of a large viral genome (more than 310 kbp) that had inserted into one of the algal chromosomes (Cock *et al.*, 2010). The inserted viral genome is closely related to the *Ectocarpus* phaeovirus EsV-1, a member of the Phycodnaviridae family, which are icosahedral viruses with internal lipid membranes and large double-stranded DNA genomes (Müller *et al.*, 1998; Wilson, Van Etten, &

Allen, 2009; see also Grimsley *et al.*, 2012, in this volume). Phaeoviruses are pandemic in several brown algal species (Müller *et al.*, 1998) and about half of the individuals in natural *Ectocarpus* populations show symptoms of viral infection (Dixon, Leadbeater, & Wood, 2000; Müller *et al.*, 2000). EsV-1 infects free-swimming zoids (spores or gametes, which lack a cell wall) and the 313 kbp viral genome then integrates into the cellular genome with the result that there is a copy in all the cells of the host as it develops (Bräutigam *et al.*, 1995; Delaroque, Maier, Knippers, & Müller, 1999; Müller, 1991). The virus remains latent in vegetative cells and viral particles are only produced in the reproductive organs (the sporangia and gametangia; Fig. 5.3) following a stimulus such as a change in light, seawater composition or temperature (Müller, Lindauer, Brüderlein & Schmitt, 1990; Müller *et al.*, 1998). Infected algae exhibit no obvious growth or developmental defects other than the partial or total inhibition of reproduction caused by the replacement of zoids by viral particles (Del Campo, Ramazanov, Garcia-Reina & Müller, 1997).

A single copy of the viral genome was found in the *Ectocarpus* genome, almost all located at a single locus on linkage group 16 (although a second, much smaller fragment was found on linkage group 24, possibly due to a translocation event that occurred after viral integration). This is consistent with previous studies in which the integrated EsV-1 was shown to behave as

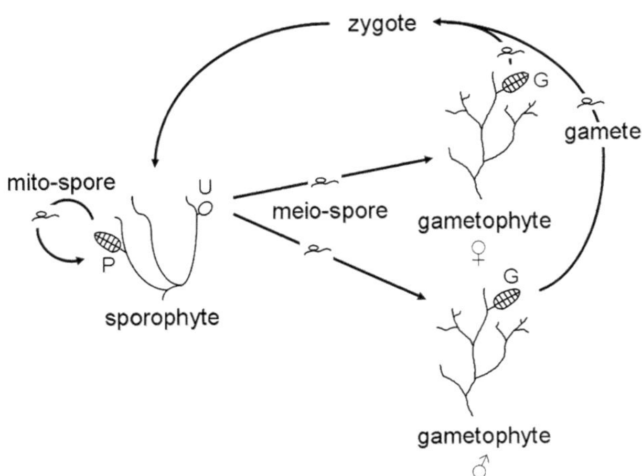

Figure 5.3 Diagram showing the alternation between the sporophyte and gametophyte generations during the life cycle of *Ectocarpus*. U, unilocular sporangium; P, plurilocular sporangium and G, plurilocular gametangium. Meiosis occurs in the unilocular sporangia to produce haploid meiospores, male and female gametes are produced in the plurilocular gametangia on the dioecious gametophytes.

a single, Mendelian locus (Bräutigam *et al.*, 1995; Delaroque *et al.*, 1999; Müller, 1991). The virus appears to have integrated as a circle, as the terminal repeat regions are adjacent in the algal chromosome. Again, this is consistent with previous reports indicating that the ends of the linear viral genome can associate to form a circle (Delaroque *et al.*, 2001).

The integration of such a large segment of foreign DNA, representing more than 1% of the total gene content of the genome, could potentially have had a significant influence on genome function. However, expression analysis showed that almost all the viral genes were silent (Cock *et al.*, 2010). The viral genes remained silent following several different stress treatments (hyperosmotic, hypoosmotic and oxidative stress) and in fertile gametophytes (where viral particles are normally produced in EsV-1-infected strains), indicating that the integrated viral genome is either defective or silenced by the host. These observations were consistent with the fact that strain Ec32 had never been observed to produce virus particles. Interestingly, small RNAs did not map preferentially to the inserted viral sequence (as had been observed for the TE-rich regions of the genome, see above), suggesting that the viral genome had been silenced by a mechanism different to that used to silence transposons. Apart from the inserted viral genome, the *Ectocarpus* genome contained very few genes that were predicted to be of viral origin, which was surprising given the pandemic levels of viral infection observed in the field (Dixon *et al.*, 2000; Müller *et al.*, 2000). *Ectocarpus* therefore may possess a mechanism that limits gene acquisition via this route.

3. EVOLUTIONARY HISTORY OF THE *ECTOCARPUS* GENOME

3.1. Evolution of Genome Gene Content

Several approaches have been used to compare the inventory of proteins encoded by the *Ectocarpus* genome (the proteome) with those of other organisms in an effort to understand the evolutionary history of the genome (Cock *et al.*, 2010). These have included comparisons with complete proteomes from other stramenopiles, blast comparisons with the National Center for Biotechnology Information database and a reconstruction of gene family loss and gain during evolution using Dollo logic (Cock *et al.*, 2010). These analyses have highlighted how different the *Ectocarpus* genome is from previously sequenced genomes. For example, more than a third (35.8%) of the *Ectocarpus* proteins matched none of the proteins in the NCBI nr_prot database, indicating that these are either entirely novel sequences or that they have diverged considerably from homologues in other species. This result underlines not only the potential of the genome as a source of new bioactivities but also the challenges associated with investigating gene function in this organism.

Analysis of gene family expansions can also provide insights into evolutionary forces that have shaped genome content. Expansions in the *Ectocarpus* genome have been particularly marked for genes involved in the cytoskeleton, flagella function, protein degradation and protein phosphorylation/dephosphorylation (http://bioinformatics.psb.ugent.be/dollo_analysis). One specificity of the *Ectocarpus* genome is the expansion of a glycosyl hydrolase family and a fucoxanthin chlorophyll-*a*/*c*-binding protein subfamily underlining the importance of carbohydrate metabolism and photosynthesis in this organism. A similar result was obtained from an analysis of the frequency of protein domains in the *Ectocarpus* genome. Domains involved in carbohydrate binding (IPR002889) and photosynthesis (IPR001344) were particularly abundant. Notch (IPR000800) and ankyrin (IPR002110 and PTHR18958) domain proteins were also significantly overrepresented. These proteins may have important roles in intercellular communication.

3.2. Endosymbiosis and Organellar Genomes

The plastids of almost all eukaryotic algae are thought to be ultimately derived from an endosymbiotic event that occurred early in the evolution of the green (Archaeplastida) lineage before the separation of the glaucophytes from the red and green algae (Reyes-Prieto, Weber & Bhattacharya, 2007; see also De Clerck, Bogaert & Leliaert, 2012, Chapter II of this volume). This event involved the capture of a cyanobacterium by a eukaryotic host cell and enslavement of the former to produce a plastid. The plastids of algae in other major eukaryotic lineages, including the chromalveolate group (to which *Ectocarpus* belongs), were then acquired by secondary, or even tertiary, endosymbiotic events involving the capture of plastid-containing algae (see Archibald, 2012, Chapter III of this volume). Both primary and secondary endosymbiosis events involved large-scale transfer of genes from the endosymbiont to the host nucleus resulting in the acquisition of a broad range of novel functions (Baurain *et al.*, 2010; Bowler *et al.*, 2008; Dorrell & Smith, 2011; Keeling, 2004). *Ectocarpus* genes that were acquired during secondary endosymbiosis can be identified by their phylogenetic affinity with red algal sequences because brown algal plastids were derived from a captured red alga. Phylogenetic analysis identified 611 such red-alga-derived (or 'red') genes in the *Ectocarpus* genome (Cock *et al.*, 2010). These genes are predicted to carry out a broad range of cellular functions indicating that the secondary endosymbiosis event not only influenced photosynthetic processes (31% of the 'red' genes were predicted to be involved in plastid function) but also represented an opportunity to optimize other cellular processes. For example, 34 mitochondrial proteins are encoded by 'red' genes (5.5% of the 'red' genes and 9.5% of proteins predicted to be targeted to the mitochondria in *Ectocarpus*), suggesting that advantageous features of the two ancestral mitochondrial proteomes may have been combined during evolution. Similarly, the 'red' genes included a glutamate/ornithine acetyltransferase and an acetylornithine aminotransferase, suggesting that the urea cycle in brown algae is partly derived from the red algal lineage.

Little is known about organelle function in brown algae at the molecular level, although the genome project has provided tools to address this aspect of brown algal biology. Both the plastid and mitochondrial genomes have been described in *Ectocarpus* (Cock *et al.*, 2010; Le Corguillé *et al.*, 2009). The mitochondrial genome is relatively small (37 kbp) and encodes only a small fraction of the mitochondrial components (3 ribosomal RNA genes, 24 tRNA genes, 35 conserved protein-coding genes and 5 putative (open reading frame (ORFs)). The majority of the mitochondrial proteins are encoded in the nucleus and transported into the mitochondria. The predicted components of this transport machinery have been analysed recently, providing several insights into the evolution of this system during the emergence of the eukaryotes (Delage *et al.*, 2011). The Hectar algorithm (Gschloessl, Guermeur & Cock, 2008) predicts that 605 proteins are encoded in the nucleus and enter the mitochondria via the protein transport system. One particularly interesting protein is the nicotinamide adenine dinucleotide (NADH) dehydrogenase (complex I) iron–sulfur protein subunit encoded by the *nad11* gene. In many protists (including oomycetes and some diatoms), the *nad11* gene is located on the mitochondrial genome, whereas in green plants, animals and fungi, the gene is nuclear. *Ectocarpus* and some other stramenopiles (*Cafeteria, Traustochytrium* and other brown algae; Oudot-Le Secq, Loiseaux-de Goër Stam, & Olsen *et al.*, 2006) possess a truncated version of this gene in the mitochondrion. A nuclear gene encoding the C-terminal part of the nad11 subunit (with a mitochondrial-targeting sequence) was found in the *Ectocarpus* genome, suggesting that the *nad11* gene was split during evolution and one part of the coding region transferred to the nucleus. The diatom *P. tricornutum* may represent an intermediate step in this process as its *nad11* gene has been split into two loci but both components are still present in the mitochondrial genome (Oudot-Le Secq & Green, 2011). The red alga *Cyanidioschyzon merolae* possesses two nuclear genes corresponding to the N-terminal and C-terminal regions of nad11. Taken together, these observations suggest that the transfer of *nad11* from the mitochondrial to the nuclear genome occurred several times during evolution, in some cases involving the sequential transfer of fragments of the gene to create two individual nuclear loci.

Dual targeting of the same protein into both the plastid and the mitochondrion has been described in land plants but the situation is more complicated in stramenopiles because a bipartite N-terminal-targeting sequence (including a signal peptide and a plastid transit peptide) is required for uptake via the four membranes of the plastid. However, at least one potential example of a dual-targeted *Ectocarpus* protein has been described, a GTPase orthologous to the bacterial MnmE (TrmE) involved in tRNA modification (Cock *et al.*, 2010). Sequence analysis suggests that dual targeting may in this case occur by the use of alternative translation initiation sites, but this will need to be confirmed experimentally.

4. INSIGHTS INTO BROWN ALGAL METABOLISM

4.1. Photosynthesis and Photosynthetic Pigments

The *Ectocarpus* light reaction and electron transport system gene complements are very similar to those of green plants, except for the lack of plastocyanin, which is usually replaced by cytochrome c_6 in chromist algae. More chlorophyll-binding protein (CBP) genes were found in the *Ectocarpus* genome than in any green plant genome studied to date (Dittami, Michel, Collen, Boyen & Tonon, 2010), suggesting that *Ectocarpus* employs a complex repertoire of CBP genes in order to cope with the exceptionally dynamic environment of the intertidal and upper subtidal zones. The 53 CBP *Ectocarpus* genes include a family of stress-induced proteins similar to the light-harvesting complex stress-related (LHCSR or LI818) CBPs. LHCSR proteins have been shown to have a photoprotective role and appear to be involved in the xanthophyll cycle-related dissipation of excess energy (Bailleul *et al.*, 2010; Gundermann & Büchel, 2007; Peers *et al.*, 2009; Zhu & Green, 2010). Interestingly, the phylogeny of LHCSR proteins suggests that they may have originated in a chlorophyll-*a*/*c*-containing organism and then been transferred horizontally to the green lineage (Dittami *et al.*, 2010).

Brown algal cells are highly pigmented, with most of the pigments playing a role in photosynthetic processes. The major brown algal pigments are chlorophyll *a*, chlorophylls c_1 and c_2, fucoxanthin, violaxanthin and β-carotene (Bjørnland & Liaaen-Jensen, 1989; Jeffrey, 1976). Some of these molecules, such as the c_1 and c_2 chlorophylls, are only found in chromalveolates and the enzymes that synthesize these pigments are unknown. Other pigments are also found in the green lineage and their biosynthesis has been studied in more detail. The *Ectocarpus* genome contains orthologues of all the genes that have been shown to be involved in the biosynthesis of chlorophyll *a*, β-carotene and violaxanthin in vascular plants and green algae (Cock *et al.*, 2010). Interestingly, the *Ectocarpus* gene encoding subunit CHL27 of the magnesium-protoporphyrin IX monomethyl ester cyclase (*acsF*) is located in the plastid. This has also recently been shown to be the case for the raphidophyte *Heterosigma akashiwo*, the xanthophyte *Vaucheria litorea* and the phaeophyte *Fucus vesiculosus* (Le Corguillé *et al.*, 2009). *CHL27* is located in the nuclear genomes of seed plants and green algae but is absent from both the nuclear and the plastid genomes of diatoms, suggesting that independent loss in the diatom lineage relative to other stramenopiles.

Light-independent (dark) protochlorophyllide oxido reductase (DPOR) allows efficient synthesis of chlorophyll in the dark or under dim light conditions (Shui *et al.*, 2009). The *Ectocarpus* plastid contains three genes (*chlB*, *chlL* and *chlN*) that encode subunits of this complex (Le Corguillé *et al.*, 2009). This is consistent with earlier observations, which indicated that species of the closely related Laminariales synthesize chlorophyll in the dark, allowing arctic species to grow during the winter (Lüning, 1990).

The carotenoid biosynthesis pathway is responsible for the production of several important molecules in chromist algae, including fucoxanthin, which is involved in the harvesting of blue light for photosynthesis and which gives these organisms their brown colouration. Diatoms and haptophytes possess two xanthophyll-based systems, the violaxanthin cycle and the diadinoxanthin cycle, which have a role in dissipating excess light energy in the plastid (Lohr & Wilhelm, 1999). These cycles are part of a biosynthetic pathway that produces fucoxanthin and both cycles are catalysed by the activity of two opposing enzymes, zeaxanthin epoxidase and violaxanthin de-epoxidase (Wilhelm *et al.*, 2006). Brown algae, in contrast, only possess the violaxanthin cycle and this was correlated with the presence of only one zeaxanthin epoxidase gene rather than the two or three copies that are usually found in diatom and haptophyte genomes (Frommolt *et al.*, 2008).

This suggested that the additional zeaxanthin epoxidase enzymes in diatoms and haptophytes might be involved in the diadinoxanthin cycle.

4.2. Carbon Metabolism

Genes for most of the enzymes involved in photosynthetic inorganic carbon fixation were found in the *Ectocarpus* genome (Cock *et al.*, 2010). The genome potentially encodes the enzymes necessary for C4 photosynthesis, consistent with the suggestion that brown algae are able to use C4 or CAM metabolism (Axelsson, 1988; Kremer & Küppers, 1977). However, the encoded proteins are not predicted to be targeted to the cellular location expected for a C4 system (Cock *et al.*, 2010). *Phosphoenolpyruvate carboxylase*, for example, is predicted to be located in the mitochondria. This is nonetheless interesting because photosynthetic activity decreases in brown algae when mitochondrial respiration is inhibited (Carr, 2005) and large numbers of mitochondria have been detected close to the cell wall in fucoid algae (Axelsson, 1988). Hence, mitochondria might play an important role in inorganic carbon uptake in the brown algae, with the initial steps of inorganic carbon fixation being partly located in the mitochondria. In addition, a recent study that combined gene analysis with extensive metabolite profiling did not provide clear support for the occurrence of an alanine/aspartate-based inducible C4-like metabolism in *Ectocarpus* (Gravot *et al.*, 2010) and suggested the presence of a classical glycolate-based photorespiration pathway in this brown alga rather than the malate synthase pathway found in diatoms.

In brown algae, excess assimilated carbon is stored as laminarin (a β-1,3-glucan with occasional β-1,6-linked branches; Read, Currie & Bacic, 1996) and the alcohol sugar mannitol (Percival & Ross, 1951). These two molecules are thought to fulfil essentially the same functions as starch and sucrose in flowering plants (Yamaguchi, Ikawa, & Nisizawa, 1966). Several potential components of the laminarin biosynthetic pathway have been identified, including putative glucose-6-phosphate isomerases, phosphoglucomutases, uridine-diphosphate (UDP)-glucose-pyrophosphorylases, β-1,3-glucan synthases and KRE6-like proteins (Michel, Tonon, Scornet, Cock & Kloareg, 2010a). Similarly, genes with predicted roles in mannitol biosynthesis and catabolism have been identified in *Ectocarpus* (Cock *et al.*, 2010) and the biochemical function of an enzyme that catalyses the first step of mannitol biosynthesis, a mannitol-1-phosphate dehydrogenase, has been investigated (Rousvoal *et al.*, 2011). The β-1,3-glucan biosynthetic pathway is thought to have been inherited vertically from the last eukaryotic common ancestor, whereas the mannitol biosynthetic pathway appears to have been acquired via a horizontal gene transfer from an actinobacterium (Michel *et al.*, 2010a).

Another unusual feature of brown algae is their cell walls, which contain large amounts of the anionic polysaccharides alginates and fucoidans (Kloareg & Quatrano, 1988). These molecules play an important role in providing the cell wall with the strength and flexibility necessary to resist the physical stresses characteristic of shoreline environments. Cell wall biosynthesis pathways, like the carbon storage pathways described above, appear to have had a complex evolutionary history (Michel, Tonon, Scornet, Cock & Kloareg, 2010b). For example, phylogenetic analysis indicates that the terminal steps of the alginate biosynthesis pathway were acquired from an actinobacterium via a horizontal gene transfer event. In contrast, the cellulose synthesis pathway appears to have been inherited from a red alga, probably via the secondary endosymbiosis event that led to the acquisition of a plastid by an ancient ancestor of the brown algae (Reyes-Prieto *et al.*, 2007). Note that, as with the carbon storage pathways, cell wall biosynthesis enzymes are starting to be characterized at the biochemical level (Tenhaken, Voglas, Cock, Neu & Hiuber, 2011).

4.3. Nitrogen Metabolism

Based on analysis of the genome sequence, *Ectocarpus* appears to be able to take up nitrogen in three different forms: ammonium, nitrate and urea (Cock *et al.*, 2010). Nitrate is presumably assimilated via nitrate reductase and nitrite reductase. Interestingly, like diatoms, *Ectocarpus* possesses both a ferredoxin and an NAD(P)H nitrite reductase. Production of a NAD(P)H nitrite reductase may allow these organisms to reduce nitrite under conditions where reduced ferredoxin is limiting, such as low light conditions. Another interesting feature that *Ectocarpus* shares with diatoms is the presence of a complete urea cycle (Allen *et al.*, 2011; Armbrust *et al.*, 2004; Cock *et al.*, 2010).

4.4. Halogen Metabolism

Some algae can accumulate halides (iodide and/or bromide) to high levels; *Laminaria digitata*, for example, can concentrate iodine to 30,000-fold the level found in the surrounding seawater. These accumulated halides have anti-oxidant activity and can also be used to produce volatile halocarbons some of which are thought to have anti-microbial activity and which may also have a significant impact on the chemistry of the atmosphere (Küpper *et al.*, 2008). *Ectocarpus* does not accumulate iodine to the same levels as *Laminaria* (0.08 mg/g dry weight of *Ectocarpus* filaments corresponding to a 1000-fold concentration compared with seawater). This difference between the two brown algae is reflected in the complement of halide metabolism genes in their genomes. Vanadium-dependent haloperoxidases (vHPO) are thought to play a central role in brown algal halogen metabolism, both with regard to halide uptake and to the production of halogenated compounds (La Barre, Potin, Leblanc & Delage, 2010). *L. digitata* possesses two large multigenic families of vanadium-dependent bromoperoxidases (vBPOs) and vanadium-dependent iodoperoxidases (Colin *et al.*, 2003, 2005), whereas *Ectocarpus* possesses only one vHPO (predicted to be an apoplastic vBPO), which is expressed at a low level during the sporophyte generation (0.1% of the *Ectocarpus* ESTs compared to the 4% vBPO sequences in *L. digitata* sporophyte ESTs). vHPOs also potentially catalyze oxidative cross-linking of cell wall polymers, an activity that is consistent with a role in spore and gamete adhesion and in cell wall strengthening (Potin & Leblanc, 2006).

In *L. digitata* sporophytes, defence responses appear to involve tightly coordinated regulation of the two distinct haloperoxidase gene families, which have probably evolved from an ancestral gene duplication (Colin *et al.*, 2005; Cosse, Potin & Leblanc, 2009). Hence, there is a marked difference between the closely related Ectocarpales and Laminariales in that a highly developed iodine-based defence metabolism has evolved in the macroscopic parenchymatous sporophytes of kelps, but this system is not present, or at least not to the same degree, in the smaller thalli of the Ectocarpales.

Interestingly, other halogen-related enzymes have been identified in the *Ectocarpus* genome. These include at least three different families (21 loci) of haloacid dehalogenase (HAD) and two haloalkane dehalogenases. The HADs belong to a large superfamily of hydrolases with diverse substrate specificity, including phosphatases and ATPases. The dehalogenase enzymes may serve to defend *Ectocarpus* against halogen-containing compounds produced as defence metabolites by kelps (Küpper *et al.*, 2008) allowing it to successfully grow as an epiphyte or endophyte on kelp thalli (Russell, 1983a, 1983b).

4.5. Uptake and Storage of Iron

Iron is an important cofactor for a broad range of enzymes involved in photosynthesis, respiration and general redox reactions. In general, iron is scarce in the marine environment, particularly in the open ocean (Bruland,

Donat, & Hutchins, 1991; Martin & Fitzwater, 1988; Wu & Luther, 1994). Analysis of the *Ectocarpus* genome has provided evidence (Cock *et al.*, 2010) that it has an iron uptake system that resembles that of strategy I plants (Moog & Bruggemann, 1994). Homologues of both FRO2, an iron chelate reductase, and natural resistance-associated macrophage proteins (NRAMP), a M^{2+}-H^{+} symporter with a preference for Fe(II) (Bauer & Bereczky, 2003; Curie & Briat, 2003; Morrissey & Guerinot, 2009), have been identified in *Ectocarpus* (Cock *et al.*, 2010). The *Ectocarpus* NRAMP homologues may be important for iron release or mobilization but no homologues of the iron carrier CCC1p were detected. Physiological studies using the bathophenanthroline disulphonic acid assay (Eckhardt & Buckhout, 1998) support the involvement of an iron chelate reductase. The iron uptake system in *Ectocarpus* would therefore appear to be similar to that of the pennate diatom *P. tricornutum* and hence different from the reductive–oxidative pathway found in the centric diatom *T. pseudonana* (Kustka, Allen, & Morel, 2007). *Ectocarpus* appears to lack homologues of any of the common iron regulatory genes including *ide, dtxR, fur* or *irr* (Bauer & Bereczky, 2003; Curie & Briat, 2003; Morrissey & Guerinot, 2009). Furthermore, while some coastal diatoms such as *P. tricornutum* have ferritin genes, ferritins have not been detected in other open-ocean taxa such as *T. pseudonana* (Marchetti *et al.*, 2009); *Ectocarpus* has no homologues of any of these proteins making it similar to *Thalassiosira* in this respect. More recently, an alternative method of iron storage in vacuoles has been elucidated in yeast and several other eukaryotes including the halotolerant alga *Dunaliella salina* (Martinoia, Maeshima, & Neuhaus, 2007; Paz, Shimoni, Weiss & Pick, 2007). Thus, at present, there is no genetically identifiable iron storage system in *Ectocarpus*. In line with this, iron K-edge x-ray absorbance spectroscopy (XAS) and Mössbauer spectroscopic analysis of *Ectocarpus* tissue showed that most of the Fe pool is present as Fe(III), most of which is coordinated in FeS clusters and FeO mineral species but, as expected not to ferritin (Lars H. Böttger, Eric P. Miller, Christian Andresen, Berthold F. Matzanke, Frithjof C. Küpper, and Carl J. Carrano, J. Exp. Bot., in press, 2012).

4.6. Lipid Metabolism

One of the interesting features of brown algal metabolism is that it includes pathways for the biosynthesis of both C18 and C20 polyunsaturated fatty acids (PUFAs), which are typical of green plant and animal lipid metabolisms, respectively. Brown algal PUFAs are probably important precursors both of oxylipins involved in defence and stress responses and of sexual pheromones (Müller, Jaenicke, Donike & Akintobi, 1971; Pohnert & Boland, 2002). In *Ectocarpus*, PUFAs appear to be synthesized in the cytoplasm from malonyl-CoA, which is itself derived from acetyl-CoA in the plastid. The most abundant PUFAs are $18{:}2n-6$, $18{:}3n-6$ and $20{:}4n-6$ for the omega 6 series, and $18{:}3n-3$ and $20{:}5n-3$ for the omega 3 series. *Ectocarpus* appears to lack docosahexaenoic acid (Schmid, Müller, & Eichenberger, 1994). A number of genes with potential roles in oxylipin biosynthesis have been identified in *Ectocarpus* (Cock *et al.*, 2010), but it was not possible to assign genes definitively to the C18 or C20 pathways based on sequence information. Sphingolipids act both as structural components of membranes and as signalling molecules in plants and mammals, and most of the genes required for the biosynthesis of these molecules have been identified in *Ectocarpus*. As far as fatty acid degradation is concerned, *Ectocarpus*, like diatoms (Armbrust *et al.*, 2004), possesses two beta-oxidation pathways, one localized in mitochondria and the other in peroxisomes.

4.7. Secondary Metabolism

Brown algae produce many phenolic compounds through the acetate–malonate pathway, and these molecules have important roles as ultraviolet (UV) protectants, adhesives, cell wall strengtheners and defence molecules (Emiliani, Fondi, Fani, & Gribaldo, 2009). Phloroglucinol is the precursor of brown algal tannins. Synthesis of this molecule in *Ectocarpus* is predicted to involve three type III polyketide synthases, enzymes that appear to be absent from oomycete or diatom genomes (Cock *et al.*, 2010). Cytochrome P450s are known to oxidatively tailor polyketide products in other organisms as seen in aflatoxin biosynthesis. *Ectocarpus* has 11 cytrochrome P450 genes and 1 pseudogene, of which only 3 have putative functions (CYP51C1 is a sterol 14-alpha demethylase, CYP97E3 and CYP97F4 are presumed carotenoid hydroxylases based on the role of the highly conserved CYP97 family in green plants). The eight remaining P450s are potential modifiers of polyketide structures or other secondary metabolites.

The shikimate pathway appears to be present in *Ectocarpus*, but some of the derivatives of this pathway that are found in terrestrial plants, such as phenylpropanoids and salicylic acid, are predicted to be absent (Cock *et al.*, 2010). *Ectocarpus* also possesses a flavonoid metabolism, a feature shared with a broad range of photosynthetic organisms including other stramenopiles, but lacks a phenylalanine ammonia-lyase (*PAL*) gene. Given that terrestrial plants appear to have acquired *PAL* following a horizontal gene transfer from a bacterial genome (Emiliani *et al.*, 2009), this suggests that the PAL enzyme was integrated into a pre-existing flavonoid metabolism in the green lineage.

5. CELLULAR PROCESSES

5.1. Receptors, Ion Channels and Signal Transduction Pathways

The *Ectocarpus* genome provided evidence for several types of sensor molecule, including molecules located both on the cell membrane and within the cytoplasm (Cock *et al.*, 2010). These included molecules that have been found in other stramenopiles, such as G-protein-coupled receptors and their associated heterotrimeric G-proteins, and molecules that have so-far appeared to be absent from chromalveolates, such as membrane-located histidine kinases. The *Ectocarpus* genome encodes three membrane-located histidine kinases, with N-terminal Mase or Chase sensor domains. Interestingly, one of these proteins is predicted to have seven trans-membrane domains and seems to be a G-protein-coupled receptor (GPCR)-histidine kinase fusion protein. One of the most remarkable discoveries in the genome was a family of 11 receptor kinases (Cock *et al.*, 2010). Both land plants and animals possess large families of similar receptor kinases and it has been suggested that the acquisition of these membrane-located signalling molecules was a key step towards the evolution of complex multicellularity in each lineage. Interestingly, animal and plant receptor kinases appear to have evolved independently and phylogenetic analysis indicated that this was also the case for the brown algal receptor kinases. Hence, the independent evolution of complex multicellularity in the plant, animal and brown algal lineages can, in each case, be correlated with the evolution of receptor kinase families.

Ectocarpus possesses at least six different types of membrane-localized ion channel, including members that are predicted to play an important role in sensing and responding to changes in the extracellular environment. For example, the ion channel proteins include a large family of transient receptor potential channels, which have been shown, in other organisms, to respond to a broad range of stimuli including temperature, light, chemicals and mechanical stress (Venkatachalam & Montell, 2007). Other families of ion channels represented in the *Ectocarpus* genome include bacterial-type, small conductance mechanosensitive channels, ionotropic glutamate receptors (which are absent from other stramenopile genomes) and an inositol triphosphate (IP3)/ryanodine-type receptor (IP3R/RyR). The presence of the latter is consistent with the observation that IP3 induces Ca^{2+} release in *Fucus serratus* embryos (Coelho *et al.*, 2002; Goddard, Manison, Tomos & Brownlee, 2000).

In animals, integrins are membrane-localized proteins involved in the transmission of mechanical signals perceived at the cell surface to the cytoskeleton (Arnaout, Goodman, & Xiong, 2007). *Ectocarpus* has three proteins that share similarity with integrin alpha subunits. Moreover, although there are no homologues of animal extracellular interacting proteins such as collagen, fibronectin and vitronectin, *Ectocarpus* does have homologues of talin and α-actinin, which are intracellular integrin partners that interact with actin microfilaments (Ziegler, Gingras, Critchley, & Emsely, 2008).

In addition to these membrane-localized molecules, *Ectocarpus* is predicted to possess the following cytosolic photoreceptors: three phytochromes, three cryptochromes (including a (6-4) family 'animal type' cryptochrome and two Cry-DASH genes) and five aureochromes (Cock *et al.*, 2010). Aureochromes are thought to be the stramenopile equivalents of phototropin blue-light receptors (Ishikawa *et al.*, 2009), explaining the absence of genes encoding the latter class of receptor in brown algal genomes.

A number of *Ectocarpus* pathogens have been described, including viruses, an oomycete (*Eurychasma dicksonii*), a chytrid (*Chytridium polysiphoniae*), a hyphochytrid (*Anisolpidium ectocarpii*) and a plasmodiophorid (*Maullinia ectocarpii*) (Charrier *et al.*, 2008; Gachon, Sime-Ngando, Strittmatter, Chambouvet and Kim, 2010). *Ectocarpus* presumably possesses sensor systems that allow it to detect the presence of these pathogens. A search for protein domains commonly found in components of land plant pathogen recognition systems failed to identify any caspase activation and recruitment domain (CARD), domain in apoptosis and interferon response (DAPIN) or Toll/Interleukin-1 receptor (TIR) domain-containing proteins. However, the genome does encode a large family of more than 250 LRR-domain-containing proteins and 15 proteins with nucleotide-binding adaptor shared by APAF-1, R proteins and CED-4 (NB-ARC) domains. A role in pathogen-specific immune responses is particularly likely for the NB-ARC-TPR proteins and a subset of about 60 of the LRR-domain loci, which encode GTPases of the ROCO family (including about 20 apparent pseudogenes). Both families exhibit evidence of rapid evolution of their ligand-binding (LRR and TPR) domains via exon shuffling (Fig. 5.4; Zambounis *et al.*, 2012). The LRR and NB-ARC domain genes are often grouped into small clusters of closely related genes and associated with probable pseudogenes in a similar fashion to the fast-evolving clusters of disease-resistance genes found in land plants (Meyers *et al.*, 2005). The predicted regulation of *Ectocarpus* LRR domain genes by microRNAs (Table 5.1) has also been observed to be a feature of land plant disease-resistant genes (Li *et al.*, 2012; Zhai *et al.*, 2011).

Ectocarpus also possesses homologues of many of the pathogenesis-related proteins, which are induced following infection or attack in land plants (Antoniw, Ritter, Pierpoint, & Van Loon, 1980; Van Loon, Rep, & Pieterse, 2006), and potential components of a programmed cell death pathway, such as metacaspases (Cock *et al.*, 2010).

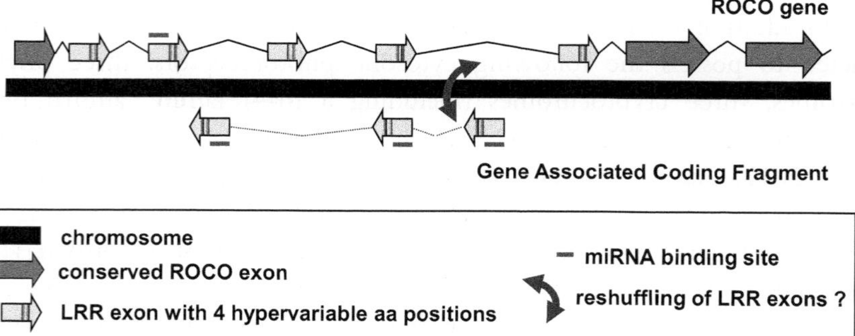

Figure 5.4 Gene structure and hypothetical post-transcriptional regulation features of the ROCO family. *Ectocarpus* ROCO proteins contain LRR domains in the N-terminal region. Each LRR motive is specified by a precisely delineated, 24-codon exon (blue arrows). Each LRR exon contains four hypervariable codons corresponding to amino acids that are exposed at the protein surface and are believed to dictate the ligand-binding specificity of the LRR domain (multicolour vertical bars in each LRR exon). These amino acid residues evolve under positive selection, suggesting a potential role of ROCO proteins in pathogen detection or immune reactions. Sequences resembling the LRR-encoding exons are also present in the introns but located on the opposite strand of the DNA. It is possible that these sequences are integrated into ROCO gene sequences as new exons following intragenic recombination events (curved blue arrow). Phylogenetic analysis also suggests that recombination occurs between members of the ROCO family, leading to swapping of LRR exons (not shown here). Moreover, the more conserved N-terminal regions of some LRR exons are predicted to be targets of miRNAs (red bars), suggesting that shuffling of LRR exons might also modulate post-transcriptional regulation. It is not yet known whether the LRR-encoding sequences on the opposite strand are transcribed but if so they could serve as alternative targets for ROCO-directed miRNAs and hence potentially modulate ROCO gene expression. See the colour plate.

Cytoplasmic signal transduction presumably involves members of the eukaryotic protein kinase (ePK) superfamily (Manning, Whyte, Martinez, Hunter, & Sudarsanam, 2002). The ePK family in *Ectocarpus* includes 258 predicted proteins representing a broad range of subfamilies (Cock *et al.*, 2010). Cytoplasmic protein kinase-mediated pathways in *Ectocarpus* also include the broadly conserved target of rapamycin (TOR) signalling pathway regulating cell growth (Durán & Hall, 2012; van Dam, Zwartkruis, Bos, & Snel, 2011). Exceptionally among stramenopiles, *Ectocarpus* lacks the central regulator of TOR kinase complexes, the small GTPase Rheb (Cock *et al.*, 2010), but this appears to have been lost several times independently in the eukaryotic evolution (van Dam *et al.*, 2011); it would be interesting to investigate whether the independent loss of Rheb leads to similar or different compensatory modifications in the TOR regulation.

A proportion of the signal transduction pathways are expected to modulate gene transcription. The *Ectocarpus* genome is estimated to encode 401 transcription-associated proteins (this includes both the transcription factors that bind directly to the DNA and the transcriptional regulators that act indirectly on transcription). *Ectocarpus* possesses eight RWP-RK domain (or nodule-inception(NIN)-like) transcription factors. Members of this family, which is completely absent from the diatom genomes studied to date, have been implicated in both nitrate signalling (Fernandez & Galvan, 2008) and gametogenesis (Lin & Goodenough, 2007) in other species.

5.2. Cell Cycle, Endoreduplication and Meiosis

Cell cycle regulation by cyclins and cyclin-dependent kinases (CDK) appears to be a universal feature across the eukaryotic tree (Cross, Buchler, Skotheim, 2011). The complement of cell cycle regulatory genes in *Ectocarpus* is more similar to that of green plants than that of animals (Cock *et al.*, 2010). This is true both for the core cyclin/CDK complex and regulators of this complex.

Spo11 creates the double-strand DNA breaks that are required to initiate recombination during meiosis (Keeney, Giroux, & Kleckner, 1997; Krogh & Symington, 2004; Lichten, 2001). Land plants possess three Spo11 homologues, Spo11-1 and Spo11-2, which are specifically required for meiosis, and Spo11-3/Top6A, which functions with Top6B as a topoisomerase and is required for endoreduplication (Hartung *et al.*, 2007; Sugimoto-Shirasu, Stacey, Corsar, Roberts, & McCann, 2002; Yin *et al.*,

2002). *Ectocarpus* is predicted to possess both a meiotic Spo11 and a Top6A homologue (together with its Top6B partner). The presence of the latter is particularly interesting, given the recent demonstration that a proportion of haploid partheno–sporophytes (produced by parthenogenetic germination of gametes) undergo endoreduplication very early in development to produce diploid individuals. As a result, these individuals can produce haploid meiospores via a normal meiotic division, allowing progression to the gametophyte generation of the life cycle (Bothwell, Marie, Peters, Cock & Coelho, 2010).

In addition to *Spo11*, *Ectocarpus* possesses most of the core meiotic genes. One notable feature of this set of genes is that, compared to other stramenopiles, *Ectocarpus* has a very complete set of *Rad51* genes, possessing at least six of the seven known eukaryotic members of the *Rad51* family (Lin, Kong, Nei & Ma, 2006), including a likely *DMC1* homologue. Completeness of the *Rad51* family has been linked to multicellularity, so this may be another factor that was important for the evolution of complex multicellularity in the brown algae.

Ectocarpus has a homologue of the DNA helicase *Mer3,* but no clear homologues of *Mms4* and *Mus81* were found (although *ERCC4*-like genes are present). This suggests that *Ectocarpus* may only have the class I, interference-sensitive crossover pathway, although it is possible that the class II, interference-insensitive pathway is present but mediated by highly divergent proteins.

5.3. Cytoskeleton, Flagella and Vesicle Trafficking

One of the remarkable features of cytokinesis in brown algae is that it has characteristics typical of both green plant and animal systems; centrosomes function as microtubule organizing centres (as in animal cells), but during cytokinesis, a structure resembling a cell plate is formed, and this extends out from the centre of the cell towards the plasma membrane (Nagasato & Motomura, 2002). This latter feature resembles cytokinesis in green plants, but no specialized phragmoplast is formed in brown algae and therefore, as expected, *Ectocarpus* does not possess genes encoding dynamin-related phragmoplastins. *Ectocarpus* does, however, possess the centrosome-associated tubulins δ and ε in addition to α-, β- and γ-tubulin.

Microtubules also play an important structural role in the flagella. Brown algal zoids have the two heteromorphic flagella typical of stramenopiles. Comparison of the *Ectocarpus* radial spoke proteins (RSP) with those of *Chlamydomonas reinhardtii, Thalassiosira pseudonana, Micromonas pusilla* and

Homo sapiens, indicated that that the complement of RSPs can be very variable (Cock *et al.*, 2010), probably depending on the mechanism of flagella bending in different species. Several flagellum- or basal body-associated proteins are encoded by Bardet–Biedl syndrome (BBS) genes, which are of considerable interest because of their implication in human disease. Seven of them are widely conserved in eukaryotes with flagella and were suggested to represent a functional module consisting of a regulatory small GTPase (BBS3 = ARL6) and its putative effector, the multisubunit complex BBSome, which probably serves as a membrane coat for trafficking to the flagellum (Jin *et al.*, 2010). The *Ectocarpus* genome encodes all the known BBS proteins that have been detected in *Chlamydomonas* and may therefore be an interesting system to study the function of these proteins. Other flagellar components encoded by the *Ectocarpus* genome include intraflagellar transport proteins and SF-assemblin (a component of system I fibers, which run parallel to flagellar root microtubules from the basal bodies; Lechtreck & Melkonian, 1991) and orthologues of the mastigoneme components Ocm1, Ocm2, Ocm3 and Ocm4 (Honda *et al.*, 2007; Yamagishi, Motomura, Nagasato, & Kawai, 2009; Yamagishi, Motomura, Nagasato, Kato, & Kawai, 2007).

In brown algae, the cortical actin network plays a crucial role both in determining the polarity of apical and axillary cell divisions and in cell wall morphogenesis (Katsaros, Karyophyllis & Galatis, 2006). *Ectocarpus* possesses a single actin gene, plus actin-associated proteins such as formin, profilin, villin, fimbrin and myosin. *Ectocarpus* also has all seven subunits of the ARP2/3-complex, which plays an important role in the reorganization of the actin cytoskeleton at the cell cortex during processes such as vesicular trafficking. There is evidence that the ARP2/3 complex has a role in mediating polarized growth in brown algae (Hable & Kropf, 2005). The Rho family of small GTPases is known to play a central role in mediating signalling directed towards the actin cytoskeleton (Etienne-Manneville & Hall, 2002). In contrast to other multicellular (and even many unicellular) eukaryotes (Brembu, Winge, Bones, & Yang, 2006), no lineage-specific expansion of the Rho family has occurred in *Ectocarpus,* which possesses only a single *Rho* gene (Cock *et al.*, 2010).

Based on ultra-structure studies, the intracellular trafficking system appears to be very active in vegetative *Ectocarpus* cells (Bouck, 1965; Oliveira & Bisalputra, 1973). It was therefore not surprising to find that the gene families associated with this process are quite complex, including, for example, 20 SNARE proteins and significant numbers of coat protein complex proteins (Cock *et al.*, 2010). However, the set of Rab GTPases (which play a key role in determining the specificity of vesicle trafficking between the diverse compartments of the endomembrane system) is not markedly expanded in *Ectocarpus* compared to unicellular relatives (diatoms and *Aureococcus anophagefferens*). This contrasts with the situation in metazoans, where there has been a significant expansion of this family, but more closely resembles the situation in green plants, where the Rab family did not undergo a marked expansion during the transition to multicellularity (Elias, Brighouse, Gabernet-Castello, Field, & Dacks, 2012). Another group of GTPases important for the function of the endomembrane system are atlastins, which are responsible for homotypic fusion of ER membranes and the generation of the tubular endoplasmic reticulum (ER) network (Hu *et al.*, 2009). Atlastins had been thought to be restricted to Metazoa, but putative orthologs have been found in the genomes of *Ectocarpus* and other stramenopile algae (Cock *et al.*, 2010).

6. FUTURE DIRECTIONS

Analysis of the *Ectocarpus* genome sequence has provided a large amount of information about the probable molecular systems that underlie a broad range of processes in brown algae. However, this information is based principally on parallels with other organisms (sequence homology) and an important challenge for the future will be to develop methods of directly investigating gene function in *Ectocarpus*. These methods need to include not only ways of disrupting gene function but also tools to analyse spatiotemporal patterns of gene expression and interactions between gene products and to assay protein activities in heterologous or *in vitro* systems. Several genetic tools are already available including protocols for UV and chemical mutagenesis, phenotypic screening methods, genetic crosses, methods for handling large populations, a large number of genetic markers, defined strains for genetic mapping and a microsatellite-based genetic map (Cock *et al.*, 2011; Coelho *et al.*, 2012; Heesch *et al.*, 2010; Peters *et al.*, 2008). At present, however, there is no reliable method to knock out or knock down gene expression in *Ectocarpus*, although a considerable effort is being invested in the development of genetic transformation and RNA interference approaches, and a targeting-induced local lesions in genomes (TILLING) population (Comai & Henikoff, 2006) is being developed for reverse genetic analysis.

As far as analysis of gene expression is concerned, an EST-based expression microarray has been developed (Dittami *et al.*, 2009), and a considerable amount of cDNA and small RNA sequence data are available (Cock *et al.*, 2010). The latter is currently being significantly expanded using new generation RNA-seq methodology. In addition, protocols have been developed for proteome and metabolome analysis (Dittami *et al.*, 2011; Ritter *et al.*, 2010).

The genome sequence data are currently being used in combination with the tools described above to explore several different aspects of brown algal biology. These include life cycle regulation (Bothwell *et al.*, 2010; Coelho *et al.*, 2011), morphogenesis (Le Bail, Billoud, Le Panse, Chenivesse, & Charrier, 2011), metabolic processes (Dittami *et al.*, 2011), interactions with the environment and with brown algal pathogens (Grenville-Briggs *et al.*, 2011; Ritter *et al.*, 2010; Zambounis *et al.*, 2012) and taxonomy (Peters *et al.*, 2010a, 2010b). In several of these studies (Coelho *et al.*, 2011; Le Bail *et al.*, 2011; Peters *et al.*, 2008), major regulatory loci have been identified through the isolation of mutants and an important aim for the future is to identify the genes affected by these genetic lesions. Phylogenetic analyses have shown that the genus *Ectocarpus* is a complex of several (cryptic) species inhabiting different geographic regions and ecological environments and which we only begin to understand (Peters *et al.*, 2010a, 2010b). Reproductive isolation between species is incomplete because interspecific crosses can be performed, although vegetative development or formation of meiospores may be impaired in the hybrids (reviewed in Stache-Crain, Müller, & Goff, 1997). Genome data from the different species of *Ectocarpus* would help to better understand the mechanisms and processes determining their interfertility, ecological niches, as well as (historical) biogeography.

Another important future objective for brown algal genomics will be to extend knowledge gained using *Ectocarpus* as a model system to other brown algae. This process should include extension of the sequencing effort to additional brown algal genomes and transcriptomes. In this respect, two approaches would be particularly interesting. First, as a group, the Ectocarpales exhibit a considerable amount of variation in terms of their life cycles, sexual systems, morphology and cell biology (Silberfeld *et al.*, 2010) and yet are expected to share a significant level of genome synteny, facilitating the use of comparative genomic approaches to investigate the genetic basis of these biological variation. Second, compared to the kelps, the Ectocarpales exhibit only a limited level of developmental complexity. It would be of considerable interest to analyse a kelp genome sequence, particularly with respect to the emergence of complex multicellularity in the brown algae.

ACKNOWLEDGEMENTS

Work on *Ectocarpus* has been supported by the Centre National de la Recherche Scientifique, the University Pierre and Marie Curie, the Inter-University Network for Fundamental Research (P6/25, BioMaGNet), the Europole Mer, the European Union Infrastructures program (project Assemble), the Interreg program France (Channel)-England (project Marinexus), the Agence Nationale de la Recherche (project Bi-cycle) and Natural Environment Research Council (United Kingdom) grants NE/F012705/1 and NE/J00460X/1.

REFERENCES

Allen, A. E., Dupont, C. L., Oborník, M., Horák, A., Nunes-Nesi, A., McCrow, J. P., et al. (2011). Evolution and metabolic significance of the urea cycle in photosynthetic diatoms. *Nature, 473*, 203–207.

Antoniw, J. F., Ritter, C. E., Pierpoint, W. S., & Van Loon, L. C. (1980). Comparison of three pathogenesis-related proteins from plants of two cultivars of tobacco infected with TMV. *Journal of General Virology, 47*, 79–87.

Archibald, J. (2012). The evolution of algae by secondary and tertiary endosymbiosis. *Advances in Botanical Research, 64*, 87–118.

Armbrust, E., Berges, J., Bowler, C., Green, B., Martinez, D., Putnam, N., et al. (2004). The genome of the diatom *Thalassiosira pseudonana*: Ecology, evolution, and metabolism. *Science, 306*, 79–86.

Arnaout, M., Goodman, S., & Xiong, J. (2007). Structure and mechanics of integrin-based cell adhesion. *Current Opinion in Cell Biology, 19*, 495–507.

Axelsson, L. (1988). Change in pH as a measure of photosynthesis by marine macroalgae. *Marine Biology, 97*.

Bailleul, B., Rogato, A., de Martino, A., Coesel, S., Cardol, P., Bowler, C., et al. (2010). An atypical member of the light-harvesting complex stress-related protein family modulates diatom responses to light. *Proceedings of the National Academy of Sciences of the United States of America, 107*, 18214–18219.

Bauer, P., & Bereczky, B. (2003). Gene networks involved in iron acqusition stategies in plants. *Agronomie, 23*, 447–454.

Baurain, D., Brinkmann, H., Petersen, J., Rodríguez-Ezpeleta, N., Stechmann, A., Demoulin, V., et al. (2010). Phylogenomic evidence for separate acquisition of plastids in cryptophytes, haptophytes, and stramenopiles. *Molecular Biology and Evolution, 27*, 1698–1709.

Belostotsky, D. (2009). Exosome complex and pervasive transcription in eukaryotic genomes. *Current Opinion in Cell Biology, 21*, 352–358.

Bjørnland, T., & Liaaen-Jensen, S. (1989). Distribution patterns of carotenoids in relation to chromophyte phylogeny and systematics. In J. C. Green, et al. (Eds.), *The chromophyte algae. Problems and perspectives* (pp. 37–61). Oxford: Clarendon Press.

Bothwell, J. H., Marie, D., Peters, A. F., Cock, J. M., & Coelho, S. M. (2010). Role of endoreduplication and apomeiosis during parthenogenetic reproduction in the model brown alga *Ectocarpus*. *New Phytologist, 188*, 111–121.

Bouck, G. (1965). Fine structure and organelle association in brown algae. *Journal of Cell Biology, 26*, 523–537.

Bowler, C., Allen, A., Badger, J., Grimwood, J., Jabbari, K., Kuo, A., et al. (2008). The *Phaeodactylum* genome reveals the evolutionary history of diatom genomes. *Nature, 456*, 239–244.

Bräutigam, M., Klein, M., Knippers, R., & Müller, D. G. (1995). Inheritance and meiotic elimination of a virus genome in the host *Ectocarpus siliculosus* (phaeophyceae). *Journal of Phycology, 31*, 823–827.

Brembu, T., Winge, P., Bones, A. M., & Yang, Z. (2006). A RHOse by any other name: A comparative analysis of animal and plant Rho GTPases. *Cell Research, 16*, 435–445.

Bruland, K. W., Donat, J. R., & Hutchins, D. A. (1991). Interactive influences of bioactive trace-metals on biological production in oceanic waters. *Limnology and Oceanography, 36*, 1555–1577.

Carr, H. (2005). *Energy balance during active carbon uptake and at excess irradiance in three marine macrophytes*. PhD thesis, Faculty of Science, Department of Botany, Stockholm University.

Carthew, R. W., & Sontheimer, E. J. (2009). Origins and Mechanisms of miRNAs and siRNAs. *Cell, 136*, 642–655.

Charrier, B., Coelho, S., Le Bail, A., Tonon, T., Michel, G., Potin, P., et al. (2008). Development and physiology of the brown alga *Ectocarpus siliculosus*: Two centuries of research. *New Phytologist, 177*, 319–332.

Cock, J., Sterck, L., Rouzé, P., Scornet, D., Allen, A., Amoutzias, G., et al. (2010). The *Ectocarpus* genome and the independent evolution of multicellularity in brown algae. *Nature, 465*, 617–621.

Cock, J. M., Peters, A. F., & Coelho, S. M. (2011). Brown algae. *Current Biology, 21*, R573–R575.

Coelho, S. M., Taylor, A. R., Ryan, K. P., Sousa-Pinto, I., Brown, M. T., & Brownlee, C. (2002). Spatiotemporal patterning of reactive oxygen production and $Ca2+$ wave propagation in *Fucus* rhizoid cells. *Plant Cell, 14*, 2369–2381.

Coelho, S. M., Godfroy, O., Arun, A., Le Corguillé, G., Peters, A. F., & Cock, J. M. (2011). *OUROBOROS* is a master regulator of the gametophyte to sporophyte life cycle transition in the brown alga *Ectocarpus*. *Proceedings of the National Academy of Sciences of the United States of America, 108*, 11518–11523.

Coelho, S. M., et al. (2012). Ectocarpus: A model organism for the brown algae. *Cold Spring Harbor Protoc*, 193–198.

Colin, C., Leblanc, C., Michel, G., Wagner, E., Leize-Wagner, E., Van Dorsselaer, A., et al. (2005). Vanadium-dependent iodoperoxidases in *Laminaria digitata*, a novel biochemical function diverging from brown algal bromoperoxidases. *Journal of Biological Inorganic Chemistry, 10*, 156–166.

Colin, C., Leblanc, C., Wagner, E., Delage, L., Leize-Wagner, E., Van Dorsselaer, A., et al. (2003). The brown algal kelp *Laminaria digitata* features distinct bromoperoxidase and iodoperoxidase activities. *Journal of Biological Chemistry, 278*, 23545–23552.

Comai, L., & Henikoff, S. (2006). TILLING: Practical single-nucleotide mutation discovery. *Plant Journal, 45*, 684–694.

Cosse, A., Potin, P., & Leblanc, C. (2009). Patterns of gene expression induced by oligoguluronates reveal conserved and environment-specific molecular defense responses in the brown alga *Laminaria digitata*. *New Phytologist, 182*, 239–250.

Cross, F. R., Buchler, N. E., & Skotheim, J. M. (2011). Evolution of networks and sequences in eukaryotic cell cycle control. *Philosophical Transactions of the Royal Society London B Biological Sciences, 366*, 3532–3544.

Curie, C., & Briat, J. F. (2003). Iron transport and signaling in plants. *Annual Review of Plant Biology, 54*, 183–206.

De Clerck, O., Bogaert, K., & Leliaert, F. (2012). Diversity and evolution of algae: Primary endosymbiosis. *Advances in Botanical Research, 64*, 55–86.

Del Campo, E., Ramazanov, Z., Garcia-Reina, G., & Müller, D. G. (1997). Photosynthetic responses and growth performance of virus-infected and noninfected *Ectocarpus siliculosus* (Phaeophyceae). *Phycologia, 36*, 186–189.

Delage, L., Leblanc, C., Nyvall Collén, P., Gschloessl, B., Oudot, M.-P., Sterck, L., et al. (2011). *In Silico* survey of the mitochondrial protein uptake and maturation systems in the brown alga *Ectocarpus siliculosus. Public Library of Science One, 6*, e19540.

Delaroque, N., Maier, I., Knippers, R., & Müller, D. (1999). Persistent virus integration into the genome of its algal host, *Ectocarpus siliculosus* (Phaeophyceae). *Journal of General Virology, 80*(Pt. 6), 1367–1370.

Delaroque, N., Müller, D., Bothe, G., Pohl, T., Knippers, R., & Boland, W. (2001). The complete DNA sequence of the *Ectocarpus siliculosus* Virus EsV-1 genome. *Virology, 287*, 112–132.

Dittami, S. M., Gravot, A., Renault, D., Goulitquer, S., Eggert, A., Bouchereau, A., et al. (2011). Integrative analysis of metabolite and transcript abundance during the short-term response to saline and oxidative stress in the brown alga *Ectocarpus siliculosus. Plant Cell and Environment, 34*, 629–642.

Dittami, S. M., Michel, G., Collen, J., Boyen, C., & Tonon, T. (2010). Chlorophyll-binding proteins revisited—A multigenic family of light-harvesting and stress proteins from a brown algal perspective. *BMC Evolutionary Biology, 10*, 365.

Dittami, et al. (2009). Global expression analysics of the brown alga *Ectocarpus siliculosus* (Phaeophyceae) reveals large-scale reprogramming of the transcriptome in response to abiotic stress. *Genome Biology, 10*, R66.

Dixon, N. M., Leadbeater, B. S. C., & Wood, K. R. (2000). Frequency of viral infection in a field population of *Ectocarpus fasciculatus* (Ectocarpales, Phaeophyceae). *Phycologia, 39*, 258–263.

Dorrell, R. G., & Smith, A. G. (2011). Do red and green make brown?: Perspectives on plastid acquisitions within chromalveolates. *Eukaryot Cell, 10*, 856–868.

Durán, R. V., & Hall, M. N. (2012). Regulation of TOR by small GTPases. *EMBO Rep, 13*, 121–128.

Eckhardt, U., & Buckhout, T. J. (1998). Iron assimilation in *Chlamydomonas reinhardtii* involves ferric reduction and is similar to Strategy I higher plants. *Journal of Experimental Botany, 49*, 1219–1226.

Elias, M., Brighouse, A., Gabernet-Castello, C., Field, M. C., & Dacks, J. B. (2012). Sculpting the endomembrane system in deep time: High resolution phylogenetics of Rab GTPases. *Journal of Cell Science, 125*, 2500–2508.

Emiliani, G., Fondi, M., Fani, R., & Gribaldo, S. (2009). A horizontal gene transfer at the origin of phenylpropanoid metabolism: A key adaptation of plants to land. *Biology Direct, 4*, 7.

Etienne-Manneville, S., & Hall, A. (2002). Rho GTPases in cell biology. *Nature, 420*, 629–635.

Fernandez, E., & Galvan, A. (2008). Nitrate assimilation in *Chlamydomonas. Eukaryot Cell, 7*, 555–559.

Frommolt, R., Werner, S., Paulsen, H., Goss, R., Wilhelm, C., Zauner, S., et al. (2008). Ancient recruitment by chromists of green algal genes encoding enzymes for carotenoid biosynthesis. *Molecular Biology and Evolution, 25*, 2653–2667.

Gachon, C. M., Sime-Ngando, T., Strittmatter, M., Chambouvet, A., & Kim, G. H. (2010). Algal diseases: Spotlight on a black box. *Trends in Plant Science, 15*, 633–640.

Goddard, H., Manison, N., Tomos, D., & Brownlee, C. (2000). Elemental propagation of calcium signals in response-specific patterns determined by environmental stimulus strength. *Proceedings of the National Academy of Sciences of the United States of America, 97*, 1932–1937.

Gravot, A., Dittami, S. M., Rousvoal, S., Lugan, R., Eggert, A., Collén, J., et al. (2010). Diurnal oscillations of metabolite abundances and gene analysis provide new insights into central metabolic processes of the brown alga *Ectocarpus siliculosus*. *New Phytologist, 188*, 98–110.

Grenville-Briggs, L., Gachon, C. M., Strittmatter, M., Sterck, L., Küpper, F. C., & van West, P. (2011). A molecular insight into algal-oomycete warfare: cDNA analysis of *Ectocarpus siliculosus* infected with the basal oomycete Eurychasma dicksonii. *PLoS One, 6*, e24500.

Grimsley, N., Thomas, R., Kegel, J., Jacquet, S., Moreau, H., & Desdevises, Y. (2012). Genomics of algal host-virus interactions. *Advances in Botanical Research, 64*, 343–378.

Gschloessl, B., Guermeur, Y., & Cock, J. (2008). HECTAR: A method to predict subcellular targeting in heterokonts. *BMC Bioinformatics, 9*, 393.

Gundermann, K., & Büchel, C. (2007). The fluorescence yield of the trimeric fucoxanthin-chlorophyll-protein FCPa in the diatom *Cyclotella meneghiniana* is dependent on the amount of bound diatoxanthin. *Photosynthesis Research, 95*, 229–235.

Hable, W., & Kropf, D. (2005). The Arp2/3 complex nucleates actin arrays during zygote polarity establishment and growth. *Cell Motility and the Cytoskeleton, 61*, 9–20.

Hartung, F., Wurz-Wildersinn, R., Fuchs, J., Schubert, I., Suer, S., & Puchta, H. (2007). The catalytically active tyrosine residues of both SPO11-1 and SPO11-2 are required for meiotic double-strand break induction in Arabidopsis. *Plant Cell, 19*, 3090–3099.

Heesch, S., Cho, G. Y., Peters, A. F., Le Corguillé, G., Falentin, C., Boutet, G., et al. (2010). A sequence-tagged genetic map for the brown alga *Ectocarpus siliculosus* provides large-scale assembly of the genome sequence. *New Phytologist, 188*, 42–51.

Hinas, A., Reimegard, J., Wagner, E. G., Nellen, W., Ambros, V. R., & Soderbom, F. (2007). The small RNA repertoire of Dictyostelium discoideum and its regulation by components of the RNAi pathway. *Nucleic Acids Research, 35*, 6714–6726.

Honda, D., Shono, T., Kimura, K., Fujita, S., Iseki, M., Makino, Y., et al. (2007). Homologs of the sexually induced gene 1 (sig1) product constitute the stramenopile mastigonemes. *Protist, 158*, 77–88.

Hu, J., Shibata, Y., Zhu, P., Voss, C., Rismanchi, N., Prinz, W., et al. (2009). A class of dynamin-like GTPases involved in the generation of the tubular ER network. *Cell, 138*, 549–561.

Huang, A., He, L., & Wang, G. (2011). Identification and characterization of microRNAs from Phaeodactylum tricornutum by high-throughput sequencing and bioinformatics analysis. *BMC Genomics, 12*, 337.

Hurst, L., Williams, E., & Pál, C. (2002). Natural selection promotes the conservation of linkage of co-expressed genes. *Trends in Genetics, 18*, 604–606.

Ishikawa, M., Takahashi, F., Nozaki, H., Nagasato, C., Motomura, T., & Kataoka, H. (2009). Distribution and phylogeny of the blue light receptors aureochromes in eukaryotes. *Planta, 230*, 543–552.

Jeffrey, S. W. (1976). The occurrence of chlorophyll c1 and c2 in algae. *Journal of Phycolology, 12*, 349–354.

Jin, H., White, S. R., Shida, T., Schulz, S., Aguiar, M., Gygi, S. P., et al. (2010). The conserved Bardet-Biedl syndrome proteins assemble a coat that traffics membrane proteins to cilia. *Cell, 141*, 1208–1219.

Katsaros, C., Karyophyllis, D., & Galatis, B. (2006). Cytoskeleton and morphogenesis in brown algae. *Annals of Botany (Lond), 97*, 679–693.

Keeling, P. (2004). Diversity and evolutionary history of plastids and their hosts. *American Journal of Botany, 91*, 1481–1493.

Keeney, S., Giroux, C., & Kleckner, N. (1997). Meiosis-specific DNA double-strand breaks are catalyzed by Spo11, a member of a widely conserved protein family. *Cell, 88*, 375–384.

Kensche, P., Oti, M., Dutilh, B., & Huynen, M. (2008). Conservation of divergent transcription in fungi. *Trends in Genetics, 24*, 207–211.

Kloareg, B., & Quatrano, R. S. (1988). Structure of the cell walls of marine algae and ecophysiological functions of the matrix polysaccharides. *Oceanography and Marine Biology Annual Review, 26*, 259–315.

Kremer, B. P., & Küppers, U. (1977). Carboxylating enzymes and pathway of photosynthetic carbon assimilation in different marine algae- evidence for the C4-pathway? *Planta, 133*, 191–196.

Krogh, B., & Symington, L. (2004). Recombination proteins in yeast. *Annual Review of Genetics, 38*, 233–271.

Kruglyak, S., & Tang, H. (2000). Regulation of adjacent yeast genes. *Trends in Genetics, 16*, 109–111.

Kumar, H., Kawai, T., & Akira, S. (2011). Pathogen recognition by the innate immune system. *International Reviews of Immunology, 30*, 16–34.

Küpper, F. C., Carpenter, L. J., McFiggans, G. B., Palmer, C. J., Waite, T. J., Boneberg, E. M., et al. (2008). Iodide accumulation provides kelp with an inorganic antioxidant impacting atmospheric chemistry. *Proceedings of the National Academy of Sciences of the United States of America, 105*, 6954–6958.

Kustka, A. B., Allen, A. E., & Morel, F. M. M. (2007). Sequence analysis and transcriptional regulation of iron acquisition genes in two marine diatoms. *Journal of Phycology, 43*, 715–729.

La Barre, S., Potin, P., Leblanc, C., & Delage, L. (2010). The halogenated metabolism of brown algae (Phaeophyta), its biological importance and its environmental significance. *Marine Drugs, 8*, 988–1010.

Le Bail, A., Billoud, B., Le Panse, S., Chenivesse, S., & Charrier, B. (2011). *ETOILE* regulates developmental patterning in the filamentous brown alga *Ectocarpus siliculosus*. *Plant Cell, 23*, 1666–1678.

Le Corguillé, G., Pearson, G., Valente, M., Viegas, C., Gschloessl, B., Corre, E., et al. (2009). Plastid genomes of two brown algae, *Ectocarpus siliculosus* and *Fucus vesiculosus*: Further insights on the evolution of red-algal derived plastids. *BMC Evolutionary Biology, 9*, 253.

Lechtreck, K., & Melkonian, M. (1991). Striated microtubule-associated fibers: Identification of assemblin, a novel 34-kD protein that forms paracrystals of 2-nm filaments in vitro. *Journal of Cell Biology, 115*, 705–716.

Li, F., et al. (2012). MicroRNA regulation of plant innate immune receptors. *Proceedings of the National Academy of Sciences U S A, 109*, 1790–1795.

Lichten, M. (2001). Meiotic recombination: Breaking the genome to save it. *Current Biology, 11*, R253–R256.

Lin, H., & Goodenough, U. (2007). Gametogenesis in the *Chlamydomonas reinhardtii* minus mating type is controlled by two genes, *MID* and *MTD1*. *Genetics, 176*, 913–925.

Lin, Z., Kong, H., Nei, M., & Ma, H. (2006). Origins and evolution of the recA/RAD51 gene family: Evidence for ancient gene duplication and endosymbiotic gene transfer. *Proceedings of the National Academy of Sciences of the United States of America, 103*, 10328–10333.

Lohr, M., & Wilhelm, C. (1999). Algae displaying the diadinoxanthin cycle also possess the violaxanthin cycle. *Proceedings of the National Academy of Sciences of the United States of America, 96*, 8784–8789.

Lüning, K. (1990). *Seaweeds: Their environment, biogeography, and ecophysiology*. New York: John Wiley & Sons, Inc.

Malone, C. D., & Hannon, G. J. (2009). Small RNAs as guardians of the genome. *Cell, 136*, 656–668.

Manning, G., Whyte, D. B., Martinez, R., Hunter, T., & Sudarsanam, S. (2002). The protein kinase complement of the human genome. *Science, 298*, 1912–1934.

Marchetti, A., Parker, M. S., Moccia, L. P., Lin, E. O., Arrieta, A. L., Ribalet, F., et al. (2009). Ferritin is used for iron storage in bloom-forming marine pennate diatoms. *Nature, 457*, 467–470.

Martin, J. H., & Fitzwater, S. E. (1988). Iron-deficiency limits phytoplankton growth in the Northeast Subarctic Pacific. *Nature, 331*, 341–343.

Martinoia, E., Maeshima, M., & Neuhaus, H. E. (2007). Vacuolar transporters and their essential role in plant metabolism. *Journal of Experimental Botany, 58*, 83–102.

Meyers, B. C., Kaushik, S., & Nandety, R. S. (2005). Evolving disease resistance genes. *Current Opinion in Plant Biology, 8*, 129–134.

Meyers, B. C., Axtell, M. J., Bartel, B., Bartel, D. P., Baulcombe, D., Bowman, J. L., et al. (2008). Criteria for annotation of plant MicroRNAs. *Plant Cell, 20*, 3186–3190.

Michel, G., Tonon, T., Scornet, D., Cock, J. M., & Kloareg, B. (2010a). Central and storage carbon metabolism of the brown alga *Ectocarpus siliculosus*: Insights into the origin and evolution of storage carbohydrates in Eukaryotes. *New Phytologist, 188*, 67–81.

Michel, G., Tonon, T., Scornet, D., Cock, J. M., & Kloareg, B. (2010b). The cell wall polysaccharide metabolism of the brown alga *Ectocarpus siliculosus*. Insights into the evolution of extracellular matrix polysaccharides in Eukaryotes. *New Phytologist, 188*, 82–97.

Moog, P. R., & Bruggemann, W. (1994). Iron reductase systems on the plant plasma-membrane—A review. *Plant Soil, 165*, 241–260.

Morrissey, J., & Guerinot, M. L. (2009). Iron uptake and transport in plants: The good, the bad, and the ionome. *Chemical Reviews, 109*, 4553–4567.

Müller, D. G. (1966). Untersuchungen zur Entwicklungsgeschichtebder Braunalge *Ectocarpus siliculosus* aus Neapel. *Planta, 68*, 57–68.

Müller, D. G. (1967). Generationswechsel, Kernphasenwechsel und Sexualität der Braunalge *Ectocarpus siliculosus* im Kulturversuch. *Planta, 75*, 39–54.

Müller, D. G. (1991). Mendelian segregation of a virus genome during host meiosis in the marine brown alga *Ectocarpus siliculosus*. *Journal of Plant Physiology, 137*, 739–743.

Müller, D. G., Gachon, C. M. M., & Küpper, F. C. (2008). Axenic clonal cultures of filamentous brown algae: Initiation and maintenance. *Cahier de Biologie Marine, 49*, 59–65.

Müller, D. G., Jaenicke, L., Donike, M., & Akintobi, T. (1971). Sex attractant in a brown alga: Chemical structure. *Science, 171*, 815–817.

Müller, D. G., Kapp, M., & Knippers, R. (1998). Viruses in marine brown algae. *Advances in Virus Research, 50*, 49–67.

Müller, K., Lindauer, A., Brüderlein, M., & Schmitt, R. (1990). Organization and transcription of Volvox histone-encoding genes: Similarities between algal and animal genes. *Gene, 93*, 167–175.

Müller, D. G., Westermeier, R., Morales, J., Reina, G. G., del Campo, E., Correa, J. A., et al. (2000). Massive prevalence of viral DNA in *Ectocarpus* (Phaeophyceae, Ectocarpales) from two habitats in the North Atlantic and South Pacific. *Botanica Marina, 43*, 157–159.

Nagasato, C., & Motomura, T. (2002). Influence of the centrosome in cytokinesis of brown algae: Polyspermic zygotes of *Scytosiphon lomentaria* (Scytosiphonales, Phaeophyceae). *Journal of Cell Science, 115*, 2541–2548.

Oliveira, L., & Bisalputra, T. (1973). Studies in the brown alga *Ectocarpus* in culture. *Journal of Submicroscopic Cytology, 5*, 107–120.

Oudot-Le Secq, M. P., & Green, B. R. (2011). Complex repeat structures and novel features in the mitochondrial genomes of the diatoms *Phaeodactylum tricornutum* and *Thalassiosira pseudonana*. *Gene, 476*, 20–26.

Oudot-Le Secq, M. P., Loiseaux-de Goër, S., Stam, W., & Olsen, J. (2006). Complete mitochondrial genomes of the three brown algae (Heterokonta: Phaeophyceae) *Dictyota dichotoma, Fucus vesiculosus* and *Desmarestia viridis*. *Current Genetics, 49*, 47–58.

Paz, Y., Shimoni, E., Weiss, M., & Pick, U. (2007). Effects of iron deficiency on iron binding and internalization into acidic vacuoles in *Dunaliella salina*. *Plant Physiology, 144*, 1407–1415.

Peers, G., Truong, T., Ostendorf, E., Busch, A., Elrad, D., Grossman, A., et al. (2009). An ancient light-harvesting protein is critical for the regulation of algal photosynthesis. *Nature, 462*, 518–521.

Percival, E. G. V., & Ross, A. G. (1951). The constitution of laminarin. Part II. The soluble laminarin of *Laminaria digitata*. *Journal of Chemical Society, 151*, 720–726.

Peters, A. F., Mann, A. D., Córdova, C. A., Brodie, J., Correa, J. A., Schroeder, D. C., et al. (2010a). Genetic diversity of *Ectocarpus* (Ectocarpales, Phaeophyceae) in Peru and northern Chile, the area of origin of the genome-sequenced strain. *New Phytologist, 188*, 30–41.

Peters, A. F., Marie, D., Scornet, D., Kloareg, B., & Cock, J. M. (2004). Proposal of *Ectocarpus siliculosus* (Ectocarpales, Phaeophyceae) as a model organism for brown algal genetics and genomics. *Journal of Phycology, 40*, 1079–1088.

Peters, A. F., Scornet, D., Ratin, M., Charrier, B., Monnier, A., Merrien, Y., et al. (2008). Life-cycle-generation-specific developmental processes are modified in the *immediate upright* mutant of the brown alga *Ectocarpus siliculosus*. *Development, 135*, 1503–1512.

Peters, A. F., van Wijk, S. J., Cho, G. Y., Scornet, D., Hanyuda, T., Kawai, H., et al. (2010b). Reinstatement of *E. crouaniorum* Thuret in Le Jolis as a third common species of *Ectocarpus* (Ectocarpales, Phaeophyceae) in western Europe, and its phenology at Roscoff, Brittany. *Phycological Research, 58*, 157–170.

Pohnert, G., & Boland, W. (2002). The oxylipin chemistry of attraction and defense in brown algae and diatoms. *Natural Products Reports, 19*, 108–122.

Potin, P., & Leblanc, C. (2006). Phenolic-based adhesives of marine brown algae. In A. M. Smith, & J. A. Callow (Eds.), *Biological adhesives*. Berlin, Heidelberg: Springer-Verlag. *151*, 105-124.

Read, S. M., Currie, G., & Bacic, A. (1996). Analysis of the structural heterogeneity of laminarin by electrospray-ionisation-mass spectrometry. *Carbohydr Research, 281*, 187–201.

Reyes-Prieto, A., Weber, A. P., & Bhattacharya, D. (2007). The origin and establishment of the plastid in algae and plants. *Annual Review of Genetics, 41*, 147–168.

Ritter, A., Ubertini, M., Romac, S., Gaillard, F., Delage, L., Mann, A., et al. (2010). Copper stress proteomics highlights local adaptation of two strains of the model brown alga *Ectocarpus siliculosus*. *Proteomics, 10*, 2074–2088.

Rousvoal, S., Groisillier, A., Dittami, S. M., Michel, G., Boyen, C., & Tonon, T. (2011). Mannitol-1-phosphate dehydrogenase activity in *Ectocarpus siliculosus*, a key role for mannitol synthesis in brown algae. *Planta, 233*, 261–273.

Russell, G. (1983a). Formation of an ectocarpoid epiflora on blades of *Laminaria digitata*. *Marine Ecology Progress Series, 11*, 181–187.

Russell, G. (1983b). Parallel growth-patterns in algal epiphytes and *Laminaria* blades. *Marine Ecology Progress Series, 13*, 303–304.

Schmid, C. E., Müller, D. G., & Eichenberger, W. (1994). Isolation and characterization of a new phospholipid from brown algae. Intracellular localiztaion and site of biosynthesis. *Journal of Plant Physiology, 143*, 570–574.

Shui, J., Saunders, E., Needleman, R., Nappi, M., Cooper, J., Hall, L., et al. (2009). Light-dependent and light-independent protochlorophyllide oxidoreductases in the chromatically adapting cyanobacterium *Fremyella diplosiphon* UTEX 481. *Plant Cell and Physiology, 50*, 1507–1521.

Silberfeld, T., Leigh, J. W., Verbruggen, H., Cruaud, C., de Reviers, B., & Rousseau, F. (2010). A multi-locus time-calibrated phylogeny of the brown algae (Heterokonta, Ochrophyta, Phaeophyceae): Investigating the evolutionary nature of the "brown algal crown radiation". *Molecular Phylogenetics and Evoltion, 56*, 659–674.

Stache-Crain, B., Müller, D. G., & Goff, L. J. (1997). Molecular systematics of *Ectocarpus* and *Kuckuckia* (Ectocarpales, Phaeophyceae) inferred from phylogenetic analysis of nuclear and plastid-encoded DNA sequences. *Journal of Phycology, 33*, 152–168.

Sugimoto-Shirasu, K., Stacey, N., Corsar, J., Roberts, K., & McCann, M. (2002). DNA topoisomerase VI is essential for endoreduplication in *Arabidopsis*. *Current Biology, 12*, 1782–1786.

Tenhaken, R., Voglas, E., Cock, J., Neu, V., & Hiuber, C. (2011). Characterization of GDP-mannose dehydrogenase from the brown alga *Ectocarpus siliculosus* providing the precursor for the alginate polymer. *Journal of Biological Chemistry, 286*, 16707–16715.

Trinklein, N., Aldred, S., Hartman, S., Schroeder, D., Otillar, R., & Myers, R. (2004). An abundance of bidirectional promoters in the human genome. *Genome Research, 14*, 62–66.

van Dam, T. J., Zwartkruis, F. J., Bos, J. L., & Snel, B. (2011). Evolution of the TOR pathway. *Journal of Molecular Evolution, 73*, 209–220.

Van Loon, L. C., Rep, M., & Pieterse, C. M. J. (2006). Significance of inducible defense-related proteins in infected plants. *Annual Review of Phytopathology, 44*, 135–162.

Vastenhouw, N., & Plasterk, R. (2004). RNAi protects the *Caenorhabditis elegans* germline against transposition. *Trends in Genetics, 20*, 314–319.

Venkatachalam, K., & Montell, C. (2007). TRP channels. *Annual Review of Biochemistry, 76*, 387–417.

Voinnet, O. (2009). Origin, biogenesis, and activity of plant microRNAs. *Cell, 136*, 669–687.

Wilhelm, C., Büchel, C., Fisahn, J., Goss, R., Jakob, T., Laroche, J., et al. (2006). The regulation of carbon and nutrient assimilation in diatoms is significantly different from green algae. *Protist, 157*, 91–124.

Wilson, W. H., Van Etten, J. L., & Allen, M. J. (2009). The Phycodnaviridae: The story of how tiny giants rule the world. *Current Topics in Microbiology and Immunology, 328*, 1–42.

Wu, J. F., & Luther, G. W. (1994). Size-fractionated iron concentrations in the water column of the Western North-Atlantic Ocean. *Limnology and Oceanography, 39*, 1119–1129.

Yamagishi, T., Motomura, T., Nagasato, C., Kato, A., & Kawai, H. (2007). A tubular mastigoneme-related protein, Ocm1, Isolated from the flagellum of a chromophyte alga, *Ochromonas danica*. *Journal of Phycology, 43*, 519–527.

Yamagishi, T., Motomura, T., Nagasato, C., & Kawai, H. (2009). Novel proteins comprising the stramenopile tripartite mastigoneme in *Ochromonas danica* (Chrysophyceae). *Journal of Phycology, 45*, 1100–1105.

Yamaguchi, T., Ikawa, T., & Nisizawa, K. (1966). Incorporation of radioactive carbon from $H^{14}CO_3^-$ into sugar constituents by a brown alga, *Eisenia bicyclis*, during photosynthesis and its fate in the dark. *Plant Cell and Physiology, 7*, 217–229.

Yin, Y., Cheong, H., Friedrichsen, D., Zhao, Y., Hu, J., Mora-Garcia, S., et al. (2002). A crucial role for the putative *Arabidopsis* topoisomerase VI in plant growth and development. *Proceedings of the National Academy of Sciences of the United States of America, 99*, 10191–10196.

Yoon, H. S., Hackett, J. D., Ciniglia, C., Pinto, G., & Bhattacharya, D. (2004). A molecular timeline for the origin of photosynthetic eukaryotes. *Molecular Biology and Evolution, 21*, 809–818.

Zambounis, A., Elias, M., Sterck, L., Maumus, F., & Gachon, C. M. (2012). Highly dynamic exon shuffling in candidate pathogen receptors. What if brown algae were capable of adaptive immunity? *Molecular Biology and Evolution, 29*, 1263–1276.

Zhai, J., Jeong, D. H., De Paoli, E., Park, S., Rosen, B. D., Li, Y., et al. (2011). MicroRNAs as master regulators of the plant NB-LRR defense gene family via the production of phased, trans-acting siRNAs. *Genes and Devlopment, 25*, 2540–2553.

Zhu, S. H., & Green, B. R. (2010). Photoprotection in the diatom *Thalassiosira pseudonana*: Role of LI818-like proteins in response to high light stress. *Biochimica et Biophysica Acta, 1797*, 1449–1457.

Ziegler, W., Gingras, A., Critchley, D., & Emsley, J. (2008). Integrin connections to the cytoskeleton through talin and vinculin. *Biochemical Society Transactions, 36*, 235–239.

CHAPTER SIX

Genomics of Volvocine Algae

James G. Umen*, Bradley J. S. C. Olson[†]

*Donald Danforth Plant Science Center, St. Louis, Missouri, USA

[†]Molecular Cellular and Developmental Biology, Ecological Genomics Institute, Division of Biology, Kansas State University, Manhattan, Kansas, USA

[1]Corresponding author: jumen@danforthcenter.org

Contents

Advances in Botanical Research, Volume 64
ISSN 0065-2296,
http://dx.doi.org/10.1016/B978-0-12-391499-6.00006-2

Abstract

Volvocine algae are a group of chlorophytes that together comprise a unique model for evolutionary and developmental biology. The species *Chlamydomonas reinhardtii* and *Volvox carteri* represent extremes in morphological diversity within the Volvocine clade. *Chlamydomonas* is unicellular and reflects the ancestral state of the group, while *Volvox* is multicellular and has evolved numerous innovations including germ-soma differentiation, sexual dimorphism and complex morphogenetic patterning. The *Chlamydomonas* genome sequence has shed light on several areas of eukaryotic cell biology, metabolism and evolution, while the *Volvox* genome sequence has enabled a comparison with *Chlamydomonas* that reveals some of the underlying changes that enabled its transition to multicellularity but also underscores the subtlety of this transition. Many of the tools and resources are in place to further develop Volvocine algae as a model for evolutionary genomics.

1. INTRODUCTION

Ever since van Leeuwenhoek first described *Volvox* collected from a roadside ditch (van Leeuwenhoek, 1700), this elegant organism and its cousins, the Volvocine algae, have fascinated biologists. Just over 300 years after their first published description, Volvocine algae entered the genomic era with genome sequencing of the unicellular species *Chlamydomonas reinhardtii* (Merchant *et al.*, 2007) followed by sequencing of *Volvox carteri* (Prochnik *et al.*, 2010). These two species are at the forefront of research in diverse areas including evolution, motility, photosynthesis and development, and it has now become possible to add an arsenal of '-omics' tools to speed progress and answer new questions. Genomics research with *Chlamydomonas* has shed light on several areas of biology by helping define conserved genes related to both plant and animal biology. The addition of the *Volvox* genome makes possible comparisons within Volvocine algae aimed at deciphering the genetic changes that underlie a major transition in biological complexity represented by differences between unicellular *Chlamydomonas* and multicellular *Volvox*. These comparisons have revealed how remarkably similar these two species are at the genome level and have prompted a reconsideration of how much change is required to drive a major evolutionary transition. In this chapter, we introduce the Volvocine algae and highlight key areas of genomics research to which they have contributed.

2. THE BIOLOGY OF THE VOLVOCALES

The Volvocales (or Volvocine algae) are a subgroup of chlorophytes (green algae) comprising dozens of species in seven major genera (Fig. 6.1). The two best-known species are unicellular *C. reinhardtii* and multicellular, *V. carteri* (Fig. 6.2). These two species represent extremes in a range of developmental complexity observed within the Volvocales (Fig. 6.2). Each of the major genera is defined by its level of organization such as colony size and morphology, cell number, cell differentiation and the degree of germ-soma specialization (Fig. 6.2). A brief introduction to the biology, taxonomy, ecology and culture resources for Volvocine algae follows.

2.1. *Chlamydomonas* is a Prototypical Volvocine Alga

C. reinhardtii has been used as a model for many decades to investigate diverse areas of biology (Grossman *et al.*, 2003, 2007; Harris, 2001, 2004; Peers & Niyogi, 2008; Stern *et al.*, 2009). Molecular genetic tools and resources have been developed for *Chlamydomonas* including nuclear and organellar transformation, insertional mutagenesis, RNA-interference (RNAi)-mediated silencing, microarrays, expressed sequence tag (EST)

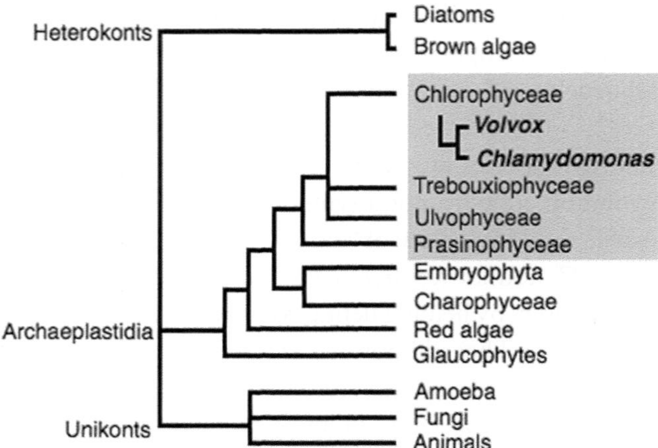

Figure 6.1 The phylogenetic relationship between the Volvocales, green algae, and other major divisions of Eukaryotes. *Chlamydomonas* and *Volvox* belong to Archaeplastidia, the group that contains primary endosymbiotic plastids and includes green algae, red algae, glaucophytes, charophytes and land plants (embryophytes). Major green algal subdivisions are the Chlorophyceae, Trebouxiophyceae, Ulvophyceae and Prasinophyceae indicated with a shaded box. Diatoms and brown algae are in a separate division, the Heterokonts. Animals, fungi and amoebae belong to the Unikonts. The unrooted cladogram is adapted from Baldauf (2003). For interpretation of the references to colour in this figure legend, the reader is referred to the online version of this book.

libraries, bacterial artificial chromosome (BAC) libraries, complementary DNA (cDNA) libraries and numerous strains and mutants (see Tables 6.1 and 6.2). The following section provides a brief overview of *Chlamydomonas*.

2.1.1. Chlamydomonas Taxonomy

C. reinhardtii belongs to the chlorophycean family Chlamydomonaceae (Fig. 6.1). The genus *Chlamydomonas* encompasses over 600 species but is now recognized as being polyphyletic and the genus will need substantial revision (Buchheim, Buchheim, & Chapman, 1997; Hoham, Bonome, Martin, & Leebens Mack, 2002; Lemieux, Turmel, & Lemieux, 1985; Morita *et al.*, 1999; Pröschold, Marin, Schlösser, Melkonian, 2001; Stern *et al.*, 2009). Molecular phylogenies place *C. reinhardtii* and a few other *Chlamydomonas* species as the closest unicellular relatives of colonial Volvocine algae that are together part of a coherent monophyletic grouping (Fig. 6.2) (Coleman, 1999; Nakada, Nozaki, & Tomita, 2010; Nozaki & Itoh, 1994; Nozaki *et al.*, 1995, 2000, 2002) with an estimated divergence time of ~200 MY (Herron *et al.*, 2009). We refer here to Volvocine algae as the group containing colonial Volvocales plus their closest unicellular relatives including *C. reinhardtii* (Fig. 6.2).

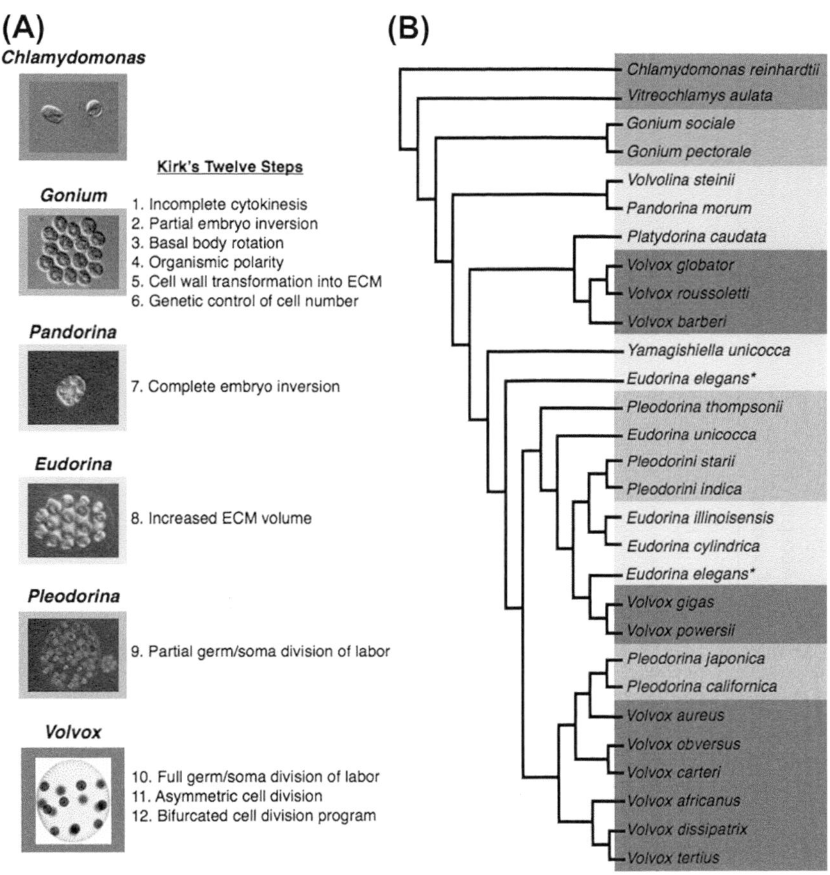

Figure 6.2 The morphology and phylogeny of the Volvocales. (A) Micrographs of Volvocine species and their relationship to each of Kirk's 12 steps of multicellular evolution (Kirk, 2005). (B) Phylogeny of selected Volvocales adapted from (Kirk, 2005; Nozaki, 2003). Species are color-coded based on morphology as indicated in panel (A). Asterisks indicate two isolates of *Eudorina elegans* that subsequently were reclassified as distinct species (Nozaki, 2003; Yamada, Miyaji, & Nozaki, 2008). See the colour plate.

2.1.2. Chlamydomonas *Cell Structure*

Each *Chlamydomonas* cell is 5–10 μM in diameter and is shaped like a pear or ovoid (Fig. 6.3). Surrounding the plasma membrane is a cell wall composed of distinct layers of crosslinked hydroxyproline-rich glycoproteins (Stern *et al.*, 2009; Woessner & Goodenough, 1994). A pair of basal bodies at the anterior end of the cell nucleates two flagella that provide motility and chemosensory functions. The basal bodies also nucleate and organize the microtubule cytoskeleton within the cell body (Dutcher, 2003; Pearson & Winey, 2009)

Table 6.1 Volvocales Culture Collections

Culture Collection	Species Available	Media	Other	Accepts New Cultures?	Web Site
Chlamydomonas collection	*Chlamydomonas reinhardtii, some other chlamydomonadaceae*	Recipes and for sale	Plasmids complementary DNA libraries, genomic libraries	Yes, all *chlamydomonadaceae*	http://chlamycollection.org/
UTEX	Some *Chlamydomonas*, many Volvocales, other chlorophytes and protists	Recipes	Genomic DNA for PCR analysis	Yes, with prior approval	http://web.biosci.utexas.edu/utex/
SAG	Many Volvocales, many other chlorophytes and protists	References		Yes, with prior approval	http://epsag.uni-goettingen.de/cgi-bin/epsag/website/cgi/show_page.cgi?kuerzel=start
CCAP	Many Volvocales, many other chlorophytes and protists	Recipes and for sale		Yes, also accepts private collections	http://www.ccap.ac.uk/index.htm
NIES	A wide array of algal strains, extensive collection of the Volvocales	Recipes		Yes, with prior approval	http://mcc.nies.go.jp/

Table 6.2 Genomic Resources for *Chlamydomonas* and *Volvox*

Resource	Description	Reference(s)	URL
General resources			
Chlamy.org	General landing page for all information about *Chlamydomonas*		http://www.chlamy.org
Chlamy Collection	Strain and plasmid stock center and general information about *Chlamydomonas*		http://www.chlamycollection.org
Sequences			
Phytozome	Latest assembly and annotations of the Chlamydomonas and Volvox genomes and comparison tools.		http://www.phytozome.net
JGI Genome Browser for *Chlamydomonas*	Version 4 of the Chlamydomonas genome including use annotations. All future assemblies will be deposited on Phytozome		http://genome.jgi-psf.org/Chlre4/Chlre4.home.html
Chlamydomonas reinhardtii minus mating-type locus	GenBank Accession GU814015	Ferris *et al.* (2010)	http://www.ncbi.nlm.nih.gov/nuccore/GU814015
Chlamydomonas reinhardtii plus mating-type locus	GenBank Accession GU814014	Ferris *et al.* (2010)	http://www.ncbi.nlm.nih.gov/nuccore/GU814014
Chlamydomonas reinhardtii chloroplast genome	GenBank Accession GU814014	Maul *et al.* (2002)	http://www.chlamy.org/chloro/default.html
Chlamydomonas reinhardtii mitochondrial genome	GenBank Accession FJ423446	Ma *et al.* (1992)	http://www.ncbi.nlm.nih.gov/nuccore/NC_001638
Chlamydomonas incerta expressed sequence tag (EST) database	GenBank Accessions GI106792973-GI106798104	Popescu *et al.*	http://megasun.bch.umontreal.ca/pepdb/pep.html

(*Continued*)

Table 6.2 Genomic Resources for *Chlamydomonas* and *Volvox*—cont'd

Resource	Description	Reference(s)	URL
JGI Genome Browser for Volvox	Version 1 of the Volvox carteri genome including user annotations. All future versions of the genome will be deposited on Phytozome		http://genome.jgi.doe.gov/Volca1/Volca1.info.html
Volvox carteri Female mating type locus	GenBank Accession GU784915	Ferris *et al.* (2010)	http://www.ncbi.nlm.nih.gov/nuccore/GU784915
Volvox carteri Male mating type locus	GenBank Accession GU784916	Ferris *et al.* (2010)	http://www.ncbi.nlm.nih.gov/nuccore/GU784916
Volvox carteri chloroplast genome	GenBank Accession GU084820.1	Smith and Lee (2009)	http://www.ncbi.nlm.nih.gov/nuccore/GU084820.1
Volvox carteri mitochondrial genome	GenBank Accession EU760701	Smith and Lee (2009)	http://www.ncbi.nlm.nih.gov/nuccore/EU760701
Libraries			
Chlamydomonas reinhardtii genomic BAC library			http://www.genome.clemson.edu
Volvox carteri female genomic BAC libraries			http://www.genome.arizona.edu/
Volvox cateri male genomic BAC libraries		Ferris *et al.* (2010)	http://www.genome.clemson.edu/
Chlamydomonas EST database		Asamizu *et al.* (2004)	http://est.kazusa.or.jp/en/plant/chlamy/EST/

Chlamydomonas 454 read database	2 Gb of 454 sequencing of *Chlamydomonas* complementary DNAs (cDNAs) from various growth conditions		http://genomes.mcdb.ucla.edu/Cre454/
Functional annotation and expression tools			
Algal functional annotation tool	Comprehensive algal genomics, annotation and expression tool	Lopez *et al.* (2011)	http://pathways.mcdb.ucla.edu/algal/index.html
GreenGenie2	Gene predictions for version 4 of the *Chlamydomonas* genome	Kwan *et al.* (2009)	http://stormo.wustl.edu/GreenGenie2/
ChlamyCyc	*Chlamydomonas centric*, integrated annotation tool that integrates with other related plant databases		http://chlamyto.mpimp-golm.mpg.de/chlamycyc/index.jsp
Plant genome database	Plant genome annotation comparison tool that includes Chlamydomonas reinhardtii		http://www.plantgdb.org/search/misc/
Plaza	Tools for comparative plant genomics	Van Bel *et al.* (2012)	http://bioinformatics.psb.ugent.be/plaza/
Proteomic databases			
Chlamydomonas flagellar proteome			http://labs.umassmed.edu/chlamyfp/protector_login.php
Plant transcription factor databases			
Plant transcription factor database	Includes both *Chlamydomonas reinhardtii* and *Volvox carteri*, version 2.0	Zhang *et al.* (2010)	http://planttfdb.cbi.edu.cn/
Plant transcription factor database	Includes *Chlamydomonas reinhardtii*, version 3.0	Pérez-Rodríguez *et al.* (2009)	http://plntfdb.bio.uni-potsdam.de/v3.0/

(*Continued*)

Table 6.2 Genomic Resources for *Chlamydomonas* and *Volvox*—cont'd

Resource	Description	Reference(s)	URL
Microarrays			
Chlamydomonas reinhardtii microarray	Agilent *Chlamydomonas reinhardtii* microarray, number 024664	Toepel *et al.* (2011)	https://earray.chem.agilent.com/earray/
Chlamydomonas reinhardtii UNIGENE cDNA array	cDNA array based on the first *Chlamydomonas* UNIGENE set superseded by the agilent arrays	Jain *et al.* (2007)	
Chlamydomonas reinhardtii oligo array	Oligonucleotide array based on the first *Chlamydomonas* EST set superseded by the agilent arrays	Eberhard *et al.* (2006)	
***Chlamydomonas* small RNA (sRNA) database**			
SiLoDb sRNA locus database	Database of sRNA locations in the *Chlamydomonas* genome		http://silodb.cmp.uea.ac.uk/

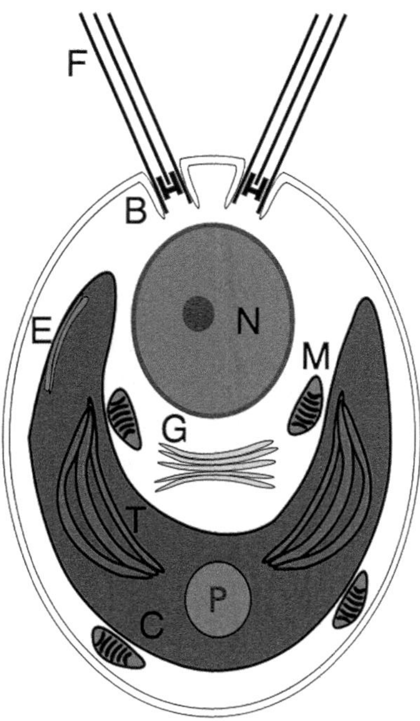

Figure 6.3 Cell structure of *Chlamydomonas reinhardtii*. Schematic representation of the subcellular components of a typical *Chlamydomonas* cell; flagella (F), basal body (B), nucleus (N), eye spot (E), golgi apparatus (G), mitochondria (M), chloroplast (C), thylakoids (T) and the pyrenoid (P). For colour version of this figure, the reader is referred to the online version of this book.

(Fig. 6.3). The flagella and basal bodies are structurally and functionally homologous to cilia and basal bodies of animal cells and *Chlamydomonas* serves as a model for investigating these organelles (Dutcher, 2003; Pedersen & Rosenbaum, 2008). The nucleus is positioned below the basal bodies, while the posterior of each cell is occupied by a single large cup-shaped chloroplast that encompasses over half of the total cell volume (Fig. 6.3). *Chlamydomonas* is of particular interest for evolutionary genomics because it has attributes of land plants (chloroplast, photosynthetic metabolism) and of animals, lower plants and protists (flagella and basal bodies) (Grossman *et al.*, 2003; Harris, 2001; Merchant *et al.*, 2007). Indeed, as described below, the unique biology of this duality is reflected in its protein-coding gene repertoire and has been exploited to identify new candidate genes that are involved in plant- and animal-specific processes.

2.1.3. Chlamydomonas Vegetative and Sexual Cycles

Chlamydomonas cells are haploid and have two mating types, *plus* and *minus*, that are specified by a mating locus (*MT*) (Ferris, Armbrust, & Goodenough,

2002; Ferris *et al.*, 2010; Goodenough, Lin, & Lee, 2007). Under nutrient replete conditions, *Chlamydomonas* cells proliferate mitotically using a modified cell cycle termed multiple fission (also called palintomy) that is shared by most other Volvocine algae and many other chlorophytes (Graham, Graham, & Wilcox, 2009; Kirk, 1995, 2005). This mode of reproduction involves a long growth period followed by a rapid series of S phases and mitoses to produce 2^n daughters (2, 4, 8, 16, . . .), where *n* is typically 1, 2, 3 or 4 in *Chlamydomonas* (Fig. 6.4). In a diurnal environment the cell cycle becomes synchronized with growth occurring during the day and division occurring at night (Harris, 2001). In *Chlamydomonas*, the division number (*n*) is variable, but in colonial Volvocine algae it is more stereotyped with variability of usually just one cycle within a species.

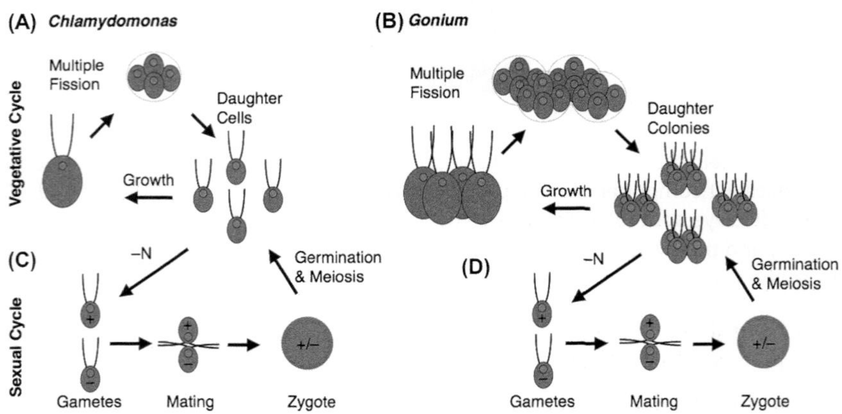

Figure 6.4 The multiple fission cell cycle and sexual cycles of *Chlamydomonas* and *Gonium*. During vegetative growth (A), *Chlamydomonas* cells may grow many fold in size with the extent of growth somewhat indeterminate. Cells then divide multiple times to produce uniform-sized daughters. Two rounds of cell division are shown in this panel, with division numbers ranging from one to four in a typical culture. *Gonium* colonies (B) follow a similar growth and division pattern as *Chlamydomonas*, but the daughter cells remain attached to each other through cytoplasmic bridges and ECM (see Fig. 6.6). The *Chlamydomonas* sexual cycle (C) is induced by lack of nitrogen (−N) that causes cells to differentiate into gametes. Gametes from each mating type (*plus* and *minus*) are similar in size. Flagellar adhesion between gametes of opposite mating type precedes cell fusion to form a diploid zygote. Upon germination, four meiotic progeny are produced that can re-enter the vegetative cycle. In *Gonium* −N also triggers gametogenesis that involves colony dissolution into unicellular gametes. Post-meiotic *Gonium* progeny are single cells that produce a new vegetative colony after their first passage through the cell cycle. For colour version of this figure, the reader is referred to the online version of this book.

When starved of nitrogen, vegetative *Chlamydomonas* cells differentiate into mating-competent gametes. While gametes are morphologically similar to vegetative cells, they express a genetic program that allows them to recognize cells of the opposite mating type and to fuse into a zygote that differentiates to become a dormant diploid zygospore. Upon return to light and nutrients, the spores undergo meiosis and germinate to yield four haploid progeny that re-enter the vegetative reproductive cycle (Goodenough, Armbrust, Campbell, & Ferris, 1995, 2007; Stern *et al.*, 2009) (Fig. 6.4).

2.1.4. *Chlamydomonas* Metabolism

Chlamydomonas cells can grow phototrophically on simple mineral nutrient media but can also grow heterotrophically using external acetate as a carbon source for respiration. Its ability to grow heterotrophically has made *Chlamydomonas* an important model for photosynthesis research since photosynthetically inactive mutants are viable when supplied with acetate (Dent *et al.*, 2005; Grossman *et al.*, 2003; Harris, 2001). Moreover, *Chlamydomonas* also has evolved anaerobic metabolic pathways that are of significant interest for understanding biological hydrogen (H_2) production (Costa & de Morais, 2011; Kruse & Hankamer, 2010; Rupprecht, 2009). Pathways for key nutrients and metabolites such as nitrogen, sulfur, phosphorous, iron, copper, starch lipids and others have been investigated extensively (Grossman *et al.*, 2003; Harris, 2001; Stern *et al.*, 2009).

2.2. Colonial Volvocine Algae Show Convergent Evolution of Form and Function

2.2.1. Overview of Colonial Volvocine Forms

The colonial Volvocine algae come in a diverse forms that can be organized based on colony size, cell type specialization and other morphological attributes. These attributes were traditionally used to group the Volvocine algae into genera and are a useful means of describing different types; but as described below and in Fig. 6.2, the morphological classification scheme is misleading with respect to phylogenetic relationships within the Volvocales where multiple instances of convergent evolution have occurred.

The simplest colonial genera is *Gonium*. Morphologically, *Gonium* looks like a group of 4, 8 or 16 *Chlamydomonas* cells that are stuck together. Indeed, each *Gonium* colony arises from a single cell that undergoes multiple fission but whose daughters remain attached after division (Hoops, Nishii, & Kirk, 2006) (Fig. 6.4). *Gonium* colonies also have polarity, with cells arranged in a slightly curved sheet with flagella on the apical surface (convex side) of the colony (Fig. 6.2). This configuration is achieved through a post-mitotic contortion termed inversion that bends the cell sheet outwards so that the flagella (that originally faced inwards after division) end up on the outside face of the colony. Radial polarity is also evident in the orientation of flagella: In 16-cell *Gonium* colonies the inner 4 cells have their flagella arranged as those in *Chlamydomonas* (with opposed beat directions), while the outer 12 cells have rotated basal bodies that allow the two flagella to beat with a more parallel stroke that is characteristic of all the larger colonial genera (Fig. 6.6). Partial inversion in *Gonium* is functionally similar to the process of full inversion in the remaining spheroidal genera that involves a complete turning inside out of the post-mitotic colony (Kirk, 2005).

The remaining genera include *Pandorina* that contain 8, 16 or 32 similar-sized cells in a compact ball with flagella pointing outwards (Fig. 6.2). *Eudorina* colonies have 16, 32 or 64 cells but are much larger than *Pandorina* due to the secretion of an extensive glycoprotein-rich extracellular matrix (ECM) in which cells are embedded. *Pleodorina* colonies are larger still (32, 64 or 128 cells) and show partial cell type specialization with some small somatic cells on the anterior of the spheroid that never grow or divide, but only provide motility. The remaining cells are flagellated and motile but continue to grow. Eventually, these cells withdraw their flagella and divide into new colonies.

Volvox is the most morphologically complex Volvocine genera in terms of size and cell-type specialization. Each spherical colony is up to 0.5 mm in diameter and contains 2000–4000 small sterile somatic cells around the outside embedded in an extensive network of ECM (Fig. 6.6). The somatic cells provide motility for the colony and secrete ECM but do not grow or reproduce. Within each colony are large reproductive cells termed gonidia that grow and undergo a morphogenesis program that is described in more detail below for the model species *V. carteri* (see Section 6.1 below).

Colonial Volvocine algae have sexual cycles similar to that of *Chlamydomonas* with the smaller colonial forms producing gametes of equal size like *Chlamydomonas*. Larger colonial genera such as *Eudorina*, *Pleodorina* and *Volvox* are anisogamous or oogamous with males producing packets of small sperm and females producing large eggs (Fig. 6.5) (Nozaki *et al.*, 2000).

2.2.2. Phylogenetic Relationships Among Volvocine Algae

Morphological criteria that led to grouping the various Volvocine species into the genera described above also suggested a simple stepwise progression in developmental complexity termed the Volvocine lineage hypothesis (Kirk, 1998; Larson, Kirk, & Kirk, 1992). In its simplest form, this hypothesis

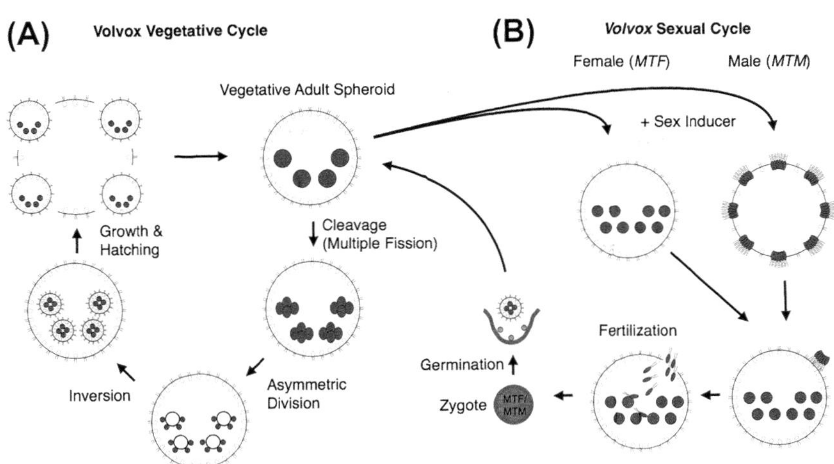

Figure 6.5 Vegetative and sexual developmental in *Volvox*. (A) The *Volvox* vegetative life cycle for males and females is identical and begins with a mature adult (upper right) whose gonidia (large green cells) undergo cleavage to begin embryogenesis. During the sixth cleavage cycle, asymmetric division occurs and leads to production of 16 large anterior cells that are destined to form the germ cells in the next generation. After a total of 12 cleavage cycles, the ~2000-celled embryo undergoes inversion to place the gonidial precursors inside the spheroid and the flagella of the somatic cells pointing outside in their final adult configuration. After a period of growth, the daughter spheroids hatch and continue to grow and mature into adults that can restart the vegetative cycle. (B) Sexual development begins when immature gonidia are exposed to sex-inducer protein. Their subsequent embryogenesis is then altered to produce egg-bearing females or sperm-packet-bearing males (Ferris *et al.*, 2010; Hallmann *et al.*, 1998). Females contain 32–48 eggs and ~2000 somatic cells, while males contain 128 somatic cells and 128 packets of 64–128 sperm. Sperm packets are released whereupon they swim as a single unit until they encounter a sexual female. Upon interacting with a sexual female, the sperm packet dissolves into individual cells that enter the female ECM through a fertilization pore and swim until an individual sperm encounters and fuses with an egg to form a diploid zygote. When a *Volvox* zygospore germinates, only one of the four meiotic products survives and differentiates into a new vegetative spheroid. See the colour plate.

predicts a nested phylogenetic tree with each new genera arising once from a less complex ancestral form. Molecular systematics reveals a more complex history in this group with multiple independent transitions between levels of organization. For example, colonials with morphology similar to *Eudorina* (Fig. 6.2) have arisen independently at least three times to create the genera now termed *Eudorina*, *Volvulina* and *Yamagishiella* (Herron *et al.*, 2009; Nozaki *et al.*, 2000). Likewise, *Volvox*-like colonial forms have evolved at least three times independently with distinct developmental patterning mechanisms that give rise to germ-soma differentiation (Desnitski, 1995).

Even more problematic for the Volvocine lineage hypothesis are trait losses such as the absence of expanded ECM in some subgroups of *Eudorina* that are morphologically closer to *Pandorina* than to other members of their clade (Fig. 6.2). While complicating the task of classification, convergent evolution within the Volvocine algae makes them appealing as a model system because it allows comparative studies of how similar developmental innovations arose or were lost independently within a well-defined clade.

2.3. Ecology of the Volvocine Algae

Volvocine algae are freshwater species that are distributed worldwide. Long distance dispersion is possible due to the production of dormant sexual zygospores that can survive for years in desiccated environments and are highly resistant to stress. Wind, attachment to animals or ingestion by animals, provides a means for the spores to travel and allows for maintenance of genetic continuity between geographically distant populations. *Chlamydomonas* is often isolated from soil samples though it can also be found in ponds and temporary bodies of water. The methods for isolating *Chlamydomonas* and other algae from soils favour germinating zygospores, so it is not clear whether significant populations of vegetative *Chlamydomonas* are present in soils where they are collected. In contrast, colonial Volvocine algae are typically isolated as vegetative forms from transient puddles, rice paddies and warm ponds. The presence of vegetative colonial Volvocine algae in temporary bodies of water implies the presence of zygospores in the soil prior to inundation as vegetative forms do not survive desiccation.

Collection method biases notwithstanding, there are several potential advantages to a colonial lifestyle in an open water environment. The first is increased size. Colonies can be resistant to grazing predators that eat unicells but which have an upper limit to the size of prey they can ingest (Bell, 1985; Bonner, 2000; Kirk, 1998). The second potential advantage of being colonial is increased efficiency in resource utilization. In general, such efficiency is thought to be of greatest benefit for multicellular species in environments that fluctuate from highly nutrient rich to nutrient poor. Under these conditions, a multicellular organism can collect and store nutrients allowing it to outcompete unicellular species whose growth is more tightly coupled to immediately available nutrients (Kirk, 1998). A third potential benefit of multicellularity is increased mobility. For example, the combined effort of 4000 somatic flagella beating on the surface of a *Volvox* spheroid supports motility at rates of ~500–900 μm/s (Solari, Kessler, & Michod, 2006; Solari, Michod, & Goldstein, 2008; Ueki *et al.*, Hallmann, 2010), while the two flagella on a *Chlamydomonas* cell allow movement at a rate of ~100–200 μm/s (Ojakian & Katz, 1973). The faster speed of a *Volvox* spheroid would allow it to travel much greater distances through a stratified water column in search of mineral nutrients. Moreover, the second advantage (efficiency of resource utilization) may be enhanced by the third advantage (motility) in the action of flagellar beating that disrupts the diffusion boundary layer around each spheroid and thereby increases the efficiency of nutrient uptake (Koufopanou, 1994; Short *et al.*, 2006).

2.4. Culture Resources for Volvocine Algae

Chlamydomonas and other Volvocine algae are maintained and distributed from any of several culture collections. Some of the major sources of Volvocine culture stocks are summarized in Table 6.1. The *Chlamydomonas* stock centre has wild-type strains, including inbred isogenic reference strains, many published and unpublished mutants and several independent wild isolates. In addition to cultures, the *Chlamydomonas* stock centre also distributes cDNA libraries and plasmids. The UTEX algal collection at the University of Texas, Austin, has a good representation of Volvocine algal strains. UTEX also maintains a comprehensive database of freshwater algal media recipes and protocols. Additional collections of Volvocine algae are SAG at the University of Gottingen in Germany, CCAP in United Kingdom and NIES in Japan. Currently, two Web-based resources provide information for Volvocine algae and other groups. Protist images (http://protist.i.hosei.ac.jp/Protist_menuE.html) has a large collection of photos and taxonomy of protists, including Volvocine algae. Algaebase (http://www.algaebase.org/) includes a comprehensive taxonomy for many Volvocine species.

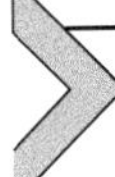

3. *CHLAMYDOMONAS* AND *VOLVOX* GENOME STRUCTURE AND CONTENT

The genome sequence of *C. reinhardtii* was published in 2007 (Merchant *et al.*, 2007) and that of *V. carteri* in 2010 (Prochnik *et al.*, 2010). These two sequences have provided new insights into the biology of Volvocine algae and have enabled a host of additional 'omics' and systems-level approaches (Castruita *et al.*, 2011; Eberhard *et al.*, 2006; Gonzalez-Ballester,

2005; Jamers, Blust, & De Coen, 2009; Rolland *et al.*, 2009; Schmidt *et al.*, 2010). The majority of genomics research on Volvocine algae has been done with *Chlamydomonas*; however, the genome of *Volvox* has already provided new perspectives on the evolution of multicellularity and of Volvocine sexual cycles (Ferris *et al.*, 2010; Prochnik *et al.*, 2010), and we can expect the area of comparative Volvocine algal genomics to be one of increasing activity and interest in the near future. In this review, we focus first on *Chlamydomonas* genomics and the insights it has provided on the evolution of a remarkably complex and sophisticated unicellular eukaryote. We then focus on the emerging area of comparative genomics of *Chlamydomonas*, *Volvox* and soon-to-be sequenced species of Volvocine algae that will be of increasing future interest.

3.1. *Chlamydomonas* Nuclear Genome Structure

The *C. reinhardtii* nuclear genome sequence is ~117 Mb distributed on 17 chromosomes. The nucleotide composition is GC biased (64% GC) as is third position codon usage (Merchant *et al.*, 2007). Protein-coding gene density is relatively high with 16.7% exonic sequence and 12.5% total repeats. *Chlamydomonas* chromosomes have genetically mappable centromeres that are not well characterized but likely contain repeats (Preuss & Mets, 2002; Merchant *et al.*, 2007). Telomeric repeats are around 300–350 bp and are composed of the sequence $(TTTTAGGG)_n$ (Petracek *et al.*, 1990).

3.1.1. The Chlamydomonas Mating-type Locus

Chlamydomonas has two mating types, *plus* and *minus*, that are governed by a mating locus with haplotypes designated MT^+ and MT^-. Although it segregates as a single Mendelian trait, *MT* is a large multigenic region of around 200–300 kb (Fig. 6.6). Sequence rearrangements (inversions and transpositions) between MT^+ and MT^- suppress recombination in this region and keep the genes within and around *MT* in linkage disequilibrium. This rearranged configuration not only keeps sex-related genes together but also keeps over a dozen housekeeping genes trapped within a non-recombining region (Umen, 2011). Thus, the *MT* locus has acquired some properties of a haploid sex chromosome and its expansion in *Volvox* is one of the most notable differences between the two algal genomes (Ferris *et al.*, 2010). The comparative genomics of the *Chlamydomonas* and *Volvox* mating loci are discussed in more detail below.

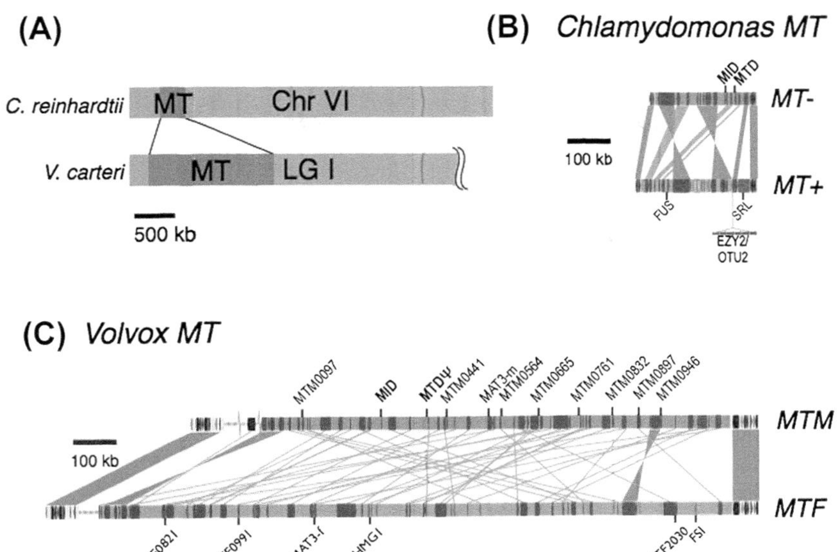

Figure 6.6 The mating-type loci of *Chlamydomonas* and *Volvox*. (A) The mating-type locus for both species is near the telomere of a syntenic chromosome (chromosome 6 in *Chlamydomonas* and linkage group I of *Volvox*). (B) MT^+ and MT^- mating haplotypes of *Chlamydomonas*. Rearrangements between the two haplotypes are indicated by gray shading. Locations of the sex determining genes *MID* and *MTD* and the sex limited gamete fusion gene *FUS1* are shown. Also indicated are the *EZY2/OTU2* regions that may be important for uniparental chloroplast inheritance (Ferris *et al.*, 2002; Goodenough *et al.*, 2007). (C) The *Volvox* male and female mating-type loci are about six times larger than *Chlamydomonas MT* and contain very little syntenic gene order between haplotypes. Several novel genes are shown as well as genes encoding homologs *MID, MTD* and the retinoblastoma tumour suppressor homolog, *MAT3*. Figure adapted from Ferris *et al.* (2010) and Umen (2011). See the colour plate.

3.1.2. Protein-Coding Genes

Various strategies have been employed for protein-coding gene predictions that are homology based or use empirical data. One of the most successful gene prediction programs for *Chlamydomonas* is AUGUSTUS that combines evidence-based and *de novo* gene model predictions (Stanke *et al.*, 2008; Stanke & Morgenstern, 2005; Stanke, Schöffmann, Morgenstern, & Waack, 2006). The *de novo* gene prediction program GreenGenie2 has been trained on a large set of known *Chlamydomonas* cDNA sequences from which it has generated high-quality gene models (Kwan *et al.*, 2009) that are available pre-formatted for processing with Cufflinks (Trapnell *et al.*, 2010), a popular tool for transcript assembly and differential gene expression analysis.

3.1.3. Genes for RNAs

Chlamydomonas encodes 259 transfer RNAs (tRNAs), over half of which are present in clusters that arose through serial gene duplication (Merchant *et al.*, 2007). Compared to other eukaryotes, *Chlamydomonas* tRNAs are relatively intron rich, with 60% intron-containing tRNA genes, and contain an unusually large number that encode a 3' terminal CCA sequence that is usually added post-transcriptionally in eukaryotes (Merchant *et al.*, 2007).

Cytosolic ribosomal RNA (rRNA) encoding gene clusters (18S, 5.8S, 28S) are found at two locations: one end of chromosome 14 and at one end of chromosome 8. A single cytosolic 5S rRNA cluster is located on chromosome 8 about 200 kb from 18S/5.8S/28S cluster with several dozen predicted protein-coding genes in between the two loci (J. Umen, unpublished observation). The highly repetitive nature of these clusters makes their exact size difficult to determine.

The nuclear genome encodes 322 small nucleolar RNAs (snoRNAs) that are involved in rRNA processing and base modifications. *Chlamydomonas* snoRNAs are notable for being clustered and for the high proportion that are encoded within spliceosomal introns of protein-coding genes (Chen *et al.*, 2008; Merchant *et al.*, 2007).

Chlamydomonas contains a complex repertoire of small 22–24 bp RNAs (sRNAs), some of which appear to be micro-RNAs (miRNAs) (Molnar *et al.*, 2007; Zhao *et al.*, 2007). The *Chlamydomonas* genome also encodes key proteins involved in sRNA biogenesis and function including Argonaute proteins and Dicer nuclease, though it does not appear to encode any RNA-dependent RNA polymerases that are potentially involved in generation or amplification of double-stranded RNAs that act as sRNA precursors (Casas-Mollano 2008; Cerutti *et al.*, 2011; Schroda, 2006). As discussed below, RNAi-based gene silencing has proven to be a powerful reverse genetics tool in Volvocine algae.

3.2. *Chlamydomonas* Organellar Genomes

The *Chlamydomonas* chloroplast genome is typical for those in the green algal/land plant lineage. It is a 200-kbp circle with two large inverted repeats that contain rRNA genes and the *psbA* gene. In total, the chloroplast encodes around 99 genes including protein-coding genes, tRNAs and rRNAs (Maul *et al.*, 2002; Grossman *et al.*, 2003). The chloroplast DNA (cpDNA) is present in around 80 copies per cell and is packaged into nucleoids containing around 10 copies each. cDNA replication occurs continuously through the cell cycle and is not synchronized with the nuclear division cycle, while chloroplast division is tightly coordinated with cytokinesis (Adams, Maple, & Møller, 2008; Chiang & Sueoka, 1967; Johnson & Porter, 1968; Turmel, Lemieux, & Lee, 1980). Though typical in size and protein-coding capacity for a chloroplast genome, the *Chlamydomonas* cpDNA is relatively rich in non-conserved short intergenic repeats that may contribute to its high rate of rearrangement (Maul *et al.*, 2002; Odom *et al.*, 2008).

The *Chlamydomonas* mitochondrial (mt) genome is a 15.8-kbp linear molecule with inverted repeats at each end (Vahrenholz *et al.*, 1993). The mt genome encodes eight proteins, mitochondrial rRNAs and three tRNAs (Grant & Chiang, 1980; Gray & Boer, 1988; Michaelis, Vahrenholz, & Pratje, 1990). The *Chlamydomonas* mt genome is notable for being linear in structure and significantly smaller than the circular mt genomes found in other algae and most plants (Boer, Bonen, Lee, & Gray, 1985; Grant & Chiang, 1980; Stern *et al.*, 2009). Other notable features of the mt genome are fragmentation of its rRNA genes that are separately transcribed as pieces that assemble non-covalently into intact mt ribosomes (Boer & Gray, 1988; Denovan-Wright, Sankoff, Spencer, & Lee, 1996) and are rapidly evolving and with respect to other mt genes cytosolic rRNAs (Popescu & Lee, 2007).

The *Chlamydomonas* organellar genomes are inherited uniparentally through post-zygotic mechanisms that are still not well understood. In meiotic progeny, cpDNA is inherited from the MT^{+} parent, while mitohondrial DNA is inherited from the MT^{-} parent (Boynton *et al.*, 1988). cpDNA inheritance involves targeted destruction of MT^{-}-derived cpDNA in early zygotes (Nishimura *et al.*, 1998; Kuroiwa, 2010) and may also involve differential replication in germinating zygotes mediated by cpDNA methylation (Umen, 2001). Loss of female mitochondrial DNA (mtDNA) appears to occur during meiosis as zygotes germinate (Aoyama *et al.*, 2006).

3.3. The *Volvox* Nuclear Genome Structure

Previous genetic mapping defined 19 linkage groups for *Volvox* that probably correspond to chromosomes (Kirk, 1998), though exact chromosome number remains to be established. The ~131 Mbp *Volvox* nuclear genome sequence assembly (version 2) contains 434 scaffolds, 100 of which are >50 kbp and contain over 98% of the sequence. Updated assemblies and annotations are available on the Phytozome Web site (Table 6.2) (Goodstein *et al.*, 2011). The extra ~14 Mbp of sequence in *Volvox* compared with *Chlamydomonas* is composed of repeats that are interspersed within and between genes (Prochnik *et al.*, 2010). A second difference between the two algae is that the *Volvox* genome is ~56% GC, while *Chlamydomonas* is ~64% GC. However, with respect to coding capacity, *Volvox* is highly similar to *Chlamydomonas* with ~15,000 predicted protein-coding genes (Prochnik *et al.*, 2010). A more detailed comparison of gene content between the two species is in Sections 5 and 6 below.

3.3.1. The Volvox Mating Locus

Volvox and other large colonials (*Pleodorina*, *Eudorina*) transitioned from an isogamous ancestral mating system to an anisogamous or oogamous one with eggs and sperm. *Volvox* sexual development occurs via a modified embryogenesis program that differs between males and females and is controlled by a mating locus with two haplotypes, *MTF* (female) and *MTM* (male).

There are striking similarities and differences between the mating loci of *Chlamydomonas* and *Volvox*. They both are relatively large regions with rearrangements that suppress recombination (Fig. 6.6). Somewhat surprisingly, the location of *MT* in both species is in the same region of a syntenic chromosome (chromosome 6 in *C. reinhardtii* and scaffold 1 in *V. carteri*), indicating that *MT* has not moved during the ~200 Mya since the *Chlamydomonas* and *Volvox* lineages last shared a common ancestor (Fig. 6.7A) (Ferris *et al.*, 2010). Unlike *Chlamydomonas MT*, *Volvox MT* is repeat rich with low gene density compared with the rest of the genome, making it similar in most respects to classical sex chromosomes (Ferris *et al.*, 2010). However, unlike diploid heteromorphic sex chromosomes, haploid sex chromosomes are not sheltered from gene loss and are less likely to lose essential genes (Bull, 1978). At >1 Mb, the *Volvox MT* region is about five times larger than *Chlamydomonas MT* and unlike the case for MT^{+}/MT^{-}, *MTF* and *MTM* contain almost no residual synteny (identical gene order) between the 70+ shared genes that they have in common (Fig. 6.6C).

3.4. The *Volvox* Organellar Genomes

Compared with *Chlamydomonas* and with most other green algae, the *V. carteri* organellar genomes are atypically large and filled with non-coding sequences, most of which are palindromic repeats (Smith & Lee, 2009). The cp genome is 420 kbp and the mt genome is 30 kbp. The cp genome is the largest and most repeat rich within the green lineage (chlorophytes and streptophytes), while the *Volvox* mt genome is also repeat rich with few genes. Other than their expanded repeat content, the organellar protein-coding genes of *Volvox* are very similar to those of *Chlamydomonas* (Smith & Lee, 2009). The increased non-coding content of all three *Volvox* genomes compared with *Chlamydomonas* (nuclear, cp and mt) is consistent with less efficient natural selection on a smaller and possibly fragmented population of *V. carteri* (Smith & Lee, 2010).

Like *Chlamydomonas*, the *Volvox* organellar genomes are inherited uniparentally, but unlike *Chlamydomonas*, both derive from the maternal parent (Adams *et al.*, 1990). The presence of the putative mating type specification gene *MID* in males of colonial Volvocine algae makes them formally equivalent to the MT^- mating type of *Chlamydomonas* that also has a *MID* gene (Ferris & Goodenough, 1997; Ferris *et al.*, 2010; Hamaji *et al.*, 2008; Nozaki *et al.*, 2006). Therefore, with respect to mating type, there has been at least one switch in mtDNA inheritance patterns in the lineage. In *Gonium pectorale* organellar DNA inheritance was found to follow a similar pattern to that of *Chlamydomonas* with cpDNA inherited from the MT^+ parent and mtDNA from the MT^- parent (Hamaji *et al.*, 2008; Setohigashi *et al.*, 2011). These findings are consistent with the *Chlamydomonas*/*Gonium* organellar inheritance patterns being ancestral, but more data from other Volvocine species will be required to confirm this idea.

Moreover, even though cpDNA is inherited maternally in all three Volvocine algal species described above, *Volvox* female *MT* is missing *EZY1*, *EZY2* and other MT^+-specific genes that are thought to be associated with uniparental cpDNA inheritance in *Chlamydomonas* (Fig. 6.6) (Armbrust, Ibrahim, & Goodenough, 1995; Ferris *et al.*, 2002; Goodenough *et al.*, 2007).The absence of these genes in *Volvox* suggests that the underlying mechanism that specifies uniparental cpDNA inheritance in the two species may differ as well. A default hypothesis for uniparental organelle DNA inheritance in oogamous species such as *Volvox* is by a passive mechanism based on differential contributions of organellar DNA from eggs versus sperm (as opposed to active elimination that occurs in *Chlamydomonas*). Nonetheless, additional active mechanisms to ensure uniparental organelle DNA inheritance may also operate in *Volvox* as they do in *Chlamydomonas* and other eukaryotic taxa (DeLuca & O'Farrell, 2012; Rawi *et al.*, 2011; Sato & Sato, 2011; Shitara *et al.*, 2000; Zouros, 2000).

3.5. Additional Volvocine Algal Genomes

Chlamydomonas and *Volvox* represent two extremes of organismal complexity in the Volvocales. Many questions about evolution in the Volvocine algae can only be answered by sequencing additional genomes of intermediate species. Projects to do so are underway and include genomes of *Gonium pectorale* and *Pleodorina starrii* (B. Olson, H. Nozaki, P. Durrand, R. Michod, personal communication).

4. MOLECULAR GENETICS AND GENOMICS TOOLS FOR VOLVOCINE ALGAE

Genomics resources for *Chlamydomonas* and *Volvox* are steadily growing and include informatics sites that provide access to browsers and computational tools as well as repositories for strains and cDNA or genomic libraries. These resources are described in Tables 6.1 and 6.2.

4.1. The *Chlamydomonas* Molecular Genetics Toolkit

4.1.1. Transformation

All three genomes (nuclear, chloroplast and mitochondrial) of *Chlamydomonas* are transformable (Kindle, 1990; Kindle *et al.*, 1989). Transformed nuclear genes insert randomly, a property that can be exploited for making tagged mutants (Dent *et al.*, 2005; Galvan *et al.*, 2007). Transformation of the nuclear genome by electroporation with either a circular or a linearized plasmid is a routine procedure typically resulting in hundreds to thousands of independent transformants (Keller, 1995; Shimogawara *et al.*, 1998). *Chlamydomonas* can also be transformed by vortexing gametolysin-treated cells with glass beads or silicon carbide whiskers (Dunahay *et al.*, 1990). Chloroplast and mitochondrial transformation occurs by homologous recombination. It is now possible to transform not just one or a few genes into the *Chlamydomonas* chloroplast but to re-engineer the entire chloroplast genome *ex vivo* and then reintroduce it to obtain gene replacements and modifications at multiple loci (O'Neill *et al.*, 2012).

4.1.2. Promoters for Nuclear Transgene Expression

Unlike the case in higher plants, heterologous promoters have so far not been successful in *Chlamydomonas*. A handful of endogenous promoters have been successfully used to drive transgene expression (Neupert *et al.*, 2012). These include an engineered hybrid promoter from the *HSP70A* and

RBCS2 genes that also contains the first intron of *RBSC2* (Lodha, Schulz-Raffelt, & Schroda, 2008; Schroda, Blöcker, & Beck, 2000; Schroda, Beck, & Vallon, 2002), the *HSP70a* and *HSP70b* promoters alone (Schroda *et al.*, 2000) and the *PSAD* promoter that is derived from an intronless gene and well-suited to drive expression of cDNAs (Fischer & Rochaix, 2001).

4.1.3. Regulated Promoters

Several regulated promoters have been described that utilize environmentally controlled transcriptional responses to drive transgene expression. These include a copper-responsive promoter from the cytochrome b_6 gene *CYC6* that can be induced using Ni^{2+} (Ferrante *et al.*, 2008; Quinn *et al.*, 2003), the nitrate inducible/ammonium repressible promoter from the nitrate reductase gene *NIT1* and the low-CO_2-inducible promoter from the carbonic anhydrase gene *CA1* (Ohresser, Matagne, & Loppes, 1997; Villand *et al.*, 1997). The activities of these promoters are coupled to environmental conditions, a property that might limit their utility. Synthetic or heterologous regulated promoters would be useful for expanding the molecular genetic and genomic toolkits of both *Chlamydomonas* and *Volvox*.

4.1.4. Selectable Markers in Chlamydomonas

Two classes of selectable markers have been utilized for nuclear transformation in *Chlamydomonas*; endogenous genes that complement auxotrophies and dominant antibiotic resistance markers (Neupert *et al.*, 2012). Complementable mutations include *nit1* (nitrate requiring) (Kindle *et al.*, 1989), *arg7* (arginine requiring) (Debuchy, Purton, & Rochaix, 1989), *nic7* (nicotinamide requiring) (Ferris, 1995) and *thi10* (thiamine requiring) (Ferris, 1995). Dominant antibiotic resistance markers that have been engineered for *Chlamydomonas* include *ble* (zeocin resistance) (Stevens, Purton, & Rochaix, 1996), *aphVIII* (paromomycin resistance) (Sizova, Fuhrmann, & Hegemann, 2001), *aph7* (hygromycin resistance) (Berthold *et al.*, 2002), *aadA* (spectinomycin resistance) (Cerutti *et al.*, 1997), *cry1-1* (emetine resistance) (Nelson *et al.*, 1994) and *ppx1* (protoporphyrinogen oxidase mutant conferring resistance to the herbicide S-23142) (Randolph-Anderson *et al.*, 1993) .

Chloroplast transformants are typically selected by either complementation of an *atpB* mutation to restore phototrophic growth (Boynton *et al.*, 1988) or with a dominant spectinomycin/kanamycin resistance gene (*aadA*) (Purton, 2007; Ramesh *et al.*, 2011). Mitochondrial transformants are selected by complementation of the respiration deficient mutant *dum1* (Boynton & Gillham, 1996; Hu *et al.*, 2011; Yamasaki *et al.*, 2005).

4.1.5. Reporter genes

Green fluorescent protein (GFP) has been codon optimized for use in *Chlamydomonas* (Fuhrmann *et al.*, 1999) and has been expressed transgenically from the nucleus, chloroplast and mitochondrial genomes (Fuhrmann *et al.*, 1999; Hu *et al.*, 2011; Purton, 2007; Ramesh *et al.*, 2011). Live cell imaging with GFP in algae is a challenge due to high background fluorescence from chlorophyll and other auto-fluorescent compounds but can be effective in cases where the GFP is well expressed or concentrated in a specific subcellular location (Diener, 2009; Fuhrmann *et al.*, 1999; Hayashi & Shinozaki, 2011; Neupert *et al.*, 2012; Yoshihara *et al.*, 2008). Luciferase reporters have also been developed for *Chlamydomonas* (Fuhrmann *et al.*, 2004; Mayfield & Schultz, 2004; Shao & Bock, 2008) and adapted for various purposes including extracellular secretion (Eichler-Stahlberg, Weisheit, Ruecker, & Heitzer, 2009) and for monitoring circadian gene expression (Matsuo *et al.*, 2008; Mayfield & Schultz, 2004; Minko *et al.*, 1999).

4.1.6. RNAi and Antisense Technology in Chlamydomonas

RNAi is an effective tool for gene silencing in *Chlamydomonas* that has all the essential components of the RNAi machinery (see Section 4.1.6). Two strategies that have proven successful for generating targeted gene knock-downs are double-strand hairpin-forming constructs (Rohr, Sarkar, Balenger, Jeong, & Cerutti, 2004) and artificial microRNAs (Molnar *et al.*, 2009; Zhao *et al.*, 2009) whose expression can also be coupled to regulated promoters (Schmollinger *et al.*, 2010).

4.2. Forward Genetics and Functional Genomics in *Chlamydomonas*

Positional cloning of *Chlamydomonas* mutants can be accomplished using Polymerase Chain Reaction (PCR) mapping makers (Kathir *et al.*, 2003; Rymarquis *et al.*, 2005) that are available from the *Chlamydomonas* resource centre (http://www.chlamycollection.org). Next generation re-sequencing-based strategies for mutant gene identification are also now feasible and should allow more rapid identification of point mutants (Dutcher *et al.*, 2012).

Insertional mutagenesis is a powerful tool for isolating tagged mutants in *Chlamydomonas* (Galván *et al.*, 2007), but its potential is even greater when combined with high-throughput methods for identifying and cataloguing insertion sites. This potential is gradually being realized. One approach for doing so involves pooling sets of mutant DNA and screening for insertions in genes of interest using PCR with a gene-specific and insert-specific pair of primers. When optimized, this method has a very high success rate (González-Ballester *et al.*, 2011; Pootakham *et al.*, 2010). A variant approach involves screening insertion mutants that have been enriched for specific phenotypes such as photoprotection or nitrogen utilization and then identifying mutants en masse by sequencing the insertion border in all the identified strains (Dent *et al.*, 2005; Gonzalez-Ballester, 2005). A next-generation approach to insertional mutant screening using bar-coded sequencing of selected pools may be on the horizon. A drawback of using insertional lines in a haploid strain is that essential genes will be highly underrepresented. A diploid insertion library is a feasible way around this problem but has yet to be generated as a reverse genetics resource.

4.3. The *Volvox* toolkit

4.3.1. Volvox *Transformation*

The *Volvox* nuclear genome can be transformed by particle bombardment using the *nitA* gene (encoding nitrate reductase) to complement a *nitA*$^-$ mutation (Gruber *et al.*, 1996; Schiedlmeier *et al.*, 1994). Dominant antibiotic resistance markers including *aphVIII* (conferring paromomycin resistance) and *ble* (conferring zeocin resistance) have also been adapted for use in *Volvox* (Hallmann & Rappel, 1999; Jakobiak *et al.*, 2004). A sex pheromone inducible promoter system has also been developed for regulated gene expression (Hallmann & Sumper, 1994). Some *Volvox* transformation vectors have also been used to successfully transform *Gonium pectorale* (Lerche & Hallmann, 2009). GFP fusion proteins expressed in *Volvox* have been used to localize proteins within the ECM (Ender *et al.*, 2002; Ishida, 2007) and in nuclei (Pappas & Miller, 2009). Gene expression can be knocked down in *Volvox* using antisense constructs as demonstrated for *GlsA* (Cheng *et al.*, 2003) and a photoreceptor, *Volvox* rhodopsin (Ebnet, 1999).

4.3.2. Forward Genetics by Transposon Tagging

It is challenging to create and maintain thousands of insertion tagged mutants in *Volvox* due to its relatively low transformation efficiency and the difficulty of maintaining so many transformed lines (Kirk, 2000). An alternative strategy for forward genetics is transposon tagging mutagenesis using the cold-inducible transposons *Idaten* (Ueki & Nishii, 2008) or *Jordan* (Miller *et al.*, 1993). This method relies on endogenous transposons that can occasionally excise to yield revertants. While successful, the method is relatively time consuming and inefficient. The development of artificial transposons would be a useful advance.

Classical genetics has been used successfully in *Volvox* to identify developmental mutants but was done in a pre-genome era with few molecular markers (Huskey *et al.*, 1979). The decreasing costs of resequencing may reopen this route for identifying important developmental regulators in *Volvox*.

5. THE *CHLAMYDOMONAS* GENOME AS A WINDOW INTO PLANT AND ANIMAL EVOLUTION

In this section, we highlight areas where genomic information from *Chlamydomonas* has shed light on eukaryotic cell biology and the evolution of the green plant lineage. While every species is in some ways specialized, *Chlamydomonas* is particularly interesting in that it does not have a reduced genome size as do marine picoalgae such as *Ostreococcus* and *Micromonas* (Derelle *et al.*, 2006; Misumi *et al.*, 2007; Palenik *et al.*, 2007; Worden *et al.*, 2009), and it has retained features that were likely present in the last common eukaryotic ancestor such as flagella and basal bodies. In addition, *Chlamydomonas* is part of the Archeaplastidia or green lineage. Archaeplastidia include a diverse group of green algae and land plants (Fig 6.1A), all of which descended from a unicellular eukaryote that underwent a primary endosymbiotic event of engulfing a cyanobacterium that subsequently evolved into the chloroplast (Chapter II of this volume; De Clerck, Bogaret, & Leliaert 2012). The tractability of *Chlamydomonas* as an experimental model makes it highly attractive for genomic studies that can then lead to experimentally testable hypotheses about gene function. Some examples of genomics-based discoveries in *Chlamydomonas* are described below.

5.1. Cell Motility and the Flagella

Long before its genome was sequenced, it was recognized that the flagella and basal bodies of *Chlamydomonas* are structurally and functionally homologous to those in other eukaryotes including animals (Preble, Giddings, & Dutcher, 1999; Silflow & Lefebvre, 2001). Indeed, a key process for flagellar biogenesis that was discovered in *Chlamydomonas*, *i*ntra*fl*agellar *t*ransport (IFT) (Kozminski *et al.*, 1993), is now widely recognized for its role in human ciliary signalling and a variety of genetic diseases (Hildebrandt *et al.*, 2011; Sharma *et al.*, 2008). *Chlamydomonas* is also a valuable resource for identifying new basal body and flagellar genes. Part of its utility for this approach stems from it having one of the best annotated sets of flagellar and basal body protein-coding genes, many of which were validated by empirical proteomics (Table 6.2) (Keller *et al.*, 2005; Merchant *et al.*, 2007; Pazour *et al.*, 2005).

Comparative approaches have been successful at identifying gene sets enriched for those encoding flagella and basal body proteins. This approach is based on the fact that higher plants, fungi and slime moulds have lost flagella and basal bodies, whereas most animals and unicellular eukaryotes have retained these organelles. By focusing on protein families where ciliated species have at least one member, but where non-ciliated species do not, new cilia or basal body proteins were identified (Avidor-Reiss *et al.*, 2004; Li *et al.*, 2004). These studies identified not only known proteins but also new ones such as *BBS5* whose human homolog is encoded by a gene associated with the genetic disease Bardet–Biedl syndrome (Li *et al.*, 2004). A *Chlamydomonas*-centric comparative genomics approach was also used to subclassify flagellar genes by identifying those that are common to organisms with different cilia subtypes. Using a similar comparative approach as above, a more comprehensive analysis of flagella genes was undertaken. Starting with a set of proteins that are present in flagellated/ciliated organisms but not in those without flagella, a more refined subset of classifications were made. For example, the nematode *elegans* has non-motile sensory cilia, and proteins found in *Chlamydomonas* and other species with motile cilia but not in *Caenorhabditis elegans* were provisionally designated as MOT genes. Similar functional comparisons were made with the moss *Physcomitrella* patens and diatom *Thalassiosira pseudonana*, each of which have structurally modified motile cilia that are missing or thought to be missing structural components found in the *Chlamydomonas* flagella that has a standard 9 + 2 microtubule doublet structure with inner and outer dynein arms and radial spokes (Merchant *et al.*, 2007).

5.2. Green Genes

Genomic analyses in *Chlamydomonas* have aided progress in understanding photosynthetic metabolism and cell biology in the green lineage. A green-lineage-specific tool based on the predicted *Chlamydomonas* proteome was used to identify gene families that are specific to photosynthetic species. Indeed, 83% of the Green Cut proteins from *Chlamydomonas* whose functions were already known were chloroplast localized (Merchant *et al.*, 2007). However, the Green Cut set also includes proteins that evolved specifically in the green lineage but may not be directly related to photosynthesis. These include proteins with predicted functions in signalling and nuclear transcription factors that may be specialized within the green lineage. Since they were first identified, a number of Green Cut protein family members were characterized and functionally verified (Grossman *et al.*, 2003; Karpowicz *et al.*, 2011) and this process will likely accelerate in the next few years as functional genomic resources such as insertional library screening are applied to *Chlamydomonas* and other green organisms.

Large-scale approaches such as Green Cut have inherent limitations that are complemented by more directed studies. For example some chloroplast import machinery proteins were missed by Green Cut, probably due to sequence divergence or incomplete gene models, but could be identified in directed searches based on screening for *Arabidopsis* Tic (translocon in the inner chloroplast envelope) and Toc (translocon in the outer chloroplast envelope) homologs (Kalanon & McFadden, 2008). Homologs of all but two *Arabidopsis* Tic and Toc proteins were identified in *Chlamydomonas*, and the two missing proteins (Tic62 and Toc64) are also missing from one or more other algal species. Comparison of the chloroplast transit peptide receptor complex proteins CrToc34 and CrToc159 indicated differences in amino acid composition that may reflect differences in composition of transit peptides between *Chlamydomonas* and higher plants (Franzen *et al.*, 1990; Kalanon & McFadden, 2008; Patron & Waller, 2007).

5.3. Selenoproteins

One interesting surprise in the *Chlamydomonas* genome was the presence of selenocysteine-containing proteins that are absent from land plants and fungi, but present in animals (Novoselov *et al.*, 2002). Selenocysteine (Sec) is inserted into translating polypeptides by a modified tRNA-Sec that recognizes the codon UGA (normally a translation terminator) whose insertion is specified by a stem loop forming sequence element (termed SECIS) in the 3' UTR of the message containing the tRNA-Sec codon (Novoselov *et al.*, 2002). *Chlamydomonas* was found to encode a single tRNA-Sec gene (Rao *et al.*, 2003) as well as the enzymes and factors needed to produce and insert tRNA-Sec into elongating polypeptide chains (Grossman *et al.*, 2007). Twelve putative selenoproteins have been identified in *Chlamydomonas*, five of which are involved in redox biochemistry (Grossman *et al.*, 2007). Five other selenoproteins have no known function but do have orthologs in mammals. This finding makes *Chlamydomonas* a potential new model for investigating selenoprotein-related biology.

5.4. Carbon Concentrating Mechanism

Inorganic carbon (Ci) is often a limiting nutrient for aquatic photosynthetic microbes such as *Chlamydomonas*. Unlike terrestrial environments with relatively stable atmospheric CO_2 concentrations, aquatic Ci exists in gaseous (CO_2) and ionic (HCO_3^-, CO_3^{2-}) forms that are not always in equilibrium with atmospheric CO_2 and whose ratios fluctuate depending on pH. Aquatic algae have evolved *c*arbon *c*oncentrating *m*echanisms (CCMs) to help overcome this limitation. CCM proteins are part of an energy-mediated transport system to move Ci from outside the cell and concentrate it in the chloroplast where it serves as the substrate for RuBisCo in photosynthetic dark reactions. CCM proteins include carbonic anhydrases (CAs) that interconvert CO_2 and HCO_3^{2-} and transporters that move HCO_3 or CO_2 across membranes. While forward genetic screens have identified many CCM proteins (Spalding, 2008; Wang, Duanmu, & Spalding, 2011), genome-wide homology searches have revealed additional candidate genes that may play a role in the CCM or related processes. These include a large repertoire of 12 CAs that localize to different subcellular compartments or the periplasmic space. They also include additional paralogs of LCI proteins (LCIA, LCIB, LCIC, LCID, LCIE) that are thought to be involved in CO_2 transport or uptake (Wang & Spalding, 2006; Grossman *et al.*, 2007) and which have homologs in other algae such as *Ostreococcus*. A second observation from the *Chlamydomonas* genome sequence is that several CCM genes cluster in a single 75-kbp region of chromosome 4 (Merchant *et al.*, 2007). These include CCP1, CCP2, LCID, LCIE and two CA paralogs, CAH1 and CAH2. Whether such clustering is connected to gene regulation remains to be determined.

EST and genome sequencing identified a medically relevant protein family, Rh, as another possible connection to CO_2 physiology. Rh is a red blood cell antigen and integral plasma membrane protein that is part of an ammonium transporter superfamily (PFAM 00909). The discovery of Rh proteins in non-metazoans such as *Chlamydomonas* was unexpected (Huang & Liu, 2001), as is their overall distribution in a few select non-metzoan taxa that include green algae (*Chlamydomonas, Coccymyxa, Chlorella*), oomycetes (*Phytophthora*), slime molds (*Dictyostelium, Polysphondylium*), Choanoflagellates (*Monosiga, Salpingoeca*), Incertae sedis (*Capsaspora*) and Heteroloboseans (*Naeglaria*). Whether Rh transports NH_4/NH_3^+ or CO_2/HCO_3 is not fully resolved; however, the *Chlamydomonas* Rh paralogs are linked to high CO_2 acclimation possibly by acting as a bidirectional CO_2 gas exchange channel (Soupene *et al.*, 2004).

5.5. Vitamin Biosynthesis

Many algae require one or more of vitamins B1 (thiamine), B7 (biotin) and B12 (cobalamin) for growth (Croft *et al.*, 2006). Vitamin auxotrophies are distributed across algal taxa in a manner that suggest multiple independent losses of genes or enzymes requiring the co-factors. *Chlamydomonas* does not require any vitamin supplements. It has biosynthetic pathways for thiamine and biotin, and its single cobalamin-dependent enzyme, the methionine synthase METH, is redundant with a cobalamin-independent enzyme, METE (Croft *et al.*, 2006). Interestingly, *V. carteri* requires B12 supplementation for growth, suggesting that it has lost its cobalamin-independent methionine synthase gene, METE (or that it has acquired another essential enzyme that is cobalamin dependent). Indeed, the sequence of *Volvox* METE has multiple frame-shift mutations and a premature stop indicating that it has become a pseudogene, thus explaining the dependence of *Volvox* on B12 (Helliwell *et al.*, 2011). Moreover, METE loss has occurred at least one other time in the Volvocine lineage. The METE gene from the 16-cell colonial species *Gonium pectorale* was also found to be a pseudogene whose loss appears to have occurred independently from that in *V. carteri* (Helliwell *et al.*, 2011); though some strains of *Gonium pectorale* are B12 independent suggesting a very recent loss of METE in some lines (Stein, 1966). The loss of METE implies that *V. carteri* and some *Gonium pectorale* strains acquire B12 environmentally and raise the question of whether they do so in specific association with other microbes.

V. carteri has orthologs of the thiamine and biotin biosynthetic enzymes identified in *Chlamydomonas* (Croft *et al.*, 2006) (J. Umen, unpublished observation), but their expression and function have not been carefully examined. The question of whether *V. carteri* truly requires biotin and thiamine supplementation merits re-examination in light of this observation.

5.6. Cell Cycle

The non-canonical multiple fission cell cycle of *Chlamydomonas* might be expected to require innovation in cell cycle control machinery compared to organisms that divide by binary fission. In addition, multiple fission is likely to have been an important factor in facilitating the evolution of colonialism in Volvocine algae (see Section 6.3.2). A genomic survey of cell cycle regulatory proteins in *Chlamydomonas* revealed a typical repertoire of eukaryotic cell division control proteins that are similar to those in land plants, though with far less gene duplication than is seen in plants (Bisova *et al.*, 2005). A similar situation was observed for another alga, the Prasinophyte *Ostreococcus tauri* (Robbens *et al.*, 2005). A few notable observations arose from these studies. First, both algae encode homologs of the plant-lineage-specific cyclin-dependent kinase (CDK) CDKB that is expressed during cell division (Bisova *et al.*, 2005; Corellou, 2005). This finding broadens the possible roles for CDKB and suggests an ancestral function that encompasses both algae and land plants. A notable feature of *Chlamydomonas* cell cycle genes is the presence of novel cyclin-dependent kinases, CDKG1 and CDKG2, which are not found in species outside of Volvocine algae and may be important for the multiple fission cycle found in the Volvocales. Genomic level analysis also confirmed the presence of key components of the retinoblastoma (RB) tumour suppressor pathway that are conserved in the green lineage and in animals but have been lost from fungi (Bisova *et al.*, 2005; Grafi *et al.*, n.d.; Robbens *et al.*, 2005; Umen & Goodenough, 2001; Xie *et al.*, 1996). Interestingly, the only cell cycle gene family in *Chlamydomonas* that underwent expansion is the D-type cyclins, which are presumed to activate CDKs that regulate RB-related proteins through phosphorylation. *Chlamydomonas* has four D-type cyclins that are diverged from each other enough to suggest an ancient duplication event. This idea is supported by the finding that *Volvox* has orthologs for each of the four D-type cyclin subfamilies that are found in *Chlamydomonas* plus additional paralogous duplications whose potential significance is discussed in Section 6e (Prochnik *et al.*, 2010). This finding indicates that a D-cyclin family expansion occurred early in the Volvocine lineage and has been maintained in two independent branches suggesting an important role for each of the D-type cyclin subtypes in Volvocine algal cell cycle control or other processes.

5.7. Scavenger Receptors

Scavenger receptor and cysteine-rich domain (SRCR) proteins are found in metazoans where they have diverse roles in the immune system (e.g. pathogen recognition by macrophages) and elsewhere such as the sea urchin sperm flagellar receptor for the egg peptide *speract* (Cardullo *et al.*, 1994; Resnick *et al.*, 1994). SRCR proteins contain a ~100 amino acid repeat module with conserved cysteine pairs that form disulfide bridges (Hohenester *et al.*, 1999). SRCR proteins are abundant in metazoans but are absent from higher plants and most algae. Remarkably, *Chlamydomonas* encodes 35 SRCR domain containing proteins, many of which were described previously (Wheeler *et al.*, 2008), and few of which have appeared in a more updated genome assembly from Phytozome (Goodstein *et al.*, 2011). *Volvox* has at least 21 SRCR domain-containing proteins, though its genome annotation is not as comprehensive as that of *Chlamydomonas*, so this family awaits a true inter-species comparison in Volvocine algae. SRCR domains are often repeated multiple times in a protein with a range of 1–11 copies in *Chlamydomonas* SRCR proteins. A notable feature of several of the *Chlamydomonas* SRCR proteins is their enormous size, with the largest containing 8671 amino acids and encoded by a gene that spans almost 60 kb. Several of the *Chlamydomonas* SRCR proteins also contain lectin domains that bind to carbohydrates and are widespread in eukaryotes (Wheeler *et al.*, 2008). The presence of these large metazoan-like adhesion molecules or receptors in *Chlamydomonas* and *Volvox* raises many interesting questions about their function(s) and provides a new avenue for investigating their evolutionary origins.

6. THE *VOLVOX* GENOME AS A WINDOW INTO MULTICELLULAR EVOLUTION

6.1. *Volvox* Development

V. carteri has a complex developmental program that involves conserved processes that are shared with *Chlamydomonas* and other features with no obvious analogs in *Chlamydomonas* (Figs 6.4 and 6.5). Insights into the evolution of *Volvox* have come from genomics and from isolation of developmental mutants (Kirk, 1998; Nishii & Miller, 2010; Prochnik *et al.*, 2010). Below, we outline key stages of *Volvox* development and discuss what genetic and genomic approaches have revealed about the origins of multicellularity in Volvocine algae. More comprehensive reviews of *V. carteri* development are also available (Kirk, 1998, 2005).

Each *V. carteri* vegetative spheroid is composed of ~2000 sterile somatic cells and 16 large reproductive cells called gonidia, all of which are embedded within a complex compartmentalized ECM that occupies 99% of the spheroid volume (Fig. 6.7). Somatic cells do not normally grow or reproduce but provide motility for the spheroid and secrete ECM. They may also help concentrate nutrients in the interior to support growth of reproductive cells (Koufopanou, 1994). The vegetative gonidial cells undergo embryogenesis involving successive cleavage divisions that follow a modified program of multiple fission in which incomplete cytokinesis leaves post-mitotic cells attached through a network of cytoplasmic bridges. At cycle 6 (32 to 64 cell stage transition), 16 cells at the anterior of the embryo divide asymmetrically to produce a large and a small cell, while the remaining cells continue to divide a total of 11 or 12 times. The 16 large cells from the asymmetric division divide one or two more times asymmetrically and then withdraw from cleavage. After embryogenesis the 16 large cells will end up differentiating into the next generation of germ cells, while the small cells become somatic. Interestingly, germ-soma differentiation is controlled by cell size rather than by partitioning of specific cell fate determinants into either of the two cell types (Kirk *et al.*, 1993). At the end of cleavage, the embryo is inside out with respect to its final configuration with gonidial cells on the outside of the spheroid and flagella of somatic cells pointing inwards. Through the remarkable process of inversion, the embryo turns itself right-side out into its adult configuration with gonidial precursor cells on the interior of the spheroid and the flagella of the somatic cells facing the exterior (Hallmann, 2006a; Kirk & Nishii, 2001; Viamontes & Kirk, 1977; Viamontes, Fochtmann, & Kirk, 1979). Juvenile spheroids remain inside their mother where they continue to grow for another day or two before hatching out to repeat the vegetative growth cycle.

In addition to its vegetative (asexual) reproductive cycle, *V. carteri* has a sexual development cycle (Fig. 6.5) that is under control of its mating locus (Fig. 6.6). While vegetative males and females of *V. carteri* are indistinguishable, the male and female sexual forms are different from each other and from vegetative *Volvox*. Sexual differentiation in *Volvox* is triggered by a glycoprotein sex-inducer produced by males that causes modified embryogenesis programs in males and females (Fig. 6.5) (Kochert, 1968; Mages, Tschochner, & Sumper, 1988; Tschochner, Lottspeich, & Sumper, 1987). Females treated with sex inducer produce sexual offspring with 32–48 egg cells instead of vegetative gonidia, while males treated with sex inducer produce sexual offspring containing a 1:1 ratio of 128 somatic cells and 128 sperm packets, each containing 64–128 sperm. Sperm packets travel as a unit and upon encountering a sexual female, break apart and tunnel inside the ECM where they find and fertilize eggs to produce diploid zygote spores. Upon germination, the spores undergoes meiosis and produce three polar bodies plus one new viable haploid progeny spheroid. While not described in Kirk's 12 steps (see below), innovations in the *Volvox* sexual cycle have been recently summarized (Ferris *et al.*, 2010; Umen, 2011).

6.2. Kirk's 12 Steps of Multicellular Evolution in the Volvocales

Kirk proposed a 12-step model of multicellular evolution in the Volvocales (Kirk, 1998, 1999, 2005) that posited the minimum changes or innovations required for a unicellular *Chlamydomonas*-like ancestor to evolve into a multicellular *V. carteri*-like spheroid (Fig. 6.2A). When broken into discrete steps, it becomes clear that half of the innovations occur in the first transition to form the simplest colonial genus exemplified by *Gonium* (Herron & Michod, 2008; Herron *et al.*, 2009; Kirk, 1998, 1999, 2005; Nishii & Miller, 2010). This finding implies that the most difficult transition is evolving the minimal functional colonial form, a finding that is partly supported by the monophyly of colonial Volvocines as a whole (i.e. only one original instance of colonialism) compared to polyphyly of several genera within the Volvocine algae (i.e. more than one instance of a transition to a more complex form) (Fig. 6.1).

Evolving *Gonium*-like colonies from *Chlamydomonas*-like unicells requires a minimum of six changes: (1) incomplete cytokinesis to maintain linkage between daughter cells (i.e. cytoplasmic bridges), (2) partial inversion to reorient post-mitotic daughters within the new colony, (3) basal body rotation by 90° in peripheral cells to align flagella (Kirk, 1998, 2005), (4) establishment of organismal polarity as evidenced by the difference in basal body and flagellar orientations of central versus peripheral cells in the colony, (5) modification of the cell wall to keep daughters attached, (6) genetic control of cell number (Fig 6.7). After the evolution of a colonial morphology typified by *Gonium*, the Volvocales show a significant increase in colony size/cell number. The spherical genera starting with *Pandorina* undergo full inversion corresponding to Kirk's step 7. *Eudorina* is representative of step 8 – expansion of the ECM. *Pleodorina* represents Kirk's step 9 – partial division of labour between germ and somatic cells. Finally, *V. carteri* represents Kirk's steps 11 and 12 – asymmetric cell divisions and a bifurcated cell division program. Notably, these last two steps are found in only one subgroup of *Volvox* species from the section Merrillosphaera. Other, species of *Volvox* develop in a different manner through progressive binary divisions that represent an alternative evolutionary solution to colony growth (Coleman, 2012; Desnitski, 1995; Kirk, 1998).

6.3. Comparing the Content of the *Chlamydomonas* and *Volvox* Genomes

It has long been assumed that an extensive new genetic toolkit would be required for multicellular evolution (King, 2004; Ruiz-Trillo *et al.*, 2007; Rokas, 2008a, 2008b). Previous genomic comparisons reinforced this idea (Putnam *et al.*, 2007; Srivastava *et al.*, 2010), though in no case had two closely related species with divergent morphology, such as

Chlamydomonas and *Volvox*, been compared (Prochnik *et al.*, 2010). On a global level, the genomes of these two algae are remarkably similar, including extensive regions of synteny (Ferris *et al.*, 2010; Prochnik *et al.*, 2010). The total protein-coding gene count for *Chlamydomonas* is ~14,516 (Merchant *et al.*, 2007), while that for the first version of the *Volvox* genome was ~14,520 genes (Prochnik *et al.*, 2010). Protein domain content is also similar between the two species: *Chlamydomonas* has 2,354 PFAM domains, while *Volvox* has 2,431 PFAM domains that are largely overlapping (Prochnik *et al.*, 2010). Of the ~15,000 genes found in *Chlamydomonas* and *Volvox*, ~64% can be assigned into protein families that are shared with other eukaryotes (Prochnik *et al.*, 2010). Of these, 80% show a 1:1 orthology between the two algae. In summary, with a few exceptions described below, large-scale comparisons of *Chlamydomonas* and *Volvox* reveal very few differences that can be tied to the substantial differences in biological organization that distinguish them.

The overall similarity between the two genomes is mirrored by results of forward genetics that identified key developmental regulators in *Volvox*. Mutants that affect somatic cell specification (*regA*) (Kirk *et al.*, 1999), asymmetric cell division (*glsA*) (Miller & Kirk, 1999) and inversion (*invA*, *invB*, *invC*) (Nishii *et al.*, 2003; Ueki & Nishii, 2008, 2009) have all been identified and cloned. Remarkably, four of the genes underlying these processes that are unique to *Volvox* have orthologs in *Chlamydomonas* (*glsA*, *invA*, *invB* and *invC*) and the fifth, *regA*, is part of a complex gene family that has members in both species (Duncan *et al.*, 2007). Moreover, two of the *Volvox* mutants, *invA* and *glsA*, can be complemented by their *Chlamydomonas* orthologs (Cheng *et al.*, 2003; Nishii *et al.*, 2003; Ueki & Nishii, 2008, 2009). Thus, it appears that many ancestral functions carried out by proteins in *Chlamydomonas* could be modified or co-opted to perform developmental functions in *Volvox*, suggesting that the genetic toolkit required for multicellularity may have been largely present in ancestral unicells (King, 2004; Prochnik *et al.*, 2010).

About 26% (1,835) of the 15,000 genes in *Chlamydomonas* and *Volvox* are only found in these algae, and are thus Volvocine specific. This latter group of Volvocine algal protein families is of special interest since they show asymmetric expansion/contraction patterns between *Chlamydomonas* and *Volvox*. These Volvocine algal protein families tend to be larger in *Volvox* than in *Chlamydomonas* (Prochnik *et al.*, 2010). Moreover, this asymmetric distribution of expansions/contractions does not extend to protein families that are shared with species outside Volvocine algae. This latter group of more ancient proteins shows similar degrees of loss/gain between the *Volvox* and *Chlamydomonas* lineages (Prochnik *et al.*, 2010).

What might be responsible for the preferential expansion/retention of Volvocine-algal-specific protein families in *Volvox*? While answers to this question are speculative, it is striking that among these protein families are two whose members are components of the *Volvox* ECM (see below), which is itself a massive elaboration of the ancestral cell wall of its *Chlamydomonas*-like ancestor. Extrapolating further, one can speculate that the entire group of Volvocine-algal-specific proteins might be highly evolvable. By definition, they have been around long enough to remain established in both algal lineages but are still young enough to be amenable to evolutionary experimentation and elaboration. Unfortunately, the functions of most Volvocine-algal-specific proteins are unknown since they are restricted to this lineage, but they may be a rich source of evolutionary plasticity and novelty that merits further examination. Supporting this idea are recent findings indicating that the origins of many metazoan proteins associated with multicellular functions (e.g. cell adhesion) extend back to their last unicellular ancestor that resembled a choanoflagellate and are also clade-specific proteins that are not found outside of holozoa (e.g. integrins, cadherins, tyrosine kinases) (Abedin & King, 2010; King *et al.*, 2008).

6.3.1. Gene Family Expansions in Chlamydomonas

Notable gene family expansions in *Volvox* relative to *Chlamydomonas* are discussed below, but there are a few instances where *Chlamydomonas* has more gene family members than *Volvox* (Prochnik *et al.*, 2010). *Chlamydomonas* has ~35 histone gene clusters, while *Volvox* has less than half this number (Prochnik *et al.*, 2010). The *Chlamydomonas* histone content is also exceptionally high compared with plants and other algae. Increased gene copy number is a possible means of increasing the rate of gene product synthesis which might be important for multiple fission where rapid successive divisions occur, but it is not clear why *Chlamydomonas* would need more histone clusters than *Volvox* since their cell division rates during mitosis are similar (Harper & John, 1986; Kirk, 1998). Proteins with ankyrin repeats were reported to be highly enriched in *Chlamydomonas* versus *Volvox* (~40 vs. 12), but a re-evaluation of protein domain content using more sensitive criteria in the new version of the *Volvox* genome (Goodstein *et al.*, 2011) shows that *Chlamydomonas* has 146 ankyrin repeat proteins versus 80 in *Volvox* (J. Umen, unpublished observation). This is still a biased ratio, but no longer appears exceptional and may be a result of genetic drift.

6.3.2. Cell Cycle Modifications

Four of Kirk's 12 steps of multicellular evolution in the Volvocales are closely or directly related to cell cycle modifications: incomplete cytokinesis (1), genetic modulation of cell number (6), asymmetric cell division (11), bifurcated cell division program (12) (Kirk, 2005). In addition, sexual development in *Volvox* entails further cell cycle modifications that include altered timing of embryonic asymmetric cell division in males and females and a set of post-embryonic germ cell divisions to produce sperm packets in males (Ferris *et al.*, 2010; Umen, 2011). As described above, *Volvox* utilizes a *Chlamydomonas*-like multiple fission cell cycle for embryonic and post-embryonic divisions. Unlike the case in *Chlamydomonas* where division number is flexible, the number of divisions in *Volvox* and other colonial Volvocine algae is more stereotyped so that cell number per colony stays relatively constant and is always a power of 2 (e.g. 4, 8, 16, 32 . . .). The triggering of asymmetric cell division at a specific embryonic cell cycle number and the subsequent bifurcated division program for germ cell and somatic cell precursors in *Volvox* is not understood but may be coupled in some manner to cell size, as is the final process of germ-soma differentiation (Kirk, 2005; Kirk *et al.*, 1993).

Forward genetics in *Chlamydomonas* identified the RB tumour suppressor pathway as a central regulator of size control during multiple fission (Fang *et al.*, 2006; Umen & Goodenough, 2001). It might be expected that this pathway is modified or elaborated in *Volvox* and perhaps other Volvocine species to accommodate alterations in colony size and/or developmental timing. While the cell division cycle protein-coding genes in *Volvox* are overall very similar to those in *Chlamydomonas*, there are two intriguing differences uncovered by genome sequencing that suggest possible modifications in *Volvox*. First, an expansion of the D1 cyclin subfamily was found in *Volvox* that has single orthologs of the other *Chlamydomonas* cyclins (D2, D3 and D4) but four cyclin D1 paralogs (Prochnik *et al.*, 2010). D-type cyclins are the activator subunits of cyclin-dependent kinases that have RB tumour suppressor-related proteins (RBRs) as their substrates and key targets for controlling cell cycle progression. The elaboration of D1 cyclins in *Volvox* may enable more fine-tuned developmental control over multiple fission than in *Chlamydomonas*, an idea that merits further investigation.

A second change in the *Volvox* RB pathway compared with *Chlamydomonas* is the incorporation of the RBR homolog *MAT3* into the *Volvox* mating locus and subsequent divergence of the male and female alleles (*vcMAT3f* and *vcMAT3m*) (Fig. 6.6) (Ferris *et al.*, 2010). The large divergence between *MAT3f* and *MAT3m* as well as their sex-regulated alternative splicing patterns suggests that this protein has acquired a role in controlling developmental patterning differences that distinguish male and female sexual forms of *Volvox*.

6.3.3. Asymmetric Cell Division

Chlamydomonas does not divide asymmetrically, so it might be expected that *Volvox* would have evolved one or more proteins that facilitate this function, perhaps through modifications to its cytoskeletal protein repertoire. On the contrary, no major differences in cytoskeletal proteins were detected between the two species (Prochnik *et al.*, 2010). Moreover, the single identified mutant that affects asymmetric division in *Volvox*, *glsA*, has a *Chlamydomonas* ortholog that complements its phenotype (Cheng *et al.*, 2003; Miller & Kirk, 1999). *glsA* encodes a DnaJ-domain protein that interacts with heat shock protein HSP70 to promote asymmetric cell division in an unknown manner (Cheng *et al.*, 2006; Cheng *et al.*, 2005). The investigation of GAR1, the *Chlamydomonas glsA* ortholog, may shed light on the ancestral role of this protein in *Chlamydomonas* cell division.

6.3.4. Inversion

Like the case for asymmetric cell division, there is no obvious correlate of inversion in *Chlamydomonas* and no candidate cytoskeletal genes that emerged from genomic comparisons as candidate mediators of this process. Forward genetic screens have identified three genes required for *Volvox* inversion – *invA*, *invB* and *invC* – that encode a kinesin, a sugar transporter, and a glycosyl transferase, respectively (Nishii *et al.*, 2003; Ueki & Nishii, 2008, 2009). *InvA* is thought to interact with the microtubule cytoskeleton at cytoplasmic bridge attachment sites to effect cell-shape changes required for inversion (Nishii *et al.*, 2003). *InvB* and *InvC* are required for expansion of the vesicle that surrounds the inverting embryo and which physically restricts inversion in the *invB* and *invC* mutants. Each of the above proteins has a highly similar ortholog in *Chlamydomonas* indicating that the pathways they control have cognate processes in *Chlamydomonas* from which they were coopted. The *invB* and *invC* functions might have ancestral roles in cell wall expansion that must occur in *Chlamydomonas* to accommodate cell growth. The potential role of the *Chlamydomonas invA* ortholog, *IAR1*, is harder to imagine but suggests the existence of cytoskeletal processes that have not yet been characterized.

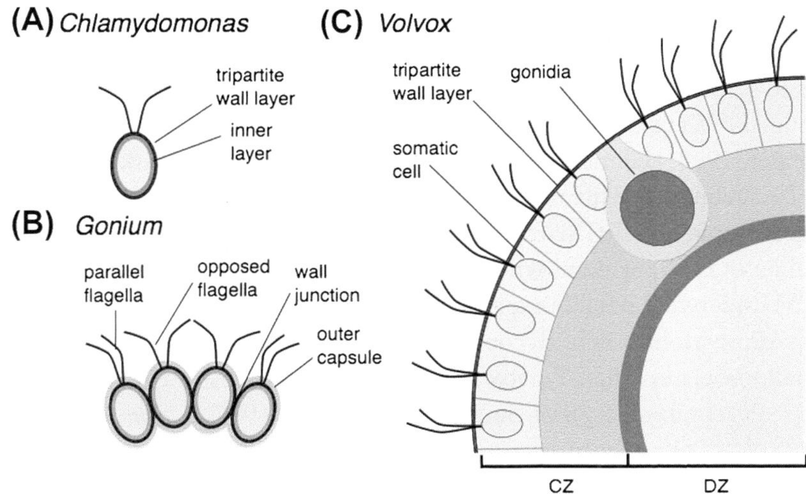

Figure 6.7 Comparison of the ECM structure of *Chlamydomonas*, *Gonium* and *Volvox*. The *Chlamydomonas* cell wall (A) is composed of an inner layer and an outer tripartite layer that is conserved with other Volvocales (indicated as a black layer around the cell). Individual *Gonium* cells (B) have an identical tripartite layer and the entire colony is surrounded by an additional outer capsule layer of ECM. Cells are held together by specialized attachments at their wall junctions. (C) *Volvox* cells are completely embedded in an expanded and compartmentalized ECM with a conserved tripartite boundary layer surrounding the entire colony (instead of individual cells). Inside the tripartite layer of the somatic cells are the cellular zones of ECM and the deep zone of the interior. For colour version of this figure, the reader is referred to the online version of this book.

6.3.5. Evolution of an Elaborated ECM

Cell–cell adhesion is a critical aspect of multicellularity (Abedin & King, 2010; Bonner, 2000). In the Volvocales, adhesion is accomplished by ECM proteins that are related to cell wall proteins from *Chlamydomonas* (Adair, Steinmetz, Mattson, Goodenough, & Heuser, 1987; Hallmann, 2006b; Kirk *et al.*, 1986). Indeed, the importance of cell–cell adhesion and ECM modifications in colonial Volvocine algae is reflected in two of Kirk's 12 steps of multicellular evolution that are changes to the ECM (Kirk, 2005). Step 5 is modification of the cell wall to maintain adhesion between post-mitotic daughters, while step 8 is expansion of the ECM that allows for colony growth in the absence of cell growth and division.

The *Chlamydomonas* cell wall is composed of three major layers of hydroxyproline-rich glycoproteins that can be divided into seven sublayers (Goodenough & Heuser, 1985). The outermost of these layers can be used to nucleate the assembly of *Volvox* cell walls demonstrating the evolutionary conservation of Volvocine cell wall structural components (Adair, 1988; Adair *et al.*, 1987). The expanded ECM in *Volvox* is organized and compartmentalized as revealed by various structural and localization studies (Fig. 6.7) (Ertl, Mengele, Wenzl, Engel, & Sumper, 1989; Godl, Hallmann, Rappel, & Sumper, 1995; Hallmann, 2006b; Hallmann & Kirk, 2000; Kirk *et al.*, 1986).

Paralleling the expansion of its ECM, the *Volvox* genome codes a much larger number of ECM-related proteins than does *Chlamydomonas*. Two families of ECM proteins whose members are known constituents of *Volvox* ECM underwent notable expansion in the *Volvox* lineage. These are a family of hydroxyproline-rich glycoproteins called pherophorins and a family of metalloproteases called Volvox metalloproteinase (VMPs) (Hallmann, 2002, 2007; Prochnik *et al.*, 2010; Sumper & Hallmann, 1997). *Chlamydomonas* has 27 predicted pheophorins versus 45 in *Volvox* and 8 VMPs versus 42 in *Volvox*.

ECM proteins in *Volvox* also play a role in the sexual cycle. The *Volvox* sex inducer is a pherophorin-related protein (Mages *et al.*, 1988) and expression of several pherophorin genes is up-regulated during sexual differentiation (Hallmann, 2006b; Prochnik *et al.*, 2010). Likewise, three related *Volvox* ECM metalloproteinases are also sexually regulated (Hallmann, 2006b; Prochnik *et al.*, 2010). The remarkable organizational properties of the *Volvox* ECM and the potential roles of ECM proteins in signalling and nutrient storage/transport are all exciting topics for future investigation.

6.3.6. *Terminally Differentiated Somatic Cells and the Origins of Volvocine Algae* RegA*-like (VARL) Genes*

Somatic cell fate in *Volvox* is controlled by *regA*, whose protein product is a nuclear localized putative transcriptional repressor (Duncan *et al.*, 2006; Duncan *et al.*, 2007; Kirk *et al.*, 1999; Meissner *et al.*, 1999; Tam *et al.*, 1991). *RegA* belongs to a family of proteins in Volvocine algae called VARL that all share a putative DNA-binding domain related to the SAND domain found in some transcriptional repressors (Duncan *et al.*, 2006, 2007). Unlike the previously described *glsA*, *invA*, *invB* and invC genes, *Volvox regA* has no clear ortholog in *Chlamydomonas*. Instead, both algae have multiple VARL paralogs that arose from a complicated pattern of duplications and/or losses in each of the two species (Duncan *et al.*, 2006, 2007). *regA* is part of a four gene paralogous duplication that is unique to *Volvox*. Interestingly, it is the only one of the four paralogs that has been identified as a developmental mutant (Kirk *et al.*, 1999). In *Chlamydomonas*, the closest VARL protein to *Volvox regA* is called RLS1 whose expression has been linked to stress and environmental responses (Nedelcu, 2009; Nedelcu & Michod, 2006), but whose function is not known. These data again point to the unicellular origins of proteins associated with multicellular development in the Volvocine algae.

6.3.7. *Chlamydomonas* and *Volvox* microRNAs are not Homologous

Both *Chlamydomonas* and *Volvox* have RNAi machinery (Cerutti *et al.*, 2011) and it might be expected that they contain homologous microRNA genes as do many higher plants and animals (Berezikov, 2011; Cuperus *et al.*, 2011). On the contrary, when *Chlamydomonas* microRNAs were compared to those predicted from *Volvox*, there was very little overlap (Zhao *et al.*, 2007), though there are caveats to this conclusion. The comparison was based on empirical data for small RNAs and predicted microRNAs in *Chlamydomonas* from studies that captured only a fraction of possible stages in the life cycle. An empirical evaluation of *Volvox* miRNAs and small RNA targets will be required to know if any underlying homology exists in the small RNA networks of the two algae. Equally important will be an improved understanding of how miRNAs and other small regulatory RNAs contribute to the developmental programs of Volvocine algae.

6.3.8. The *Volvox* Mating Locus and Evolution of Sexual Dimorphism

As discussed above, the sexual cycles of *Chlamydomonas* and *Volvox* are under the control of a mating locus (*MT*) whose location is similar in the two species, but whose size, content and evolutionary dynamics are very different (Ferris *et al.*, 2010; Umen, 2011). The structure of the *MT* region from *Volvox* and *Chlamydomonas* is described in Section 6.3. Besides its relative expansion and loss of residual synteny between shared genes (those with an allele in both mating types), the divergence rates of *Volvox* mating locus genes are up to two orders of magnitude higher than those in *Chlamydomonas* (Ferris *et al.*, 2010). Whereas shared genes in *Chlamydomonas MT* retain high nucleotide identity (typically 99% in coding regions and >90% in introns), shared genes in *Volvox* have coding region identities as low as 40–50% and typically have no similarity in their intron sequences. This large divergence is the result of completely blocked recombination that can be traced back through speciation events in sister species of *Volvox carteri* (Ferris *et al.*, 2010). Paradoxically, these differences with *Chlamydomonas MT* makes *Volvox MT* look older even though it is part of a presumably younger and more recently derived lineage (Charlesworth & Charlesworth, 2010; Umen, 2011). One hypothesis that needs testing is whether cryptic recombination might occur in *Chlamydomonas MT* that acts to maintain homogeneity between shared genes in the two *MT* haplotypes (Umen, 2011). The effect of long-term isolation of female and male alleles of *Volvox* mating locus genes is divergence in both primary sequence and expression patterns, with a remarkable number of genes having acquired sex-regulated expression including female or male biases and up- or down-regulation during sexual differentiation (Ferris *et al.*, 2010). Thus, even shared genes in the *Volvox* mating locus have diverged enough to acquire male- or female-specific functions.

Another remarkable feature of *Volvox MT* is the presence of numerous male-limited and female-limited genes that do not have an allelic counterpart in the opposite mating type (Fig. 6.6). Of 15 such genes (5 female specific and 10 male specific), only 2, *VcMID* and *VcMTD*, have any similarity to *Chlamydomonas* genes and only 1, encoding an HMG-box protein, shows similarity to genes in any other species (Ferris *et al.*, 2010). The number of 'new' genes in the *Volvox* mating locus stands in contrast to the paucity of *Volvox*-specific autosomal genes (Prochnik *et al.*, 2010) (and see below). This finding implies that the *Volvox* mating locus has evolved into a hotbed of evolutionary experimentation (Ferris *et al.*, 2010), an idea that might be confirmed by obtaining sequences from mating loci of related Volvocine species and additional wild isolates of *V. carteri*.

VcMID and VcMTD are in male *MT* and have homologs *CrMID* and *CrMTD* in the *Chlamydomonas* MT^{-} locus. *MID* and *MTD* homologs have been found in MT^{-} strains of *Gonium pectorale* (Hamaji *et al.*, 2008) and *MID* has also been found in males of *P. starrii* (Nozaki *et al.*, 2006). Mid proteins are related to RWP-RK domain putative transcription factors, while Mtd proteins are Volvocine algal specific. In *Chlamydomonas*, *MID* is a key sex determining gene whose presence or absence dictates differentiation as *minus* or *plus* gametes (Ferris & Goodenough, 1997). Mtd appears to play an auxiliary role in augmenting the expression of *MID* (Lin & Goodenough, 2007). Functional studies of *MID* have not been reported outside of *Chlamydomonas*, but two observations suggest that sexual differentiation in *Volvox* will involve additional factors besides *VcMID* and *VcMTD*. First, *VcMID* messenger RNA (mRNA) is expressed constitutively in both vegetative and sexually induced males and is not controlled by sex inducer or by nitrogen starvation as is the case for *MID* orthologs in *Chlamydomonas*, *Gonium* and *Pleodorina* (Ferris & Goodenough, 1997; Hamaji *et al.*, 2008; Lin & Goodenough, 2007; Nozaki *et al.*, 2006). Second, *VcMTD* appears to be a pseudogene in *Volvox* with no start codon anywhere near the 5' end of its mRNA (Ferris *et al.*, 2010). These observations suggest that other genes in *MT* or elsewhere contribute to sexual differentiation in *Volvox*.

7. FUTURE PERSPECTIVES

Genomics research with Volvocine algae is still far from reaching its potential. While the unexpectedly similar genome sequences of *Chlamydomonas* and *Volvox* have not led to a simple list of genes that underlie each major trait difference that distinguishes them, they have provided a framework for more in-depth studies. Genomics has also refocused investigations into the origin of complexity towards the unicellular species *Chlamydomonas* that is remarkably sophisticated and appears to possess most of the genetic toolkit that was required for Volvocine algae to become multicellular. Critical for understanding how Volvocine algae and their genomes evolve will be sequencing additional species. Such sequencing will allow important genomic attributes such as gene gains and losses to be polarized in order to determine the direction of change within the group. Given the complex evolutionary history within Volvocine algae (Fig. 6.2), additional genomes will provide insights into how convergent traits evolved in different subgroups. As genomes from closely related Volvocine algal subgroups are compared, the key genetic changes that distinguish them may come into focus. The decreasing cost of sequencing will soon make it possible to do population genomics on Volvocine algae that will begin to shed light on how population structure and natural variation relate to long-term evolutionary change and speciation. Low-cost genome sequencing is an enabling technology for large-scale forward genetics and functional genomics that are of increasing importance for *Chlamydomonas* and will soon be so for *Volvox* and other species. Sequencing is also revolutionizing the ability to rapidly create new model systems from previously unstudied species. Efforts to develop additional Volvocine algal models such as *Gonium pectorale* (whose genome is currently being sequenced) will greatly expand the utility of Volvocine algae as a premier model for evolution and development.

REFERENCES

Abedin, M., & King, N. (2010). Diverse evolutionary paths to cell adhesion. *Trends in Cell Biology, 20*, 734–742.

Adair, W. (1988). Organization and in vitro assembly of the *Chlamydomonas reinhardtii* cell wall. *Self-assembling architecture*. (pp 25–51), Alan R. Liss, Inc.

Adair, W. S., Steinmetz, S. A., Mattson, D. M., Goodenough, U. W., & Heuser, J. E. (1987). Nucleated assembly of *Chlamydomonas* and *Volvox* cell walls. *Journal of Cell Biology, 105*, 2373–2382.

Adams, C. R., Stamer, K. A., Miller, J. K., McNally, J. G., Kirk, M. M., & Kirk, D. L. (1990). Patterns of organellar and nuclear inheritance among progeny of two geographically isolated strains of *Volvox carteri*. *Current Genetics, 18*, 141–153.

Adams, S., Maple, J., & Møller, S. G. (2008). Functional conservation of the MIN plastid division homologues of *Chlamydomonas reinhardtii*. *Planta, 227*, 1199–1211.

Aoyama, H., Hagiwara, Y., Misumi, O., Kuroiwa, T., & Nakamura, S. (2006). Complete elimination of maternal mitochondrial DNA during meiosis resulting in the paternal inheritance of the mitochondrial genome in *Chlamydomonas* species. *Protoplasma, 228*, 231–242.

Armbrust, E. V., Ibrahim, A., & Goodenough, U. W. (1995). A mating type-linked mutation that disrupts the uniparental inheritance of chloroplast DNA also disrupts cell-size control in *Chlamydomonas*. *Molecular Biology of the Cell, 6*, 1807–1818.

Asamizu, E., Nakamura, Y., Miura, K., Fukuzawa, H., Fujiwara, S., Hirono, M., et al. (2004). Establishment of publicly available cDNA material and information resource of Chlamydomonas reinhardtii (*Chlorophyta*) to facilitate gene function analysis. *Phycologia, 43*, 722–726.

Avidor-Reiss, T., Maer, A. M., Koundakjian, E., Polyanovsky, A., Keil, T., Subramaniam, S., et al. (2004). Decoding cilia function: defining specialized genes required for compartmentalized cilia biogenesis. *Cell, 117*, 527–539.

Baldauf, S. (2003). The deep roots of eukaryotes. *Science, 300*, 1703.

Bell, G. (1985). The origin and early evolution of germ cells, as illustrated by the Volvocales. In H. Halvorson, & A. Monroy (Eds.), *The Origin and Evolution of Sex,* 1984 Wood's Hole Symposium. (pp 221–256) New York: Liss.

Berezikov, E. (2011). Evolution of microRNA diversity and regulation in animals. *Nature Reviews Genetics, 12*, 846–860.

Berthold, P., Schmitt, R., & Mages, W. (2002). An engineered Streptomyces hygroscopicus aph 7" gene mediates dominant resistance against hygromycin B in *Chlamydomonas reinhardtii*. *Protist, 153*, 401–412.

Bisova, K., Krylov, D. M., & Umen, J. G. (2005). *Genome-wide annotation and expression profiling of cell cycle regulatory genes in* Chlamydomonas reinhardtii, *137*, 475–491.

Boer, P. H., Bonen, L., Lee, R. W., & Gray, M. W. (1985). Genes for respiratory chain proteins and ribosomal RNAs are present on a 16-kilobase-pair DNA species from *Chlamydomonas reinhardtii* mitochondria. *Proceedings of the National Academy of Sciences of the United States of America, 82*, 3340–3344.

Boer, P. H., & Gray, M. W. (1988). Scrambled ribosomal RNA gene pieces in *Chlamydomonas reinhardtii* mitochondrial DNA. *Cell, 55*, 399–411.

Bonner, J. (2000). *First signals, the evolution of multicellular development.* Princeton, NJ: Princeton University Press.

Boynton, J., Gillham, N., Harris, E., Hosler, J., Johnson, A., Jones, A., et al. (1988). Chloroplast transformation in *Chlamydomonas* with high velocity microprojectiles. *Science, 240*, 1534–1538.

Boynton, J. E., & Gillham, N. W. (1996). Genetics and transformation of mitochondria in the green alga *Chlamydomonas*. *Methods in Enzymology, 264*, 279–296.

Buchheim, M. A., Buchheim, J. A., & Chapman, R. L. (1997). Phylogenyof the VLE-14 Chlamyomonas (Chlorophyceae) group: a study of 18S rRNA gene sequences. *Journal of Phycology, 33*, 1024–1030.

Bull, J. (1978). Sex chromosomes in haploid dioecy: a unique contrast to Muller's theory for diploid dioecy. *American Naturalist, 112*, 245–250.

Cardullo, R. A., Herrick, S. B., Peterson, M. J., & Dangott, L. J. (1994). Speract receptors are localized on sea urchin sperm flagella using a fluorescent peptide analog. *Developmental Biology, 162*, 600–607.

Casas-Mollano, J. A., Rohr, J., Kim, E.-J., Balassa, E., van Dijk, K., & Cerutti, H. (2008). Diversification of the core RNA interference machinery in *Chlamydomonas reinhardtii* and the role of DCL1 in transposon silencing. *Genetics, 179*, 69–81.

Castruita, M., Casero, D., Karpowicz, S. J., Kropat, J., Vieler, A., Hsieh, S. I., et al. (2011). Systems biology approach in *Chlamydomonas* reveals connections between copper nutrition and multiple metabolic steps. *The Plant Cell, 23*, 1273–1292.
Cerutti, H., Johnson, A. M., Gillham, N. W., & Boynton, J. E. (1997). Epigenetic silencing of a foreign gene in nuclear transformants of *Chlamydomonas*. *Plant Cell, 9*, 925–945.
Cerutti, H., Ma, X., Msanne, J., & Repas, T. (2011). RNA-mediated silencing in algae: biological roles and tools for analysis of gene function. *Eukaryotic Cell, 10*, 1164–1172.
Charlesworth, D., & Charlesworth, B. (2010). Evolutionary biology: the origins of two sexes. *Current Biology, 20*, R519–R521.
Chen, C. L., Chen, C. J., Vallon, O., Huang, Z. P., Zhou, H., & Qu, L. H. (2008). Genomewide analysis of box C/D and box H/ACA snoRNAs in *Chlamydomonas reinhardtii* reveals an extensive organization into intronic gene clusters. *Genetics, 179*, 21–30.
Cheng, Q., Fowler, R., Tam, L.-W., Edwards, L., & Miller, S. M. (2003). The role of GlsA in the evolution of asymmetric cell division in the green alga *Volvox carteri*. *Development Genes and Evolution, 213*, 328–335.
Cheng, Q., Hallmann, A., Edwards, L., & Miller, S. M. (2006). Characterization of a heat-shock-inducible hsp70 gene of the green alga *Volvox carteri*. *Gene, 371*, 112–120.
Cheng, Q., Pappas, V., Hallmann, A., & Miller, S. M. (2005). Hsp70A and GlsA interact as partner chaperones to regulate asymmetric division in *Volvox*. *Developmental Biology, 286*, 537–548.
Chiang, K.-S., & Sueoka, N. (1967). Replication of chloroplast DNA in *Chlamydomonas* reinhardi during vegetative cell cycle: its mode and regulation. *Proceedings of the National Academy of Sciences of the United States of America, 57*, 1506.
Coleman, A. W. (1999). Phylogenetic analysis of "Volvocacae" for comparative genetic studies. *Proceedings of the National Academy of Sciences of the United States of America, 96*, 13892–13897.
Coleman, A. (2012). A comparative analysis of the Volvocaceae (Chlorophyta). *Journal of Phycology, 48*, 491–513.
Corellou, F. (2005). Atypical regulation of a green lineage-specific B-type cyclin-dependent kinase. *Plant Physiology, 138*, 1627–1636.
Costa, J. A. V., & de Morais, M. G. (2011). The role of biochemical engineering in the production of biofuels from microalgae. *Bioresource Technology, 102*, 2–9.
Croft, M. T., Warren, M. J., & Smith, A. G. (2006). Algae need their vitamins. *Eukaryotic Cell, 5*, 1175–1183.
Cuperus, J. T., Fahlgren, N., & Carrington, J. C. (2011). Evolution and functional diversification of MIRNA genes. *Plant Cell, 23*, 431–442.
Debuchy, R., Purton, S., & Rochaix, J.-D. (1989). The argininosuccinate lyase gene of *Chlamydomonas reinhardtii*: an important tool for nuclear transformation and for correlating the genetic and molecular maps of the ARG7 locus. *The EMBO Journal, 8*, 2803.
De Clerck, O., Bogaret, K., & Leliaert, F. (2012). Diversity and evolution of algae: primary endosymbiosis. *Advances in Botanical Research*. 64, 55–86.
DeLuca, S. Z., & O'Farrell, P. H. (2012). Barriers to male transmission of mitochondrial DNA in sperm development. *Developmental Cell, 22*, 660–668.
Denovan-Wright, E. M., Sankoff, D., Spencer, D. F., & Lee, R. W. (1996). Evolution of fragmented mitochondrial ribosomal RNA genes in *Chlamydomonas*. *Journal of Molecular Evolution, 42*, 382–391.
Dent, R. M., Haglund, C. M., Chin, B. L., Kobayashi, M. C., & Niyogi, K. K. (2005). Functional genomics of eukaryotic photosynthesis using insertional mutagenesis of *Chlamydomonas reinhardtii*. *Plant Physiology, 137*, 545–556.
Derelle, E., Ferraz, C., Rombauts, S., Rouze, P., Worden, A. Z., Robbens, S., Partensky, et al.. (2006). Genome analysis of the smallest free-living eukaryote *Ostreococcus tauri*

unveils many unique features. *Proceedings of the National Academy of Sciences of the United States of America, 103*, 11647–11652.

Desnitski, A. (1995). A review on the evolution of development in *Volvox*-morphological and physiological aspects. *European journal of protistology, 31*, 241–247.

Diener, D. (2009). Chapter 6-Analysis of Cargo Transport by IFT and GFP Imaging of IFT in Chlamydomonas. In Pazour, S. M. K. A. G. J. (Ed.) (pp. 111–119). Methods in Cell Biology: Academic Press.

Dunahay, T. G., Adler, S. A., & Jarvik, J. W. (1997). Transformation of microalgae using silicon carbide whiskers. *Methods in Molecular Biology, 62*, 503–509.

Duncan, L., Nishii, I., Harryman, A., Buckley, S., Howard, A., Friedman, N. R., et al. (2007). The VARL gene family and the evolutionary origins of the master cell-type regulatory gene, regA, in *Volvox carteri*. *Journal of Molecular Evolution, 65*, 1–11.

Duncan, L., Nishii, I., Howard, A., Kirk, D., & Miller, S. M. (2006). Orthologs and paralogs of regA, a master cell-type regulatory gene in *Volvox carteri*. *Current Genetics, 50*, 61–72.

Dutcher, S. K. (2003). Elucidation of basal body and centriole functions in *Chlamydomonas reinhardtii*. *Traffic, 4*, 443–451.

Dutcher, S. K., Li, L., Lin, H., Meyer, L., Giddings, T. H., Jr., Kwan, A. L., et al. (2012) *Whole-genome sequencing to identify mutants and polymorphisms in Chlamydomonas reinhardtii*, 2. 15–22.

Eberhard, S., Jain, M., Im, C.-S., Pollock, S., Shrager, J., Lin, Y., et al. (2006). Generation of an oligonucleotide array for analysis of gene expression in *Chlamydomonas reinhardtii*. *Current Genetics, 49*, 106–124.

Ebnet, E. (1999). *Volvox*rhodopsin, a light-regulated sensory photoreceptor of the spheroidal green alga *Volvox carteri*. *Plant Cell, 11*, 1473–1484.

Eichler-Stahlberg, A., Weisheit, W., Ruecker, O., & Heitzer, M. (2009). Strategies to facilitate transgene expression in *Chlamydomonas reinhardtii*. *Planta, 229*, 873–883.

Ender, F., Godl, K., Wenzl, S., & Sumper, M. (2002). Evidence for autocatalytic cross-linking of hydroxyproline-rich glycoproteins during extracellular matrix assembly in *Volvox*. *Plant Cell, 14*, 1147–1160.

Ertl, H., Mengele, R., Wenzl, S., Engel, J., & Sumper, M. (1989). The extracellular matrix of *Volvox carteri*: molecular structure of the cellular compartment. *Journal of Cell Biology, 109*, 3493–3501.

Fang, S.-C., de los Reyes, C., & Umen, J. G. (2006). Cell size checkpoint control by the retinoblastoma tumor suppressor pathway. *PLoS Genetics, 2*, e167.

Ferrante, P., Catalanotti, C., Bonente, G., & Giuliano, G. (2008). An optimized, chemically regulated gene expression system for *Chlamydomonas*. *PLoS ONE, 3*, e3200.

Ferris, P., & Goodenough, U. (1997). Mating type in *Chlamydomonas* is specified by mid, the minus-dominance gene. *Genetics, 146*, 859–869.

Ferris, P. J. (1995). Localization of the nic-7, ac-29 and thi-10 genes within the mating-type locus of *Chlamydomonas reinhardtii*. *Genetics, 141*, 543–549.

Ferris, P. J., Armbrust, E. V., & Goodenough, U. W. (2002). Genetic structure of the mating-type locus of *Chlamydomonas reinhardtii*. *Genetics, 160*, 181–200.

Ferris, P. J., Olson, B., de Hoff, P. L., Douglass, S., Casero, D., Prochnik, S. E., et al. (2010). Evolution of an expanded sex-determining locus in *Volvox*. *Science, 328*, 351–354.

Fischer, N., & Rochaix, J. D. (2001). The flanking regions of PsaD drive efficient gene expression in the nucleus of the green alga *Chlamydomonas reinhardtii*. *Molecular and General Genomics, 265*, 888–894.

Franzén, L.-G., Rochaix, J.-D., & Heijne von, G. (1990). Chloroplast transit peptides from the green alga *Chlamydomonas reinhardtii* share features with both mitochondrial and higher plant chloroplast presequences. *FEBS Letters, 260*, 165–168.

Fuhrmann, M., Hausherr, A., Ferbitz, L., Schödl, T., Heitzer, M., & Hegemann, P. (2004). Monitoring dynamic expression of nuclear genes in *Chlamydomonas reinhardtii* by using a synthetic luciferase reporter gene. *Plant Molecular Biology, 55*, 869–881.

Fuhrmann, M., Oertel, W., & Hegemann, P. (1999). A synthetic gene coding for the green fluorescent protein (GFP) is a versatile reporter in *Chlamydomonas reinhardtii. Plant Journal, 19*, 353–361.

Galván, A., González-Ballester, D., & Fernández, E. (2007). Insertional mutagenesis as a tool to study genes/functions in *Chlamydomonas. Advances in Experimental Medicine and Biology, 616*, 77–89.

Godl, K., Hallmann, A., Rappel, A., & Sumper, M. (1995). Pherophorins: a family of extracellular matrix glycoproteins from *Volvox* structurally related to the sex-inducing pheromone. *Planta, 196.*

Gonzalez-Ballester, D. (2005). Functional genomics of the regulation of the nitrate assimilation pathway in *Chlamydomonas. Plant Physiology, 137*, 522–533.

González-Ballester, D., Pootakham, W., Mus, F., Yang, W., Catalanotti, C., Magneschi, L., et al. (2011). Reverse genetics in *Chlamydomonas*: a platform for isolating insertional mutants. *Plant Methods, 7*, 24.

Goodenough, U., Lin, H., & Lee, J.-H. (2007). Sex determination in *Chlamydomonas. Seminars in Cell and Developmental Biology, 18*, 350–361.

Goodenough, U. W., Armbrust, E. V., Campbell, A. M., & Ferris, P. J. (1995). Molecular genetics of sexuality in *Chlamydomonas. Annual Review of Plant Physiology and Plant Molecular Biology, 46*, 21–44.

Goodenough, U. W., & Heuser, J. E. (1985). The *Chlamydomonas* cell wall and its constituent glycoproteins analyzed by the quick-freeze, deep-etch technique. *Journal of Cell Biology, 101*, 1550–1568.

Goodstein, D. M., Shu, S., Howson, R., Neupane, R., Hayes, R. D., Fazo, J., et al. (2011). Phytozome: a comparative platform for green plant genomics. *Nucleic Acids Research, 40*, D1178–D1186.

Grafi, G., Burnett, R. J., Helentjaris, T., Larkins, B. A., DeCaprio, J. A., Sellers, W. R., et al. A maize cDNA encoding a member of the retinoblastoma protein family: involvement in endoreduplication.

Grafi, G., Burnett, R. J., Helentjaris, T., Larkins, B. A., DeCaprio, J. A., Sellers, W. R., et al. (1996). A maize cDNA encoding a member of the retinoblastoma protein family: involvement in endoreduplication. *Proceedings of the National Academy of Sciences USA, 93*, 8962–8967.

Graham, L. E., Graham, J. M., & Wilcox, L. W. (2009). *Algae*. Benjamin Cummings: The University of California.

Grant, D., & Chiang, K.-S. (1980). Physical mapping and characterization of *Chlamydomonas* mitochondrial DNA molecules: their unique ends, sequence homogeneity, and conservation. *Plasmid, 4*, 82–96.

Gray, M. W., & Boer, P. H. (1988). Organization and expression of algal (*Chlamydomonas reinhardtii*) mitochondrial DNA. *Philosophical Transactons of the Royal Society Lond, B, Biological Sciences, 319*, 135–147.

Grossman, A., Harris, E., Hauser, C., Lefebvre, P., Martinez, D., Rokhsar, D., et al. (2003). *Chlamydomonas reinhardtii* at the Crossroads of Genomics. *Eukaryotic Cell, 2*, 1137.

Grossman, A. R., Croft, M., Gladyshev, V. N., Merchant, S. S., Posewitz, M. C., Prochnik, S., et al. (2007). Novel metabolism in *Chlamydomonas* through the lens of genomics. *Current Opinion in Plant Biology, 10*, 190–198.

Gruber, H., Kirzinger, S. H., & Schmitt, R. (1996). Expression of the *Volvox* gene encoding nitrate reductase: mutation-dependent activation of cryptic splice sites and intron-enhanced gene expression from a cDNA. *Plant Molecular Biology, 31*, 1–12.

Hallmann, A. (2002). Extracellular matrix and sex-inducing pheromone in *Volvox*. *International Review of Cytology, 227*, 131–182.

Hallmann, A. (2006a). Morphogenesis in the family Volvocaceae: different tactics for turning an embryo right-side out. *Protist, 157*, 445–461.

Hallmann, A. (2006b). The pherophorins: common, versatile building blocks in the evolution of extracellular matrix architecture in Volvocales. *Plant Journal, 45*, 292–307.

Hallmann, A. (2007). A small cysteine-rich extracellular protein, VCRP, is inducible by the sex-inducer of *Volvox carteri* and by wounding. *Planta, 226*, 719–727.

Hallmann, A., Godl, K., Wenzl, S., & Sumper, M. (1998). The highly efficient sex-inducing pheromone system of *Volvox*. *Trends in Microbiology, 6*, 185–189.

Hallmann, A., & Kirk, D. L. (2000). The developmentally regulated ECM glycoprotein ISG plays an essential role in organizing the ECM and orienting the cells of *Volvox*. *Journal of Cell Science, 113*(24)), 4605–4617.

Hallmann, A., & Rappel, A. (1999). Genetic engineering of the multicellular green alga *Volvox*: a modified and multiplied bacterial antibiotic resistance gene as a dominant selectable marker. *Plant Journal, 17*, 99–109.

Hallmann, A., & Sumper, M. (1994). Reporter genes and highly regulated promoters as tools for transformation experiments in *Volvox carteri*. *Proceedings of the National Academy of Sciences of the United States of America, 91*, 11562–11566.

Hamaji, T., Ferris, P. J., Coleman, A. W., Waffenschmidt, S., Takahashi, F., Nishii, I., et al. (2008). Identification of the minus-dominance gene ortholog in the mating-type locus of *Gonium pectorale*. *Genetics, 178*, 283–294.

Harper, J., & John, P. (1986). Coordination of division events in the *Chlamydomonas* cell cycle. *Protoplasma, 131*.

Harris, E. H. (2001). *Chlamydomonas* as a model organism. *Annual Review of Plant Physiology and Plant Molecular Biology, 52*, 363–406.

Harris, E. H. (2004). *Introduction to* Chlamydomonas*, the molecular biology of chloroplasts and mitochondria in* Chlamydomonas. Dordrecht: Kluwer Academic Publishers.

Hayashi, Y., & Shinozaki, A. (2012). Visualization of microbodies in Chlamydomonas reinhardtii. *Journal of Palnt Research, 125*, 579–586.

Helliwell, K. E., Wheeler, G. L., Leptos, K. C., Goldstein, R. E., & Smith, A. G. (2011). Insights into the evolution of vitamin B12 auxotrophy from sequenced algal genomes. *Molecular Biology and Evolution, 28*, 2921–2933.

Herron, M. D., Hackett, J. D., Aylward, F. O., & Michod, R. E. (2009). Triassic origin and early radiation of multicellular volvocine algae. *Proceedings of the National Academy of Sciences of the United States of America, 106*, 3254–3258.

Herron, M. D., & Michod, R. E. (2008). Evolution of complexity in the volvocine algae: transitions in individuality through Darwin's eye. *Evolution, 62*, 436–451.

Hildebrandt, F., Benzing, T., & Katsanis, N. (2011). Ciliopathies. *New England Journal of Medicine, 364*, 1533–1543.

Hoham, R. W., Bonome, T. A., Martin, C. W., & Leebens Mack, J. H. (2002). A combined 18S rDNA and rbcL phylogenetic analysis of Chloromonas and *Chlamydomonas* (Chlorophyceae, Volvocales) emphasizing snow and other cold-temperature habitats. *Journal of Phycology, 38*, 1051–1064.

Hohenester, E., Sasaki, T., & Timpl, R. (1999). Crystal structure of a scavenger receptor cysteine-rich domain sheds light on an ancient superfamily. *Nature Structural Biology, 6*, 228–232.

Hoops, H. J., Nishii, I., & Kirk, D. L. (2000). Cytoplasmic Bridges in *Volvox* and Its Relatives. In Madame Curie Bioscience Database [Internet]. Austin (TX): Landes Bioscience-. Available from: http://www.ncbi.nlm.nih.gov/books/NBK6424/.

Hoops, H., Nishii, I., & Kirk, D. (2006). Cytoplasmic bridges in *Volvox* and its relatives. In F. Baluška, D. Volkmann, & P. Barlow (Eds.), *Cell-cell channels*. Landes Bioscience.

Hu, Z., Zhao, Z., Wu, Z., Fan, Z., Chen, J., Wu, J., et al. (2011). Successful expression of heterologous egfp gene in the mitochondria of a photosynthetic eukaryote *Chlamydomonas reinhardtii*. *Mitochondrion, 27*, 716–721.

Huang, C.-H., & Liu, P. Z. (2001). New insights into the Rh superfamily of genes and proteins in erythroid cells and nonerythroid tissues. *Blood Cells, Molecules, and Diseases, 27*, 90–101.

Huskey, R., Griffin, B., Cecil, P., & Callahan, A. (1979). A preliminary genetic investigation of *Volvox carteri*. *Genetics, 91*, 229–244.

Ishida, K. (2007). Sexual pheromone induces diffusion of the pheromone-homologous polypeptide in the extracellular matrix of *Volvox carteri*. *Eukaryotic Cell, 6*, 2157–2162.

Jain, M., Shrager, J., Harris, E. H., Halbrook, R., Grossman, A. R., Hauser, C., et al. (2007). EST assembly supported by a draft genome sequence: an analysis of the *Chlamydomonas reinhardtii* transcriptome. *Nucleic Acids Research, 35*, 2074–2083.

Jakobiak, T., Mages, W., Scharf, B., Babinger, P., Stark, K., & Schmitt, R. (2004). The bacterial paromomycin resistance gene, aphH, as a dominant selectable marker in *Volvox carteri*. *Protist, 155*, 381–393.

Jamers, A., Blust, R., & De Coen, W. (2009). Omics in algae: paving the way for a systems biological understanding of algal stress phenomena? *Aquatic Toxicology, 92*, 114–121.

Johnson, U. G., & Porter, K. R. (1968). Fine structure of cell division in *Chlamydomonas* reinhardi. Basal bodies and microtubules. *Journal of Cell Biology, 38*, 403–425.

Kalanon, M., & McFadden, G. I. (2008). The chloroplast protein translocation complexes of *Chlamydomonas reinhardtii*: a bioinformatic comparison of Toc and Tic components in plants, green algae and red algae. *Genetics, 179*, 95–112.

Karpowicz, S. J., Prochnik, S. E., Grossman, A. R., & Merchant, S. S. (2011). The GreenCut2 resource, a phylogenomically derived inventory of proteins specific to the plant lineage. *Journal of Biological Chemistry, 286*, 21427–21439.

Kathir, P., LaVoie, M., Brazelton, W. J., Haas, N. A., Lefebvre, P. A., & Silflow, C. D. (2003). Molecular map of the *Chlamydomonas reinhardtii* nuclear genome. *Eukaryotic Cell, 2*, 362–379.

Keller, L. C., Romijn, E. P., Zamora, I., Yates, J. R., & Marshall, W. F. (2005). Proteomic analysis of isolated *Chlamydomonas* centrioles reveals orthologs of ciliary-disease genes. *Current Biology, 15*, 1090–1098.

Keller, L. R. (1995). Electroporation of DNA into the unicellular green alga *Chlamydomonas reinhardtii*. *Methods in Mololecular Biology, 55*, 73–79.

Kindle, K. L. (1990). High-frequency nuclear transformation of *Chlamydomonas reinhardtii*. *Proceedings of the National Academy of Sciences of the United States of America, 87*, 1228–1232.

Kindle, K. L., Schnell, R. A., Fernández, E., & Lefebvre, P. A. (1989). Stable nuclear transformation of *Chlamydomonas* using the *Chlamydomonas* gene for nitrate reductase. *Journal of Cell Biology, 109*, 2589–2601.

King, N. (2004). The unicellular ancestry of animal development. *Developmental Cell, 7*, 313–325.

King, N., Westbrook, M. J., Young, S. L., Kuo, A., Abedin, M., Chapman, J., et al. (2008). The genome of the choanoflagellate Monosiga brevicollis and the origin of metazoans. *Nature, 451*, 783–788.

Kirk, D. (1995). Asymmetric division, cell size and germ-soma specification in *Volvox*. *Seminars in Developmental Biology, 6*, 369–379.

Kirk, D. (1998). *Volvox: molecular-genetic origins of multicellularity and cellular differentiation.* New York: Cambridge University Press, Cambridge, U.K.
Kirk, D. (1998). Volvox*: molecular-genetic origins of multicellularity and cellular differentiation.* Cambridge University Press.
Kirk, D. L. (1999). Evolution of multicellularity in the volvocine algae. *Current Opinion in Plant Biology, 2*, 496–501.
Kirk, D. L. (2000). *Volvox* as a model system for studying the ontogeny and phylogeny of multicellularity and cellular differentiation. *Journal of Plant Growth Regulation, 19*, 265–274.
Kirk, D. L. (2005). A twelve-step program for evolving multicellularity and a division of labor. *Bioessays, 27*, 299–310.
Kirk, D. L., Birchem, R., & King, N. (1986). The extracellular matrix of *Volvox*: a comparative study and proposed system of nomenclature. *Journal of Cell Science, 80*, 207–231.
Kirk, D. L., & Nishii, I. (2001). *Volvox carteri* as a model for studying the genetic and cytological control of morphogenesis. *Development, Growth and Differentiation, 43*, 621–631.
Kirk, M. M., Ransick, A., McRae, S. E., & Kirk, D. L. (1993). The relationship between cell size and cell fate in *Volvox carteri*. *Journal of Cell Biology, 123*, 191–208.
Kirk, M. M., Stark, K., Miller, S. M., Müller, W., Taillon, B. E., Gruber, H., et al. (1999). regA, a *Volvox* gene that plays a central role in germ-soma differentiation, encodes a novel regulatory protein. *Development, 126*, 639–647.
Kochert, G. (1968). Differentiation of reproductive cells in *Volvox carteri*. *Journal of Protozoology, 15*, 438–452.
Koufopanou, V. (1994). The evolution of soma in the Volvocales. *American Naturalist,* 907–931.
Kozminski, K. G., Johnson, K. A., Forscher, P., & Rosenbaum, J. L. (1993). A motility in the eukaryotic flagellum unrelated to flagellar beating. *Proceedings of the National Academy of Sciences of the United States of America, 90*, 5519–5523.
Kruse, O., & Hankamer, B. (2010). Microalgal hydrogen production. *Current Opinion in Biotechnology, 21*, 238–243.
Kuroiwa, T. (2010). The replication, differentiation, and inheritance of plastids with enlphasis on the concept of organelle nuclei. *International Review of Cytology, 128*, 1–33.
Kwan, A. L., Li, L., Kulp, D. C., Dutcher, S. K., & Stormo, G. D. (2009). Improving gene-finding in *Chlamydomonas reinhardtii*:GreenGenie2. *BMC Genomics, 10*, 210.
Larson, A., Kirk, M. M., & Kirk, D. L. (1992). Molecular phylogeny of the volvocine flagellates. *Molecular Biology and Evolution, 9*, 85–105.
Lemieux, B., Turmel, M., & Lemieux, C. (1985). Chloroplast DNA variation in *Chlamydomonas* and its potential application to the systematics of this genus. *BioSystems, 18*, 293–298.
Lerche, K., & Hallmann, A. (2009). Stable nuclear transformation of *Gonium pectorale*. *BMC Biotechnology, 9*, 64.
Li, J. B., Gerdes, J. M., Haycraft, C. J., Fan, Y., Teslovich, T. M., May-Simera, H., et al. (2004). Comparative genomics identifies a flagellar and basal body proteome that includes the BBS5 human disease gene. *Cell, 117*, 541–552.
Lin, H., & Goodenough, U. W. (2007). Gametogenesis in the *Chlamydomonas reinhardtii* minus mating type is controlled by two genes, MID and MTD1. *Genetics, 176*, 913–925.
Lodha, M., Schulz-Raffelt, M., & Schroda, M. (2008). A new assay for promoter analysis in *Chlamydomonas* reveals roles for heat shock elements and the TATA box in HSP70A promoter-mediated activation of transgene expression. *Eukaryotic Cell, 7*, 172–176.

Lopez, D., Casero, D., Cokus, S. J., Merchant, S. S., & Pellegrini, M. (2011). Algal Functional Annotation Tool: a web-based analysis suite to functionally interpret large gene lists using integrated annotation and expression data. *BMC Bioinformatics, 12*, 282.

Ma, D.-P., King, Y.-T., Kim, Y., Luckett, W. S., Jr., Boyle, J. A., & Chang, Y.-F. (1992). Amplification and characterization of an inverted repeat from the *Chlamydomonas reinhardtii mitochondrial genome. Gene, 119*, 253–257.

Mages, H. W., Tschochner, H., & Sumper, M. (1988). The sexual inducer of *Volvox carteri*. Primary structure deduced from cDNA sequence. *FEBS Letters, 234*, 407–410.

Matsuo, T., Okamoto, K., Onai, K., Niwa, Y., Shimogawara, K., & Ishiura, M. (2008). A systematic forward genetic analysis identified components of the *Chlamydomonas* circadian system. *Genes and Development, 22*, 918–930.

Maul, J. E., Lilly, J. W., Cui, L., Depamphilis, C. W., Miller, W., Harris, E. H., et al. (2002). The *Chlamydomonas reinhardtii* plastid chromosome: islands of genes in a sea of repeats. *Plant Cell, 14*, 2659–2679.

Mayfield, S. P., & Schultz, J. (2004). Development of a luciferase reporter gene, luxCt, for Chlamydomonas reinhardtii chloroplast. *Plant Journal, 37*, 449–458.

Meissner, M., Stark, K., Cresnar, B., Kirk, D. L., & Schmitt, R. (1999). *Volvox* germline-specific genes that are putative targets of RegA repression encode chloroplast proteins. *Current Genetics, 36*, 363–370.

Merchant, S. S., Prochnik, S. E., Vallon, O., Harris, E. H., Karpowicz, S. J., Witman, G. B., et al. (2007). The *Chlamydomonas* genome reveals the evolution of key animal and plant functions. *Science, 318*, 245–250.

Michaelis, G., Vahrenholz, C., & Pratje, E. (1990). Mitochondrial DNA of *Chlamydomonas reinhardtii*: the gene for apocytochrome b and the complete functional map of the 15.8 kb DNA. *Molecular Genetics and Genomics, 223*.

Miller, S. M., & Kirk, D. L. (1999). glsA, a *Volvox* gene required for asymmetric division and germ cell specification, encodes a chaperone-like protein. *Development, 126*, 649–658.

Miller, S. M., Schmitt, R., & Kirk, D. L. (1993). Jordan, an active *Volvox* transposable element similar to higher plant transposons. *Plant Cell, 5*, 1125–1138.

Minko, I., Holloway, S. P., Nikaido, S., Carter, M., Odom, O. W., Johnson, C. H., et al. (1999). Renilla luciferase as a vital reporter for chloroplast gene expression in *Chlamydomonas. Molecular and General Genetics, 262*, 421–425.

Misumi, O., Yoshida, Y., Nishida, K., Fujiwara, T., Sakajiri, T., Hirooka, S., et al. (2007). Genome analysis and its significance in four unicellular algae, *Cyanidioshyzon merolae, Ostreococcus tauri, Chlamydomonas reinhardtii*, and *Thalassiosira pseudonana. Journal of Plant Research, 121*, 3–17.

Molnar, A., Bassett, A., Thuenemann, E., Schwach, F., Karkare, S., Ossowski, S., et al. (2009). Highly specific gene silencing by artificial microRNAs in the unicellular alga *Chlamydomonas reinhardtii. Plant Journal.*

Molnár, A., Schwach, F., Studholme, D. J., Thuenemann, E. C., & Baulcombe, D. C. (2007). miRNAs control gene expression in the single-cell alga *Chlamydomonas reinhardtii. Nature, 447*, 1126–1129.

Morita, E., Abe, T., Tsuzuki, M., Fujiwara, S., Sato, N., Hirata, A., et al. (1999). Role of pyrenoids in the CO2-concentrating mechanism: comparative morphology, physiology and molecular phylogenetic analysis of closely related strains of *Chlamydomonas* and Chloromonas (Volvocales). *Planta, 208*, 365–372.

Nakada, T., Nozaki, H., & Tomita, M. (2010). Another origin of coloniality in volvocaleans: the phylogenetic position of Pyrobotrys arnoldi (Spondylomoraceae, Volvocales). *Journal of Eukaryotic Microbiology, 57*, 379–382.

Nedelcu, A. M. (2009). Environmentally induced responses co-opted for reproductive altruism. *Biology Letters, 5*, 805–808.

Nedelcu, A. M., & Michod, R. E. (2006). The evolutionary origin of an altruistic gene. *Molecular Biology and Evolution, 23*, 1460–1464.

Nelson, J. A., Savereide, P. B., & Lefebvre, P. A. (1994). The CRY1 gene in *Chlamydomonas reinhardtii*: structure and use as a dominant selectable marker for nuclear transformation. *Molecular and Cellular Biology, 14*, 4011–4019.

Neupert, J., Shao, N., Lu, Y., & Bock, R. (2012). Genetic transformation of the model green alga *Chlamydomonas reinhardtii*. *Methods in Molecular Biology, 847*, 35–47.

Nishii, I., & Miller, S. M. (2010). *Volvox*: simple steps to developmental complexity? *Current Opinion in Plant Biology, 13*, 646–653.

Nishii, I., Ogihara, S., & Kirk, D. L. (2003). A kinesin, invA, plays an essential role in *volvox* morphogenesis. *Cell, 113*, 743–753.

Nishimura, Y., Higashiyama, T., Suzuki, L., Misumi, O., & Kuroiwa, T. (1998). The biparental transmission of the mitochondrial genome in *Chlamydomonas reinhardtii* visualized in living cells. *European Journal of Cell Biology, 77*, 124–133.

Novoselov, S. V., Rao, M., Onoshko, N. V., Zhi, H., Kryukov, G. V., Xiang, Y., et al. (2002). Selenoproteins and selenocysteine insertion system in the model plant cell system, *Chlamydomonas reinhardtii*. *EMBO Journal, 21*, 3681–3693.

Nozaki, H. (2003). Origin and evolution of the genera *Pleodorina* and *Volvox* (Volvocales). *Biologia* 425–431.

Nozaki, H., & Itoh, M. (1994). Phylogenetic relationships within the colonial Volvocales (Chlorophyta) inferred from cladistic analysis based on morphological data. *Journal of Phycology, 30*, 353–365.

Nozaki, H., Itoh, M., Sano, R., Uchida, H., Watanabe, M., & Kuroiwa, T. (1995). Phylogenetic relationships within the colonial Volvocales (Chlorophyta) inferred from rbcL gene sequence data. *Journal of Phycology, 31*, 970–978.

Nozaki, H., Misawa, K., Kajita, T., Kato, M., Nohara, S., & Watanabe, M. M. (2000). Origin and evolution of the colonial volvocales (Chlorophyceae) as inferred from multiple, chloroplast gene sequences. *Molecular Phylogenetics and Evolution, 17*, 256–268.

Nozaki, H., Mori, T., Misumi, O., Matsunaga, S., & Kuroiwa, T. (2006). Males evolved from the dominant isogametic mating type. *Current Biology, 16*, R1018–R1020.

Nozaki, H., Takahara, M., Nakazawa, A., Kita, Y., Yamada, T., Takano, H., et al. (2002). Evolution of rbcL group IA introns and intron open reading frames within the colonial Volvocales (Chlorophyceae). *Molecular Phylogenetics and Evolution, 23*, 326–338.

O'Neill, B. M., Mikkelson, K. L., Gutierrez, N. M., Cunningham, J. L., Wolff, K. L., Szyjka, S. J., et al. (2012). An exogenous chloroplast genome for complex sequence manipulation in algae. *Nucleic Acids Research, 40*, 2782–2792.

Odom, O. W., Baek, K.-H., Dani, R. N., & Herrin, D. L. (2008). *Chlamydomonas* chloroplasts can use short dispersed repeats and multiple pathways to repair a double-strand break in the genome. *Plant Journal, 53*, 842–853.

Ohresser, M., Matagne, R. X. F., & Loppes, R. (1997). Expression of the arylsulphatase reporter gene under the control of the nit1 promoter in *Chlamydomonas reinhardtii*. *Current Genetics, 31*, 264–271.

Ojakian, G. K., & Katz, D. F. (1973). A simple technique for the measurement of swimming speed of *Chlamydomonas*. *Experimental Cell Research, 81*, 487–491.

Palenik, B., Grimwood, J., Aerts, A., Rouze, P., Salamov, A., Putnam, N., et al. (2007). The tiny eukaryote *Ostreococcus* provides genomic insights into the paradox of plankton speciation. *Proceedings of the National Academy of Sciences of the United States of America, 104*, 7705.

Pappas, V., & Miller, S. M. (2009). Functional analysis of the *Volvox carteri* asymmetric division protein GlsA. *Mechanisms of Development, 126*, 842–851.

Patron, N. J., & Waller, R. F. (2007). Transit peptide diversity and divergence: a global analysis of plastid targeting signals. *Bioessays, 29*, 1048–1058.

Pazour, G. J., Agrin, N., Leszyk, J., & Witman, G. B. (2005). Proteomic analysis of a eukaryotic cilium. *Journl of Cell Biology, 170*, 103–113.

Pearson, C. G., & Winey, M. (2009). Basal body assembly in ciliates: the power of numbers. *Traffic, 10*, 461–471.

Pedersen, L. B., & Rosenbaum, J. L. (2008). Chapter two intraflagellar transport (IFT): role in ciliary assembly, resorption and signalling. *Current Topics in Developmental Biology, 85*, 23–61.

Peers, G., & Niyogi, K. K. (2008). Pond scum genomics: the genomes of *Chlamydomonas* and *Ostreococcus*. *Plant Cell, 20*, 502–507.

Perez-Rodriguez, P., Riano-Pachon, D. M., Correa, L. G. G., Rensing, S. A., Kersten, B., & Mueller-Roeber, B. (2009). PlnTFDB: updated content and new features of the plant transcription factor database. *Nucleic Acids Research, 38*, D822–D827.

Petracek, M. E., Lefebvre, P. A., Silflow, C. D., & Berman, S. D. (1990). *Chlamydomonas* telomere sequences are A+T-rich but contain three consecutive G-C base pairs. *Proceedings of the National Academy of Sciences of the United States of America, 87*, 8222.

Pootakham, W., González-Ballester, D., & Grossman, A. R. (2010). Identification and regulation of plasma membrane sulfate transporters in *Chlamydomonas*. *Plant Physiology, 153*, 1653–1668.

Popescu, C. E., Borza, T., Bielawski, J. P., & Lee, R. W. (2006). Evolutionary rates and expression level in chlamydomonos. *Genetics, 172*, 1567–1576.

Popescu, C. E., & Lee, R. W. (2007). Mitochondrial genome sequence evolution in *Chlamydomonas*. *Genetics, 175*, 819–826.

Preble, A. M., Giddings, T. H., Jr., & Dutcher, S. K. (1999). Basal bodies and centrioles: their function and structure. *Current Topics in Developmental Biology, 49*, 207–233.

Preuss, D., & Mets, L. (2002). Summaries of national science foundation-sponsored Arabidopsis 2010 projects and national science foundation-sponsored plant genome projects that are generating Arabidopsis resources for the community. *Plant Physiology, 129*, 421–422.

Prochnik, S. E., Umen, J., Nedelcu, A. M., Hallmann, A., Miller, S. M., Nishii, I., et al. (2010). Genomic analysis of organismal complexity in the multicellular green alga *Volvox carteri*. *Science, 329*, 223–226.

Pröschold, T., Marin, B., Schlösser, U. G., & Melkonian, M. (2001). Molecular phylogeny and taxonomic revision of *Chlamydomonas* (Chlorophyta). I. Emendation of *Chlamydomonas* Ehrenberg and Chloromonas Gobi, and description of Oogamochlamys gen. nov. and Lobochlamys gen. nov. *Protist, 152*, 265–300.

Purton, S. (2007). Tools and techniques for chloroplast transformation of *Chlamydomonas*. *Advances in Experimental Medicine and Biology, 616*, 34–45.

Putnam, N. H., Srivastava, M., Hellsten, U., Dirks, B., Chapman, J., Salamov, A., et al. (2007). Sea anemone genome reveals ancestral eumetazoan gene repertoire and genomic organization. *Science, 317*, 86–94.

Quinn, J. M., Kropat, J., & Merchant, S. (2003). Copper response element and Crr1-dependent Ni2+-responsive promoter for induced, reversible gene expression in *Chlamydomonas reinhardtii*. *Eukaryotic Cell, 2*, 995–1002.

Ramesh, V., Bingham, S., & Webber, A. (2011). A simple method for chloroplast transformation in *Chlamydomonas reinhardtii*. *Methods in Molecular Biology, 684*, 313–320.

Randolph-Anderson, B. L., Boynton, J. E., Gillham, N. W., Harris, E. H., Johnson, A. M., Dorthu, M.-P., et al. (1993). Further characterization of the respiratory deficient dum-1 mutation of *Chlamydomonas reinhardtii* and its use as a recipient for mitochondrial transformation. *Molecular Genetics and Genomics, 236-236*, 235–244.

Rao, M., Carlson, B. A., Novoselov, S. V., Weeks, D. P., Gladyshev, V. N., & Hatfield, D. L. (2003). *Chlamydomonas reinhardtii* selenocysteine tRNA[Ser]Sec. *RNA, 9*, 923–930.

Rawi, Al, S., Louvet-Vallée, S., Djeddi, A., Sachse, M., Culetto, E., Hajjar, C., et al. (2011). Postfertilization autophagy of sperm organelles prevents paternal mitochondrial DNA transmission. *Science, 334*, 1144–1147.

Resnick, D., Pearson, A., & Krieger, M. (1994). The SRCR superfamily: a family reminiscent of the Ig superfamily. *Trends in Biochemical Sciences, 19*, 5–8.

Robbens, S., Khadaroo, B., Camasses, A., Derelle, E., Ferraz, C., Inzé, D., et al. (2005). Genome-wide analysis of core cell cycle genes in the unicellular green alga *Ostreococcus tauri*. *Molecular Biology and Evolution, 22*, 589–597.

Rohr, J., Sarkar, N., Balenger, S., Jeong, B.-R., & Cerutti, H. (2004). Tandem inverted repeat system for selection of effective transgenic RNAi strains in *Chlamydomonas*. *Plant Journal, 40*, 611–621.

Rokas, A. (2008a). The molecular origins of multicellular transitions. *Current Opinion in Genetics and Development, 18*, 472–478.

Rokas, A. (2008b). The origins of multicellularity and the early history of the genetic toolkit for animal development. *Annual Review of Genetics, 42*, 235–251.

Rolland, N., Atteia, A., Decottignies, P., Garin, J., Hippler, M., Kreimer, G., et al. (2009). *Chlamydomonas* proteomics. *Current Opinion in Microbiology, 12*, 285–291.

Ruiz-Trillo, I., Burger, G., Holland, P. W. H., King, N., Lang, B. F., Roger, A. J., et al. (2007). The origins of multicellularity: a multi-taxon genome initiative. *Trends in Genetics, 23*, 113–118.

Rupprecht, J. (2009). From systems biology to fuel—*Chlamydomonas reinhardtii* as a model for a systems biology approach to improve biohydrogen production. *Journal of Biotechnology, 142*, 10–20.

Rymarquis, L. A., Handley, J. M., Thomas, M., & Stern, D. B. (2005). Beyond complementation. Map-based cloning in *Chlamydomonas reinhardtii*. *Plant Physiology, 137*, 557–566.

Sato, M., & Sato, K. (2011). Degradation of paternal mitochondria by fertilization-triggered autophagy in C. elegans embryos. *Science, 334*, 1141–1144.

Schiedlmeier, B., Schmitt, R., Müller, W., Kirk, M. M., Gruber, H., Mages, W., et al. (1994). Nuclear transformation of *Volvox carteri*. *Proceedings of the National Academy of Sciences of the United States of America, 91*, 5080–5084.

Schmidt, B. J., Lin-Schmidt, X., Chamberlin, A., Salehi-Ashtiani, K., & Papin, J. A. (2010). Metabolic systems analysis to advance algal biotechnology. *Biotechnology Journal, 5*, 660–670.

Schmollinger, S., Strenkert, D., & Schroda, M. (2010). An inducible artificial microRNA system for *Chlamydomonas reinhardtii* confirms a key role for heat shock factor 1 in regulating thermotolerance. *Current Genetics, 56*, 383–389.

Schroda, M. (2006). RNA silencing in *Chlamydomonas*: mechanisms and tools. *Current Genetics, 49*, 69–84.

Schroda, M., Beck, C. F., & Vallon, O. (2002). Sequence elements within an HSP70 promoter counteract transcriptional transgene silencing in *Chlamydomonas*. *Plant Journal, 31*, 445–455.

Schroda, M., Blöcker, D., & Beck, C. F. (2000). The HSP70A promoter as a tool for the improved expression of transgenes in *Chlamydomonas*. *Plant Journal, 21*, 121–131.

Setohigashi, Y., Hamaji, T., Hayama, M., Matsuzaki, R., & Nozaki, H. (2011). Uniparental Inheritance of Chloroplast DNA Is Strict in the Isogamous Volvocalean *Gonium*. *PLoS ONE, 6*, e19545.

Shao, N., & Bock, R. (2008). A codon-optimized luciferase from Gaussia princeps facilitates the in vivo monitoring of gene expression in the model alga *Chlamydomonas reinhardtii*. *Current Genetics, 53*, 381–388.

Sharma, N., Berbari, N. F., & Yoder, B. K. (2008). Ciliary dysfunction in developmental abnormalities and diseases. *Current Topics in Developmental Biology, 85*, 371–427.

Shimogawara, K., Fujiwara, S., Grossman, A., & Usuda, H. (1998). High-efficiency transformation of *Chlamydomonas reinhardtii* by electroporation. *Genetics, 148*, 1821–1828.

Shitara, H., Kaneda, H., Sato, A., Inoue, K., Ogura, A., Yonekawa, H., et al. (2000). Selective and continuous elimination of mitochondria microinjected into mouse eggs from spermatids, but not from liver cells, occurs throughout embryogenesis. *Genetics, 156*, 1277–1284.

Short, M. B., Solari, C. A., Ganguly, S., Powers, T. R., Kessler, J. O., & Goldstein, R. E. (2006). Flows driven by flagella of multicellular organisms enhance long-range molecular transport. *Proceedings of the National Academy of Sciences of the United States of America, 103*, 8315–8319.

Silflow, C. D., & Lefebvre, P. A. (2001). Assembly and Motility of Eukaryotic Cilia and Flagella. Lessons from *Chlamydomonas reinhardtii*. *Plant Physiology, 127*, 1500–1507.

Sizova, I., Fuhrmann, M., & Hegemann, P. (2001). A Streptomyces rimosus aphVIII gene coding for a new type phosphotransferase provides stable antibiotic resistance to *Chlamydomonas reinhardtii*. *Gene, 277*, 221–229.

Smith, D. R., & Lee, R. W. (2009). The mitochondrial and plastid genomes of *Volvox carteri*: bloated molecules rich in repetitive DNA. *BMC Genomics, 10*, 132.

Smith, D. R., & Lee, R. W. (2010). Low nucleotide diversity for the expanded organelle and nuclear genomes of *Volvox carteri* supports the mutational-hazard hypothesis. *Molecular Biology and Evolution, 27*, 2244–2256.

Solari, C. A., Kessler, J. O., & Michod, R. E. (2006). A hydrodynamics approach to the evolution of multicellularity: flagellar motility and germ-soma differentiation in volvocalean green algae. *American Naturalist, 167*, 537–554.

Solari, C. A., Michod, R. E., & Goldstein, R. E. (2008). *Volvox* barberi, the fastest swimmer of the Volvocales (Chlorophyceae). *Journal of Phycology, 44*, 1395–1398.

Soupene, E., Inwood, W., & Kustu, S. (2004). Lack of the rhesus protein Rh1 impairs growth of the green alga *Chlamydomonas reinhardtii* at high CO2. *Proceedings of the National Academy of Sciences of the United States of America, 101*, 7787–7792.

Spalding, M. H. (2008). Microalgal carbon-dioxide-concentrating mechanisms: *Chlamydomonas* inorganic carbon transporters. *Journal of Experimental Botany, 59*, 1463–1473.

Srivastava, M., Simakov, O., Chapman, J., Fahey, B., Gauthier, M. E. A., Mitros, T., et al. (2010). The Amphimedon queenslandica genome and the evolution of animal complexity. *Nature, 466*, 720–726.

Stanke, M., Diekhans, M., Baertsch, R., & Haussler, D. (2008). Using native and syntenically mapped cDNA alignments to improve de novo gene finding. *Bioinformatics, 24*, 637–644.

Stanke, M., & Morgenstern, B. (2005). AUGUSTUS: a web server for gene prediction in eukaryotes that allows user-defined constraints. *Nucleic Acids Research, 33*, W465–W467.

Stanke, M., Schöffmann, O., Morgenstern, B., & Waack, S. (2006). Gene prediction in eukaryotes with a generalized hidden Markov model that uses hints from external sources. *BMC Bioinformatics, 7*, 62.

Stein, J. (1966). Growth and mating of *Gonium pectorale* (Volvocales) in defined media. *Journal of Phycology, 2*, 23–28.

Stern, D. B., Witman, G., & Harris, E. H. (2009). The *Chlamydomonas* sourcebook (Second Edition. ed.). Academic Press. Oxford; Burlington, MA; San Diego, CA.

Stevens, D. R., Purton, S., & Rochaix, J. D. (1996). The bacterial phleomycin resistance geneble as a dominant selectable marker in*Chlamydomonas*. *Molecular Genetics and Genomics, 251*, 23–30.

Sumper, M., & Hallmann, A. (1997). Biochemistry of the extracellular matrix of *Volvox*. *International Review of Cytology, 180*, 51–85.

Tam, L., Stamer, K. A., & Kirk, D. L. (1991). Early and late gene expression programs in developing somatic cells of *Volvox carteri**1. *Developmental Biology, 145*, 67–76.

Toepel, J., Albaum, S. P., Arvidsson, S., Goesmann, A., la Russa, M., Rogge, K., et al. (2011). Construction and evaluation of a whole genome microarray of *Chlamydomonas reinhardtii*. *BMC Genomics, 12*, 579.

Trapnell, C., Williams, B. A., Pertea, G., Mortazavi, A., Kwan, G., van Baren, M. J., et al. (2010). Transcript assembly and quantification by RNA-Seq reveals unannotated transcripts and isoform switching during cell differentiation. *Nature Biotechnology, 28*, 511–515.

Tschochner, H., Lottspeich, F., & Sumper, M. (1987). The sexual inducer of *Volvox carteri*: purification, chemical characterization and identification of its gene. *EMBO Journal, 6*, 2203–2207.

Turmel, M., Lemieux, C., & Lee, R. W. (1980). Net synthesis of chloroplast DNA throughout the synchronized vegetative cell-cycle of *Chlamydomonas*. *Current Genetics, 2*, 229–232.

Ueki, N., Matsunaga, S., Inouye, I., & Hallmann, A. (2010). How 5000 independent rowers coordinate their strokes in order to row into the sunlight: phototaxis in the multicellular green alga *Volvox*. *BMC Biology, 8*, 103.

Ueki, N., & Nishii, I. (2008). Idaten is a new cold-inducible transposon of *Volvox carteri* that can be used for tagging developmentally important genes. *Genetics, 180*, 1343–1353.

Ueki, N., & Nishii, I. (2009). Controlled enlargement of the glycoprotein vesicle surrounding a *volvox* embryo requires the InvB nucleotide-sugar transporter and is required for normal morphogenesis. *Plant Cell, 21*, 1166–1181.

Umen, J., & Goodenough, U. (2001). Control of cell division by a retinoblastoma protein homolog in *Chlamydomonas*. *Genes and Development, 15*, 1652.

Umen, J. G. (2001). Chloroplast DNA methylation and inheritance in *Chlamydomonas*. *Genes and Development, 15*, 2585–2597.

Umen, J. G. (2011). Evolution of sex and mating loci: an expanded view from Volvocine algae. *Current Opinion in Microbiology, 14*, 634–641.

Vahrenholz, C., Riemen, G., Pratje, E., Dujon, B., & Michaelis, G. (1993). Mitochondrial DNA of *Chlamydomonas reinhardtii*: the structure of the ends of the linear 15.8-kb genome suggests mechanisms for DNA replication. *Current Genetics, 24*, 241–247.

Van Bel, M., Proost, S., Wischnitzki, E., Movahedi, S., Scheerlinck, C., van de Peer, Y., & Vandepoele, K. (2012). Dissecting Plant Genomes with the PLAZA Comparative Genomics Platform. *Plant Physiol, 158*, 590–600.

van Leeuwenhoek, A. (1700). Part of a letter from Mr Antony van Leeuwenhoek, concerning the worms in sheeps livers, gnats, and animalcula in the excrements of frogs. *Philosophical Transactions, 22*, 509–518.

Viamontes, G. I., Fochtmann, L. J., & Kirk, D. L. (1979). Morphogenesis in *Volvox*: analysis of critical variables. *Cell, 17*, 537–550.

Viamontes, G. I., & Kirk, D. L. (1977). Cell shape changes and the mechanism of inversion in *Volvox*. *Journal of Cell Biology, 75*, 719–730.

Villand, P., Eriksson, M., & Samuelsson, G. (1997). Carbon dioxide and light regulation of promoters controlling the expression of mitochondrial carbonic anhydrase in *Chlamydomonas reinhardtii*. *Biochemical Journal, 327*(Pt 1), 51–57.

Wang, Y., Duanmu, D., & Spalding, M. H. (2011). Carbon dioxide concentrating mechanism in *Chlamydomonas reinhardtii*: inorganic carbon transport and CO2 recapture. *Photosynthesis Research, 109*, 115–122.

Wang, Y., & Spalding, M. H. (2006). An inorganic carbon transport system responsible for acclimation specific to air levels of CO2 in *Chlamydomonas reinhardtii*. *Proceedings of the National Academy of Sciences of the United States of America, 103*, 10110–10115.

Wheeler, G. L., Miranda-Saavedra, D., & Barton, G. J. (2008). Genome analysis of the unicellular green alga *Chlamydomonas reinhardtii* indicates an ancient evolutionary origin for key pattern recognition and cell-signaling protein families. *Genetics, 179*, 193–197.

Woessner, J., & Goodenough, U. (1994). Volvocine cell walls and their constituent glycoproteins: an evolutionary perspective. *Protoplasma, 181*, 245–258.

Worden, A. Z., Lee, J.-H., Mock, T., Rouze, P., Simmons, M. P., Aerts, A. L., et al. (2009). Green evolution and dynamic adaptations revealed by genomes of the marine picoeukaryotes micromonas. *Science, 324*, 268–272.

Xie, Q., Sanz-Burgos, A. P., Hannon, G. J., & Gutiérrez, C. (1996). Plant cells contain a novel member of the retinoblastoma family of growth regulatory proteins. *EMBO Journal, 15*, 4900–4908.

Yamada, T. K., Miyaji, K., & Nozaki, H. (2008). A taxonomic study of *Eudorina* unicocca (Volvocaceae, Chlorophyceae) and related species, based on morphology and molecular phylogeny. *European Journal of Phycology, 43*, 317–326.

Yamasaki, T., Kurokawa, S., Watanabe, K. I., Ikuta, K., & Ohama, T. (2005). Shared molecular characteristics of successfully transformed mitochondrial genomes in *Chlamydomonas reinhardtii. Plant Molecular Biology, 58*, 515–527.

Yoshihara, C., Inoue, K., Schichnes, D., Ruzin, S., Inwood, W., & Kustu, S. (2008). An Rh1-GFP fusion protein is in the cytoplasmic membrane of a white mutant strain of *Chlamydomonas reinhardtii. Molecular Plant, 1*, 1007–1020.

Zhao, T., Li, G., Mi, S., Li, S., Hannon, G. J., Wang, X.-J., et al. (2007). A complex system of small RNAs in the unicellular green alga *Chlamydomonas reinhardtii. Genes and Development, 21*, 1190–1203.

Zhao, T., Wang, W., Bai, X., & Qi, Y. (2009). Gene silencing by artificial microRNAs in *Chlamydomonas. Plant Journal, 58*, 157–164.

Zhang, H., Jin, J., Tang, L., Zhao, Y., Gu, X., Gao, G., et al. (2010). PlantTFDB 2.0: update and improvement of the comprehensive plant transcription factor database. *Nucleic Acids Research, 39*, D1114–D1117.

Zouros, E. (2000). The exceptional mitochondrial DNA system of the mussel family Mytilidae. *Genes and Genetic Systems, 75*, 313–318.

CHAPTER SEVEN

Genomics and Genetics of Diatoms

Thomas Mock* and Linda K. Medlin†
*School of Environmental Sciences, University of East Anglia, Norwich Research Park, Norwich NR47TJ, UK
†University of Pierre and Marie Curie, CNRS, Observatoire Océanologique, LOMIC, UMR 7621, BP 44, Banyuls sur Mer 66651, France
Corresponding author: E-mail: medlin@obs-banyuls.fr and E-mail: t.mock@uea.ac.uk

Contents

Abstract

Diatoms are unicellular eukaryotes with nano-patterned silica cell walls and they contribute about 20% of global primary production. Their beautiful shells and significance for life on our planet already caused scientific interest many centuries ago. However, the development of genetics and genomics-enabled technology about two decades ago and their application to diatom research has caused a step change in our understanding of diatom evolution, biology and ecology. In contrast to plants and green algae, which were derived from primary endosymbiosis, diatom evolution seems to be based on secondary endosymbiosis involving green and red algae as

Advances in Botanical Research, Volume 64
ISSN 0065-2296,
http://dx.doi.org/10.1016/B978-0-12-391499-6.00007-4

endosymbionts and a heterotrophic exosymbiont that is believed to have provided the ability to use silicate for an external cell wall made of silica. This review will discuss how results obtained by the application of genetics and genomics have impacted our understanding of diatoms. We will provide evidence for how their complex evolution has shaped key features of their biology and their global distribution by adaptive diversification to very different habitats.

1. INTRODUCTION

The diatoms are unique among the major eukaryotic algae because of their silicified cell walls (frustules), which consist of two overlapping thecae, each in turn consisting of a valve and a number of hoop-like or segmental girdle bands (Round, Crawford, & Mann, 1990). This feature makes them one of the most easily recognizable algal groups both in living and fossil assemblages (Fig. 7.1). One of the earliest known, well preserved and diversified deposits of fossil diatoms is dated from the early Albian (Lower Cretaceous) and was recovered from what is now the Weddell Sea, Antarctica (Gersonde & Harwood, 1990). The diatom frustules in that deposit bear little resemblance to modern diatoms in their morphology, although it is clear that, at that time, diatoms were very highly developed and diversified. From the time of their origin until today, the diatoms have been a rapidly evolving group, adapting to many different kinds of habitats and it is clear that the morphology and the ecology of the group have changed dramatically over time. In this chapter, we review the evolution of the diatoms drawing in evidence from genomic studies to provide insights into which genes have shaped their evolution into undoubtedly one of the most successful and adaptable algal groups today.

2. DIVERSITY OF DIATOMS: INTEGRATING MORPHOLOGICAL AND MOLECULAR DATA

2.1. Morphological Insights into Diatom Diversity and Evolution

Diatoms belong to the Kingdom Heterokonta, which are those organisms with two heterodynamic flagella, one covered with tripartite hairs and the other smooth (Van den Hoek, Mann, & Jahns, 1995). Pigmented heterokonts are chlorophyll $a + c$ containing algae and are the last divergence in the heterokont phylogeny (Cavalier-Smith & Chao, 2006). In the emergence of the pigmented heterokonts, the diatoms and their sister group the

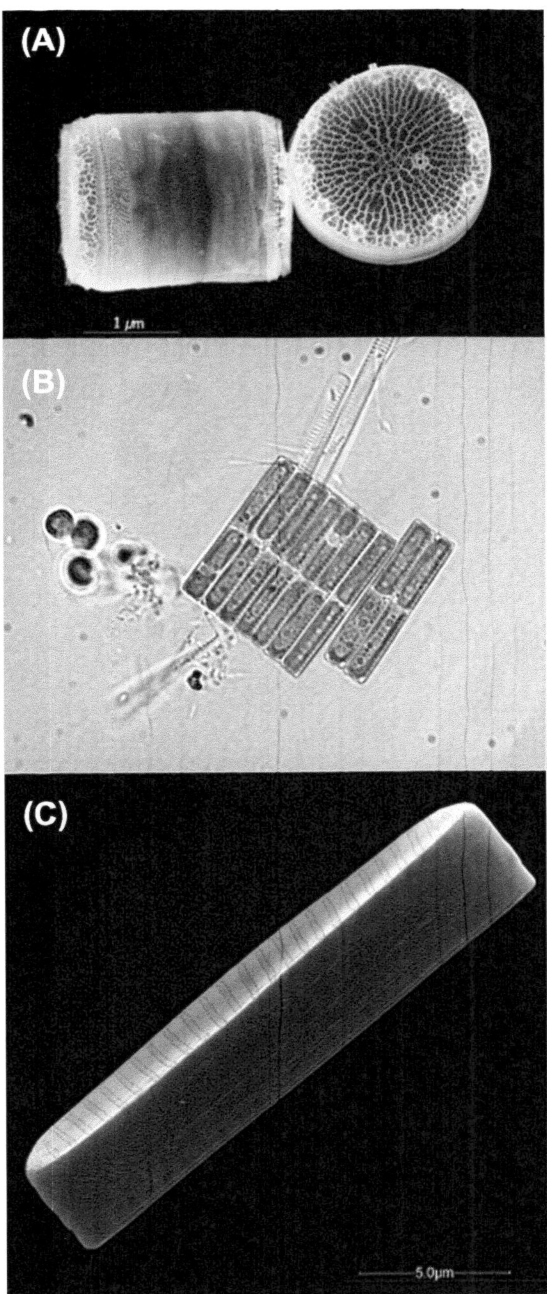

Figure 7.1 (A) Scanning electron micrograph (SEM) of the silica cell wall from the marine diatom *Thalassiosira pseudonana*. Courtesy of Nils Kröger, Technische Universität Dresden, Germany. (B) Light micrograph of the polar marine diatom *Fragilariopsis cylindrus* forming ribbons. (C) SEM of a single silica shell from *F. cylindrus. Courtesy of Gerhard Dieckmann (B and C), Alfred-Wegener Institute Bremerhaven, Germany.* For colour version of this figure, the reader is referred to the online version of this book.

bolidomonads are usually recovered as a basal divergence from all remaining heterokont algae (Cavalier-Smith & Chao, 2006). In the diatoms, the flagellar apparatus is reduced or absent and only the spermatozoids of the oogamous 'centric' diatoms are flagellated. The sperm are uniflagellate (Von Stosch, 1950), lacking all trace of the smooth posterior flagellum. Despite this loss of the smooth flagellum, the diatoms are clearly heterokonts.

Most diatomists have long assumed that there are two types of diatoms: the centrics and the pennates, which can be distinguished by their pattern centres or symmetry, mode of sexual reproduction, and plastid number and structure (Round *et al.*, 1990). These two groups of diatoms are well known to most aquatic and cell biologists, and each term conveys a distinct image of a particular type of diatom cell. However, these terms are only descriptive and have no taxonomic value because molecular analysis has shown that there are two groups of centric diatoms and within the pennate diatoms, there are two groups of araphid pennate diatoms. Nevertheless despite the non-monophyly of the centrics, the oogamous centric diatoms, with radially symmetrical valve ornamentation and numerous discoid plastids, are distinct from the isogamous bilaterally symmetrical pennate diatoms with generally fewer, plate-like plastids.

A historical review of the classification of the diatoms has been done by Williams (2007). Since 1896, centric and pennate diatoms have been separated either as two classes or orders. However, in 1990, Round *et al.* recognized three classes: Coscinodiscophyceae (centric diatoms), Fragilariophyceae (araphid pennate diatoms) and Bacillariophyceae (raphid pennate diatoms), which gave equal status to the centrics as to the raphid pennate diatoms with a double-slit opening (raphe) in the cell wall for movement and to the araphid pennate diatoms without this slit. Molecular data have seriously challenged this view of the diatoms. Recently, Medlin and colleagues (literature summarized by Medlin & Kaczmarska, 2004) have divided the diatoms into two groups on the basis of molecular sequence data. 'Clade 1' in Medlin's early molecular work (Medlin, Williams, & Sims, 1993) contains those centric diatoms with radial symmetry of valve shape and structure. 'Clade 2' consists of two groups, the first of which contains the bi- or multipolar centrics and the radial Thalassiosirales ('Clade 2a') and the second, the pennates ('Clade 2b') (Fig. 7.3). Medlin and Kaczmarska (2004) formally recognized these two clades based on the orphological and cytological support for these clades reviewed in Medlin, Kooistra, and Schmid (2000). Clades 1 and 2 are now known at the subdivision level as the Coscinodiscophytina and Bacillariophytina, respectively, and Clades 1, 2a, and 2b are now recognized at the class level, as the Coscinodiscophyceae, Mediophyceae and Bacillariophyceae (Medlin & Kaczmarska 2004). These classes are not recovered as monophyletic if alignments are not performed using the secondary structure of the ribosomal RNA (rRNA) genes as a guide or if single outgroups are used (Medlin, 2010; Medlin, Sato, Mann, & Kooistra, 2008; Medlin *et al.*, 1993; Sato, 2008).

The best morphological support for the two clades comes from the position of the labiate process, the position of the cribrum or covering in the chambered holes or areolae in the silica cell wall, the presence of a structure inside the annulus or starting point of valve silicification and the type of interlocking mechanisms used to link the cells into chains. Labiate processes in the Coscinodiscophytina, when present, are located in a ring around the valve perimeter or scattered over the valve face but never in the annulus. Several genera possess spines at the edge of the valve face, which interlock or interdigitate with similar structures on the valve face of the adjacent sister cell to link cells firmly in chains. External cribra are found in these diatoms with loculate areolae, with one possible exception of *Endictya*, which is certainly misclassified because it has eccentric areolation. There is a central structure in the mediophycean valves that may be a labiate process, a strutted process or a sternum. Internal cribra dominate in the mediophycean diatoms, especially if the valves are loculate as in the Thalassiosirales and *Triceratium*. One possible exception is *Eupodiscus* whose loculate areolae have external cribra. In the simple poroid mediophycean valves, the cribrum appears to be at the internal valve surface, e.g. *Lampriscus*. Chain formation in the bipolar mediophycean diatoms tends to be less robust than in the Coscinodiscophytina. To a lesser extent, there is some support for the two groups in the internal arrangement of some organelles. The Bacillariophytina have a perinuclear arrangement of the Golgi apparatus, whereas in the Coscinodiscophytina, the Golgi stacks arranged in Golgi–endoplasmic reticulum–mitochondrion (G–ER–M) units (available data summarized in Medlin *et al.*, 2000, and Medlin & Kaczmarska, 2004) have been documented in a few genera but it is not known how widespread this feature is among the radial centrics. There are some exceptions in each group (see Schmid, 1988).

The best independent support for three classes not associated with the structure of the cell wall comes from the auxospore structure (Fig. 7.2), the specialized zygote of the diatoms that swells to restore the cells to their original cell size, which has diminished with progressive vegetative division

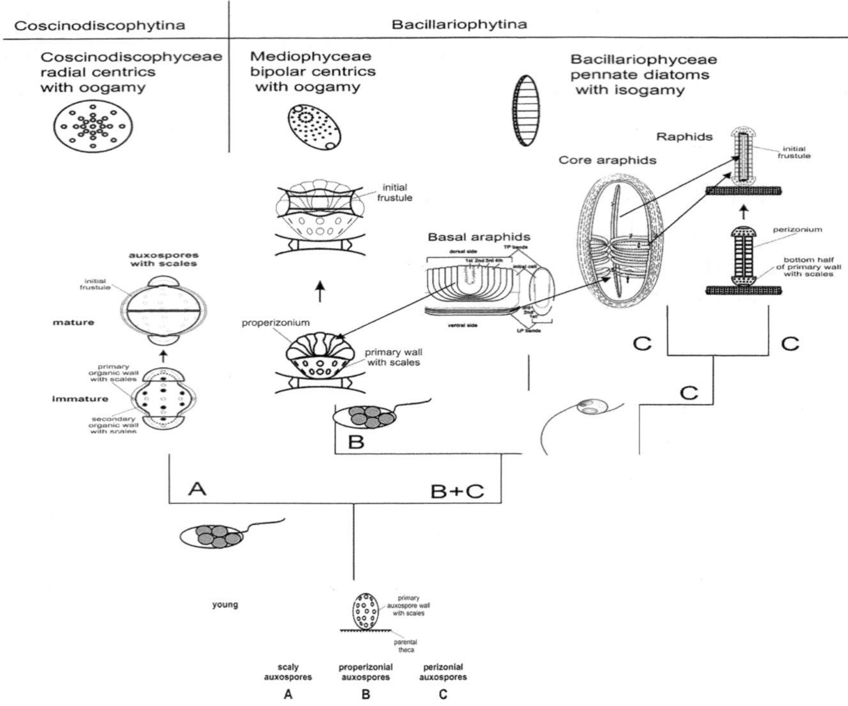

Figure 7.2 Summary of the phylogeny of the diatoms using the auxospore and the sperm cells as the character defining the lower branches in the clade. *Redrawn from Medlin and Kaczmarska (2004) and Medlin and Sato (2009).* For colour version of this figure, the reader is referred to the online version of this book.

(Kaczmarska, Ehrman, & Bates, 2001; Medlin & Kaczmarska, 2004). The class Coscinodiscophyceae has only scales and isodiametric auxospores that can swell in all directions to make a final radial cell. The class Mediophyceae also has scales but they have in addition closed hoops or bands (a properizonium) that restricts the swelling of the auxospores to bipolar or multipolar directions. This auxospore is termed an anisodiametric auxospore. Anisodiametric auxospores are found in the class Bacillariophyceae but these form a complex tubular perizonium, consisting of transverse hoops and longitudinal bands. Among the Mediophyceae, the Thalassiosirales have retained the scaly isodiametric auxospores of the Coscinodiscophyceae, but they are in the Mediophyceae because they possess a process in the valve centre or annulus, a perinuclear arrangement of the Golgi bodies and an internal cribrum in the loculate areolae. Among the pennates, the araphids have two different types of auxospores, which correlates with their two clades in the molecular tree, and the raphids have only one type of auxospore. The two araphid diatom clades have been termed basal and core araphids (Medlin & Sato, 2009; Sato, 2008). The basal araphids have both a properizonium and a perizonium, whereas the core araphids have only a perizonium.

2.1. Molecular Data and Classification of Diatoms

Two different means of calibrating molecular trees: (1) by fossil dates for the entire clade (Kooistra & Medlin, 1996) and (2) by biomarker compounds for the clade containing *Rhizosolenia* (Sinninghe-Damsté *et al.*, 2004) have been used to calculate the rate of evolution in coscinodiscophycean diatoms. Both suggest that these diatoms are evolving very quickly (1% per 21.5 and 14 Ma for the rRNA gene, respectively) and this could explain why the morphology of the diatoms changes so rapidly across the Cretaceous, between Lower and Upper Cretaceous floras before the pennate diatoms diversify. L. K. Medlin and S. Sato (unpublished data) and Sato (2008) using four genes and the program, Multidivtime, with designated maximum and minimum divergence times of 250 and 190 Ma, respectively, found much older divergence times for all the classes, especially the pennates (Fig. 7.3). Their clock using a maximum likelihood tree as input suggests that the radial

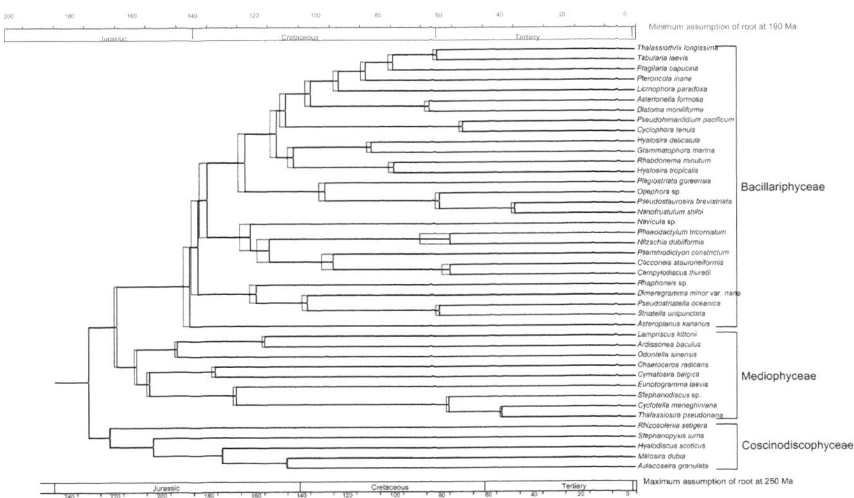

Figure 7.3 Molecular timescale for diatoms using a maximum likelihood analysis of four genes with a maximum inference of the age of the root at 190 Ma (grey) and 250 Ma (black) tree-based four taxa outgroup rooting data set. Prior of 250 and 190 Ma for the root of diatoms and respective time scales are indicated above and below the tree topology.

centrics, Class Coscinodiscophyceae, emerged from 240 to 180 Ma and the bipolar centrics, Class Mediophyceae, diverged from the Class Bacillariophyceae at 183 to 238 Ma (minimum to maximum, respectively). These results also indicate that the early divergence of the pennates into three major clades, basal araphid, core araphid and raphid diatoms, took place over a very short period. By the end of the Cretaceous in all analyses, all major clades (approximate orders or families) of araphid diatoms appeared. Thus, the molecular diversification of the diatoms appears to be much earlier than the first appearance of these taxa in the fossil record at 180 Ma. However, modern diversifications of the genera in these lineages usually coincide with the first appearances of the extant genera.

2.2. How Does the Rate of Evolution in Diatom Genomes Compare to Other Groups?

The genome sequence of *Phaedoactylum tricornutum* and a comparative approach with *Thalassiosira pseudonana* revealed for the first time how molecular diversification took place on a genome-wide level between the pennate and centric lineages. The bipolar centric diatom *T. pseudonana* belongs to the group of Mediophyceae and the pennate diatom *P. tricornutum* belongs to Bacillariophyceae. Both groups have been diverging for about 90 Ma years (see above). Over this period of time, *T. pseudonana* and *P. tricornutum* diverged about 45% based on the percentage of amino acid identity of 4267 orthologous gene pairs (Bowler *et al.*, 2008). This degree of divergence is similar to other unicellular eukaryotes. For instance, *Saccharomyces cerevisiae* and the related yeast *Kluyveromyces lactis* show the same divergence over a similar period of evolutionary time. The two sequenced *Ostreococcus* strains (*Ostreococcus tauri* and *Ostreococcus lucimarinus*) from the group of marine Prasinophytes have a divergence of about 25% in amino acid identity over their orthologous protein-coding genes (Palenik *et al.*, 2007). However, both strains diverged over a shorter period of time compared to the two diatoms and two yeast species. A slightly higher divergence though was observed between the two sequenced *Micromonas* strains (Worden *et al.*, 2009). A molecular divergence of approximately 0.5% per Ma in unicellular eukaryotes is relatively fast compared to their multicellular counterparts. For instance, the divergence between *Homo sapiens* and the puffer fish *Takifugu rubripes* is with about 40%, similar to the divergence between the two sequenced diatoms from above but humans and fish have been diverging for the last 550 Ma (Bowler *et al.*,

2008). Thus, their molecular divergence is only about 0.07% over a Ma. According to this comparison, microbial eukaryotes diverge faster, which might be related to a higher mutation rate, larger effective population size (depending on the distribution of fitness effects of new mutations) and shorter generation times. The mutation rate is not an abiotic process with random variation at a uniform rate across the genome (Bromham, 2011). The per-base mutation rate of DNA is modulated by biological features of the organism. Such features include accuracy of DNA replication and the efficiency of DNA repair. Furthermore, mutation rates can vary between ecosystems with different physical constrains (e.g. temperature, UV radiation, solar irradiance) and the length of a lifetime of an organisms (e.g. age). Mutation rates in a biological context are heritable and therefore can vary between species and even strains. In asexual microbes, such as *T. pseudonana* and *P. tricornutum*, mutation rates can be shaped by selection. A mutation that increases the mutation rate in large populations of unicellular eukaryotes may generate novel traits that are selected for if the selection pressure is strong enough. In multicellular organisms with a smaller population size, advantageous mutations are rare and any mutated alleles would become unlinked from the beneficial mutations they generate through sexual reproduction (Bromham, 2011). Their longer life histories (Mann, 1988) add to a reduced diversification by a longer time span for reproduction compared to unicellular eukaryotes.

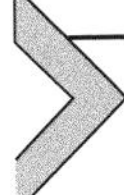

3. ENDOSYMBIOSIS AND THE EVOLUTION OF DIATOM PLASTIDS

3.1. The Complexity of Plastid Evolution in Algae

The origins of the plastids have remained one of the most intriguing and well-researched topics in biology because photosynthesis has played such a fundamental role in shaping the biosphere (see reviews in Bhattacharya & Medlin, 1995; Yoon, Hackett, Pinto, & Bhattacharya, 2002; Keeling, 2004). The first photosynthetic eukaryote is believed to have arisen when a cyanobacterium was engulfed by a heterotrophic host cell and converted into a plastid. This so-called primary endosymbiosis gave rise to the green algae, the red algae and the glaucophyte algae (we refer to chapter II of this volume for a review on the evolution of algae after primary endosymbiosis, De Clerck, Bogaret, & Leliaert, 2012). A secondary endosymbiotic event then occurred, in which a second heterotrophic host engulfed one of the primary endosymbiotic algae. Euglenoid and chlororachniophyte algae were the result of the secondary endosymbiosis of a engulfed a green alga, whereas if it was a red alga that was engulfed, then the new eukaryotic cell became a cryptophyte, haptophyte, heterokont or dinoflagellate alga.

All these endosymbiotic events that began to take place about 2000 Ma since eukaryotes arose were accompanied by extensive gene loss but also transfers from the endosymbiont (nucleus, mitochondria and plastid) to the host nucleus. The latter significantly shaped eukaryotic genomes and contributed to the chimeric nature of many algal lineages but especially the nature of diatom genomes. However, most of the genes from the endosymbionts got lost as indicated by a comparison between genomes of mitochondria and plastids of recent eukaryotes with free-living α-proteobacteria and cyanobacteria, which gave rise to present-day mitochondria and plastids, respectively. The marine α-proteobacterium, HTCC2150, for instance, was isolated from surface ocean water and has a genome size of 3.58 Mb encoding 3667 protein-coding genes (Kang *et al.*, 2010). However, many mitochondria from various algae and plants have no more than about 50 protein-coding genes (Timmis, Ayliffe, Huang, & Martin, 2004). A very similar picture arises when comparing between cyanobacterial genomes and plastids. The free-living surface ocean cyanobacterium, *Prochlorococcus*, strain MED4, has a genome size of 1.66 Mb, which encodes 1716 genes (Rocap *et al.*, 2003) but plastid DNA from algae and plants rarely encode more than 200 protein-coding genes (Timmis *et al.*, 2004).

The most important step towards the development of autotrophic eukaryotes, such as algae and plants, was the evolution of plastids. Comparative genomics between free-living cyanobacteria and plastid genomes revealed that the most significant loss in genes had taken place in the non-photosynthetic-related genes, which makes sense if the host aimed to use the novel companion as a light-driven power station in addition to mitochondria (Timmis *et al.*, 2004). However, some of the endosymbiont's nuclear genes that were related either to photosynthesis or respiration in mitochondria were incorporated into the host nuclear genome. Plastids evolved after turning a free-living bacterium into a mitochondrion. The most ancestral plastid is assumed to have had at least about 200 protein-coding genes. The closest plastid in present-day algae to a cyanobacterial ancestor is that of the glaucophytes, a small group of freshwater algae (Price *et al.*, 2012). Chloroplasts of glaucophytes are known as cyanelles that are believed to be a relict of the endosymbiotic origin of plastids from cyanobacteria. They still have a peptidoglycan layer surrounding them as do many bacteria. The plastid genome of the glaucophyte *Cyanophora paradoxa* encodes 136 genes (Price *et al.*, 2012). More significant losses, however, took place later on especially when chlorophyte, prasinophyte and land plants evolved (the so-called green lineage). For instance, the chloroplast genome of the green alga *Chlorella vulgaris* has 78 genes encoded and the chloroplast genome of the land plant *Arabidopsis thaliana* only 31 genes (Green, 2011; Marin *et al.*, 2002). Many of the NADH dehydrogenase genes were lost, which were involved in shuttling electrons from an electron donor via flavin mononucleotide (FMN) and iron–sulphur (Fe-S) centres, to quinone in the photosynthetic electron transport chain. Another example of significant gene loss in the green lineage is the group of *ycf* proteins. Many of them have unknown function but some of them are known to be involved in assembly of photosystems. Interestingly, plastid genomes of red algae have the highest number of genes with the lowest number of losses as compared to members of the green lineage and those algae evolved by secondary endosymbiosis (Oudot-Le Secq *et al.*, 2007).

More complex is plastid evolution in so-called red lineage containing the groups of cryptophyte, haptophyte, heterokont or dinoflagellate algae whose plastid arose from the secondary endosymbiosis that took place subsequently after the primary endosymbiosis. The chromalveolate hypothesis (Cavalier-Smith, 1999) proposed that the engulfed photosynthetic eukaryote was a red alga, which gave rise to all extant chromists (stramenopiles [e.g. diatoms], haptophytes [e.g. coccolithoporids], cryptophytes) and alveolates (ciliates, dinoflagellates, apicomplexans).

Many of these taxa derived from multiple endosymbiotic events retained not only parts of the plastid genome from the engulfed photosynthetic eukaryote but sometimes also from the nucleus, which resulted in a very complex pattern of gene relocation between endosymbiont and host (We refer to chapter III of this volume for a detailed review of secondary and tertiary symbiosis in algae, Archibald, 2012). Those algae that did not loose the nuclear genome from the endosymbiont most likely transferred genes over a significant period of time to the host nucleus. Evidence for that comes from the group of cryptophytes and chlorarachniophytes, which reduced the nucleus of the endosymbiont to a nucleomorph. The nucleomorph genome from both groups of organisms reveals strong compaction and a very limited number of genes ($\leq$464 protein-coding genes) (Douglas *et al.*, 2001). The intergenic space is greatly reduced and genes sometimes overlap. These features suggest that nucleomorph genomes have been streamlined over time by reduction of intergenic space, gene loss or transfer to the nucleus.

A similar streamlining process can be observed in many endosymbionts but also parasites that live in a constant environment. Those organisms basically outsource their metabolism to the host, which can lead to significant host specificity. Genes in nucleomorph genome mostly encode genes for nucleomorph functions (e.g. replication). However, some of them still encode for genes with plastid function, which is remnant of gene relocation from the plastid to nuclear genome in the endosymbiont at a time after primary endosymbiosis but before secondary endosymbiosis took place. The other algal lineages have either transferred the genes from the nucleus of the engulfed photosynthetic eukaryote to the nucleus of the host organism or they have lost them.

Many of the transferred genes either from the chloroplast or nucleus of the endosymbiont can still be found in recent genomes of diatoms, coccolithophores and dinoflagellates. For instance, several chloroplast genes (*ftsH*, *petF*, *cbbX*, *secA*) have strong homologues in diatom coccolithophore nuclear genomes (Oudot-Le-Secq *et al.*, 2007) indicating a successful transfer to the host nucleus with re-targeting to the plastid. There are also cases where one copy of a gene (e.g. *psb28*) is still encoded in the plastid genome but a second copy has been transferred to the nucleus. This appears to be a gene transfer where the chloroplast copy has not yet been deleted (evolution in process). The *acP* (acyl carrier protein) gene has three different copies in *P. tricornutum*. One in the plastid genome and two in the nuclear genome. One of the nuclear copies is targeted to the plastid and the other one to the mitochondria. Thus, even redirection to a different cellular compartment can take place and therefore increase the variation in metabolic complexity in these organisms.

3.2. Green or Red Plastids in Diatoms?

Evidence that the primary endosymbiosis resulting in the green, red and glaucophyte algae happened only once is fairly conclusive (Yoon, Hackett, Ciniglia, Pinto, & Bhattacharya, 2004; Yoon *et al.*, 2002). However, such evidence that the secondary endosymbiosis happened only once is very controversial and the number of secondary endosymbioses is continually debated in the literature with the most recent evidence heavily weighted in favour of the secondary endosymbiosis having happened multiple times (Baurain *et al.*, 2010). Moreover, it is clear that an early timing of this event (1.3 Ga) as determined by Yoon *et al.* (2004) does not match the fossil record of the diatoms, dinoflagellates and coccolithophores (Lipps, 1970) that are the modern components of the red algal secondary endosymbiotic event. Clearly, the host lineages did not take immediate advantage of their newly acquired photosynthetic function. All of the early divergences in the heterokont tree are heterotrophic and the cells appear to have lost the plastid from their secondary endosymbiosis. There is a final divergence in this lineage of all the autotrophic golden brown and brown algae, which include the diatoms. To complicate further this endosymbiosis story, there is also growing molecular evidence that some of the lineages with the red algal plastid, especially the diatoms, may have originally had a green plastid, which was later exchanged for a red algal one (Frommolt *et al.*, 2008; Moustafa Beszteri, Maier, Valentin, & Bhattacharya, 2009; Petersen, Teich, Brinkmann, & Cerff, 2007). It is presumed that both types of plastids were present in the cell at the early origin of secondary endosymbiosis. If this hypothesis is found to be true, then some ecological event is likely to have driven the elimination of the green plastid in favour of the red one. None of the molecular clocks with appropriate sister groups to root the diatom origin have placed their origin earlier than 250 Ma, which likely corresponds to the Permian–Triassic (PT) extinction. Thus, the heterokont algae to which the diatoms belong likely radiated at the PT boundary as did the dinoflagellates and the haptophytes (Medlin, 2008, 2011), although at different taxonomic levels. Each of the three groups were at different points in their radiation and it is likely that each group engulfed a red alga separate from the other groups and then rapidly diversified, albeit at different taxonomic levels (Medlin, 2011). The volcanic eruption in China at the PT boundary, which caused the most massive extinction event that the world has known (Erwin, 1990), caused the ocean to become anoxic and ocean trace metal chemistry to change. This abrupt change in the ocean chemistry likely gave the host plants with a red algal plastid the adaptive advantage that they needed to radiate (Falkowski, Katz, *et al.*, 2004; Falkowski, Schofield, Katz, van de Schootbrugge, & Knoll, 2004) because red and green algal plastids differ greatly in their need for certain trace metals. The abundance of Fe after the PT boundary favours the growth of the red algal plastid, which has the Fe-containing cytochrome C6 in its photosynthetic electron carrier complex instead of the Cu-containing plastocyanin found in the photosystems of the green algae (Falkowski, Katz, et al., 2004; Falkowski, Schofield, et al., 2004). With the many open niches that appeared at the end of the PT extinction event, algae bearing red plastids would have been able to evolve rapidly at that time (Medlin, Kooistra, Gersonde, Sims, & Wellbrock, 1997). Thus, it is most likely that the PT mass extinction event is the event that triggered the elimination of the green algal plastid from the cell in favour of the red plastid, which resulted in the radiation of the heterokont algae and other members of the modern phytoplankton. Molecular clocks using in-group dates (Medlin, 2011) instead of outgroup dates (Yoon *et al.*, 2002) support the radiation of the major groups of the modern plankton at this time, which adds further support to the hypothesis that the secondary endosymbiosis happened more than once.

This scenario of the double secondary endosymbiosis was uncovered with the genome analysis of the two diatoms. The reason for that was a lack of genome sequences from marine members of the green lineage (e.g. Prasinophytes) at the time when the comparative genome analysis was done with *T. pseudonana* and *P. tricornutum*. Moustafa *et al.* (2009) discovered a large proportion of green genes with phylogenetic affiliation to prasinophytes (e.g. *Ostreococcus* and *Micromonas*) of which only some were shared with Chlorophyta and Streptophyta. Thus, a comparative genome analysis to non-marine members of the green lineage was insufficient to discover these genes. The authors suggested that about 70% of genes derived either from red or green genomes were actually of green lineage provenance, which is in contradiction to the chromalveolate hypothesis. Interestingly some of these diatom green genes were also shared with apicomplexans, plastid-lacking ciliates and the haptophyte *Emiliania huxleyi*. These data suggested that a large proportion of these green genes were of ancient provenance and therefore pre-dated the split of cryptophytes and haptophytes from other chromalveolates. These data provided some evidence that a prasinophyte-like endosymbiont is the common ancestor of chromalveolates.

3.3. Potential Mechanisms for Gene Transfer from Plastids to the Nucleus

Comparative genomics studies clearly have revealed that gene transfers have occurred at several times in the past and indicate that this process is continuing. The time necessary for an endosymbiotic gene transfer has been estimated by reverse genetic approaches in higher plants and green algae. Gene transfer from the chloroplast to the nucleus is at least six orders of magnitudes slower in green algae than in higher plants (Lister, Bateman, Purton, & Howe, 2003). The presence of much fewer chloroplasts in unicellular organisms as compared to higher plants might explain why. If chloroplast lysis and subsequent genome degradation is required for endosymbiotic gene transfer, then organisms with only very few chloroplasts might have problems to survive. However, there are some large unicellular algae, such as the radial centric diatom, *Coscinodiscus wailesii*, which have many hundreds of chloroplasts in a single cell. Thus, it remains to be seen whether endosymbiotic genes transfer is always low in eukaryotic unicellular microalgae.

Although there is no clear evidence yet, there is assumed that DNA can be transferred as 'bulk DNA' and 'complementary DNA (cDNA) intermediates' (Thorsness and Weber, 1996). Bulk DNA transfer seems likely because whole-organelle DNA molecules that are >100-kb long are found recombined into eukaryotic chromosomes that still contain the organelle introns, transfer RNA and non-coding regions. Possible 'cDNA intermediates' have been identified in higher plant nuclear genomes. Evidence for 'cDNA intermediates' comes from mitochondrial protein-coding genes found in nuclear genomes of higher plants (Adams, Qui, Stoutemyer, & Palmer, 2002). These genes in flowering plants often have introns and RNA editing. Copies of these genes have been found in the plant nucleus but without the organelle-specific introns and the edited sites. Thus, the assumption is cDNA intermediates were involved in the transfer from the mitochondria to the nucleus. However, conclusive experimental evidence still is missing for this hypothesis (Timmis *et al.*, 2004).

If gene transfer from the organelles to the nucleus is still ongoing, why should there be any genes left in the chloroplasts or mitochondria? The nucleus, in general, is the safer place because there is no production of radical oxygen species as in the chloroplast and there is sexual recombination. However, some genes seem to prefer to stay in the organelles and not to transfer to the nucleus. The theory of 'redox control' explains the retention of genes in the chloroplast genome by the need to control the expression of genes that encode components of the electron transport chain so that they can be synthesized when they are needed to maintain the redox balance (Allen, 2003). However, there is also experimental evidence of redox control of nuclear genes. Nevertheless, the types of genes that plastids and mitochondria retain fit well with the redox-control hypothesis (Allen, 2003).

4. THE SILICA FRUSTULE: DIATOM'S HALLMARK

4.1. Origin of the Silica Frustule

The diatoms are one of the major contributors to this radiation of the heterokont algae and have continued to rise in importance ever since the

PT extinction. Today, the diatoms are found in almost all aquatic and most wet terrestrial habitats (Not *et al.*, 2012 in this volume). Current hypotheses of diatom origins tend to agree that the pre-diatom or 'Ur-diatom' developed from a scaly ancestor, in shallow marine environments and were tychoplanktonic, that is being only occasionally swept into the plankton (see review in Sims, Mann, & Medlin 2006). The early stages of the auxospore, the specialized zygote of the diatoms, have a covering of silica scales in many genera (Round et al., 1990). The multipolar and pennate groups of diatoms have additional bands added to the developing auxospores but in all of the studies on sexual reproduction to date with electron microscopy, the initial stages of all auxospores of all diatoms is a rounded cell covered with silica scales. Several heterokont algae, namely the Dictyochophyceae, Synurophyceae, Chrysophyceae, Parmophyceae, and Xanthophyceae (Van den Hoek *et al.*, 1995; Graham & Wilcox, 2000) also produce silica structures, either as resting stages or as part of their vegetative cell. The ability to metabolize silica was probably inherited from the heterotrophic heterokont ancestor because these groups are spread across the heterokont phylogeny (Medlin, Kooistra, Potter, Saunders, & Andersen, 1997). Scales are present on the reproductive stage of the Labyrinthuloides, which are earlier divergences in the heterokont lineage (Medlin, Kooistra, Potter, et al., 1997). Phylogenetic analyses have documented that the closest sister group to the diatoms are the Bolidophyceae, uniflagellated picoplankters (Guillou *et al.*, 1999). Recently the Parmales, which are characterized as siliceous cyst-like cells, have been shown to be imbedded inside the Bolidomonads. So, the Bolidomonads appear to have a non-siliceous flagellated stage and a siliceous non-flagellate stage. Both stages have been isolated and maintained in culture and there is no existing evidence as to how the stages are interrelated.

In Harwood et al. (2004), suggested that diatoms arose in terrestrial habitats because a new early diatom deposit (175 Ma) had been found in Korea that is terrestrial in origin. Because the Bolidophyceae, their true sister group, are an exclusively marine group of picoplankton, the idea that the diatoms originated on land would appear to conflict with molecular data. To try to resolve this proposed terrestrial origin, Medlin (2002, 2004) has proposed a scenario in which 'Ur–diatoms', abundant as non-silicifying unicells in coastal waters, could have become stranded in isolated tidal pools as eustatic seas retreated after flooding the continents. Over considerable time periods, these large saline pools would begin to dry up and the unicellular, flagellated Ur-diatom, if they survived the desiccation, had to adapt to a semi-terrestrial habitat. The ability to metabolize silica and the production of thick silica walls could have evolved at this time as protection against desiccation (and higher salinity) and to put the cells into a temporary resting state (Medlin 2002, 2004) until the areas were re-flooded. Thus, a simple naked bi-flagellate cell initially evolved or retained silicon metabolism, which prevented the cell from ageing and thus aided its survival by placing it in a prolonged resting state while it was stranded in the tidal pools. Medlin (2002) has reviewed the literature that shows that mammalian cells grown on a silica substratum are placed in a temporary resting state and she proposed that this same benefit of being placed in a temporary resting state was the force driving the ancestral cell to both the diatoms (and the bolidomonads) to metabolize silica. Pascher (1921) originally proposed the evolution of the diatom vegetative cell from a resting cell stage and this idea was later expanded by Mann and Marchant (1989) who proposed that the Parmales could be a close relative of the diatoms long before it was known that the Parmales were bolidomonads. The naked cell stage of the bolidomonads is a reminder of what the ancestral cells of this lineage must have looked like. The fact that both the naked stage and the silicified stage of the bolidomonads can be cultured independently is further evidence for such a scenario for the evolution of silica to have some merit.

4.2. The Genetic and Genomic Basis for Acquisition of Silicate and Biochemical Formation of the Frustule

Genes encoding for proteins for silicic acid transport, deposition and morphogenesis most likely evolved in the exosymbiont and not in either of the two (red and green) endosymbionts because there are no silicifying algal species in the green and red lineage. However, there are many heterokonts that silicify, such as chrysophytes, synurophytes, silicoflagellates and heliozoans as detailed above.

Silicification and deposition of silica takes place in a specialized vesicle (silicon deposition vesicle [SDV]) that was discovered about 50 years ago by pioneering work of Drum and Pankratz (1964). Later microscopic work carried out mainly by Volcani, Schmid and Pickett-Heaps have identified more detailed processes showing how the SDV aids silicon deposition in relation to the formation of the components of the silica shell and in relation to cell division (e.g. Nakajima & Volcani, 1969, 1970; Pickett-Heaps, Schmid, & Tippit, 1984; Schmid, 1979). These early microscopic studies already indicated the complexity of silica deposition. Recent studies with the fluorescent dyes, PDMPO and Rhodamine 123 that visualize the process of silica deposition *in vivo*, added exciting new details to what we already knew about the SDV and frustule formation in general (Descles *et al.*, 2008). Morphogenesis of valves and girdle bands is highly synchronized with cell division and therefore the formation of SDVs and deposition of silica. The valve SDVs form immediately after cytokinesis until the full valve has been synthesized and released via exocytosis in each of the sibling cells. Shortly after the sibling cells have been separated, formation of girdle-band SDVs take place to enable the protoplast to expand until the genetically determined cell size has been obtained (Kroeger & Poulsen, 2008). No one has yet been able to isolate the SDV to identify most of the molecules involved in silicification and morphogenesis, although templates for the valve pattern have now been isolated (Scheffel, Poulsen, Shian, & Kroeger, 2011).

However, biochemical studies on isolated components of diatom cell walls or membranes have revealed the first enzymes involved in uptake of silica and silicification. The following key proteins could be identified by applying the above-mentioned approaches: silicic acid transporter proteins (SITs), frustulins, pleuralins, silaffins, p150 proteins, and silacidins. It took more than 40 years to identify this handful of proteins after Volcani and co-workers isolated the first proteins from acid hydrolysates of purified diatom cell walls (Nakajima and Volcani, 1969, 1970). The publication of the genomes of *T. pseudonana* (Armbrust *et al.*, 2004) and *P. tricornutum* (Bowler *et al.*, 2008), however, opened entirely new opportunities to identify formerly unknown genes and proteins involved in the formation of the silica shell and to identify differences between centric and pennate diatoms. The first comparative analysis of genes related to silicification was done on diatom silicon transporters (Thamatrakoln, Alverson, & Hildebrand, 2006). However, this comparative study was still done on gene amplification from many different diatoms and was not based on Expressed sequence tags (EST) libraries or any other kind of genomic resources from diatoms. Nevertheless, it revealed first more global insights into the diversity of this key group of proteins. The overall percent identity of all amplified SITs in this study was only 19% for overlapping regions of 286 amino acids, which is at the border (20%) of the safe zone (>20%) and midnight zone (<20%), where homologous relationships cannot be reliably determined (Rost, 1999). Generally, there was a higher similarity among SITs within the groups of centric and pennate diatoms than between both groups. SITs from pennate diatoms even share coiled-coil motifs that are absent in SITs from centric diatoms. However, their role is unclear. SITs seem to evolve by either dimerization of two monomers or by internal gene duplication. Many diatom species have paralogs that are more closely related to each other than to any SITs from other diatom species, which is indicative of gene duplication. Some SITs follow the 18S phylogeny of diatoms; others are highly divergent as the tpSIT3 from *T. pseudonana*. These divergent SITs are either ancient SIT types that have had a long time to evolve or they have evolved more rapidly than other paralogs from the same genome, suggesting that they may have a distinct function from other SITs (Thamatrakoln *et al.*, 2006).

By comparing proteins directly involved in silica precipitation and nanostructure formation, we certainly are in the midnight zone of protein sequence identity because there is even less identity compared to diatom SITs. This uncertainty in identifying more of those proteins (e.g. silaffins) and new ones by homology-dependent screening methods was the reason why the first genome-enabled approaches on silicification in diatoms were based on experiments for proteome and whole-transcriptome profiling. Frigeri, Radabaugh, Haynes, and Hildebrand (2006) published the first diatom proteome targeted on cell wall proteins of *T. pseudonana* for which the genome only became available 2 years before this study have been published. Their tandem mass spectrometry analysis on enriched cell wall fractions identified 31 proteins of which 13 had patterns of gene expression similar to silaffins under synchronized cell division. Interestingly, neither silaffins nor silacidins, cingulins and silicon transporters were among the identified proteins. However, two proteins involved in cytoskeleton formation have been indentified (myosin and dynein), one of the p150 proteins and a glutamate acetyltransferase. The latter is involved in the synthesis of polyamines that aid precipitation of silica. The SDV interacts with the cytoskeleton and it might therefore be possible that the identified myosin and dynein proteins are involved in movement of moulding of the SDV. The group of p150 proteins was discovered under copper stress as a group of thiourea extractable cell wall proteins of apparent molecular mass of 150 kDa (p150) (Davis, Hildebrand, & Palenik, 2005). Immunofluorescence experiments with *T. pseudonana* revealed that these proteins are associated to the girdle-band regions of the cell wall and the proteins seem to be there not only under copper stress. However, their precise function and role in cell wall synthesis is still unclear.

In 2008, Mock *et al.* published the first diatom whole-genome transcriptome study to identify new genes involved in silicification. Growth experiments under many different conditions were chosen to identify only those genes responsive to the availability of silicic acid in the growth medium. Unlike the study of Frigeri *et al.* (2006), a silaffin (ProtID 11366) was identified among many other genes significantly (p-value $\leq$ 0.05) upregulated under silicate (4.3-fold) but also under iron limitation (4.9-fold) but none of the other conditions (N, T, CO_2 – limitation) tested. The majority of other genes in this co-regulated cluster under Si and Fe limitation but also those specifically induced only under Si limitation had no known function but many of the proteins showed secretorial signal sequences, transmembrane-spanning domains and were enriched in amino acids S, T, K, R or P. Thus, it was likely that those newly identified proteins were involved in cell wall, membrane and vesicle-bound transport processes. The most intriguing and least anticipated result from this study, however, was the co-regulation of the *T. pseudonana* transcriptome under silicate and iron limitation (Mock *et al.*, 2008). There was some evidence of a link between iron and silicate coming from ecological and physiological studies with diatoms before but not based on any genetic data. These data suggested for the first time that there seem to be pathways in common for the biological formation of the silica shell and for dealing with iron limitation. Alternatively, it was assumed that iron may serve as a required cofactor for silicon bioprocesses. Unfortunately, no new insights have been gained about the link between silicon and iron since the publication of that study, indicating that deciphering of this iron-silicon link may be more difficult than previously thought. However, a transcriptome study with *P. tricornutum* about 1 year later (Sapriel *et al.*, 2009) revealed a very similar link between iron and silicon in *P. tricornutum*, suggesting that this link might be common in many diatoms (Sapriel *et al.*, 2009). Evolutionary and functional insights into both kinds of metabolism are highly needed to reveal how the most characteristic and unique biology of diatoms has shaped to success of this class of organisms.

Post-genomic screening methods unbiased with respect to homology-based identification of proteins involved in silicification were very successful in indentifying novel metabolism related to the silica cell wall of diatoms. Subsequent targeted biochemical and molecular biological analyses of the newly identified proteins need now to unravel how these proteins contribute to the formation of the intricate silica structures of diatoms. A more biased approach with respect to the identification of novel silaffin-like proteins revealed how genomics in combination with targeted biochemical approaches is able to discover proteins with novel functions but which are structurally still similar to silaffins (Scheffel *et al.*, 2011). A homology-independent screening method was used in this study to retrieve proteins from the *T. pseudonana* genome that share structural similarities to known silaffins. They were called cingulins because of their association with the cingulum (girdle-band region of cell wall). The only search criteria were a) an amino acid domain (>100 amino acid) containing ≥18% serine and ≥10% lysine residues and b) an N-terminal ER signal peptide. This screen retrieved 89 hits including the previously identified silaffin proteins. Six of those with previously unknown functions were the cingulins. Their over-expression in *T. pseudonana* revealed an association to the girdle bands. Further analyses showed that cingulins are integral components of a silica-forming organic matrix (microrings). Remarkably and never observed before, these microrings direct silica morphogenesis, suggesting that these pre-assembled protein-based templates are important components for silica morphogenesis. This study shows how genomics and bioinformatics lead to new insights into unknown biology when combined with biochemical approaches. The next step now seems to be using similar biochemical approaches but combined with unbiased post-genomics methods (e.g. proteomics, transcriptomics) to be able to discover the function of novel proteins involved in silica morphogenesis. As revealed by all the latest genomics-enabled studies, more than 50% of the identified proteins potentially involved in silica morphogenesis have no known function. Thus, exciting biology waits to be discovered for new evolutionary and functional insights into diatom silica metabolism.

5. SEX AND THE DIATOM LIFE CYCLE

Modern diatoms grow by sliding apart of the two cell walls or thecae, with simultaneous addition of new girdle elements to the inner, younger theca. Throughout this process, the thecae remain intimately linked, forming a single integrated cell wall. Over time, the average size of the diatom cell reduces by the thickness of one valve width, the so-called McDonald-Pfitzer rule (Pfitzer, 1871). Ultimately, the diatom cell becomes so reduced in size that it will die. The diatoms have evolved an expandable zygote that restores the original large cell size and the process of cell size reduction with each division starts over. Unless the diatom undergoes sexual reproduction, it will die. Once sexual reproduction occurs, a zygote swells in size to restore the maximum cell size to the diatom. The link between size reduction, sexual reproduction and auxospore formation may not be universal in diatoms as it has been thought to be, and even where present, it may not be as strong as has historically been assumed. Because it has been known for many years (see Geitler, 1932; Wiedling, 1948) that some pennate diatoms are able to avoid size reduction and can even expand during cell division (see Crawford, 1981, for discussion). Lewis (1984) has pointed out that the cell division mechanism per se is an inadequate explanation for the evolution of the size reduction–restitution cycle. Size reduction can be avoided if the girdle widens as it elongates during the cell cycle. Some diatoms appear to be entirely asexual. After 20 years of observations of *Caloneis amphisbaena* populations in Edinburgh lakes, Mann (in Sims *et al.*, 2006) never saw sexual stages in this species and no evidence of size reduction. In cases like this and the *Nitzschia* species studied by Wiedling (1948), it is plain that an 'anomalous' life cycle has secondarily evolved (Chepurnov, Mann, Sabbe, & Vyverman, 2004) because other members of the same genus and related genera have a 'normal' life cycle, with sexually produced auxospores. However, because there are exceptions that do not undergo size reduction–restitution, Lewis (1984) argued that size reduction must have adaptive significance and cannot be a mere by-product of the wall structure and cell division mechanism. He suggested was a chronometer for sex, allowing diatoms to spread the high costs of sexual reproduction over several or many years (see also Mann, 1989). Certainly, diatoms do not undergo sexual reproduction at regular intervals because of the time needed to reduce in cell size to enter the size window needed to induce sexual reproduction.

Diatoms are diploid organisms and few other microalgae are also uniquely diploid. Through gametogenesis, haploid gametes are formed. The centric diatoms (both the Coscinodiscophyceae and the Mediophyceae) have oogamous reproduction but the pennate diatoms (Bacillariophyceae) have isogamous reproduction. In oogamy, the female gametangium will produce one large egg and the male gametangia will produce up to eight sperm. In isogamy or anisogamy, one gametangium will form from one up to four gametes, depending on the species and plus (ameoboid motile gametes) and minus (non-motile gametes) mating types must first find one another before meiosis is initiated. It has been documented that in the plankton, when sexual reproduction among centrics occurs, many centric cells undergo sexual reproduction at the same time (Crawford, 1995; Amato *et al.*, 2007). If the sexual life cycle of the diatoms were uncertain during its early stages of evolution, then the close proximity of mating types would have been beneficial (Crawford & Sims, 2007). Kooistra, Forlani, and De Stefano (2008) have argued that isogamy in the plankton is not as successful as oogamy and that this may explain why there are so few planktonic pennate diatoms. A change from oogamy in centric diatoms to isogamy or anisogamy in raphid pennates is therefore understandable as the diatoms evolve into actively motile cells with the evolution of the raphe. What is curious is that this change in mode of sexuality predated the evolution of the raphe, taking place instead in the araphid pennate diatoms and in particular in the basal araphid group, which bears both properizonial bands of the bi-polar centrics and the perizonial bands of the raphe pennates (Medlin & Sato, 2009) and likely an filamentous microtubule-containing appendage that pulls the sex cells together (Sato, Beakes, Idei, Nagumo, & Mann, 2011).

In all diatoms, sexual reproduction can only be induced below a cell size threshold (Chepurnov *et al.*, 2004; Rozumek, 1968). When the cells enter the reproductive size window, sexual reproduction in the homothallic centric diatoms is entirely triggered by environmental factors (e.g. light intensity, photoperiod, temperature or salinity). In contrast, in the predominately allogamous pennates, external factors are less important and cell–cell interactions between sexually compatible female and male cells is more commonly the factor that determines if and when sexual reproduction takes place (Chepurnov *et al.*, 2002, 2005; Geitler, 1932; Mann, 1989; Roshchin, 1994; Sato *et al.*, 2011). This difference obviously places different evolutionary pressures on the evolution of two groups and is also likely a reflection of the habitats that the two groups thrive.

The sexual life cycle of diatoms is highly diverse but generally comprises four different phases: growth, sex, quiescence and cell death. All diatoms known so far have a diplontic life cycle, where 2N cells undergo cell division and thus growth but not 1N cells (e.g. Round et al., 1990). Thus, it seems that this life cycle type is strongly conserved in all diatoms, although there is a huge and largely uncovered diversity of sexual reproduction. Centric diatoms are oogamous, producing eggs and uniflagellate sperm, whereas both non-motile araphid and motile raphid pennate diatoms are morphologically isogamous and non-flagellate. The pennate lineage is assumed to have evolved from centric diatoms. Molecular phylogenies with pennate diatoms have indicated that the motile raphids diverged from among the araphid diatoms. Thus, the pennate diatoms have lost to produce flagellated sperm to increase the chance of encounter between male and female gametes, which, is critical in the diluted aquatic environment. However, the timing of flagellum loss is unknown. A new appendage has now been discovered in two different araphid genera, which appears to be involved in manoeuvring the sex cells into close proximity (Sato *et al.*, 2011; Davidovich et al., 2012). Sato *et al.* (2011) revealed that *Pseudostaurosia trainorii* that belongs to the core araphid pennate lineage and has male gametes with sticky microtubule-based 'threads' to catch and draw eggs. The production and release of female pheromones induce the production of male gametes, which has never been observed in diatoms before. The induction of sex in many centric diatoms is controlled by cell size. When a centric diatom divides asexually, the new cell is formed within the older cell, which causes the new cell to have a smaller size. Thus, asexual cell division leads to cell size reduction. When a critical size has been reached, original cell size is restored by induction of gametes that fuse to form an auxospore that enlarges and hosts the formation of a large vegetative cell and the size therefore is restored. Sex in pennate diatoms is usually induced by stressful conditions, such as nutrient starvation, changes in temperature and/or light. However, not all diatoms reproduce sexually. Many diatom species, centrics and pennate, appear to have lot the ability for sexual reproduction because sexual stages have never been observed. Two of them are *T. pseudonana* and *P. tricornutum* for which no sexual stages have ever been reported. These two diatom species and many others like them are capable of asexual size restoration, a mechanism of unknown biology. Loss of sex may be related to the energetic costs related to it and comparing the ecology of asexual and obligate sexual diatom species would provide new valuable insights into evolutionary forces that shape the occurrence of sex in diatoms. Especially meiosis comes at a high cost because of its slow progression and higher rate of loss and mutations compared to mitosis. However, genetic recombination leads to new genotypes and phenotypes and will reduce deleterious mutations (e.g. Haag & Roze, 2007). Thus, it remains to be seen which kind of evolutionary forces have caused some diatom to abandon sex in favour of asexual reproduction and asexual size restitution. Genomics and post-genomics might be able to provide some answers but the whole field of evolution of sex and the diatom life cycle still is largely unexplored in relation to similar studies in animals and plants. Only very few molecular studies have been conducted over the last few years to unravel the molecular underpinnings of the diatom life cycle and so far we have not much gained from whole-genome sequencing projects with diatoms. One interesting example though was the discovery of the SIG protein family specifically expressed centric diatom sperm by applying small-scale transcriptome approach (Armbrust, 1999). The SIG proteins encode for the flagellar mastigonemes in centric diatom sperm (Honda *et al.*, 2007). However, large-scale transcriptome sequencing currently is underway with several different diatoms and their life cycle stages (W. Vyverman, personal communication). Unlike in diatoms, whole-genome sequencing projects with several members from the class of Prasinophytes (e.g. *Ostreococcus* and *Micromonas*) have revealed a set of genes specific to meiosis, which were conserved between plants, animals and fungi (Derelle *et al.*, 2006; Worden *et al.*, 2009). Sexual life cycle stages have not been observed in these microalgae either but these data suggest the existence of unknown sexual phases. The most comprehensive study so far on life cycle stages and sex in any kind of marine microalgae has been conducted with 1 and 2N cells from the prymnesiophyte *E. huxleyi* (Von Dassow *et al.*, 2009). This comparative whole-genome transcriptome approach revealed many life cycle-specific genes. For instance, Ca^{2+}, H^+ and bicarbonate transporters were strongly over-represented in 2N cells, indicating involvement in calcification. Flagellated 1N cells are not calcifying and therefore do not express these genes. This study gives an example on how genome-enabled tools might be able to provide new information about the functional role of different life cycle stages in marine phytoplankton linked to biogeochemical processes (e.g. carbon cycle). This information is key to decipher how environmental conditions select life cycle traits, an area that has barely been explored for diatoms.

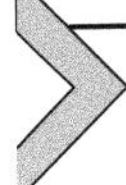

6. EVOLUTION INTO THE PLANKTON

6.1. Coastal versus Open Ocean

With further sea-level changes after the Cretaceous, the early diatoms would have recolonized coastal waters, but their newly developed heavy silica walls would have made them too heavy for a planktonic existence and restricted them to a benthic, nearshore existence. Gradually, the cell walls became less robust and the cells became lighter and there was a strong evolutionary change to cells that remained planktonic. Finkel, Katz, Wright, Schofield, & Falkowski (2005) have noted that there is a decreasing size in the diatoms since the Miocene, which is when the diversification of the Thalassiosirales occurs. This order contains the most diverse and speciose members of the marine plankton. The widespread distribution of the order Thalassiosirales was one major reason to choose *T. pseudonana* for the first diatom genome diatom-sequencing project. The strain sequenced was isolated from Moriches Bay coastal waters at Long Island, New York. Many different strains of this species have been isolated not only from other coastal areas around the globe but also from the open ocean. Thus, it seems that the species *T. pseudonana* has been diverged into different ecotypes adapted to their specific aquatic environment. Re-sequencing of some of these strains is ongoing (M. S. Parker, personal communication) to identify mutations (e.g. single-nucleotide polymorphism (SNPs), insertion or deletion (INDELS)) that may reveal which of the ca. 11,000 genes might be under selection for adaptation to environmental conditions. Recent phylogenetic analyses of morphological and molecular data sets of *T. pseudonana* showed that a) *T. pseudonana* marks an early divergence in a major freshwater radiation by diatoms and b) as a species, *T. pseudonana* is likely ancestrally freshwater (Alverson, Beszteri, Julius, & Theriot, 2011). According to these findings, *T. pseudonana* with all its strains and ecotypes therefore seem to represent recent recolonizations of higher saline habitats. Thus, it is likely that most of the physiological traits of *T. pseudonana* and the genes underlaying those traits might differ from those of strictly marine diatoms. There is no evidence that the other two diatoms for which we have the genome available are of freshwater ancestry but both of them have also been isolated from coastal waters. However, the freshwater input in the Southern Ocean in general is limited compared to most other oceans and not very different between coastal and open-ocean waters because of sea-ice formation and meltdown and the impact of glaciers and ice bergs. The fact that the Southern Ocean is an high-nutrient low chlorophyll (HNLC) area adds a further attribute of an open ocean and the genome of *Fragilariopsis cylindrus* will reveal how this species has evolved to cope with nutrient conditions characteristic for 35% of the open ocean. But so far most evidence for adaptation to open- versus coastal-ocean conditions have not come from comparative genomics approaches but physiological and biochemical studies comparing coastal with open-ocean diatom species. The most important environmental difference between coastal and open oceans is the nutrient concentration in the water (Sarmiento, Gruber, Brezinski, & Dunne, 2004). This is why most studies have focused on understanding the adaptation strategies of diatoms to thrive under high or low nutrient concentrations. However, the freshwater ancestry of the *T. pseudonana* clade with several re-sequenced strains and a worldwide distribution from the coastal to open ocean would make this species the ideal test bed for investigating the adaptation to the salty marine system. Nutrients in coastal waters are usually higher with some exceptions being subtropical desert-like coasts with almost no vegetation and water (nutrient) runoff and permanently frozen shelf-ice coasts of the Antarctic continent. However, some areas on the Antarctic continent, such as coastal waters around the Antarctic Peninsula, seem to be influenced by glacial meltwater particularly over the summer months. These meltwater blooms play a critical role for water column stratification and fertilization (iron) and therefore impact the development of phytoplankton (Dierssen, Smith, & Vernet, 2002). Most of the other coastal waters have higher nutrient concentrations compared to their offshore open-ocean ecosystems because of stronger mixing and therefore release of nutrients from the sediments and runoff of nutrient-rich water from land. The latter is particularly pronounced in industrialized areas with large rivers entering the ocean. Water in these rivers is often loaded with nutrients from agricultural fertilizers (e.g., N, P). But there is also natural fertilization of coastal waters: coastal upwelling of deep nutrient-rich water (Bruland, Rue, & Smith, 2001). However, the degree of upwelling can vary considerably because of changes in meteorological and oceanographic conditions. This kind of variable ecosystem with pulses of high nutrient concentrations is the preferred ecosystem for many different diatoms. It is preferred because diatoms have evolved to be successful in dynamic marine environments characterized by strong seasonality and mixing. The key adaptations to thrive in these environments are a) their ability to take up nutrients quickly after they become available, b) to store nutrients either in vacuoles or by storage molecules (e.g. ferritin), c) by fast growth, d) metabolic flexibility and e) by a resilience to periods of starvation (e.g., Smetacek, 1999).

The two most important nutrients in terms of bloom formation and global distribution of diatoms are nitrogen and iron. Significant amounts of nitrogen are needed for the synthesis of amino and nucleic acids and the most important forms of nitrogen take up from the environment are NO_3, NO_2, NH_4 and also organic forms of nitrogen (e.g. amino acids). Diatom genomes revealed many different transporters for uptake of inorganic and organic forms of nitrogen and there are many reports of storage of inorganic nitrogen in vacuoles of diatoms, which can occupy the majority of the cell volume (Armbrust *et al.*, 2004; Bowler *et al.*, 2008). Thus, diatoms seem to be very versatile in using different forms of nitrogen and storage enables them to extend their mitotic reproduction beyond depletion of nitrogen in the environment. Competitive advantage seems also to come from flexibility in terms of how nitrogen compounds are being metabolized. The genome sequence of *T. pseudonana* has revealed the presence of a metazoan-type urea cycle in diatoms (Armbrust *et al.*, 2004). The key enzyme carbamoyl phosphate synthetase III is located inside the mitochondria and responsible for incorporating NH_4 and HCO_3 into carbomoyl phosphate (Allen *et al.*, 2011). Carbomoyl phosphate feeds into the urea cycle that is mainly responsible for the production of proline, spermidine and urea.

Proline is a major osmolyte in some diatoms, such as the psychrophile *F. cylindrus*, and spermidine is required for silica precipitation during cell wall formation. Urea, the waste product of N-metabolism in metazoans, may severe as storage of nitrogen in diatoms to successfully cope with episodic nitrogen availability in surface ocean waters. Alternatively, planktonic organisms' ability to substitute non-phosphate lipids for phosphate lipids in their membranes under phosphate limitations (Van Mooy *et al.*, 2009) has provided clues to the unknown function of the urea cycle in diatoms with its side production of ornithine in the urea cycle, which is one of the compounds used in the non-phosphate lipids. The enzyme urease breaks down urea into CO_2 and NH_4. Glutamine synthetase in mitochondria and/ or plastids will catalyse the ATP-dependent condensation of glutamate with NH_4 produced by the urease to yield glutamine that can feed into the tricarboxylic acid cycle (TCA) cycle. The urea cycle in diatoms, in fact, is coupled to various intermediates of the TCA cycle and the glutamine synthase/glutamate synthase cycle (Allen *et al.*, 2011). Recently, a proteomics study by Hockin, Mock, Mulholland, Kopriva, and Malin (2012) has shown that the central carbon and especially TCA cycle metabolism response of *T. pseudonana* to nitrogen starvation might differ considerably from that of green algae and higher plants. The TCA cycle under N-limitation seems to be a hub for providing carbon skeletons, for nitrogen reassimilation, or in the diversion of excess carbon into fatty acid biosynthesis. Thus, both, the urea and TCA cycle seem to serve as hubs for redistribution and reassimilation of inorganic carbon and nitrogen in diatoms and therefore contribute to episodic events of nitrogenous compounds in the ocean. However, both cycles are present in all diatoms studied so far and it therefore remains to be seen how differences in the regulation of gene expression and/ or protein activity contribute to adaptation of nitrogen-rich coastal versus nutrient-poor open-ocean habitats. Bacteria, however, follow a different strategy to cope with reduced concentrations of nitrogen in open-ocean habitats. Amino acids sequences from open-ocean bacteria seem to be reduced in nitrogen, but increased in average mass compared to coastal-ocean microorganisms (Grzymski & Dussaq, 2012). N-cost minimization, especially of highly expressed proteins, reduces the total cellular N budget by 2.7–10%. This minimization in combination with reduction in genome size is an evolutionary adaptation to nitrogen limitation. Beginning to sequence open-ocean diatom species and their entire communities will reveal whether a similar N-cost minimization exists also in diatoms. A further discussion about how the environment may shape genomic architecture (base composition, amino acid content and genome size) is discussed in chapter X of this volume (Toulza, Blanc-Mathieu, Gourbiere, & Piganeau, 2012).

Iron and other metals, such as cadmium, are available in much higher concentrations in coastal than open-ocean waters because of closer contact with metal-rich sediments and coastal freshwater runoff. In fact, phytoplankton in about 35% of the surface ocean is limited by the availability of iron, which is the reason why many studies in the past have focussed on how phytoplankton is able to cope with iron limitation (Martin *et al.*, 1994; Pitchford & Brindley, 1999). Some oceans with low-iron concentrations are enriched with nitrogen (e.g., Southern Ocean, Equatorial Pacific, North Pacific) because phytoplankton is not able to consume all the nitrogen that is available. Thus, iron in these ocean areas is the scarcest resource (Liebig's law of the minimum). However, sometimes there is co-limitation with silicate as in the Equatorial Pacific (Marchetti *et al.*, 2010). Those areas of the Ocean are therefore called HNLSiLC (High-NO_3, low-$Si(OH)_4$, low-chlorophyll a regions). Interestingly, a genome-wide transcriptome analysis with *T. pseudonana* under five different growth limitations identified a common set of 84 genes induced under both silicon and iron limitations, suggesting that silicon metabolism share pathways in common with iron, or alternatively, that iron may serve as a required cofactor for proteins involved in silicon metabolism (Mock *et al.*, 2008). Many different iron fertilization experiments have shown that massive phytoplankton blooms can be induced by addition of iron to seawater.

The genome sequence of *F. cylindrus* isolated from the Southern Ocean so far is the only publically available diatom genome sequence from and iron-limited ocean. Annotation of its genome is ongoing and comparative genomics to the coastal diatoms, *T. pseudonana* and *P. tricornutum*, will reveal the molecular underpinnings of adaptation to iron limitation. However, some key studies on iron-related physiology in diatoms already provided significant insights into adaptation to low-iron concentrations. The majority of iron in diatoms is required for electron transport and associated photosynthetic redox reactions in the chloroplasts. Laboratory studies on model diatoms have shown that there are fundamental differences between coastal and oceanic diatoms in their photosynthetic architecture (Strzepek & Harrison, 2004) and use of metal cofactors for electron transport (Peers & Price, 2006). Ocean diatoms had a fivefold lower photosystem I and up to sevenfold lower cytochrome b6f complex concentrations than the studied coastal diatoms. These changes decrease the cellular iron requirements of the oceanic diatoms because both protein complexes require iron for electron transport. However, they do not impact photosynthetic rates but might influence the ability to acclimate to rapid fluctuations of the light intensity as is characteristic for turbid coastal waters (Peers & Price, 2006). Another strategy to reduce the requirement for iron is substituting iron by a metal with similar electrochemical properties, such as copper. Like iron, copper is scarce in the open ocean but this switch reduces the requirement for iron and copper is still relatively more abundant in the open ocean (Peers & Price, 2006). The oceanic mediophycean diatom, *Thalassiosira oceanica*, as well as *F. cylindrus* from the Southern Ocean have plastocyanin that is a copper-containing electron carrier for photosynthesis, which was only known to exist in organisms with chlorophyll *b* and cyanobacteria. Those organisms were most likely the source of the diatom genes acquired via horizontal gene transfer. Coastal diatoms, such as *Thalassiosira weissflogii*, lack plastocyanin but have the Fe-containing homologue cytochrome C6 as do all organisms whose plastid lineage is from the original red algal endosymbiont. These results suggest a strong selection pressure imposed by iron limitation (e.g. the Permian–Triassic Extinction. A recent comparative metatranscriptomics study (Marchetti *et al.*, 2012) with iron-limited and iron-replete diatom communities from the North-Eastern Pacific Ocean (Station Papa at 50°N, 145 °W) revealed that these communities keep expressing genes encoding non-iron-containing proteins even after iron addition, which confirms their evolutionary adaptation to low-iron condition in seawater. Other phytoplankton groups, such as haptophytes, which are less well adapted to low iron in seawater, switch to iron-containing proteins after iron addition. It seemed that diatoms use the newly acquired iron towards metabolic pathways involved in nitrate assimilation (nitrate reductase in diatoms requires iron as a cofactor) and growth to be able to outcompete other phytoplankton groups by a more efficient use of iron for cell division and thus bloom formation. This study also observed a potential role of bacteria-like diatom rhodopsins in dealing with low-iron conditions because these genes were very abundant in iron-limited diatom communities. Bacteria-like rhodopsins in diatoms were first discovered in the genome of *F. cylindrus* and they are also present in diverse dinoflagellates (autotrophic and heterotrophic) and haptophytes (Slamovits, Okamoto, Burri, James, & Keeling, 2011). Their function in these organisms is still unclear but most of them have homology to light-driven proton pumps in bacteria. Thus, it is most likely they have been acquired by horizontal gene transfer. However, it is very unlikely that all these proteins from different marine algal species are involved in dealing with low-iron conditions but it seems that at least some isoforms might help to cope with iron stress. Only isolation of carefully selected bacteria-like diatom rhodopsins and molecular and physiological characterization in heterologous hosts will reveal the true function of these proteins and therefore their significance for diatoms. It is interesting to note that this metatranscriptome study did not support the significance of the iron storage molecule ferritin for open-ocean pennate diatom communities. Ferritin in oceanic pennate diatoms was discovered in 2009 (Marchetti *et al.*, 2009) and was suggested as a key molecule to outcompete algae without this iron storage capability (e.g. coastal diatoms or non-ferritin-containing open-ocean diatoms). Ferritin-containing oceanic diatoms (e.g., *Pseudo-nitzschia granii*) underwent several more cell divisions in the absence of iron than non-ferritin-containing diatoms (e.g., *T. oceanica*). The transcript levels of ferritin in *P. granii* were positively correlated with iron concentration in the growth medium (Marchetti *et al.*, 2009). However, ferritin transcripts were neither detected in iron-limited pennate diatom communities containing *P. granii* nor after iron addition that specifically caused enrichment in transcripts from *P. granii* including those from iron transporters (Marchetti *et al.*, 2012). Thus, the metatranscriptome data based on natural communities (Marchetti *et al.*, 2012) suggest that the competitive strength of low-iron-adapted open-ocean diatom communities lies in reduction of iron-containing proteins and not so much in their ability to store iron by ferritin.

6.2. Polar versus Temperate and Tropical Oceans

Comparing polar with temperate and especially tropical diatom communities might reveal new insights into how global climate change impact phytoplankton communities. Global warming has led to a significant reduction of sea-ice coverage in the Arctic Ocean over the last 50 years (National Snow and Ice Data Center, United States) with consequences for the earth system as a whole. Of special interest are marine diatom communities, which are the basis of the entire Arctic food web supporting large stocks of fish, contributing significantly to carbon cycling and emission of climate active trace gases (e.g. dimethyl sulfide) (Hobson & Welch, 1992). It is expected that many sea-ice phytoplankton species would not be able to adapt because the predicted environmental changes will occur on a time scale too fast for evolutionary processes (Arrigo, van Dijken, & Pabi, 2008). Thus, it is more likely that species well adapted to the low-temperature Arctic environment (e.g. psychrophiles) will be replaced by intruders from lower latitudes outside the Arctic Circle, a process that may already be underway (Greene, Pershing, Cronin, & Cecis, 2008). Despite the severity of current climate changes in polar and especially the Arctic Ocean caused by global warming, there is a significant lack of fundamental data about phylogenetic and functional diversity in eukaryotic phytoplankton communities from polar seawater and sea ice in comparison to communities from temperate and tropic seawater. The sea-ice community in the Arctic is more diverse than that in the Antarctic. There are two different types of communities in the Artic, associated with annual ice and multiyear ice. With the increasing loss of multiyear ice, this entire community will be lost.

Diatom abundance and diversity in geological times scales was largely influenced by temperature. Opening of the Drake Passage about 34 Ma years ago led to a major climatic shift, whereby permanent polar ice developed on the Antarctic continent, resulting in a major climate cooling (Falkowski & Oliver, 2007). The last ice house occurred about 250 Ma years ago during the Permian period. This change in ocean thermal structure was accompanied by a rapid increase in diatom diversity (Falkowski & Oliver, 2007). Also today, diatoms dominate phytoplankton communities in cooler temperate and polar oceans with strong seasonality and mixing, which reflect their evolutionary adaptation. These diatoms are usually well adapted to a wide range of temperatures because their aquatic systems are characterized by thermal fluctuations all year around, except polar waters with permanently cold temperatures (Huertas, Rouco, Lopez-Rodas, & Costas, 2011). These polar diatom communities often lack the genetic redundancy required to withstand an environmental change, such as warming. Many of the polar diatoms are facultative or obligate psychrophiles and one of them from the Southern Ocean (*F. cylindrus*) has been sequenced to reveal its molecular underpinnings of adaptation to polar marine conditions.

A comparative analysis of the *F. cylindrus* genome sequence is still ongoing but many other physiological studies already revealed key mechanisms that separated polar diatoms from their temperate counterparts and therefore gave insights into how they have evolved to polar marine habitats. The genome of *F. cylindrus*, several EST libraries from this species and the mediophycean centric, *Chaetoceros neogracile*, has revealed the presence of a novel group of ice-binding proteins (IBPs), some of which also reduce the freezing temperature of seawater and therefore possibly act as antifreeze proteins (Bayer-Giraldi, Uhlig, John, Mock, & Valentin, 2010). Members of this protein family have not been identified in any non-polar diatom yet and phylogenetic analyses have revealed that they have been acquired by horizontal gene transfer from polar bacteria and fungi. The whole group of IBPs is very diverse and only some of their genes can be induced by freezing temperature. Diatom IBPs are very widespread and they have even been found in genomes of animals that feed on diatoms (copepods) (Kiko, 2010), which indicates the significance of this group of proteins for survival at and below freezing temperatures of sea water.

Photosynthesis of polar diatoms seems exceptionally well adapted to freezing temperatures unlike in non-polar diatoms and especially plants that usually stop growing at freezing temperature. Photosynthetic electron transport, CO_2 fixation and growth was not much depressed at $-1\,^{\circ}C$ compared to $+7\,^{\circ}C$ in *F. cylindrus* given enough time for acclimation (months). However, shifting temperatures from $+\ 4\,^{\circ}C$ to $+10\,^{\circ}C$ in *C. neogracile*, which does not seem to be much, caused growth depression, changed pigment composition and downregulation of photosynthesis genes based on RNA profiling (Hwang, Jung, & Jin, 2008). From these data, it seems that only a slight rise in temperature might have severe consequences for at least some polar diatoms. Those temperature sensitive diatoms will be the first to get replaced by intruders from lower latitudes outside polar oceans. However, our current knowledge about the molecular evolution and adaptation of polar phytoplankton is very limited and only based on a handful of investigations. We therefore have an urgent need to intensify our efforts because global warming has already begun to impact the climatically most susceptible areas on our planet with severe consequences for all life on earth.

7. CONCLUSIONS

Genetic and especially genomic and the first post-genomic studies on diatoms have led to a step change in our understanding how this globally significant group of organisms has evolved and is adapted to the marine habitat. Descriptive observations from the past have been combined with mechanistic insights into diatom biology and evolution gained from molecular approaches. Especially, the recently developed molecular tools for reverse genetics will make diatoms as tractable as almost any other model groups in biology, such as plants or yeast. However, the adaptability of different diatoms species is significantly higher than between species of many other organismal groups. Thus, the major challenge in studies to come is to disentangle specific adaptations between different diatom species from what makes a diatom a diatom. Knowledge gained from this work will lead to the next step change in diatom biology.

REFERENCES

Adams, K. L., Qui, Y.-L., Stoutemyer, M., & Palmer, J. D. (2002). Punctuated evolution of mitochondrial gene content: high and variable rates of mitochondrial gene loss and transfer to the nucleus during angiosperm evolution. *Proceedings of the National Academy of Sciences of the United States of America, 99*, 380–395.

Allen, A. E., Dupont, C. L., Oborník, M., Horák, A., Nunes-Nesi, A., McCrow, J. P., et al. (2011). Evolution and metabolic significance of the urea cycle in photosynthetic diatoms. *Nature, 473*, 203–207.

Allen, J. F. (2003). The function of genomes in bioenergetic organelles. *Philosophical Transactions of the Royal Society B Biological Sciences, 358*, 19–38.

Alverson, A. J., Beszteri, B., Julius, M. L., & Theriot, E. C. (2011). The model marine diatom *Thalassiosira pseudonana* likely decended from a freshwater ancestor in the genus *Cyclotella*. *BMC Evolutionary Biology, 11*, 125.

Amato, A., Kooistra, W. H. C. F., Hee, J., Ghiron, L., Mann, D. G., Pröschold, T., et al. (2007). Reproductive isolation among sympatric cryptic species in marine diatoms. *Protist, 158*, 193–207.

Archibald, J. (2012). The evolution of algae by secondary and tertiary endosymbiosis. *Advances in Botanical Research, 64*, 87–118.

Armbrust, E. V. (1999). Identification of a new gene family expressed during the onset of sexual reproduction in the centric diatom *Thalassiosira weissflogii*. *Applied and Environmental Microbiology, 65*, 3121–3128.

Armbrust, E. V., Berges, J. A., Bowler, C., Green, B. R., Martinez, D., Putnam, N. H., et al. (2004). The genome of the diatom *Thalassiosira pseudonana*: ecology, evolution, and metabolism. *Science, 306*, 79–86.

Arrigo, K. A., van Dijken, G., & Pabi, S. (2008). Impact of a shrinking Arctic ice cover on marine primary production. *Geophysical Research Letters, 35*, L19603.

Baurain, D., Brinkmann, H., Petersen, J., Rodriguez-Ezpeleta, N., Stechmann, A., Demoulin, V., et al. (2010). Phylogenomic evidence for separate acquisition of plastids in cryptophytes, haptophytes, and stramenopiles. *Molecular Biology and Evolution, 27*, 1698–1709.

Bhattacharya, D., & Medlin, L. K. (1995). The phylogeny of plastids. A review based on comparisons of small subunit ribosomal RNA coding regions. *Journal of Phycology, 31*, 489–498.

Bayer-Giraldi, M., Uhlig, C., John, U., Mock, T., & Valentin, K. (2010). Antifreeze proteins in polar sea ice diatoms: diversity and gene expression in the genus *Fragilariopsis*. *Environmental Microbiology, 12*, 1041–1052.

Bromham, L. (2011). The genome as a life history character: why rate of molecular evolution varies between mammal species. *Philosophical Transactions of the Royal Society: Biological Sciences, 366*, 2503–2513.

Bruland, K. W., Rue, E. L., & Smith, G. J. (2001). Iron and macronutrients in California coastal upwelling regimes: implications for diatom blooms. *Limnology and Oceanography, 46*, 1661–1674.

Bowler, C., Allen, A. E., Badger, J. H., Grimwood, J., Jabbari, K., Kuo, A., et al. (2008). The *Phaeodactylum* genome reveals the dynamic nature and multi-lineage evolutionary history of diatom genomes. *Nature, 456*, 239–244.

Cavalier-Smith, T. (1999). Principles of protein and lipid targeting in secondary symbiogenesis: euglenoid, dinoflagellate, and sporozoan plastid origins and the eukaryotic family tree. *Journal of Eukaryotic Microbiology, 46*, 347–366.

Cavalier-Smith, T., & Chao, E. E.-Y. (2006). Phylogeny and megasystematics of phagotrophic heterokonts (Kingdom Chromista). *Journal of Molecular Evolution, 62*, 388–420.

Chepurnov, V. A., et al. (2005). Sexual reproduction, mating system, chloroplast dynamics and abrupt cell size reduction in *Pseudo-nitzschia pungens* from the North Sea (Bacillariophyta). *European Journal of Phycology, 40*, 379–395.

Chepurnov, V. A., Mann, D. G., Sabbe, K., & Vyverman, W. (2004). Experimental studies on sexual reproduction in diatoms. *International Revue Cytology, 237*, 91–154.

Chepurnov, V. A., Mann, D. G., Vyverman, W., Sabbe, K., & Danielidis, D. B. (2002). Sexual reproduction, mating system and protoplast dynamics of *Seminavis* (Bacillariophyceae). *Journal of Phycology, 38*, 1004–1019.

Crawford, R. M. (1981). Some considerations of size reduction in diatom cell walls. In R. Ross (Ed.), *Proceedings of the 6th symposium on recent and fossil diatoms* (pp. 253–265). Koenigstein, Germany: O. Koeltz.

Crawford, R. M. (1995). The role of sex in the sedimentation of a marine diatom bloom. *Limnology* and *Oceanography, 40*, 200–204.

Crawford, R. M., & Sims, P. A. (2007). Some principles of chain formation as evidenced by the early diatom fossil record. *Nova Hedwigia Beiheifte, 133*, 171–186.

Davis, A. K., Hildebrand, M., & Palenik, B. (2005). A stress-induced protein associated with the gridle band region of the diatom *Thalassiosira pseudonana* (Bacillariophyta). *Journal of Phycology, 41*, 577–589.

Davidovich, N. A., Kaczmarska, I., Karpov, S. A., Davidovich, O. I., Michael, L., MacGillivary, M. L., & Mather, L. (2012). Mechanism of male gamete motility in araphid pennate diatoms from the genus. *Tabularia* (Bacillariophyta) *Protist, 163*, 480–494.

De Clerck, O., Bogaret, K., & Leliaert, F. (2012). Diversity and evolution of algae: primary endosymbiosis. *Advances in Botanical Research, 64*, 55–86.

Derelle, E., Ferraz, C., Rombauts, S., Rouzé, P., Worden, A. Z., Robbens, S., et al. (2006). Genome analysis of the smallest free-living eukaryote Ostreococcus tauri unveils many unique features. *Proceedings of the National Academy of Sciences of the United States of America, 103*, 11647–11652.

Douglas, S., Zauner, S., Fraunholz, M., Beaton, M., Penny, S., Deng, L.-T., et al. (2001). The highly reduced genome of an enslaved algal nucleus. *Nature, 410*, 1091–1096.

Descles, J., Vartanian, M., El Harrak, A., Quinet, M., Bremond, N., Sapriel, G., et al. (2008). New tools for labeling silica in living diatoms. *New Phytologist, 177*, 822–829.

Dierssen, H. M., Smith, R. C., & Vernet, M. (2002). Glacial meltwater dynamics in coastal waters west of the Antarctic peninsula. *Proceedings of the National Academy of Sciences of the United States of America, 99*, 1790–1795.

Drum, R. W., & Pankratz, H. S. (1964). Pyrenoids, raphes and other fine structure in diatoms. *American Journal of Botany, 51*, 405–418.

Erwin, D. H. (1990). The end Permian mass extinction. *Annual Review of Ecology and Systematics, 21*, 69–91.

Falkowski, P. G., Katz, M. E., Knoll, A. J., Quigg, A., Raven, J. A., Schofield, O., et al. (2004). The evolution of the modern phytoplankton. *Science, 305*, 354–359.

Falkowski, P. G., & Oliver, M. J. (2007). Mix and match: how climate selects phytoplankton. *Nature Reviews Microbiology, 5*, 813–819.

Falkowski, P. G., Schofield, O., Katz, M. E., van de Schootbrugge, B., & Knoll, A. (2004). Why is the land green and the ocean red? In H. Thierstein, & J. Young (Eds.), *Coccolithophores – from molecular processes to global impact* (pp. 429–453) Amsterdam: Elsevier.

Finkel, Z. V., Katz, M., Wright, J., Schofield, O., & Falkowski, P. (2005). Climatically-driven evolutionary change in the size structure of diatoms over the Cenozoic. *Proceedings of the National Academy of Sciences of the United States of America, 102*, 8927–8932.

Frigeri, L. G., Radabaugh, T. R., Haynes, P. A., & Hildebrand, M. (2006). Identification of proteins from a cell wall fraction of the diatom *Thalasiosira pseudonana* – insights into silica structure formation. *Molecular and Cellular Proteomics, 5*, 182–193.

Frommolt, R., Werner, S., Paulsen, H., Goss, R., Wilhelm, C., Zauner, S., et al. (2008). Ancient recruitment by chromists of green algal genes encoding enzymes for carotenoid biosynthesis. *Molecular Biology and Evolution, 25*, 2653–2657.

Geitler, L. (1932). Der Formwechsel der pennaten Diatomeen. *Archiv für Protistenkunde, 78*, 1–226.

Gersonde, R., & Harwood, D. M. (1990). Lower Cretaceous diatoms from ODP leg 113 site 693 (Weddell Sea) II, vegetative cells. InBarker, P F., & Kennett, J. P., et al. (Eds.) (1990). *Proceedings of the ocean drilling program scientific results*, Vol. 113 (pp. 365–403), College Station, TX Texas A&M University Publishers.

Graham, L. E., & Wilcox, L. W. (2000). *Algae*. London, UK: Prentice-Hall.

Greene, C. H., Pershing, A. J., Cronin, T. M., & Ceci, N. (2008). Arctic climate change and its impacts on the ecology of the North Atlantic. *Ecology, 89*, 24–38.

Green, B. R. (2011). Chloroplast genomes of photosynthetic eukaryotes. *The Plant Journal, 66*, 34–44.

Grzymski, J. J., & Dussaq, A. M. (2012). The significance of nitrogen cost minimization in proteomes of marine microorganisms. *The ISME Journal, 6*, 71–80.

Guillou, L., Chrétiennot-Dinet, M.-J., Medlin, L. K., Claustre, H., Loiseaux-de Goër, S., & Vaulot, D. (1999). *Bolidomonas*, a new genus with two species belonging to a new algal class, the Bolidophyceae (Heterokonta). *Journal of Phycology, 35*, 368–381.

Haag, C. R., & Roze, D. (2007). Genetic load in sexual and asexual diploids, segregation, dominance and genetic drift. *Genetics, 176*, 1663–1678.

Harwood, D. M., Chang, K. H., & Nikolaev, V. A. (2004). Late Jurassic to earliest Cretaceous diatoms from Jasong Synthem, Southern Korea, evidence for a terrestrial origin. In A. Witkowski, T. Radziejewska, B. Wawrzyniak-Wydrowska, G. Daniszewska-Kowalczyk, & M. Bok (Eds.), *Abstracts, 18th international diatom symposium* (p. 81). Miedzyzsroje: Poland and University of Szczecin.

Hobson, K. A., & Welch, H. E. (1992). Determination of trophic relationships within a high Arctic marine food web using delta13C and delta 15N analysis. *Marine Ecology Progress Series, 84*, 9–18.

Hockin, N. L., Mock, T., Mulholland, F., Kopriva, S., & Malin, G. (2012). The response of diatom central carbon metabolism to nitrogen starvation is different from that of green algae and higher plants. *Plant Physiology, 158*, 299–312.

Honda, D., Shono, T., Kimura, K., Fujita, S., Iseki, M., & Makino, A. (2007). Homologs of the sexually induced gene 1 (sig1) product constitute the stramenopile mastigonemes. *Protist, 158*, 77–88.

Huertas, E., Rouco, M., Lopez-Rodas, V., & Costas, E. (2011). Warming will affect phytoplankton differently, evidence through a mechanistic approach. *Proceedings of the Royal Society B Biological Sciences, 278*, 3534–3543.

Hwang, Y.-S., Jung, G., & Jin, E. (2008). Transcriptome analysis of acclamatory responses to thermal stress in Antartic algae. *Biochemcial and Biophysical Rsearch Communications, 367*, 635–641.

Kaczmarska, I., Ehrman, J. M., & Bates, S. S. (2001). A review of auxospore structure, ontogeny and diatom phylogeny. In A. Economou-Amilli (Ed.), *Proceedings of the 16th international diatom symposium* (pp. 153–168). Greece: University of Athens.

Keeling, P. J. (2004). Diversity and evolutionary history of plastids and their hosts. *American Journal of Botany, 91*, 1481–1493.

Kang, I., Oh, H-M., Vergin, K. L., Giovannoni, S. T., & Cho, J-C (2010). Genome sequence of the marine alphaproteobacterium HTCC2150, assigned to the *Roseobacter clade. Journal of Bacteriology, 192*, 6315–6316.

Kiko, R. (2010). Acquisition of freeze protection in a sea-ice crustacean through horizontal gene transfer? *Polar Biology, 33*, 543–556.

Kooistra, W. H. C. F., & Medlin, L. K. (1996). The evolution of the diatoms (Bacillariophyta) IV. A reconstruction of their age from small subunit rRNA coding regions and the fossil record. *Molecular Phylogenetics and Evolution, 6*, 391–407.

Kooistra, W. H. C. F., Forlani, G., & De Stefano, M. (2008). Adaptation of araphid pennate diatoms to a planktonic existed. *Marine Ecology, 30*, 1–15.

Kroeger, N., & Poulsen, N. (2008). Diatoms – from cell wall biogenesis to nanotechnology. *Annual Review of Genetics, 42*, 83–107.

Lewis, W. M. (1984). The diatom sex clock and its evolutionary significance. *American Naturalist, 123*, 73–80.

Lipps, J. H. (1970). Plankton evolution. *Evolution, 24*, 1–22.

Lister, D. L., Bateman, J. M., Purton, S., & Howe, C. J. (2003). DNA transfer from chloroplast to nucleus is much rarer in *Chlamydomonas* than in tobacco. *Gene, 316*, 33–38.

Mann, D. G. (1988). Why didn't Lund see sex in *Asterionella*? A discussion of the diatom life cycle in nature. In F. E. Round (Ed.), *Algae and the aquatic environment* (pp. 383–412). Bristol, UK: Biopress.

Mann, D. G. (1989). On auxospore formation in *Caloneis* and the nature of *Amphiraphia* (Bacillariophyta). *Plant Systematics and Evolution, 163*, 43–52.

Mann, D. G., & Marchant, H. J. (1989). The origin of the diatom and its life cycle. In J. C. Green, B. S. C. Leadbeater, & W. L. Diver (Eds.), *The Chromophyte algae, problems and perspectives* (pp. 307–323). Oxford, UK: Clarendon Press.

Marchetti, A., Schruth, D. M., Durkin, C. A., Parker, M. S., Kodner, R. B., Berthiaume, C. T., et al. (2012). Comparative metatranscriptomics identifies molecular bases for the physiological responses of phytoplankton to varying iron availability. *Proceedings of the National Academy of Sciences of the United States of America, 109*, 1828–1829.

Marchetti, A., Varela, D. E., Lance, V. P., Johnson, Z., Palmucci, M., Giordano, M., et al. (2010). Iron and silicic acid effects on phytoplankton productivity, diversity, and chemical composition in the central equatorial Pacific Ocean. *Limnology and Oceanography, 55*, 11–29.

Marchetti, A., Parker, M. S., Moccia, L. P., Lin, E. O., Arrieta, A. L., Ribalet, F., et al. (2009). Ferritin is used for iron storage in bloom-forming marine pennate diatoms. *Nature, 457*, 468–470.

Marin, W., Rujan, T., Richly, E., Hansen, A., Cornelsen, S., Lins, T., et al. (2002). Evolutionary analysis of Arabidopsis, cyanobacterial, and chloroplast genomes reveals plastid phylogeny and thousands of cyanobacterial genes in the nucleus. *Proceedings of the National Academy of Sciences of the United States of America, 99*, 12246–12251.

Martin, J. H., Coale, K. H., Johnson, K. S., Fitzwater, S. E., Gordon, R. M., Tanner, S. J., et al. (1994). *Nature, 371*, 123–129.

Medlin, L. K. (2002). Why silica or better yet why not silica? Speculations as to why the diatoms utilise silica as their cell wall material. *Diatom Research, 17*, 453–459.

Medlin, L. K. (2004). Comment in reply to Schmid (2003). The evolution of the silicified diatom cell wall revisited. *Diatom Research, 19*, 345–351.

Medlin, L. K. (2008). Molecular clocks and inferring evolutionary milestones and biogeography in the microalgae. In H. Okada, S. F. Mawatari, N. Suzuki, & P. Gautam (Eds.), *Origin and evolution of natural diversity* (pp. 31–42), Sapporo, Japan Hokkaido University Press.

Medlin, L. K. (2010). Pursuit of a natural classification of diatoms: an incorrect comparison of published data. *European Journal of Phycology, 45*, 155–166.

Medlin, L. K. (2011). The Permian Triassic Extinction forces the radiation of the modern phytoplankton. *Phycologia, 50*, 684–693.

Medlin, L. K., & Kaczmarska, I. (2004). Evolution of the diatoms: V. Morphological and cytological support for the major clades and a taxonomic revision. *Phycologia, 43*, 245–270.

Medlin, L. K., Kooistra, W. C. H. F., Gersonde, R., Sims, P., & Wellbrock, U. (1997). Is the origin of diatoms related to the end-Permian mass extinction. *Nova Hedwigia Festschrift für U. Geissler, 65*, 1–11.

Medlin, L. K., Kooistra, W. H. C. F., Potter, D., Saunders, G. W., & Andersen, R. A. (1997). Phylogenetic relationships of the 'golden algae' (haptophytes, heterokont chromophytes) and their plastids. In D. Bhattacharya (Ed.), *Origins of algae and their plastids* (pp. 187–219). Wien, Germany: Springer-Verlag.

Medlin, L. K., Kooistra, W. H. C. F., & Schmid, A. M. M. (2000). A review of the evolution of the diatoms – a total approach using molecules, morphology and geology. In A. Witkowski, & J. Sieminska (Eds.), *The origin and early evolution of the diatoms, fossil, molecular and biogeographical approaches* (pp. 13–35). Cracow, Poland: Szafer Institute of Botany, Polish Academy of Science.

Medlin, L. K., & Sato, S. (2009). The biological reality of the core and basal groups of araphid diatoms. *Diatom Research, 24*, 503–508.

Medlin, L. K., Sato, S., Mann, D. G., & Kooistra, W. C. H. F. (2008). Molecular evidence confirms sister relationship of *Ardissonea, Climacosphenia* and *Toxarium* within the bipolar centric diatoms (Mediophyceae, Bacillariophyta). *Journal of Phycology, 44*, 1340–1348.

Medlin, L. K., Williams, D. M., & Sims, P. A. (1993). The evolution of the diatoms (Bacillariophyta), I. Origin of the group and assessment of the monophyly of its major divisions. *European Journal of Phycology, 28*, 261–275.

Mock, T., Samanta, M. P., Iverson, V., Berthiaume, C., Robison, M., Holtermann, K., et al. (2008). Whole-genome expression profiling of the marine diatom *Thalassiosira pseudonana* indentifies genes involved in silicon bioprocesses. *Proceedings of the National Academy of Sciences of the United States of America, 105*, 1579–1584.

Moustafa, A., Beszteri, B., Maier, U. G., Valentin, K., & Bhattacharya, D. (2009). Genomic footprints of a cryptic plastid endosymbiosis in diatoms. *Science, 324*, 1724–1726.

Nakajima, T., & Volcani, B. E. (1969). 3,4-Dihydroxyproline – a new amino acid in diatom cell wall. *Science, 164*, 1400–1401.

Nakajima, T., & Volcani, B. E. (1970). ε-N-trimethyl-L-δ-hydroxine phosphate and its nonphosphorylated compound in diatom cell walls. *Biochemical and Biophysical Research Communications, 39*, 28–33.

Not, F., Siano, R., Kooistra, W. H. C. F., Simon, N., Vaulot, D., & Probert, I. (2012). Diversity and ecology of eukaryotic marine phytoplankton. *Advances in Botanical Research, 64*, 1–53.

Oudot-Le Secq, M. P., Grimwood, J., Shapiro, H., Armbrust, E. V., Bowler, C., & Green, B. R. (2007). Chloroplast genomes of the diatoms Phaeodactylum tricornutum and Thalassiosira pseudonana, comparison with other plastid genomes of the red lineage. *Molecular Genetics and Genomics, 277*, 427–439.

Palenik, B., Grimwood, J., Aerts, A., Rouzé, P., Salamov, A., Putnam, N., et al. (2007). The tiny eukaryote Ostreococcus provides genomic insights into the paradox of plankton speciation. *Proceedings of the National Academy of Sciences of the United States of America, 104*, 7705–7710.

Pascher, A. (1921). Über die einstimmungen zwischen Diatomeen, Heterokonten und Chrysomonaden. *Berliner Deutsch Botanische Gesellschaft, 39*, 236–240.

Petersen, J., Teich, R., Brinkmann, H., & Cerff, R. (2007). A "green" phosphoribulokinase in complex algae with red plastids, evidence for a single secondary endosymbiosis leading to haptophytes, cryptophytes, heterokonts, and dinoflagellates. *Journal of Molecular Evolution, 62*, 43–57.

Pfitzer, E. (1871). Untersuchungen fiber Bau und Entwicklung der Bacillariaceen (Diatomaceen). *Bototanische Abhandlungen, 2*, 1–189.

Round, F. E., Crawford, R. M., & Mann, D. G. (1990). *The diatoms, biology and morphology of the genera*. Cambridge, UK: Cambridge University Press.

Roshchin, A. M. (1994). *Zhiznennye Tsikly Diatomovykh Vodoroslej*. Kiev, Ukraine: Naukova Dumka. 170 pp.
Peers, G., & Price, N. M. (2006). Copper-containing plastocyanin used for electron transfer by an oceanic diatom. *Nature, 18*, 341–344.
Pickett-Heaps, J. D., Schmid, A.-M. M., & Tippit, D. H. (1984). Cell division in diatoms. *Protoplasma, 120*, 132–154.
Price, D. C., Chan, C. X., Yoon, H. S., Yang, E. C., Qiu, H., Weber, A. P., et al. (2012). *Cyanophora paradoxa* genome elucidates origin of photosynthesis in algae and plants. *Science, 335*, 843–847.
Pitchford, J. W., & Brindley, J. (1999). Iron limitation, grazing pressure and oceanic high-nutrient-low-chlorophyll (HNLC) regions. *Journal of Plankton Research, 21*, 525–547.
Rocap, G., Larimer, F. W., Lamerdin, J., Malfatti, S., Chain, P., Ahlgren, N. A., et al. (2003). Genome divergence in two *Prochlorococcus* ecotypes reflects oceanic niche differentiation. *Nature, 424*, 1042–1047.
Rost, B. (1999). Twilight zone of protein sequence alignments. *Protein Engineering, 12*, 85–94.
Rozumek, K.-E. (1968). Der Einflub der Umweltfaktoren Licht und Temperatur auf die Ausbildung der Sexualstadien bei der pennaten Diatomee *Rhabdonema adriaticum* Kutz. *Beitrage Biologie Pflanzen, 44*, 365–388.
Scheffel, A., Poulsen, N., Shian, S., & Kroeger, N. (2011). Nanopatterned protein microrings from a diatom that direct silica morphogenesis. *Proceedings of the National Academy of Sciences of the United States of America, 108*, 3175–3180.
Sapriel, G., Quinet, M., Heijde, M., Jourdren, L., Tanty, V., Luo, G., et al. (2009). *PloS One, 4*, e7458.
Sarmiento, J. L., Gruber, N., Brezinski, M. A., & Dunne, J. P. (2004). High-latitude controls of thermocline nutrients and low latitude productivity. *Nature, 427*, 56–60.
Sato, S. (2008). *Phylogeny of araphid diatoms, inferred from morphological and molecular data.* PhD Dissertation, University of Bremen. http://elib.suub.uni-bremen.de/diss/docs/00011057.pdf
Sato, S., Beakes, G., Idei, M., Nagumo, T., & Mann, D. G. (2011). Novel sex cells and evidence for sex pheromones in diatoms. *PLoS One, 6*, e26923.
Schmid, A.-M. M. (1988). The special Golgi–ER–mitochondrium unit in the diatom genus *Coscinodiscus*. *Plant Systematics and Evolution, 158*, 211–233.
Schmid, A.-M. M. (1979). The development of structure in the shells of diatoms. *Nova Hedwegia, 64*, 219–236.
Sims, P. A., Mann, D. G., & Medlin, L. K. (2006). Evolution of the diatoms, insights from fossil, biological and molecular data. *Phycologia, 45*, 361–402.
Sinninghe-Damsté, J. S., Muyzer, G., Abbas, B., Rampen, S. W., Masse, G., Allard, W. G., et al. (2004). The rise of the rhizosolenoid diatoms. *Science, 304*, 584–587.
Slamovits, C. H., Okamoto, N., Burri, L., James, E. R., & Keeling, P. J. (2011). A bacterial proteorhodopsin proton pump in marine eukaryotes. *Nature Communications*. doi: 10.1038/ncomms1188.
Smetacek, V. (1999). Diatoms and the ocean carbon cycle. *Protist, 150*, 25–32.
Strzepek, R. F., & Harrison, P. J. (2004). Photosynthetic architecture differs in coastal and oceanic diatoms. *Nature, 431*, 689–692.
Timmis, J. N., Ayliffe, M. A., Huang, C. Y., & Martin, W. (2004). Endosymbiotic gene transfer, organelle genomes forge eukaryotic chromosomes. *Nature Review Genetics, 5*, 123–135.
Thamatrakoln, K., Alverson, A. J., & Hildebrand, M. (2006). Comparative sequence analysis of diatom silicon transporters, toward a mechanistic model of silicon transport. *Journal of Phycology, 42*, 822–834.

Thorsness, P. E., & Weber, E. R. (1996). Escape and migration of nucleic acids between chloroplasts, mitochondria, and the nucleus. *International Review of Cytology, 165*, 207–234.

Toulza, E., Blanc-Mathieu, R., Gourbiere, S., & Piganeau, G. (2012). Environmental genomics of microbial algae: power and challenges of metagenomics. *Advances in Botanical Research, 64*, 379–423.

Van den Hoek, C., Mann, D. G., & Jahns, H. M. (1995). *Algae, an introduction to phycology*. Cambridge, UK: Cambridge University Press.

Van Mooy, B. A. S., Fredricks, H. F., Pedler, B. E., Dyhrman, A. T., Karl, D. M., Koliizek, M., et al. (2009). Phytoplankton in the ocean use non-phosphorus lipids in response to phosphorus scarcity. *Nature, 458*, 69–72.

Von Dassow, P., Ogata, H., Probert, I., Wincker, P., Da Silva, C., Audic, S., et al. (2009). Transcriptome analysis of functional differentiation between haploid and diploid cells of *Emiliania huxleyi*, a globally significant photosynthetic calcifying cell. *Genome Biology, 10*, R114.

Von Stosch, H. (1950). Oogany in a centric diatom. *Nature, 165*, 531–532.

Williams, D. M. (2007). Classsification and diatom systematics, the past, the present and the future. In J. Brodie, & J. Lewis (Eds.), *Unravelling the algae* (pp. 57–91). Boca Raton, FL: CRC Press..

Wiedling, S. (1948). Beiträge zur Kenntnis der vegetativen Vermehrung der Diatomeen. *Botaniska Notiser, 1948*, 322–354.

Worden, A. Z., Lee, J. H., Mock, T., Rouzé, P., Simmons, M. P., Aerts, A. L., et al. (2009). Green evolution and dynamic adaptations revealed by genomes of the marine picoeukaryotes *Micromonas. Science, 324*, 268–272.

Yoon, H. S., Hackett, J. D., Pinto, G., & Bhattacharya, D. (2002). The single ancient origin of chromist plastids. *Proceedings of the National Academy of Sciences of the United States of America, 99*, 15507–15512.

Yoon, H. S., Hackett, J., Ciniglia, C., Pinto, G., & Bhattacharya, D. (2004). A molecular timeline for the origin of photosynthetic eukaryotes. *Molecular Biology and Evolution, 21*, 809–818.

CHAPTER EIGHT

Microalgae, Functional Genomics and Biotechnology

Jean-Paul Cadoret[1],*, Matthieu Garnier*, and Bruno Saint-Jean*
*Ifremer, Laboratoire Physiologie et Biotechnologie des Algues, rue de l'île d'Yeu BP 21105 44311 Nantes cedex 3, France
[1]Corresponding author: E-mail: jean.paul.cadoret@ifremer.fr

Contents

Abstract

Microalgae have been studied for decades, but a new wave of research has recently begun as part of the search for renewable and sustainable energy sources. For economic optimization, microalgal biomass is being considered as a whole (main products and co-products) in an overall 'biorefinery' concept. Applications of microalgae cover a broad spectrum, including the food and (livestock) feed industries, bioenergy, cosmetics, healthcare and environmental restoration or protection. In the field of biotechnology, the access to genomic data is playing a growing role. As the cost of sequencing strategies has fallen, studies of gene function at the transcript, protein and biosynthesis pathway levels have multiplied. Notably, sequencing and mass spectrometry technologies are used to delineate the pathways of lipid synthesis, which will

Advances in Botanical Research, Volume 64
ISSN 0065-2296,
http://dx.doi.org/10.1016/B978-0-12-391499-6.00008-6

be valuable for the future application of microalgae in the biotechnology and biofuel industries. Another field making an applied use of genomics is the 'cell factory' approach, which uses the cell to manufacture (express) natural or recombinant proteins for diverse purposes. In this chapter, we present a vision of the potential future of genomics in the biotechnology of microalgae from several points of view.

1. INTRODUCTION

Microalgae in biotechnology are presently the focus of an unprecedented surge in interest and investment worldwide. Over recent decades, research predicted the explosion of attention this field would attact following the U.S. Aquatic Species Program (Sheehan, Dunahay, Benemann, & Roessler, 1998), as microalgae can provide a new source of vegetal material. They offer complementary products to land plants and higher manipulability, but as the consequence of their large phylogenetic spread (reviewed in Chapter I of this volume), they have vast unknown metabolic potential because most species are, as yet, unexamined.

Driven by the giants of the energy industry, the race to develop mass microalgal production capacity started about 5 years ago, fuelled by hundreds of millions of U.S. dollars targeting the production of renewable biofuels. The challenges we face today are to adapt and improve existing methods, develop new processes and achieve a drastic reduction in costs. The objective is to use this green biomass in its entirety and not only for energy production. The potential is huge and the fields of study numerous, offering very high added value in the areas of new energy (oil, hydrogen and fermentation), healthcare (pigments, enzymes and secondary metabolites), food (human or animal), environmental management (depuration and assimilation mechanisms) and industry (recovery of silica, enzymes or pigments). Here, we have chosen to focus our presentation on the world of microalgae, their broad fields of application, the advances in genomics for biotechnologies and some of the bottlenecks that need to be overcome.

1.1. Microalgae

We use the term *microalgae* to cover a heterogeneous group of single-celled photosynthetic organisms, including photosynthetic eukaryotes and photosynthetic prokaryotes like *Prochlorococcus* and *Synechococcus*, which are of major global importance and considered as key players among phytoplanktonic organisms in oligotrophic oceanic areas. It would be vastly overambitious to attempt to cover the biotechnological potential of the entire aquatic photosynthetic world in one book chapter, so this review will address only the genomics and biotechnology of eukaryotic microalgae.

Depending on environmental conditions such as salinity, light, temperature, pH and nutrient concentrations, the size and appearance of microalgae can change profoundly, making their identification difficult without molecular tools. The estimated number of described species ranges between 40,000 and 60,000, but estimations of the number of undescribed species range from hundreds of thousands to millions of species spread over the globe (Norton, Melkonian, & Andersen, 1996, Sastre & Posten, 2010). In comparison, only 250,000 land plant species have been recorded. Half of the world's oxygen is produced via microalgal photosynthesis. Microalgae contribute up to 50% of all aquatic productivity and 25% of global productivity (Raven & Falkowski, 1999). They are the foundation of the aquatic food chain and have colonized nearly all biotopes from the polar ice to deserts and hot springs. They have adapted to extreme environments, living in salt marshes, acidic environments or conditions with very low light. Through their presence on the surface of the oceans, which cover 70% of the earth, they play a major role in global climate regulation, as a machine that transforms CO_2 into organic matter (Raven & Falkowski, 1999).

Ancestors of the present day cyanobacteria invented photosynthesis as far back as 3.6 billion years ago (Gould, Waller, & McFadden, 2008) and the primary endosymbiotic event at the origin of all photosynthetic eukaryotes can be traced to 1.8 billion years ago (Finazzi, Moreau, & Bowler, 2010). The number and the diversity of algal species offer a whole new field of research when considering their potential commercial applications and biotechnology. Although progress still needs to be made on culture techniques, algal production systems on scales from a few litres up to cubic metre volumes, in photobioreactors or open ponds, are now a reality at the industrial level. Microalgae have clear advantages over land plants. Their photosynthetic yields are slightly better than those of land plants (Wijffels, Barbosa, & Eppink, 2010) and the fact that they live in an aqueous medium gives them direct access to their nutrients and explains why they display higher growth productivity. As an example, the productivity of classic crops in Europe is around 1–2 $g/m^2/day$ (dry weight), whereas the microalgae in small and medium-sized enterprises on the Atlantic coast produce around 10 $g/m^2/day$. Additionally, aqueous cultures in marine water offer the advantage of using land unsuitable for food crops, avoiding the much-publicized dilemma between 'food and fuel'. Other differences between land plants and microalgae that could give microalgae the advantage include the possibility of performing continuous cultures in photobioreactors with a high level of control, the potential to couple microalgal production with the disposal of effluents that provide nutritive components, the attractive idea of using industrial CO_2 sources and the saving of freshwater by cultivation of microalgae in seawater. The opportunity to cultivate in photobioreactors offers the additional possibility of adjusting and adapting culture conditions in real time, allowing growers to react instantaneously to the culture situation. The biological diversity of microalgae provides an exceptional range of adaptability and represents a vast potential as a source of food and feed, biomaterial, original molecules and applications in the broad field of biotechnology. Gene transfer of the means to produce selected molecules by genetic engineering will provide a complementary production method for novel compounds.

1.2. Applications of Microalgae

The current and forthcoming applications of microalgae are numerous and diverse, including food, feed, healthcare, industry and energy. Although the use of cyanobacteria in food dates back many hundreds of years, advances in this area were made in the 20th century (Habib, Huntington, & Hasan, 2008). The market for microalgae as food and food supplements is dominated by the Cyanobacteria *Spirulina platensis* (also called *Arthrospira platensis*), the Chlorophyta *Chlorella* sp., and in France, the diatom *Odontella aurita*. In addition, the green microalga *Dunaliella salina* is used for its beta-carotene, *Haematococcus pluvialis* for astaxanthin and the Cyanobacteria *Aphanizomenon flos-aquae* as a dietary supplement. Investigation is still needed on the use of other microalgae as food, requiring effort to be made for the acceptance of these alternative sources. For example, cookies made from the Haptophyta *Isochrysis galbana*, rich in omega-3, have already been produced (Gouveia *et al.*, 2008).

The area in which microalgae were first mass produced was aquaculture. Phytoplanktonic organisms are an essential food for the rearing of molluscs and fish, especially to feed the early life stages of bivalves, for which microalgae must be provided as live food. A large production capacity is devoted to this activity worldwide. Although around 40 microalgal species are used in this way, the number routinely grown is closer to a dozen. The technology and skills developed as part of this culture are important for the future of microalgal biotechnology. Microalgae could become an important source of land animal feed. The most common species used for this are

Spirulina, *Chlorella* and *Scenedesmus*. In chicken farming, it is reported that the incorporation of 5–10% microalgae in the diet has an effect on the colour of the meat and egg yolk (Becker, 2007). The potential substitution of fish oil with algae oil has also been discussed (AbuGhazaleh, Potu, & Ibrahim, 2009).

Algae also offer several benefits in the field of human healthcare. Land plants and animals lack the enzymes to synthesise polyunsaturated fatty acids (PUFAs) longer than 18 carbon atoms. Long-chain PUFAs like gamma-linolenic, arachidonic (AA), eicosapentaenoic (EPA) and docosahexaenoic acid (DHA), produced by microalgae, accumulate in most marine animals. Sufficient consumption of such fatty acids could have beneficial effects on human health. The oil from the stramenopile *Schizochytrium* sp. (permitted as a food ingredient) contains 35–45% DHA. In comparison, most conventional oils rich in omega-3 (walnut oil, canola oil) contain about 10% alpha-linolenic acid, the precursor of omega-3. The production of these PUFAs will undoubtedly be a major challenge in the coming years.

Algal pigments, such as carotenoids, are already commercially exploited but are also the subject of intensive research. The most popular among these are beta-carotene, alpha-carotene, lutein, lycopene and zeaxanthin. Even though the main supply of astaxanthin to colour salmon is 95% of synthetic origin, natural sources such as the green microalga *H. pluvialis* are authorized in Japan and Canada (Lorenz & Cysewski, 2000). Among other marine pigments of interest, the phycobiliproteins are a very unusual class identified in microalgae. First commercialized in clinical and immunological analysis, broader uses in industry and therapy are envisioned (Sekar & Chandramohan, 2008).

The uptake of oxygen by organisms can cause the formation of dangerous derivatives, including singlet oxygen and free radicals. These forms of highly reactive oxygen species (ROS) play an important role in various chronic diseases (cancer, atherosclerosis, osteoarthritis, Parkinson's, etc.) or acute reactions (inflammation, septic shock, etc.). However, ROS production can also be used as a means of therapy in human health. Indeed, photodynamic therapy (PDT) is an innovative discipline calling for photosensitive molecules with a tumour tropism that react to light and destroy the surrounding tissues by ROS production. Only a few drugs are presently in use for PDT. Less than a dozen molecules have been identified so far and none are, as yet, considered very efficient. It is, however, a promising field as our laboratory was able to identify a group of molecules from microalgae that is 30 times more efficient than the best commercial gold standard (T. Patrice, J. P. Cadoret, L. Picot, R. Kaas, and J. B. Berard, unpublished work).

The polysaccharides extracted from the red microalga *Porphyridium purpureum* have been proven to have antiviral activity on cell lines as well as *in vivo* in rabbits (Huheihel, Ishanu, Tal, & Arad, 2002). Indeed, red algae have been studied for their polysaccharide contents both for health (Matsui, Muizzuddin, Arad, & Marenus, 2003) and industry applications (Gourdon *et al.*, 2008). Apart from structural polysaccharides, some microalgae synthesize exopolysaccharides. These polymeric compounds form a hydrophilic and polyanionic matrix, retain water and trap cations, allowing the microalgae to resist desiccation. These properties suggest that the algae could be useful for biotechnological applications in environmental fields through the detoxification of biotopes polluted by heavy metals (Pb, As, Hg, Cd) and in the recovery of some metals such as gold and uranium. The physico-chemical characteristics of polysaccharides—particularly their rheological, lubricant and flocculent properties—have been suggested for various applications.

A few hundred microalgae are classified as dangerous due to their toxin production. Among the 90 recorded species, 70 belong to the dinoflagellate group. The potential applications of these toxins in human healthcare have been reviewed by Camacho *et al.* (2006). Characteristics such as the anti-fouling properties of microalgae could be exploited produce a range of 'biogenic' products (Bhadury & Wright, 2004).

Some algal extracts are considered emollients and are incorporated into anti-aging creams to prevent wrinkles and stimulate collagen synthesis; their ultraviolet (UV) protection properties are also being researched. Although many of the marketing claims about algal bioproducts still need to be proven, business prospects justify the interest shown in this field. *Arthrospira* and *Chlorella* are again those involved in the anti-aging and regenerative products (Spolaore, Joannis-Cassan, Duran, & Isambert, 2006). However, while many applications of microalgae are already in existence, genomics is opening up still more opportunities.

1.3. Genomics and Microalgae

The rise of next-generation sequencing (NGS) technologies, accompanied by a sharp fall in their cost, has led to the acquisition of important genomic data on microalgae since the 1990s. The pace of the availability of microbial genomes is obviously increasing with NGS technologies and in addition to the 14 nuclear genomes available (see Chapters II and III of this volume for a review), the gene repertoire of many additional species is now available through transcriptomics, as discussed below. Due to its phylogenetic proximity to land plants and because many molecular tools are available, the Chlorophyta *Chlamydomonas reinhardtii* was chosen as a model among photosynthetic organisms and the sequencing of its entire nuclear genome completed in 2007 (Merchant *et al.*, 2007). Comparative phylogenomic analyses have provided insight into the evolution of plants and animals, allowing genes to be associated with photosynthesis and flagellar functions, and links established between ciliopathy and the composition and function of flagellae (Umen & Olson 2012 in this volume). Over the past decade, many post-genomics and genetic tools have been used on this species, including microarrays, antibodies, RNA interference (RNAi) and genetic transformation. These approaches have enabled the exploration of metabolic pathways and biological processes such as responses to stress, the circadian clock (Matsuo & Ishiura, 2011), photosynthetic electron transport chains (Hermsmeier, Schulz, & Senger, 1994), mechanisms of carbon concentration (Yamano & Fukuzawa, 2009) and flagellar assembly (Iomini, Till, & Dutcher, 2009). In addition, proteomic studies have provided major research contributions in the areas of photosynthesis, molecular biology and evolution (Muhlhaus, Weiss, Hemme, Sommer, & Schroda, 2011; Rolland *et al.*, 2009). The other alga species sequenced were chosen due to their ecological role, phylogenetic distribution or harmful nature. Sequencing provided extensive information on the evolution of these species, helped to identify metabolic pathways and specific genes and clarified processes involved in the cycles of iron, calcium, silica, urea and nitrogen. In addition, sequence data provide essential references for matching with post-genomic investigations, including transcriptomic and proteomic analyses.

The gene content of microalgae is only beginning to be explored. Microalgal genomes can be structurally complex and sizes range from 12.6 Mbp for the Chlorophyta *Ostreococcus tauri* and 168 Mbp for the Haptophyta *Emiliania huxleyi* to an estimated 10,000 Mbp for the Dinophyta *Karenia brevis* (see Chapter XI for a discussion of genome size variations in algae). These large genome sizes can preclude full-genome sequencing, thus enforcing the use of transcriptome sequencing to build gene catalogues. Many authors have made this choice, although aware of the risk of neglecting non-transcribed sequences. Among the species studied in this way, we can mention the Ochrophyta *Pseudochattonella farcimen*, which is associated with fish mortalities (Dittami *et al.*, 2011), green microalgae

Chlorella vulgaris UTEX 395 (Guarnieri *et al.*, 2011), *D. salina* (Zhao *et al.*, 2011) and *Dunaliella tertiolecta* (Rismani-Yazdi, Haznedaroglu, Bibby, & Peccia, 2011) and the coccolithophore *E. huxleyi* (Von Dassow *et al.*, 2009). Transcriptomic data have been used for phylogenomics and opened the way for functional post-genomics approaches to the study of physiology, environmental adaptation, life cycles, metabolism and signal transduction pathways. Several major projects for transcriptome sequencing are currently underway (Table 8.1). One example is the 'Marine Microbial Eukaryotic Transcriptome Project', which aims to sequence the transcriptomes of approximately 750 samples expected to represent hundreds of species and strains with key ecological roles and evolutionary importance in the tree of microeukaryotes (http://marinemicroeukaryotes.org/). To date, 39 microbial algal transcriptomes have been sequenced (Table 8.1). In order to establish a reference database from ecologically and phylogenetically relevant photosynthetic protists for the 'Tara Oceans expedition', the 'Prometheus project' is proposing to sequence about 30 species of ecological or phylogenetic importance (http://oceans.taraexpeditions.org) (Karsenti *et al.*, 2011).

We can therefore hope, in a few months or years, to have a very large number of new transcriptomic and genomic data for algae. The development of genomics has already made a major contribution to fundamental research on photosynthetic eukaryotes in the fields of functional biology, global ecology and the evolution of organisms. These data will accelerate the commercialization of alga-derived compounds by providing a framework for hypothesis-based strain improvement programs built on an improved fundamental understanding of the specific pathways and regulation of networks. These studies are also the source of new biotechnologies that will be presented in the following sections.

2. BIOTECHNOLOGY AND MICROALGAE

For 2011, a search using the two keywords 'microalgae' and 'biotechnology' returned 51 publications in Web of Science database. More than a third of these were on energy and biofuels. In second position, around 20% of the papers deal with different cultivation and extraction techniques. Cell factories, i.e. the production of recombinant proteins, came in third position, with a number of technical advances in *Chlamydomonas* sp.

Table 8.1 Ongoing Microalgae Transcriptomic Projects

Phylum	Species	Strain	Status
Bacillariophyta	*Asterionellopsis glacialis*	1712	Assembly & annotation
Bacillariophyta	*Chaetoceros sp.*		Assembly & annotation
Bacillariophyta	*Corethron hystrix*	308	Assembly & annotation
Bacillariophyta	*Cylindrotheca closterium*		Assembly & annotation
Bacillariophyta	*Grammatophora oceanica*	410	Assembly & annotation
Bacillariophyta	*Melosira sp.*	CCMP 2643	Sequencing
Bacillariophyta	*Navicula transitans*	80	Assembly & annotation
Bacillariophyta	*Odontella sp.*		Assembly & annotation
Bacillariophyta	*Odontella sinensis*	Grunow 1884	Sequencing
Bacillariophyta	*Skeletonema costatum*	1716	Sequencing
Bacillariophyta	*Stephanopyxis turris*	CCMP 815	Sequencing
Chlorarachniophyta	*Lotharella oceanica*	CCMP622	Assembly & annotation
Chlorarachniophyta	*Lotharella globosa*	LEX01	Assembly & annotation
Chlorarachniophyta	*Lotharella amoebiformis*	CCMP2058	Assembly & annotation
Chlorarachniophyta	*Bigelowiella natans*	CCMP 2755	Assembly & annotation
Chlorophyta	*Crustomastix stigmata*	CCMP3273	Sequencing
Chlorophyta	*Dolichomastix tenuilepis*	CCMP3274	Sequencing
Chlorophyta	*Micromonas sp.*	CCMP2099	Assembly & annotation
Chlorophyta	*Nephroselmis pyriformis*	CCMP717	Assembly & annotation
Chlorophyta	*Pyramimonas parkeae*	CCMP725	Assembly & annotation
Chlorophyta	*Tetraselmis sp.*	GSL018	Sequencing
Cryptophyta	*Chroomonas mesostigmatica*	CCMP1168	Assembly & annotation
Cryptophyta	*Cryptomonas paramecium*	CCAP977/2a	Assembly & annotation
Cryptophyta	*Goniomonas pacifica*	CCMP1869	Sequencing
Cryptophyta	*Guillardia theta*	CCMP2712	Assembly & annotation

(*Continued*)

Table 8.1 Ongoing Microalgae Transcriptomic Projects—cont'd

Phylum	Species	Strain	Status
Cryptophyta	*Hemiselmis andersenii*	CCMP644	Sequencing
Dinophyta	*Alexandrium minutum*	CCMP113	Sequencing
Dinophyta	*Crypthecodinium cohnii*	Seligo	Sequencing
Dinophyta	*Karenia brevis*	SP3	Sequencing
Dinophyta	*Oxyrrhis marina*	CCMP788	Assembly & annotation
Dinophyta	*Oxyrrhis marina*	CCMP1795	Sequencing
Dinophyta	*Symbiodinium kawagutii*	CCMP2468	Sequencing
Euglenophyta	*Eutreptiella gymnastica*	NIES-381	Assembly & annotation
Haptophyta	*Hyalolithus neolepis*	TMR5	Sequencing
Ochrophyta	*Dinobryon sp.*	UTEXLB2267	Assembly & annotation
Ochrophyta	*Ochromonas sp.*	CCMP 1393	Assembly & annotation
Rhodophyta	*Rhodosorus marinus*	769	Assembly & annotation

Source: From http://marinemicroeukaryotes.org/project_organisms.

2.1. Microalgal Lipids as Biofuel and Food

2.1.1. *Algal Lipid Synthesis: The Contribution of Genomic Data*

Compared with land plants, the lipid composition of algae shows great specificity, such as the presence of long-chain PUFAs or the species-specific absence of phosphatidylcholine and phosphatidylserine in the membranes, replaced by diacylglyceryltrimethylhomoserine (Guschina & Harwood, 2006). In addition, for many algal species, high-energy reserves of tri-acylglycerol (TAG) accumulate in large amounts in lipid droplets in response to different types of stress or nutrient deficiency. TAG represents >50% of the algal dry weight and serve for membrane synthesis or carbon storage (Hu *et al.*, 2008), making it possible to obtain oil yields 10 times higher per hectare than with land plant species. Recent soaring oil prices, diminishing world reserves and the environmental damage associated with fossil fuel consumption have led to increased interest in using algae as an alternative and renewable feedstock for fuel production. The development of the microalgal biodiesel industry depends primarily on the reduction of production costs and one strategy to achieve this is to increase lipid productivity. This explains the large investment being placed in such technology and demonstrates why most genomics work on algae is aimed at describing and orienting their lipid metabolism (Norsker, Barbosa, Vermue, & Wijffels, 2011).

Many studies have been conducted on land plants to understand their mechanisms of lipid synthesis and the development of reserves in their seeds. It was reported that environmental conditions (nutrients, salinity, light, etc.) affect microalgal fatty acid accumulation (for a review, see Hu *et al.*, 2008). However, molecular mechanisms that trigger and control the accumulation of storage lipids in microalgae are poorly understood. Genomic data have allowed the identification of new enzymes and helped to show how lipid pathways interrelate with energy and carbohydrate metabolism (Wallis & Browse, 2010). Until recently, the molecular mechanisms involved in regulatory pathways in algae were still poorly understood. With genomic data and genetic tools available for the green microalga *C. reinhardtii*, lipid metabolism has been mainly studied in this species and overviews of these findings can be found in several papers (Guschina & Harwood, 2006; Khozin-Goldberg & Cohen, 2011; Moellering & Benning, 2010). Many genes of *C. reinhardtii* involved in fatty acids and TAG metabolism have been identified based on their orthological relationships to fungi and land plants. In green microalgae, starch synthesis shares common carbon precursors with lipid synthesis. In *C. reinhardtii*, it has been shown that shunting of carbon precursors from the starch synthesis pathway may facilitate carbon partitioning into the fatty acid synthesis pathway resulting in enhanced production of TAG (Li, Han, Hu, Dauvillee *et al.*, 2010). Identification of genes and biosynthetic pathways implicated in lipid biosynthesis is usually made using starchless mutants. With regard to the metabolism of TAGs, genomic data have shown conservation of the main biosynthetic pathways between microalgae and seed plants. Briefly, fatty acids are synthesized in the chloroplasts in which acetyl-CoA carboxylase (ACCase) provides the malonyl-CoA substrate for the biosynthesis of fatty acids thanks to the fatty acid synthase, a multifunctional enzymatic complex (Guschina & Harwood, 2006). Free fatty acids are then either used for the synthesis of membrane lipids or exported to the endoplasmic reticulum for the biosynthesis of TAGs. This synthesis involves the sequential transfer of acyl groups from acyl-CoA to different positions of glycerol-3-phosphate. Most acyltransferases and a phosphatases involved have been identified in the genome of *C. reinhardtii* (Merchant, Kropat, Liu, Shaw, & Warakanont, 2011). Nevertheless, significant differences from land plants were observed in the TAG pathways of *C. reinhardtii*, such as the absence of the extra-plastidic lysophosphatidyl acyltransferase in the genome and the presence of new enzymes that are, as yet, poorly characterized (Hu *et al.*, 2008). Most recently, an alternative chloroplast pathway of TAG synthesis was identified in *C. reinhardtii* (Fan, Andre, & Xu, 2011). TAG accumulates in lipid droplets in which proteomics techniques revealed the importance of a major lipid droplet protein (MLDP). Miller *et al.* (2010) used 454 and Illumina technologies for transcriptomic analysis and showed how nitrogen deprivation redirects lipid metabolism. In brief, genomic and post-genomic data have allowed lipid metabolism pathways and regulation to be characterized in the Chlorophyta *C. reinhardtii*. However, this alga is not an oleaginous species. With the great diversity that exists among algae, specific studies are now being conducted on lipid-accumulating species in numerous laboratories around the world.

2.1.2. Algal Lipids as Biofuel

Very recently, several studies have used post-genomics to study the lipid metabolism of high-oil-content algae. This illustrates a real drive in the exploration of the functional metabolism of oleaginous algae. In 2011, Rismani-Yazdi *et al.* (2011) published the NGS and transcriptome annotation of a non-model member of the Chlorophyta: *D. tertiolecta.*

Genes-encoding key enzymes were identified by homology and metabolic pathways involved in the biosynthesis and catabolism of fatty acids, TAG and starch were reconstructed (Rismani-Yazdi *et al.*, 2011). A few months later, similar work was reported in a strain of the oil-producing green alga *Botryococcus braunii* (Baba, Ioki, Nakajima, Shiraiwa, & Watanabe, 2011). In parallel, proteomic approaches have identified new proteins involved in the storage of TAG in the lipid droplets of the Chlorophyta *H. pluvialis* (Peled *et al.*, 2011). Guarnieri *et al.* (2011) reported a comprehensive proteomic and transcriptomic investigation of lipid accumulation in the unsequenced green alga *C. vulgaris* UTEX 395. The authors presented the first utilization of a *de novo* assembled transcriptome as a search model for proteomic analysis. The regulation of fatty acid and TAG biosynthetic pathways was analyzed under nitrogen limitation. This oleaginous species is extensively studied due to its relatively fast growth rate, its value as both a food supplement and a potential biofuel feedstock and its ability to produce high-economic value molecules and to remediate heavy metals from wastewater. For these reasons, the genome of the Chlorophyta *Chlorella variabilis* NC64A was previously sequenced by Blanc *et al.* (2010). However, difficulties were encountered in identifying proteins by comparing data with strains of species from the same phylum. The researchers pointed out the importance of having unique sequence data to study species and strains of interest (Guarnieri *et al.*, 2011).

Although lipid biosynthesis pathways have been studied in several species, very few studies focus on the regulation of these pathways. Given the induction of TAG biosynthesis by different stresses, it is likely that the mechanisms for the regulation of TAG synthesis differ between algae and seed plants, as the latter produce oil during a specific phase of their life cycle and in specialized tissues. The means of regulation are presently of great interest, as these are the key to engineering algal crop production without causing weakening through nutrient stress. Although transcriptomics offer a wealth of information on gene expression, the processes of messanging RNA (mRNA) splicing, ribosome recruitment and post-translational regulations of proteins are not well understood in algae and transcriptomic analysis does not adequately define the control points for metabolic regulation.

By providing insight into the mechanisms underpinning lipid metabolic processes, results can be of use for the genetic manipulation of organisms to enhance the production of feedstock for commercial microalgal biofuels. By 1996, Dunahay and co-authors were able to overexpress ACCase in the diatom *Cyclotella cryptica*, which is a key enzyme in the biosynthesis of fatty acids (Dunahay, Jarvis, Dais, & Roessler, 1996). However, no increase in the amount of lipid was observed. In expressing recombinant thioesterases to enhance the expression of shorter chain length fatty acids, Radakovits, Jinkerson, Darzins, & Posewitz (2010) were able to improve the level of lauric and myristic acids in the diatom *Phaeodactylum tricornutum*. This creates an advantage for biofuel feedstock because biodiesel made from saturated short or medium chain length fatty acids has a relatively low cloud point and is resistant to oxidation. In addition, several studies have shown metabolic shifts in starchless mutants of *C. reinhardtii* in favour of an overexpression of TAG (Moellering, Miller, & Benning, 2009; Li, Han, Hu, Sommerfeld, & Hu, 2010; Wang, Ullrich, Joo, Waffenschmidt, & Goodenough, 2009). In a starchless mutant, Moellering *et al.* (2010) inhibited the expression of MLDPs by RNAi, which not only increased the size of the lipid globules but also resulted in decreased growth. In contrast, the fatty acid content of a starchless selected mutant of *Chlorella pyrenoidosa* was doubled without detriment to its growth characteristics (Ramazanov & Ramazanov, 2006). This suggests that it is possible to improve the productivity of microalgae using lipid selection strategies. To date, the genomic data available on the selected species are still patchy, and reverse genetic tools are completely absent in these species. We also lack genetic information on the molecular mechanisms leading to these beneficial mutations. The exponential increase of genomic and post-genomic technology should enable biologists to acquire data, and reverse genetic tools should improve our understanding of the metabolism of these lipids and demonstrate ways in which these processes can be improved.

Recently, we put one of the first varietal selection strategies into action in our laboratory. We used successive rounds of UV mutation and cell sorting to improve the TAG production of the Haptophyta *I. galbana* affinis Tahiti (a strain related to the *I. galbana* strain), a species that offers numerous advantages for lipid production. This approach, which does not create genetically modified organisms (GMOs), allowed us to obtain a strain that accumulates twice the amount of neutral lipids as the original without affecting the growth rate (Rouxel, Bougaran, Doulin-Grouas, Dubois, & Cadoret, 2011). This strategy quickly improved the performance of an unsequenced selected species, so similar strategies will now be tried on other species and other valuable molecules. From now on, the acquisition of transcriptomic and proteomic data will be used to identify genes and molecular processes involved in the increase of lipid accumulation.

2.1.3. *Algal Lipids as Feed and Food*

Apart from the high importance of TAG from algae, the identification of enzymes involved in the synthesis of PUFAs, such as the long-chain PUFAs AA, EPA and DHA, is of great interest due to the health benefits they offer. Production of PUFAs involves a consecutive series of desaturations and elongations of the fatty acyl chain. Until recently, numerous authors isolated and characterized lipid metabolism and enzymes using biochemical technologies. These studies are reported in a review by Guschina and Harwood (2006). Over the past few years, authors have used genomic data to understand the biosynthetic pathways of PUFAs. Because of the putative role of PUFAs in the virulence of the fish pathogen *P. farcimen*, Dittami and co-authors analyzed the expressed sequence tags (ESTs) of this species. Focusing their attention on PUFA metabolic pathways, they identified new specific desaturases related to this virulence (Dittami *et al.*, 2011). In the same way, the ESTs of *Myrmecia incisa*, a green coccoid freshwater microalga rich in AA, were analyzed and a putative new elongase was identified (Yu, Liu, Li, & Zhou, 2011). Pan *et al.* (2011) sequenced the genome of the high PUFA content species of Heterokonta *Nannochloropsis oceanica* using next-generation Illumina sequencing technologies. Sequence similarity-based investigation identified new elongase- and desaturase-encoding genes involved in the biosynthesis of long-chain PUFAs, which provide the genetic basis of its rich EPA content.

To date, major lipid primary metabolism has been well studied in model species, but regulation pathways, catabolism and secondary metabolic pathways of lipids are complex and rarely studied. Many metabolites of lipids have high biotechnological potential. The control of lipid metabolism, which is highly regulated, is of great interest as a means of increasing the lipid yields in culture. Furthermore, strategies using random mutations and strain selection have succeeded in increasing the lipid content of selected strains, but without a clear understanding of the mechanisms involved. This demonstrates that there are still many gaps in the knowledge that would help us to optimize lipid production from algae. Genomic and post-genomic studies on a variety of microalgae will provide the basis for identifying metabolic and signalling pathways.

2.2. Bioactive Natural Products

Commercial applications of microalgae include their use as natural sources of valuable macromolecules, such as carotenoids and phycocolloids. Due to the exceptionally high diversity of the different groups and the low level of exploration carried out so far, algae are a burgeoning reservoir of high added value compounds. During the last decade, full genome analysis unveiled numerous new natural products in bacteria and fungi. Indeed, it appears that many of their genomes contain more gene clusters coding for the biosynthesis of natural products than natural products isolated from these same species (Winter, Behnken, & Hertweck, 2011). Similar results have been observed in microalgae. For example, *in silico* analysis of the Heterokonta *Aureococcus anophagefferens* genome revealed the presence of five berberine bridge enzymes involved in the synthesis of toxic isoquinoline alkaloids, although this type of alkaloid had never been previously identified in this harmful species (Gobler *et al.*, 2011). Genomic exploration of microalgae appears to be a promising way to discover new bioactive products. To date, the analysis of available genomes has aided the identification of pathways to known compounds, thereby greatly facilitating regulatory and functional investigations. The search for enzymes involved in the biosynthesis of polyketides, isoprenoids, non-ribosomal peptides, oxylipins and alkaloids was conducted *in silico* by looking for homologous genes of land plants in sequenced genomes of microalgae (for review, see Sasso, Pohnert, Lohr, Mittag, & Hertweck, 2011). Although some pathways have been elucidated, there are still many gaps in our knowledge of the metabolism of the secondary metabolites. For example, isoprenoids comprise numerous bioactive molecules such as sterols, phytohormones, phytol, prenylated quinones and carotenoids, which have numerous qualities of interest for biotechnology. While the common first steps of the synthesis of isoprenoid compounds have been well described (Lohr, Schwender, & Polle, 2012), very little is known about the biosynthesis of secondary isoprenoids except for the carotenoids. The genetic basis of the biosynthetic pathways of sterols and carotenoids in algae has been examined in detail by phylogenomics across several phyla of algae in order to gain insight into the evolution and diversity of photosynthetic eukaryotes (Cui, Wang, & Qin, 2011; Desmond & Gribaldo, 2009; Frommolt *et al.*, 2008) (see Archibald, 2012; De Clerck, Bogaret, & Leliaert, 2012; Not *et al.*, 2012, in this volume). This has led to the identification of genes in organisms where pathways had not been identified before and demonstrated the steps by which more new enzymes could be discovered. The induction and regulation of astaxanthin and carotenoid biosynthesis in Chlorophyta such as *Sphaerella lacustris* or *D. salina* has received considerable attention owing to the increasing use of secondary carotenoids as a source of pigmentation for fish in aquaculture and their potential as free-radical quenching drugs in cancer prevention. In aiming to identify the proteins involved in the regulation and biosynthesis of astaxanthin, comparative proteomics and transcriptomics were applied to the chlorophytes *H. pluvialis* and *Haematococcus lacustris* (re-named *S. lacustris*) under nitrogen starvation and irradiance stress (Eom, Lee, & Jin, 2005; Kim *et al.*, 2006; Tran, Park, Hong, & Lee, 2009) and the regulated genes identified. These genes putatively play a role in signal transduction from stress to the cellular defence system and activate the biosynthesis of astaxanthin. Complementary in-depth analysis should confirm the significance of these results. These genes include potential targets to increase the expression of astaxanthin.

Overall, it is clear that our understanding of secondary metabolism and its regulation is still rudimentary. Secondary metabolites include a large number of natural bioactive products, many of which are unknown. *In silico* genome analyses are a key to the identification of new metabolic and signalling pathways. Post-genomics can be applied to identify physiological conditions that lead the expression of new pathways and so identify hitherto undetected metabolites.

2.3. Molecular Farming

The extraction of natural substances remains the main source of supply for a large number of pharmaceutical molecules. However, since it is possible to identify the genes responsible for building a protein molecule, they can be introduced into cultured cells, which then become cell factories, making millions of copies of the desired product. This strategy—the expression of molecules with high added value in recombinant cell systems—offers extraordinary opportunities for the development of a very promising biotechnology market (estimated to be worth up to several tens of billions of dollars, depending on the information source) (Gasdaska, Spencer, & Dickey, 2003; Schmidt, 2004). The production systems available are bacteria, yeasts and animal or plant cells, which are genetically modified to produce insulin, growth hormones, monoclonal antibodies and other therapeutic proteins. Each system has advantages and disadvantages relating to factors such as cost, production safety, ease of extraction, purification and complexity of producing the molecules. Some solutions, however, combine a number of benefits, putting them in a strong position for the future of this industry.

Microalgae have several advantages over other expression systems for the production of recombinant proteins, such as (1) a high growth rate (they commonly double their biomass within 24 h), (2) easy cultivation at a low production cost (they only require water and nutrients), (3) the possibility of performing post-transcriptional and post-translational modification as in other eukaryotic expression systems and (4) photobioreactor culture methodologies that prevent transgenes from escaping into the environment, which is a potential risk when using land plants (Janssen, Tramper, Mur, & Wijffels, 2003).

Several interesting reviews on transgenic tools describe the use of microalgae as a platform for production of recombinant proteins (Bozarth, Maier, & Zauner, 2009; Hallmann, Amon, Godl, Heitzer, & Sumper, 2007; Potvin & Zhang, 2010; Walker, Collet, & Purton, 2005). Here, we focus on recent progress and results on transgenic microalgae technology for the production of therapeutic recombinant proteins and discuss the contribution of genomic studies for the optimization of genetic manipulation in microalgae.

2.3.1. Transgenic Microalgae as a Platform for Biopharmaceutical Proteins

In this section, we provide a review of biopharmaceutical proteins expressed in microalgae systems according to their intracellular cell localization (chloroplastic or nuclear). The interest in the N-glycosylation of pharmaceutical proteins will also be discussed.

Although no recombinant protein produced by transgenic algae is yet available on the market, some therapeutic proteins have been successfully produced using microalgae, mainly the Chlorophyta *C. reinhardtii* for which suitable transgenic tools and genomic data are available (for all three genomes: nuclear, chloroplastic and mitochondrial). Mayfield's group has done considerable work on the chloroplastic expression of recombinant protein in *C. reinhardtii* (Rasala & Mayfield, 2011). Indeed, the majority of microalgal therapeutic proteins have been produced by chloroplasts (Table 8.2). Recombinant protein can accumulate to much higher levels in the transgenic chloroplast than when expressed by the nuclear genome because plastids lack disadvantages such as gene-silencing mechanisms (Bock, 2007). Indeed, expression of foreign proteins remains very low for reasons that are not yet fully understood (Potvin & Zhang, 2010). The chloroplast of *C. reinhardtii* has been used to produce a range of recombinant proteins, including reporters such as glucuronidase, luciferase, green fluorescent

Table 8.2 Biopharmaceutical Proteins Expressed In Microalgae

Gene Expressed	Function	Host Species and Cell Localization	Expression Level Achieved	Application	Source
HSV8-lsc	Mammalian antibody	*Chlamydomonas reinhardtii*, Chloroplast	Detectable	Pharmaceutical	Mayfield *et al.* (2003)
CTB-VP1	Cholera toxin B subunit fused to foot and mouth disease VP1	*Chlamydomonas reinhardtii*, Chloroplast	3% TSP	Vaccine	Sun *et al.* (2003)
HSV8-scFv	Classic single-chain antibody	*Chlamydomonas reinhardtii*, Chloroplast	0.5% TSP	Pharmaceutical	Mayfield *et al.* (2005)
hMT-2	Human metallothionine-2	*Chlamydomonas reinhardtii*, Chloroplast	Detectable	Pharmaceutical, UV-protection	Zhang *et al.* (2006)
hTRAIL	Human tumor necrosis factor-related apoptosis-inducing ligand (TRAIL)	*Chlamydomonas reinhardtii*, Chloroplast	~0.67% TSP	Pharmaceutical	Yang *et al.* (2006)
M-SAA	Bovine mammary-associated serum amyloid	*Chlamydomonas reinhardtii*, Chloroplast	~5% TSP	Therapeutics, oral delivery	Manuell *et al.* (2007)
CSFV-E2	Swine fever virus E2 viral protein	*Chlamydomonas reinhardtii*, Chloroplast	~2% TSP	Vaccine	He *et al.* (2007)
hGAD65	Diabetes-associated anutoantigen human glutamic acid decarboxylase 65	*Chlamydomonas reinhardtii*, Chloroplast	~0.3% TSP	Diagnostics and therapeutics	Wang *et al.* (2008)

(*Continued*)

Table 8.2 Biopharmaceutical Proteins Expressed In Microalgae—cont'd

Gene Expressed	Function	Host Species and Cell Localization	Expression Level Achieved	Application	Source
83K7C	Full-length IgG1 human monoclonal antibody against anthrax protective antigen 83	*Chlamydomonas reinhardtii*, Chloroplast	0.01% dry algal biomass	Therapeutics	Tran *et al.* (2009)
IgG1	Murine and human antibodies (LC and HC)	*Chlamydomonas reinhardtii*, Chloroplast	Detectable	Therapeutics	Tran *et al.* (2009)
VP28	White spot syndrome virus protein 28	*Chlamydomonas reinhardtii*, Chloroplast	~10.5% TSP	Vaccine	Surzycki *et al.* (2009)
CTB-D2	D2 fibronectin-binding domain of Staphylococcus aureus fused with the cholera toxin B subunit	*Chlamydomonas reinhardtii*, Chloroplast	0.7% TSP	Oral vaccine	Dreesen *et al.* (2010)
10NF3, 14FN3	Domains 10 and 14 of human fibronectin	*Chlamydomonas reinhardtii*, Chloroplast	14FN3: 3% TSP 10FN3: detectable	Therapeutics	Rasala *et al.* (2010)
M-SAA-Interferon β1	Multiple sclerosis treatment fused to M-SAA	*Chlamydomonas reinhardtii*, Chloroplast	Detectable	Therapeutics	Rasala *et al.* (2010)
Proinsulin	Blood sugar level-regulating hormone, type I diabetes treatment	*Chlamydomonas reinhardtii*, Chloroplast	Detectable	Therapeutics	Rasala *et al.* (2010)
VEGF			2% TSP	Therapeutics	Rasala *et al.* (2010)

	Human vascular endothelial growth factor isoform 121	*Chlamydomonas reinhardtii*, Chloroplast			
HMGB1	High mobility group protein B1	*Chlamydomonas reinhardtii*, Chloroplast	2.5% TSP	Therapeutics	Rasala *et al.* (2010)
NP-1	Rabbit neutrophil peptide-1	*Chlorella ellipsoidea*, nuclear	Detectable	Antimicrobial	Chen *et al.* (2001)
ARS2-crEpo-his6	Human erythropoietin fused to ARS2 export sequence w/6xhis tag	*Chlamydomonas reinhardtii*, Nuclear	100 μg/L culture	Pharmaceutical, protein export	Eichler-Stahlberg *et al.* (2009)
CL4mAb and HBsAg	Human antibody CL4mAB and the Hepatitis B surface antigen (HBsAg)	*Phaeodactylun tricornutum*, Nuclear	CL4mAb: 8.7% TSPHBsAg: 0.7% TSP	Vaccine	Hempel, Lau *et al.* (2011)
mEPO	Murine Erythropoietin	*Phaeodactylun tricornutum*, Nuclear	300 μg/L culture	Therapeutics	Carlier A (unpublished work)

TSP: Total Soluble Proteins.
Source: Modified from Specht *et al.* (2010). Recent successes in therapeutic protein production in algae.

protein (GFP), industrial enzymes, vaccines and therapeutic enzymes (Rasala & Mayfield, 2011).

The first therapeutic protein expressed by transgenic microalgae was produced at Mayfield's laboratory using chloroplast transformation in the green microalga *C. reinhardtii*. In the study of Mayfield, Franklin, & Lerner (2003), the entire imunnoglobulin A heavy chain protein (HSV8-lcs) fused to the variable region of the light chain was expressed and accumulated as a soluble protein able to bind to the herpes virus protein. Nevertheless, the expression yield was too low (detectable only) for commercial use, even though several regulation sequences (promoters) were tested. Regulation sequence aspects will be examined in the next section. This previous study was completed by the chloroplastic expression of a single chain fragment variable antibody (HSV8-scFv) that accumulated to 0.25% of total soluble protein (TSP) (Mayfield & Franklin, 2005). In their next study, the same team successfully increased the accumulation of a bioactive mammalian protein, bovine mammary-associated serum amyloid A (M-SAA), to 5% of TSP with by chloroplasts using different promoter sequences and an interesting strategy consisting of replacing an endogenous gene by the expression cassette (Manuell *et al.*, 2007). Recently, a full-length human monoclonal antibody was expressed in the chloroplast of *C. reinhardtii*, proving that this eukaryotic green alga is capable of synthesising and assembling a full-length antibody in transgenic chloroplasts (Tran, Zhou, Pettersson, Gonzalez, & Mayfield, 2009). More recently, a study was conducted to examine the versatility of algal chloroplasts for the expression of seven different therapeutic proteins: human erythropoietin (EPO), the 10th and 14th human fibronectin type III domains (14FN3 and 10FN3), human interferon β1, the human vascular endothelial growth factor isoform, the high mobility group protein (HMGB1) and the human proinsulin. Of the seven proteins tested, four were successfully expressed in transgenic chloroplasts to above 2% of TSP (Rasala *et al.*, 2010). However, no detectable expression was shown for EPO or interferon β1. Like Mayfield's group, other research groups have successfully shown that the chloroplast of *C. reinhardtii* is a perfect platform to produce recombinant proteins at an economically viable cost (Wang *et al.*, 2008; Yang *et al.*, 2006; Zhang, Shen, & Ru, 2006). In addition to therapeutic proteins, some vaccines have been successfully produced in algal chloroplasts. Indeed, a fusion protein between the foot and mouth disease virus VP1 and the cholera toxin B subunit (as mucosal adjuvant) was reported to accumulate to 3% of TSP in transgenic algal chloroplasts (Sun *et al.*, 2003). This fusion protein retained both specific ganglioside-binding affinity and antigenic function. A classical swine fever virus E2 recombinant protein was also successfully expressed in chloroplast to around 2% of TSP and observed to have immunological activity (He *et al.*, 2007). Surzycki *et al.* (2009) reported a strong expression of the white spot syndrome VP28 protein by chloroplasts to around 10.5% of TSP. Moreover, in this study, the authors attempted to determine factors affecting the level of recombinant protein expression, which will be covered in the next section. Recently, Dreesen, Charpin-El Hamri, & Fussenegger (2010) reported the oral immunization of mice by transgenic algae expressing (to 0.7% of TSP) the *Staphylococcus aureus* fibronectin-binding domain D2 fused to the cholera toxin B subunit.

It is important to reiterate that all these studies were carried out using transgenic chloroplasts of the green algae *C. reinhardtii*. To our knowledge, there are no reports of biopharmaceutical protein expression by transgenic chloroplasts in other microalgae.

Although it is estimated that most of the therapeutic human antibodies used in therapy do not require glycosylation, other therapeutic proteins require the correct glycosylation pattern to function properly (Dove, 2002). Nevertheless, nuclear expression of therapeutic proteins remains limited because of some problems in reducing yield expression (Potvin & Zhang, 2010). Transgenic microalgal technologies are still in their infancy and the therapeutic proteins expressed by the nuclear genome are still rare in microalgae.

Initial work has been done by Hawkins and Nakamura (1999) to produce human growth hormone in the extracellular medium of *Chlorella sorokiniana* and *C. vulgaris* C-27. In a subsequent study, growth hormone of sole was produced and expressed as a stable product in *C. ellipsoidea* (since renamed *Chloroidium ellipsoideum*). Soles fed with these transgenic microalgae increased in size by 25% (Kim *et al.*, 2002). Another research team has shown the efficient expression and biological activity of rabbit neutrophile peptide-1 in *C. ellipsoideum* cells (Chen, Wang, Sun, Zhang, & Li, 2001).

Recently, Dauvillée *et al.* (2010) expressed a nuclear protein corresponding to the *Plasmodium* antigens that fuse to granule-bound starch synthase (GBSS), a protein involved in the starch matrix of plants and algae. The C-terminal domains from apical major antigen (AMA1) or major surface protein (MSP1) fused to GBSS were both efficiently expressed in nuclear cells and targeted starch particles in the chloroplasts, taking advantage of the transit peptide on the GBSS protein. Although expressed in the nucleus, these fusion proteins directly targeted starch granules, avoiding post-translation modification such as N-glycosylation. Immunogenicity tests for both fusion proteins were successfully performed in mice (Dauvillee *et al.*, 2010).

More recently, diatoms have also been used as cell factories to produce recombinant proteins. Diatoms are an algal group of great ecological importance (Mock & Medlin 2012). Their contribution to global CO_2 fixation represents around 40% of marine carbon production. Diatoms like *P. tricornutum* represent an interesting subject for a variety of biotechnological applications, and this species has become a model organism for the diatoms (Bowler *et al.*, 2008; Hempel, Bozarth *et al.*, 2011, Hempel, Lau, Klingl, & Maier, 2011; Siaut *et al.*, 2007). Indeed, its whole genome has been sequenced and molecular tools for functional genomics are available (Maheswari, Mock, Armbrust, & Bowler, 2009; Siaut *et al.*, 2011). To date, diatoms have not been employed for expression of any biopharmaceutical proteins, but a research team has recently reported the first stable expression of a full-length human antibody and the respective antigen in *P. tricornutum* (Hempel, Lau *et al.*, 2011). In this study, the antibody and respective antigen were both expressed and accumulated within the endoplasmic reticulum (ER) using the ER retention signal. Interestingly, while the same expression vector and molecular tools were used for the expression of both these recombinant proteins, different expression levels were observed for the antibody (7.8% of TSP) and antigen (0.7% of TSP). This result confirms that not all foreign proteins are equally expressed (Potvin & Zhang, 2010).

At our laboratory, we became interested by the potential of microalgae as a means to produce therapeutic proteins (Cadoret *et al.*, 2008). This interest led to the creation of a private company by our laboratory: Algenics. Algenics is the first privately owned European biotechnology company focusing on innovative uses of microalgae to produce recombinant biotherapeutics. Using microalgae as a platform for recombinant proteins, our laboratory filed a patent on the production of glycosylated proteins in microalgae (Cadoret *et al.*, 2009). Recently, as proof of the concept, we successfully produced another therapeutic protein, murine erythropoietin (mEPO), in the diatom *P. tricornutum* (A. Carlier, M. Bardor, P. Lerouge, P. Delavault, B. Saint-Jean, A. Gerard and J. P. Cadoret, unpublished work). The data show that recombinant mEPO accumulates to around of 0.05% of TSP (or 300 μg/L). This recombinant EPO is glycosylated and able to bind the human EPO receptor *in vitro* with the same affinity. These results, combined with Hempel's data, confirm the high potential of diatoms to express biopharmaceutical proteins.

This last result corroborates the expression specificity of some foreign proteins according to cell localization and/or algal taxon. Indeed, no detection of recombinant EPO has been reported in Chlorophyta *C. reinhardtii* transgenic chloroplasts (Rasala *et al.*, 2010). In contrast, Eichler-Stahlberg, Weisheit, Ruecker, & Heitzer (2009) observed a minor accumulation of recombinant EPO up to around 100 μg/L in nuclear expression by *C. reinhardtii* cells. Thus, EPO protein accumulates differently and at different expression levels according to cell localization or species.

To conclude, many efforts have been made to produce biopharmaceutical proteins at a level sufficient to be economically viable, but extensive research to optimize microalgae as cell factories still needs to be done. Recent success in microalgal transgenesis and input from genomic data will allow a response to the growing demand for biopharmaceutical molecules. However, microalgae can also provide compounds other than pharmaceutical proteins. Indeed, an interesting study has recently been reported that used microalgae to produce industrial products such as bioplastic: Hempel and co-workers (2011) expressed three prokaryotic enzyme genes in the diatom *P. tricornutum* to produce poly-3-hydroxybutyrate (PHB). These genes (i.e. a ketolase, an acetoacetyl-CoA reductase and a PHB synthase) are able to synthesize PHB from acetyl-CoA in diatom cells up to a level of 10.6% of algal dry weight.

Of the post-translational modifications encountered in eukaryotic proteins, N-glycosylation is the most prevalent of those that appear essential for biological functions (biological activity, short half-life). Moreover, glycosylation is of particular interest for biopharmaceutical proteins since more than 70% of biopharmaceuticals are glycoproteins. Glycosylation capability is an advantage for any system used to produce biopharmaceuticals. This pathway is currently well understood among the different production systems available today, such as cultured mammalian, yeast and plant cells. Plants have N-glycosylation capability similar to mammalian cells. However, N-glycosylation patterns processed in plant cells differ from those of humans and other mammals. In plants, N-linked glycans contain β(1,2)-xylose and α(1,3)-fucose instead of the α(1,6)-fucose found in mammals. These plant-specific glycans are considered to be potentially antigenic and/or allergenic epitopes (Bakker *et al.*, 2001). Several strategies have been studied to remove the antigenic potential of plant-specific glycans. One simple approach is aglycosylation to obtain recombinant protein with no N-glycosylation by mutating the N-glycosylation sites of expressed genes (Conley, Mohib, Jevnikar, & Brandle, 2009). This approach is effective if the biological activity is not affected by aglycosylation.

Another approach consists of retaining the foreign protein in the ER using KDEL/HDEL (i.e Lys/His-Asp-Glu-Leu) polypeptide retention signals to avoid plant-specific glycan residues such as β-(1,2)-xylose and α-(1,3)-fucose (Gomord *et al.*, 2004; Ko *et al.*, 2003; Petruccelli *et al.*, 2006). Indeed, glycosylation processing in the ER is conserved between the plant and animal kingdoms and restricted to high mannose-type N-glycans, whereas the further glycosylation process in the Golgi apparatus, where additional glycans are added for glycan maturation, is highly diverse. Another approach to eliminating plant-specific glycan residues is to knock out the gene expression of glycosyltransferases involving β-(1,2)-xylosylation and α-(1,3)-fucosylation (Gomord *et al.*, 2004). However, in addition to eliminating plant-specific sugar, humanization of N-glycosylation is also essential for the production of authentic glycosylated recombinant proteins in plants. The strategy to humanize plant N-glycans consists of expressing mammalian glycosyltransferases, which would complete N-glycan maturation, in plants (Bakker *et al.*, 2001).

So far, little information regarding the glycosylation of microalgae is available and it is interesting, both from a purely scientific point of view and for biotechnological applications, to determine their capacity for this process. Our laboratory published the first *in silico* N-glycosylation study in microalgae. Using the genomic data available for *P. tricornutum*, we identified specific genes coding enzymes involved in the N-glycosylation pathway in diatoms (Baiet *et al.*, 2011). Moreover, by structural analyses of N-linked glycans, this study also demonstrated that *P. tricornutum* proteins carry mainly high mannose-type N-glycans. Interestingly, other recent biochemical studies have reported the existence of special glycosyltransferase and glycosylation pathways, unique to the red alga *Porphyridium* sp. (Levy-Ontman *et al.*, 2011).

The emergence of genomic data in microalgae will provide the opportunity to perform comparative genomic studies and to dissect biosynthetic pathways such as N-glycosylation.

Recently, we initiated new studies to evaluate the N-glycosylation pathway of microalgae representing different phyla: green and red microalgae, glaucophytes, alveolates, stramenopile and haptophytes. This study will help us to determine how this specific process evolved within the eukaryotes. Moreover, demonstrating that microalgae are a suitable alternative system for the production of biopharmaceuticals requires the demonstration of their N-glycosylation capability.

2.3.2. Genomic Strategies for Optimising Recombinant Protein Expression

In this section, we report three strategies commonly used to optimize recombinant protein accumulation in microalgae.

2.3.2.1. Translation optimization by codon usage bias

Specific variations in codon usage are often cited as one of the major factors impacting protein expression level. The presences of rare codons that are correlated with low levels of their endogenous transfer RNA species in the cell can reduce the translation rate of target mRNA. The classical strategy to bypass this problem is to redesign genes to increase their expression level. For this, two approaches have been attempted, both of which require choosing from a vast number of possible DNA sequences. The first approach consists of assigning the most abundant codon of the host of a given amino acid in the target sequence. The second uses translation tables based on the frequency distribution of the codons in an entire genome or for a range of highly expressed genes. This approach was successfully used in *C. reinhardtii* to improve the expression level of foreign proteins such as GFP in the nucleus (a 5-fold increase) (Fuhrmann, Oertel, & Hegemann, 1999) and chloroplasts (increased up to 80-fold) (Franklin, Ngo, Efuet, & Mayfield, 2002). Similar studies using a codon-optimized human antibody gene or luciferase reporter gene confirmed that codon bias play an important role in protein accumulation in chloroplasts of *C. reinhardtii* (Mayfield & Schultz, 2004; Mayfield *et al.*, 2003).

The nuclear and chloroplastic genome of *C. reinhardtii* may exhibit different codon bias, and thus, adjustment of codons in foreign gene sequences is necessary to obtain a high rate of protein production. To overcome this issue, the codon adaptation index (CAI) is used as a quantitative tool to predict the expression level of transgenes based on their codon usage. Several molecular software programs are available to determine and optimize codon usage. A list of these programs is given in Villalobos, Ness, Gustafsson, Minshull, & Govindarajan (2006).

This approach, which consists of optimizing the codon usage of transgenes, was successfully used in the green alga *C. reinhardtii* and diatom *P. tricornutum*. Specific codon usage is a field that will benefit from the contribution of future microalgal genomic and transcriptomic sequences.

2.3.2.2. Identification of promoter sequences

Genome data are also necessary to identify functional sequences such as promoter, 5′ and 3′- untranslated region (UTR) sequences that regulate the gene expression rate. These sequences are specific for each gene and microalgal strain. Due to the presence of plastid and nuclear genomes in microalgae, there are different types of promoter sequences according to cell localization. Plastid transgenes are expressed under the control of an endogenous promoter and 5′ and 3′-UTR. Overall, promoter sequence control transcription and 5′-UTR mediate mRNA stability, and translation initiation and 3′-UTR regulate stability and act in the termination of transcription. The same sequences were found for nuclear promoters, but other regulated sequences such as intron sequences are also involved in the regulation of nuclear gene expression. Previous studies identified sequences within the 5′-UTR that were involved in RNA stability and used as a means to increase recombinant protein synthesis. For a comprehensive review of chloroplast translation regulation, see Marin-Navarro, Manuell, Wu, & Mayfield (2007).

Concerning chloroplastic transformation in microalgae, the green alga *C. reinhardtii* has been intensively studied. Among chloroplastic promoters for the expression of foreign proteins (Table 8.3), the endogenous *atpA*, *psbD*, *rbcL* and *psbA* promoters are generally used (Hallmann *et al.*, 2007; Specht, Miyake-Stoner, & Mayfield, 2010). An excellent study performed by Barnes *et al.* (2005) reported the effect of various promoters and UTRs on recombinant proteins in the chloroplast of *C. reinhardtii*. Using different combinations of chimeric proteins corresponding to the promoters and 5′-UTRs of chloroplast genes, *atpA*, *rbcL*, *psbA*, *psbD* and *16S* rRNA, fused to the GFP reporter and followed by 3′-UTR of either gene, they observed different protein accumulation levels. Moreover, they showed that mRNA accumulation is, in general, proportional to protein accumulation. Also, according to chimeric construction, they observed that the 5′-UTR sequence had a significant impact on recombinant protein production, while 3′-UTR had little effect. The highest level of reporter protein was found using the *atpA* or *psbD* promoter and 5′-UTR, while a minor protein accumulation level was observed under control of *rbcL* and *psbA* and no expression was seen using the *16S* rRNA promoter and 5′-UTR (Barnes *et al.*, 2005).

Interestingly, the *psbA* promoter fused with its 5′-UTR was actually the most used (Manuell *et al.*, 2007; Surzycki *et al.*, 2009). Recently, Rasala, Muto, Sullivan, & Mayfield (2011) reported a high recombinant protein expression level with the *psbA* promoter in comparison to the levels reached with the *atpA* promoter. It remains unclear why certain regulatory elements induce a high expression level in some genes but not in others

Table 8.3 Promoter Used For Microalgae Genetic Transformation

Host Species of Microalgae	Promoter of Gene and its Product	Cell Expression Localization	Source of Promoter	Source
Chlamydomonas reinhardtii	*arg7*, arginosuccinate lyase	Nuclear	*Chlamydomonas reinhardtii*	Debuchy, Purton, and Rochaix (1989)
	35S, cauliflower mosaic virus 35S	Nuclear	Cauliflower mosaic virus	Brown, Sprecher, and Keller (1991) Tang, Qiao, and Wu (1995) Kumar *et al.*, 2004
	RbcS2, rubisco small subunit 2	Nuclear	*Chlamydomonas reinhardtii*	Auchincloss, Loroch, & Rochaix (1999) Fuhrmann *et al.*(1999) Sizova, Fuhrmann, and Hegemann (2001) Stevens, Rochaix, and Purton (1996) Nelson and Lefebvre (1995) Kovar, Zhang, Funke and Weeks (2002) Cerutti, Johnson, Gillham, and Boynton (1997) Cordero *et al.* (2011)
	HSP70, heat shock protein 70 (fused to other promoter)	Nuclear	*Chlamydomonas reinhardtii*	Schroda *et al.* (2000) Eichler-Stahlberg *et al.* (2009)

(*Continued*)

Table 8.3 Promoter Used For Microalgae Genetic Transformation—cont'd

Host Species of Microalgae	Promoter of Gene and its Product	Cell Expression Localization	Source of Promoter	Source
	Nos, nopaline synthase	Nuclear	*Agrobacterium tumefaciens*	Hall, Taylor, and Jones (1993)
	Nit1, nitrate assimilation 1	Nuclear	*Chlamydomonas reinhardtii*	Ohresser, Matagne, and Loppes (1997) Llamas, Igeno, Galvan, and Fernandez (2002)
	Cop, chlamyopsin	Nuclear	*Chlamydomonas reinhardtii*	Fuhrmann *et al.* (1999)
	TubA1, alpha-tubulin	Nuclear	*Chlamydomonas reinhardtii*	Kozminski, Diener, and Rosenbaum (1993)
	β2-tubulin	Nuclear	*Chlamydomonas reinhardtii*	Blankenship and Kindle (1992) Berthold, Schmitt, and Mages (2002)
	CabII-1, chlorophyl-ab binding	Nuclear	*Chlamydomonas reinhardtii*	Blankenship and Kindle (1992)
	pcy1, plastocyanin	Nuclear	*Chlamydomonas reinhardtii*	Quinn and Merchant (1995)
	atpC, gamma-subunit of chloroplast ATPase	Nuclear	*Chlamydomonas reinhardtii*	Quinn and Merchant (1995)
	psaD, photosystem I complex protein	Nuclear	*Chlamydomonas reinhardtii*	Fischer and Rochaix (2001)
	atpA, alpha subunit of adenosine triphosphate	Chloroplast	*Chlamydomonas reinhardtii*	Sun *et al.* (2003)
	psbD, photosystem II D1	Chloroplast	*Chlamydomonas reinhardtii*	Manuell *et al.* (2007)

	RbcL, ribulose bisphosphate carboxylase large subunit	Chloroplast	*Chlamydomonas reinhardtii*	Dreesen *et al.* (2010)
	psbA, photosystem II psbA	Chloroplast	*Chlamydomonas reinhardtii*	Rasala, Muto *et al.* (2011)
Dunaliella salina	*Ubi1- Ω*, ubiquitin-Ω	Nuclear	*Zea mays*	Geng, Wang, Wang, Li, and Sun (2003)
	35S, cauliflower mosaic virus 35S	Nuclear	Cauliflower mosaic virus	Tan, Qin, Zhang, Jiang, and Zhao (2005) Sun *et al.* (2005) Feng, Xue, Liu, and Lu (2009) Wang, Xue *et al.* (2007)
	NR, Nitrate reductase	Nuclear	*Dunaliella salina*	Li *et al.* (2007) Li *et al.* (2008)
	RbcS2, rubisco small subunit	Nuclear	*Dunaliella salina*	Sun *et al.* (2005)
Dunaliella bardawil	*35S*, cauliflower mosaic virus 35S	Nuclear	Cauliflower mosaic virus	Anila *et al.* (2011)
Chlorella ellipsoida	*35S*, cauliflower mosaic virus 35S	Nuclear	Cauliflower mosaic virus	Jarvis and Brown (1991)
	Ubi1- Ω, ubiquitin-Ω	Nuclear	*Zea mays*	Chen *et al.* (2001)
	RbcS2, rubisco small subunit 2	Nuclear	*Chlamydomonas reinhardtii*	Kim *et al.* (2002)
Chlorella sorokiniana	*NR*, nitrate reductase	Nuclear	*Chlorella* sp.	Dawson, Burlingame, and Cannons (1997)

(*Continued*)

Table 8.3 Promoter Used For Microalgae Genetic Transformation—cont'd

Host Species of Microalgae	Promoter of Gene and its Product	Cell Expression Localization	Source of Promoter	Source
Chlorella vulgaris	*35S*, cauliflower mosaic virus 35S	Nuclear	Cauliflower mosaic virus	Cha, Yee, and Aziz (2011) Chow and Tung (1999) Wang, Xue *et al.* (2007)
	NR, nitrate reductase	Nuclear	*Phaeodactylum tricornutum*	Niu *et al.* (2011)
Platymonas subcodiformis (Tetraselmis)	*CMV*, cytomegalovirus	Nuclear	Cytomegalovirus	Cui *et al.* (2010)
Nannochloropsis sp	*HSP70*, heat shock protein 70 / *RbcS2*, rubisco small subunit 2	Nuclear	*Chlamydomonas reinhardtii*	Chen *et al.* (2008) Li and Tsai (2008)
	35S, cauliflower mosaic virus 35S	Nuclear	Cauliflower mosaic virus	Cha, Chen *et al.* (2011)
	VCP, violaxanthin/ chlorophyl binding protein	Nuclear	*Nannochloropsis sp*	Kilian, Benemann, Niyogi, and Vick (2011)
Haematococcus pluvialis	*SV40*, simian virus	Nuclear	simian virus	Teng *et al.* (2002)
	pds, phytoene desaturase		*Haematococcus pluvialis*	Steinbrenner and Sandmann (2006)
	35S, cauliflower mosaic virus 35S	Nuclear	Cauliflower mosaic virus	Kathiresan *et al.* (2009)
Volvox carteri	*NR*, nitrate reductase	Nuclear	*Volvox carteri*	Schiedlmeier *et al.* (1994)

Gonium pectorale	*psD*, photosystem I complex protein / *HSP70*, heat shock protein 70		*Chlamydomonas reinhardtii*	Lerche and hallmann (2009)
Closterium peracerosum-strigosum litorrale	*HSP70* heat shock protein 70 / *Cab*, chlorophyl-ab binding Ch a/b-binding protein		*Closterium peracerosum-strigosum litorrale*	Abe, Hiwatashi, Ito, Hasebe, and Sekimoto (2008) Abe *et al.* (2011)
Lotharella amoebiformis	*RbcS2*, rubisco small subunit 2	Nuclear	*Lotharella amoebiformis*	Hirakawa *et al.* (2008)
Cyclotella criptyca	*Acc1*, acetylCoA carboxylase	Nuclear	*Cyclotella criptyca*	Dunahay *et al.* (1995)
Navicula saprophila	*Acc1*, acetylCoA carboxylase	Nuclear	*Cyclotella criptyca*	Dunahay *et al.* (1995)
Phaeodactylum tricornutum	*fcpA/B/C/E*, fucoxanthin chlorophyll	Nuclear	*Phaeodactylum tricornutum*	Apt *et al.* (1996)
	fcpF, fucoxanthin chlorophyll	Nuclear	*Phaeodactylum tricornutum*	Falciatore *et al.* (1999)
	fcpA, fucoxanthin chlorophyll	Nuclear	*Phaeodactylum tricornutum*	Zaslavskaia and Lippmeier (2000)
	cah, carbonic anyhdrase	Nuclear	*Phaeodactylum tricornutum*	Harada and Matsuda (2005)
	CMV, cytomegalovirus; *PRSV-LTR*, rous sarcoma virus; *35S*, cauliflower mosaic virus 35S	Nuclear	Cytomegalovirus; Rous sarcoma virus; Cauliflower mosaic virus	Sakaue *et al.* (2008)

(*Continued*)

Table 8.3 Promoter Used For Microalgae Genetic Transformation—cont'd

Host Species of Microalgae	Promoter of Gene and its Product	Cell Expression Localization	Source of Promoter	Source
	fcpA, fucoxanthin chlorophyll	Nuclear	*Phaeodactylum tricornutum*	Coesel *et al.* (2009)
	fcp, fucoxanthin chlorophyll and *NR*, nitrate reductase	Nuclear	*Cylindrotheca fusiformis*	Miyagawa *et al.* (2009)
Cylindrotheca fusiformis	Pδ, frustulin α3	Nuclear	*Cylindrotheca fusiformis*	Fischer, Robl, Sumper, and Kroger (1999)
	NR, nitrate reductase	Nuclear	*Cylindrotheca fusiformis*	Poulsen and Kroger (2005)
Thalassiosira pseudonana	*fcp*, fucoxanthin chlorophyll	Nuclear	*Thalassiosira pseudonana*	Poulsen *et al.* (2006)
Thalassiosira weissflogii	*fcpB*, fucoxanthin chlorophyll	Nuclear	*Thalassiosira pseudonana*	Falciatore *et al.* (1999)
Chaetoceros sp.	*pTpNR* (nitrate reductase de *Thallassiosira psudomana*)	Nuclear	*Thalassiosira pseudonana*	Miyagawa-Yamaguchi *et al.* (2011)

Amphidinium spp. *Symbiodinium microadriaticum*	*35S*, cauliflower mosaic virus 35S	Nuclear	Cauliflower mosaic virus	Ten Lohuis and Miller (1998)
Cyanidioschyzon merolae	UMP synthase	Nuclear	*Cyanidioschyzon merolae*	Minoda, Sakagami, Yagisawa, Kuroiwa, and Tanaka (2004)
	β-tubulin	Nuclear	*Cyanidioschyzon merolae*	Ohnuma, Yokoyama, Inouye, Sekine, and Tanaka (2008)
	cat, catalase	Nuclear	*Cyanidioschyzon merolae*	Ohnuma *et al.* (2009)
	apcC, phycocyanin-associated protein	Nuclear	*Cyanidioschyzon merolae*	Watanabe, Ohnuma, Sato, Yoshikawa, and Tanaka (2011)
Porphyidium sp.	*AHAS*, acetohydroxyacid synthase	Nuclear	*Porphyidium sp.*	Lapidot, Raveh, Sivan, Arad, and Sapira (2002)
Euglena gracilis	*psbA*, photosystem II complex protein	Chloroplast	*Euglena gracilis*	Doetsch, Favreau, Kuscuoglu, Thompson, and Hallick (2001)

(Marin-Navarro *et al.*, 2007). The *psbA* promoter and 5′-UTR are the most studied but essentially require a *psbA*-deficient genetic background for high foreign protein accumulation (Rasala & Mayfield, 2011).

Other exogenous promoter sequences have been used in *C. reinhardtii* chloroplasts. Kato, Kolenic, & Pardini (2007) showed the functionality of the inducible system of the *lac* operon of *Escherichia coli* in *C. reinhardtii* chloroplasts. At the same time, a riboswitch was reported to act as a translational regulatory factor in *C. reinhardtii* (Croft, Moulin, Webb, & Smith, 2007). Finally, all the data suggest that the translation mechanism and mRNA accumulation are primarily controlled by the promoter and 5′-UTR and that the choice of these sequences is a critical factor to consider for each protein of interest in order to achieve high yields of recombinant proteins. On the other hand, to our knowledge, no chloroplast transformation has been reported for microalgae other than *C. reinhardtii* and the unicellular flagellate protist *Euglena gracilis*, leaving the way open for research to study this mechanism of expression in other microalgae.

Concerning nuclear promoters, several studies have been performed in different taxa of microalgae using endogenous, exogenous and synthetic promoters (Table 8.3). The most widely used constitutive promoter in the chlorophyte group is *RbcS* (RuBisCO small subunit). Interestingly, some endogenous promoters of *C. reinhardtii* can be used in other Chlorophyta algae. Indeed, the *C. reinhardtii RbcS* promoter has been successfully used in the green alga *D. salina* (Sun *et al.*, 2005), *C. ellipsoideum* formerly *Chlorella ellipsoidea* (Kim *et al.*, 2002), the Heterokonta *Nannochloropsis oculata* (Chen, Li, Huang, & Tsai, 2008; Li, Xue, Yan, Liu, & Liang, 2008) and the Chlorarachniophyta *Lotharella amoebiformis* (Hirakawa, Kofuji, & Ishida, 2008). A chimeric promoter using heat shock protein A (*HSP70A*) fused to *psaD* was also successfully used in *C. reinhardtii* (Fischer & Rochaix, 2001; Schroda, Blocker, & Beck, 2000) and more recently in the multicellular alga *Gonium pectorale* (Lerche & Hallmann, 2009). The same strategy, using *HSP70* fused to *CAb* (chlorophyll-binding protein), was reported in the charophyte *Closterium peracerosum* (Abe *et al.*, 2011). Usual plant promoters, such as the cauliflower mosaic virus *35S* with *Ubiquitin-Ω*, have been also tested in some microalgae (Chen *et al.*, 2001; Jarvis & Brown, 1991; Kumar, Misquitta, Reddy, Rao, & Rajam, 2004; Wang, Wang, Su, & Gao, 2007, Wang, Xue et al., 2007), and recently, the *35S* promoter demonstrated efficiency in the diatom *P. tricornutum* (Sakaue, Harada, & Matsuda, 2008), Chlorophytes *Haematococcus* sp. (Kathiresan, Chandrashekar, Ravishankar, & Sarada, 2009) and *Dunaliella bardawil* (Anila, Chandrashekar, Ravishankar, &

Sarada, 2011) and the Heterokonta *Nannochloropsis* sp. (Cha, Chen, Yee, Aziz, & Loh, 2011). Moreover, inducible promoters have been chosen for some algae. Indeed, the gene expression under the control of the nitrate reductase promoter is switched off when cells are grown in the presence of ammonium and becomes switched on when cells are transferred to a medium-containing nitrate. This approach was reported for the diatom *Cylindrotheca fusiformis* (Poulsen & Kroger, 2005) and recently in *C. vulgaris* (strain not reported) using the *NR* cassette (promoter and 5′ and 3′-UTR of nitrate reductase) of the diatom *P. tricornutum* (Niu *et al.*, 2011). A similar strategy was applied in *P. tricornutum* using the endogenous *NR* cassette (Hempel, Bozarth *et al.*, 2011, Hempel, Lau *et al.*, 2011) and exogenous *NR* cassette from the diatom *Cylindrotheca fusiformis* (Miyagawa *et al.*, 2009). Miyagawa-Yamaguchi *et al.* (2011) reported the same approach in another diatom, *Chaetoceros* sp., using the *NR* cassette of the diatom *Thalassiosira pseudonana*. The study performed by Niu *et al.* (2011) is particularly interesting as the diatom *NR* cassette was shown to be functional in green algae, suggesting that this type of inducible promoter could be universally employed across diverse species of algae.

In contrast to plastid promoters, several studies have been performed on nuclear promoters in diatoms. So far, unlike in green algae, the RuBisCO small subunit gene of diatoms is encoded by the chloroplast genome and its promoter is not adapted for nuclear transformation. Other promoters were identified from genomic and transcriptomic data from diatoms. Early studies reported protein expression using the acetyCoA carboxylase (*Acc1*) promoter in diatoms *C. cryptica* and *Navicula saprophila* (Dunahay, Jarvis, & Roessler, 1995). Members of the family of light-inducible fucoxanthin chlorophyll (*Fcp*) promoters have also been used to produce foreign protein in diatoms *P. tricornutum* (Apt, Kroth-Pancic, & Grossman, 1996; Falciatore, Casotti, Leblanc, Abrescia, & Bowler, 1999; Zaslavskaia & Lippmeier, 2000) and *Thalassiosira* sp. (Falciatore *et al.*, 1999; Poulsen, Chesley, & Kroger, 2006). In contrast to the *NR* promoter, the *Fcp* promoter appears to be more specific to the host as, for example, the *Fcp* promoter of *P. tricornutum* is not functional in *Cylindrotheca fusiformis* (Poulsen & Kroger, 2005).

Some use of virus promoters other than *35S* has also been reported in algae, such as mammalian cytomegalovirus *CMV* in *P. tricornutum* (Sakaue *et al.*, 2008), and recently in the Chlorophyta *Platymonas subcordiformis* (Cui, Wang, Jiang, Bian, & Qin, 2010), as well as the Rous sarcoma virus in the diatom *P. tricornutum* (Sakaue *et al.*, 2008).

Surprisingly, the use of microalgal virus sequences in algal expression constructs to enhance gene expression has still not been explored. To date, several algal viruses have been identified and their full genomes sequenced in some microalgal taxa, specifically chlorophytes, dinoflagellates, diatoms and haptophytes (for reports and reviews on this topic, see Nagasaki, 2008; Nissimov *et al.*, 2011; Schroeder, Oke, Malin, & Wilson, 2002; Van Etten & Dunigan, 2012; Wilson, Van Etten, & Allen, 2009). To date, algal viruses represent a largely unexplored source of genetic elements for engineering algae and land plants. This approach has previously been used in both monocotyledonous and dicotyledonous land plants as well as in bacteria (Mitra, Higgins, & Rohe, 1994). Another study reported the functionality of a translation enhancer element from the *Chlorella* virus in the plant *Arabidopsis thaliana* (Nguyen, Falcone, & Graves, 2009).

Another strategy to increase the yield of recombinant proteins consists of adding intronic sequences to the expression vector to act as an endogenous enhancer. Indeed, while regulation of gene expression occurs at the post-transcriptional level in the plastid, it appears that most regulation occurs both at the transcriptional and at the translational levels in the nucleus (Marin-Navarro *et al.*, 2007). The introns are non-encoding sequences but can affect the expression of genes by alternative splicing or through the regulation of transcription. In *C. reinhardtii*, Lumbreras and Purton (1998) reported that the insertion of endogenous introns from heterologous genes increases the expression level. Recently, a similar approach has been used to increase the expression level of the *Renilla*-luciferase gene reporter in *C. reinhardtii* (Eichler-Stahlberg *et al.*, 2009). However, the way in which the introns affect the expression level is still unclear.

2.3.2.3. The challenges of transgene silencing and proteolysis

Transgene silencing is another problem for high-yield recombinant protein expression in plants and algae, but different strategies exist to overcome this obstacle. Indeed, gene silencing can function as a protective system against pathogens or viruses (Specht *et al.*, 2010). Plant virus-encoded suppressors of RNA silencing are useful tools for counteracting silencing, but their wide application in transgenic plants is limited because their expression often causes harmful developmental effects. To our knowledge, this approach has not yet been attempted in microalgae. Recently, another strategy to prevent transgene silencing was reported in *C. reinhardtii* using a process of UV mutation and selection by antibiotic resistance on a selective medium (Neupert, Karcher, & Bock, 2009).

To date, most efforts to improve recombinant protein accumulation in plants or algae have focused on increasing protein expression. Moreover, proteolysis is also one of the factors that can affect the yield of recombinant protein accumulated and also lead to difficulties in purification due to degraded forms or non-functional protein (Doran, 2006; Surzycki *et al.*, 2009). However, proteolytic enzymes are essential for the degradation of misfolding or incorrectly processed endogenous proteins. Some strategies have been attempted to minimize foreign protein degradation in plants and microalgae, like producing recombinant proteins in other cell compartments that have an environment with less proteolytic activity. Indeed, for nuclear-expressed protein, the targeting of the ER using the HDEL or KDEL retention signal prevents the degradation of the foreign protein. A similar approach has been successfully used by our laboratory to express recombinant EPO in diatoms. Another approach used in plants consists of concomitantly producing protease inhibitor to neutralize endogenous protease (Doran, 2006).

3. FUTURE OUTLOOK

In this chapter, we tried to provide an overview of the principal applications of microalgae and show how genomics and post-genomics can improve their uses in biotechnology. Here, we mainly focused on some of the most popular applications. Without wishing to suggest that they are less important, we chose not to make a detailed review of other applications such as environmental biomarkers, silica synthesis from diatoms or hydrogen and methane production for energy. In any case, the future of microalgal biotechnology will depend on several steps, including domestication and a search for new intrinsic species characteristics, steps for which the contributions of 'omics' technologies will be invaluable.

3.1. Domestication

A strategy comparable to that used for the domestication of crops is now making its way into the world of microalgae. This is a matter of selecting favourable mutations and finding markers that will help select the desired traits. In agriculture, the cross-breeding of species and selection of strains was conducted empirically for thousands of years before Mendel's laws provided a scientific basis for species improvement. For sexual reproduction, considerable work remains to be done in microalgae. Knowledge of reproductive strategies is of major importance for maintaining strains, cultivating them on a long-term basis in continuous culture or envisaging selection strategies. Some algal groups have become the subject of increased attention concerning their reproduction strategies and sexual behaviour, including the diatoms (Chepurnov, Chaerle, Roef, Meirhaeghe, & Vanhoutte, 2011). In the Coccolithophore *E. huxleyi*, different morphotypes associated with different forms of ploidy have been observed and studied by transcriptomic analysis. This work revealed mechanisms involved in functional differentiation without proving that sexual reproduction occurs (Von Dassow *et al.*, 2009). To date, there is too little knowledge to envisage the improvement of strains by cross-breeding and selection through sexual reproduction, so future studies in this direction will be of great interest.

Mutation followed by selection for favourable phenotypes has been used for crop plants, and some promising strategies are now beginning to emerge for algae. This domestication route calls for induced mutations and subsequent selection. Bonente, Formighieri, Morosinotto, and Bassi (2011) identified the major relevant points for the selection of H_2-producing *Chlamydomonas* sp., namely a reduction of photosynthetic antenna size, an alteration of photosystem II to manipulate the oxygen concentration and a maximized electron flow towards hydrogenase. This strategy was thought to enhance carotenoid levels. Early studies involved *D. salina* and the selection of beta-carotene-rich strains (Shaish, Ben-Amotz, & Avron, 1991). These were followed by a search for hyperproductive variants sorted by flow cytometry (Mendoza *et al.*, 2008), and recently, there have been improvements in lutein production in the microalga *C. sorokiniana* (Cordero, Couso, Leon, Rodriguez, & Angeles Vargas, 2011). This strategy was implemented to enhance overall lipid contents or EPA and DHA, in particular, with the Haptophytes *I. galbana* (Molina Grima *et al.*, 1995) and *Pavlova lutheri* (Meireles, Guedes, & Malcata, 2003), the Heterokonta *N. oculata* (Chaturvedi & Fujita, 2006) or the chlorophyte *D. salina* (Mendoza *et al.*, 2008). In this context, the availability of a reliable marker, such as Nile Red or BODIPY for staining lipid bodies, greatly helps in the selection process. In the Haptophyte *I. galbana* affinis Tahiti, this strategy allowed our laboratory to select an improved strain that could stably produce twice the amount of TAG compared to its wild-type counterpart (Rouxel *et al.*, 2011).

We have reported that the improvement of microalgae for biotechnology uses will come through the domestication of strains and this approach has already been initiated. Far from being in conflict, the different approaches ('natural' vs. 'GMO') are complementary. Synthetic biology, synthetic genomics and genome engineering are disruptive technologies. Indeed, the development of 'GMO' strategies is very promising for applications with very high added value such as the production of drugs or antibodies. However, taking into account the environmental risks arising with such transgenic species and societal pressure against their use, their culture will have to be performed in confined and controlled conditions. Their use for energy and food (large outdoor cultures) therefore seems somewhat inappropriate. The completely opposite point of view is that, given the immeasurable biodiversity of algae, the ideal alga for a given application is probably available in nature. Although this perspective is somewhat optimistic, the exploration of biodiversity was the source of the algae presently in use and will doubtless continue to be in the future. Screening this diversity will enable us to identify new, more efficient strains with new features, some of which may have uses that have not yet been imagined. This does not preclude subsequent domestication to improve these species for use in biotechnology.

3.2. Working Towards a New Algal Metabolism, Enzymes and Compounds

As seen earlier in this chapter, the implementation of genomic and post-genomic approaches is now largely underway in the world of microalgae. In parallel, ecological approaches in metagenomics have only been seen very recently. These will hopefully lead to the identification of a large number of presently unknown microalgae and, consequently, to new gene networks, enzymes and metabolic pathways. Due to the wide variety of microalgae and difficulties in cultivating certain of them, many metabolic pathways have remained out of our reach until the present. Metagenomics aims to analyze all the genomic data in a given ecosystem without a strain isolation and cultivation step. This allows access to unknown mechanisms of potential biotechnological interest. In this situation, the 'sequencing campaigns' on research cruises (Karsenti *et al.*, 2011) will offer new and valuable insight in the field of microalgal genomics. Chapter X of this volume provides a review on the power and challenges of metagenomics for microbial algae (Toulza, Blanc-Mathieu, Gourbiere, & Piganeau 2012). Conversely, metagenomics will be greatly aided by new methods like single cell genome analysis (Ebenezer, Medlin, & Ki, 2011), which can improve methods of isolation and cultivation of new algae.

Among the wide variety of metabolic pathways conceivable across the diversity of microalgae, particular attention should be paid to metabolism from extremes environments. Like bacteria, although to a lesser extent, some microalgal species live under severe physicochemical pressures such as high salinity, extreme temperatures from below 0 °C to over 50 °C, alkaline or acidic waters or very high irradiance. Additionally, some strains have been isolated downstream from industrial sites such as acid mine drainage or in waters rich in contaminants such as metals (for review, see Das *et al.*, 2009). Extremophiles offer numerous advantages including (1) the absence of contaminants in open door cultures subjected to physicochemical pressure, (2) their potential adaptation to industrial environments such as presence of toxins, radioactive elements or extreme pH and consequently their potential use for the biocatalysis of effluents and (3) their ability to produce enzymes with biotechnological applications. Proteomics have been carried out to highlight the adaptation mechanisms in the halophilic species *D. salina* (Katz, Waridel, Shevchenko, & Pick, 2007; Liska, 2004) for which the genome sequence will soon be available. Transcriptomics and comparative genomics have dealt with the very high biochemical versatility of thermoacidophilic *Galdieria sulphuraria* (Barbier *et al.*, 2005; Weber *et al.*, 2004). Psychrophilic species have been identified, such as *Fragilariopsis cylindrus*, *Xanthonema* sp., *Koliella antarctica* and *Chlamydomonas* sp. ICE-L. However, the culture of psychrophiles is far from being technically mastered, making post-genomic approaches difficult. In such cases, metagenomics would be an appropriate solution. Overall, efforts are still needed to isolate and cultivate extremophiles, and genomics will provide a source of new applications.

3.3. Algal Pathogens: Looking Towards the Future

Like land plants, phytoplankton are susceptible to diseases and parasitism, which impact their population dynamics and use in commercial industry. Interactions between bacteria and microalgae in the environment and in cultures are numerous, and bacteria can have beneficial or negative effects on the growth of microalgae. For a review, see Fukami, Nishijima, and Ishida (1997). Numerous algicidal bacteria have been identified in the ocean and their influence on algal bloom dynamics has been demonstrated (Mayali & Azam, 2004). Although they have not yet been associated with real economic losses in cultures, the experience in production of other marine species suggests that diseases will likely appear in parallel with the expansion of the industry. Viruses are extremely abundant in seawater and are believed to be significant pathogens to photosynthetic protists. They are known to affect the regulation of eukaryotic phytoplankton population densities. Since the discovery of the very high abundance of viruses in the marine environment, researchers have highlighted their possible ecological significance. To date, more than 40 viruses infecting marine microalgae have been isolated and characterized to different extents (Nagasaki, 2008). Several studies have focused on the relationship between eukaryotic microalgae and their viruses (for review, see Nagasaki, 2008). Without going into great detail, it is interesting to note that most algae can use various strategies of resistance to their viruses, but the mechanisms involved are not yet clearly understood (Morin, 2008; Thomas *et al.*, 2011). The next chapter of this volume, 'Genomics of Algal Host-Virus Interactions', reviews algal host–virus interactions (Grimsley *et al.*, 2012). Finally, it will be interesting to compare the emergence of pathogens such as plant viruses during this new agricultural revolution and increase in demand for algal culture by industry. Microalgal cultivation remains a niche market in almost all countries, but the increasing interest in sustainable biofuel sources has triggered a high investment in culture facilities all over the world. Consequently, intensive algal aquaculture using open pond systems for the mass culture of microalgae might favour disease outbreaks. Gachon, Sime-Ngando, Strittmatter, Chambouvet, and Kim (2010) suggest that development towards intensive macroalgal production correlates with more damaging disease outbreaks. The best example is the case of edible red macroalgae *Porphyra* sp., which represents a very valuable industry in Asia. The market is estimated to be worth about $1.5 billion worldwide and has reported losses of about 10% of annual production due to oomycete pathogens, although these outbreaks can even lead to losses of 25–40% in some cases. To overcome this problem, sea farmers can use chemical treatments, but their use in large doses could have a real impact on ecosystems as well as on production costs. A better understanding of the relationships between pathogens and microalgae would be useful to identify causes and possible solutions to overcome disease epidemics. Prophylactic and microbial flora management of cultures will probably be a key to the durability of production in the coming years. Varietal selection of microalgal strains for resistance to a large range of pathogens is one strategy that could increase the resistance of microalgae cultures. In any case, in the context of intensive microalgal production, we must anticipate future epidemics that will affect algal culture yields.

4. CONCLUSIONS

Due to the huge amount of diversity among microalgae, their applications have a very bright future. As seen in this chapter, the uses of microalgae are numerous and there is work for many research teams in many fields of specialization.

Genomics and post-genomics have led to new areas of research and development and to the modernization of our view of biology. The increase in sequencing capacities will soon face a data tsunami, a fantastic amount of data that will soon be generated by fast low-cost sequencing methods. However, storage, calculation power, annotations and access to this information now pose a limit to its optimal exploration. Data mining and conversion of data into biological knowledge will be an important challenge in coming years. The confirmation of all the *in silico* analyses and discoveries will require a return to experimental testing, and the association of molecular data with biological functions will become vital work in the future (Lopez, Casero, Cokus, Merchant, & Pellegrini, 2011).

The culture of microalgae for biomass production dates back to the 1940s when it started in the United States before spreading to Europe, Japan and Israel (Grobbelaar, 2010). Since then, work has continued all over the world at different speeds, with irregular publication rates. Some of the early work still forms the basis for the today's revival of the microalgal trend (Sheehan *et al.*, 1998). The dramatic increase in the world population concerns about the ecological equilibrium, pollution, the world energy demand and failing supplies of oil and coal have all led to a more 'bio'-orientated attitude, meaning a general increase in the attention paid to 'renewable' resources. From a global perspective, in the context of a demographic crisis, the major issues in the coming years will be to provide everyone with access to water, food, education and healthcare. In a world of limited resources (energy, clean water, arable land) and increasing anthropogenic pressure on the environment, the development of biotechnological processes to provide renewable energy, new molecules and molecular farming and cleaner industrial processes is one of the key challenges. Marine microalgae possess assets that make them suitable for some of these applications. While land plants are the subject of numerous programs aiming to use vegetal organisms for the so-called 'green chemistry', the algae, particularly microalgae, are expected to participate in this race in a complementary way. Although far less studied than their terrestrial counterparts, microalgae offer an, as yet, untapped diversity and manipulability that explain the enthusiasm and investment from around the world. Phytoplankton research is being revisited and enriched by modern techniques like molecular biology and biocomputing, and the 'omics' technologies offer new insights into their biology. The young generation of students will have the chance, at the strictly scientific level, to be present when the majority of the genomes are still to be sequenced, the transcriptome is unknown and even the reproductive strategies or size of the genomes are undefined. The earth still hides a tremendous amount of original biology, including much that concerns microalgae. Their discovery, study, analysis and use will serve applications in all imaginable fields.

REFERENCES

Abe, J., Hiwatashi, Y., Ito, M., Hasebe, M., & Sekimoto, H. (2008). Expression of exogenous genes under the control of endogenous HSP70 and CAB promoters in the *Closterium peracerosum*-strigosum-littorale complex. *Plant and Cell Physiology, 49*, 625–632.

Abe, J., Hori, S., Tsuchikane, Y., Kitao, N., Kato, M., & Sekimoto, H. (2011). Stable nuclear transformation of the *Closterium peracerosum*-strigosum-littorale complex. *Plant and Cell Physiology, 52*, 1676–1685.

AbuGhazaleh, A. A., Potu, R. B., & Ibrahim, S. (2009). Short communication: The effect of substituting fish oil in dairy cow diets with docosahexaenoic acid-micro algae on milk composition and fatty acids profile. *Journal of Dairy Science, 92*, 6156–6159.

Anila, N., Chandrashekar, A., Ravishankar, G. A., & Sarada, R. (2011). Establishment of *Agrobacterium tumefaciens*-mediated genetic transformation in *Dunaliella bardawil*. *European Journal of Phycology, 46*, 36–44.

Apt, K. E., Kroth-Pancic, P. G., & Grossman, A. R. (1996). Stable nuclear transformation of the diatom *Phaeodactylum tricornutum*. *Molecular and General Genetics, 252*, 572–579.

Archibald, J. (2012). The evolution of algae by secondary and tertiary endosymbiosis. *Advances in Botanical Research, 64*, 87–118.

Auchincloss, A. H., Loroch, A. I., & Rochaix, S. D. (1999). The argininosuccinate lyase gene of *Chlamydomonas reinhardtii*: Cloning of the cDNA and its characterization as a selectable shuttle marker. *Molecular and General Genetics, 261*, 21–30.

Baba, M., Ioki, M., Nakajima, N., Shiraiwa, Y., & Watanabe, M. M. (2011). Transcriptome analysis of an oil-rich race: A strain of *Botryococcus braunii* (BOT-88-2) by de novo assembly of pyrosequencing cDNA reads. *Bioresource Technology, 109*, 282–286.

Baiet, B., Burel, C., Saint-Jean, B., Louvet, R., Menu-Bouaouiche, L., Kiefer-Meyer, M. C., et al. (2011). N-glycans of *Phaeodactylum tricornutum* diatom and functional characterization of Its N-acetylglucosaminyltransferase I enzyme. *The Journal of Biological Chemistry, 286*, 6152–6164.

Bakker, H., Bardor, M., Molthoff, J. W., Gomord, V., Elbers, I., Stevens, L. H., et al. (2001). Galactose-extended glycans of antibodies produced by transgenic plants. *Proceedings of the National Academy of Sciences of the United States of America, 98*, 2899–2904.

Barbier, G., Oesterhelt, C., Larson, M. D., Halgren, R. G., Wilkerson, C., Garavito, R. M., et al. (2005). Comparative genomics of two closely related unicellular thermoacidophilic red algae, *Galdieria sulphuraria* and *Cyanidioschyzon merolae*, reveals the molecular basis of the metabolic flexibility of *Galdieria sulphuraria* and significant differences in carbohydrate metabolism of both algae. *Plant Physiology, 137*, 460–474.

Barnes, D., Franklin, S., Schultz, J., Henry, R., Brown, E., Coragliotti, A., et al. (2005). Contribution of 5′- and 3′-untranslated regions of plastid mRNAs to the expression of *Chlamydomonas reinhardtii* chloroplast genes. *Molecular Genetics and Genomics, 274*, 625–636.

Becker, W. (2007). Microalgae in human and animal nutrition. In A. Richmond (Ed.), *Handbook of microalgal culture. Biotechnology and apllied phycology* (pp. 312–351). Oxford: Blackwell Science.

Berthold, P., Schmitt, R., & Mages, W. (2002). An engineered *Streptomyces hygroscopicus* aph 7″ gene mediates dominant resistance against hygromycin B in *Chlamydomonas reinhardtii. Protist, 153*, 401–412.

Bhadury, P., & Wright, P. C. (2004). Exploitation of marine algae: Biogenic compounds for potential antifouling applications. *Planta, 219*, 561–578.

Blanc, G., Duncan, G., Agarkova, I., Borodovsky, M., Gurnon, J., Kuo, A., et al. (2010). The Chlorella variabilis NC64A genome reveals adaptation to photosymbiosis, coevolution with viruses, and cryptic sex. *The Plant Cell, 22*, 2943–2955.

Blankenship, J. E., & Kindle, K. L. (1992). Expression of chimeric genes by the light-regulated Cabii-1 promoter in chlamydomonas-reinhardtii—A Cabii-1/Nit1 gene functions as a dominant selectable marker in a Nit1-Nit2-strain. *Molecular and Cellular Biology, 12*, 5268–5279.

Bock, R. (2007). Plastid biotechnology: Prospects for herbicide and insect resistance, metabolic engineering and molecular farming. *Current Opinion in Biotechnology, 18*, 100–106.

Bonente, G., Formighieri, C., Morosinotto, T., & Bassi, R. (2011). Domestication of wild unicellular algae for growth in photobioreactors. *FEBS Journal, 278*, 64–65.

Bowler, C., Allen, A. E., Badger, J. H., Grimwood, J., Jabbari, K., Kuo, A., et al. (2008). The Phaeodactylum genome reveals the evolutionary history of diatom genomes. *Nature, 456*, 239–244.

Bozarth, A., Maier, U. G., & Zauner, S. (2009). Diatoms in biotechnology: Modern tools and applications. *Applied Microbiology and Biotechnology, 82*, 195–201.

Brown, L. E., Sprecher, S. L., & Keller, L. R. (1991). Introduction of exogenous DNA into *Chlamydomonas reinhardtii* by electroporation. *Molecular and Cellular Biology, 11*(4), 2328–2332.

Cadoret, J.-P., Bardor, M., Lerouge, P., Cabigliera, M., Henriquez, V., & Carlier, A. (2008). Les microalgues: Usines cellulaires productrices de molécules commerciales recombinantes. *Medecine Science, 24*, 375–382.

Cadoret, J.-P., Carlier, A., Burel, C., Maury, F., Bardor, M., & Lerouge, P. (2009). *Production of glysosylated proteins in microalgae.* PCT/EP/2009/051672.

Camacho, F. G., Rodríguez, J. G., Mirón, A. S., García, M. C. C., Belarbi, E. H., Chisti, Y., et al. (2006). Biotechnological significance of toxic marine dinoflagellates. *Biotechnology Advances, 25*, 176–194.

Cerutti, H., Johnson, A. M., Gillham, N. W., & Boynton, J. E. (1997). A eubacterial gene conferring spectinomycin resistance on *Chlamydomonas reinhardtii*: Integration into the nuclear genome and gene expression. *Genetics, 145*, 97–110.

Cha, T. S., Chen, C. F., Yee, W., Aziz, A., & Loh, S. H. (2011). Cinnamic acid, coumarin and vanillin: Alternative phenolic compounds for efficient *Agrobacterium*-mediated transformation of the unicellular green alga, *Nannochloropsis sp. Journal of Microbiological Methods, 84*, 430–434.

Cha, T. S., Yee, W., & Aziz, A. (2011). Assessment of factors affecting *Agrobacterium*-mediated genetic transformation of the unicellular green alga, *Chlorella vulgaris*. *World Journal of Microbiology & Biotechnology*. doi: 10.1007/s11274-011-0991-0.

Chaturvedi, R., & Fujita, Y. (2006). Isolation of enhanced eicosapentaenoic acid producing mutants of *Nannochloropsis oculata* ST-6 using ethyl methane sulfonate induced mutagenesis techniques and their characterization at mRNA transcript level. *Phycological Research, 54*, 208–219.

Chen, Y., Wang, Y. Q., Sun, Y. R., Zhang, L. M., & Li, W. B. (2001). Highly efficient expression of rabbit neutrophil peptide-1 gene in *Chlorella ellipsoidea* cells. *Current Genetics, 39*, 365–370.

Chen, H. L., Li, S. S., Huang, R., & Tsai, H. J. (2008). Conditional production of a functional fish growth hormone in the transgenic line of *Nannochloropsis oculata* (Eustigmatophyceae). *Journal of Phycology, 44*, 768–776.

Chepurnov, V. A., Chaerle, P., Roef, L., Meirhaeghe, A., & Vanhoutte, K. (2011). Classical breeding in diatoms: scientific background and practical perspectives. In J. Seckbach, & P. Kociolek (Eds.), *The diatom world* (pp. 167–194). The Netherlands: Springer.

Chow, K. C., & Tung, W. L. (1999). Electrotransformation of *Chlorella vulgaris*. *Plant Cell Reports, 18*, 778–780.

Coesel, S., Mangogna, M., Ishikawa, T., Heijde, M., Rogato, A., Finazzi, G., et al. (2009). Diatom PtCPF1 is a new cryptochrome/photolyase family member with DNA repair and transcription regulation activity. *EMBO Reports, 10*, 655–661.

Conley, A. J., Mohib, K., Jevnikar, A. M., & Brandle, J. E. (2009). Plant recombinant erythropoietin attenuates inflammatory kidney cell injury. *Plant Biotechnology Journal, 7*, 183–199.

Cordero, B. F., Couso, I., Leon, R., Rodriguez, H., & Angeles Vargas, M. (2011). Enhancement of carotenoids biosynthesis in *Chlamydomonas reinhardtii* by nuclear transformation using a phytoene synthase gene isolated from *Chlorella zofingiensis*. *Applied Microbiology and Biotechnology, 91*, 341–351.

Croft, M. T., Moulin, M., Webb, M. E., & Smith, A. G. (2007). Thiamine biosynthesis in algae is regulated by riboswitches. *Proceedings of the National Academy of Sciences of the United States of America, 104*, 20770–20775.

Cui, Y. L., Wang, J. F., Jiang, P., Bian, S. G., & Qin, S. (2010). Transformation of *Platymonas (Tetraselmis) subcordiformis* (Prasinophyceae, Chlorophyta) by agitation with glass beads. *World Journal of Microbiology & Biotechnology, 26*, 1653–1657.

Cui, H., Wang, Y., & Qin, S. (2011). Molecular evolution of lycopene cyclases involved in the formation of carotenoids in eukaryotic algae. *Plant Molecular Biology Reporter, 29*, 1013–1020.

Das, B. K., Roy, A., Koschorreck, M., Mandal, S. M., Wendt-Potthoff, K., & Bhattacharya, J. (2009). Occurrence and role of algae and fungi in acid mine drainage environment with special reference to metals and sulfate immobilization. *Water Research, 43*, 883–894.

Dauvillee, D., Delhaye, S., Gruyer, S., Slomianny, C., Moretz, S. E., d'Hulst, C., et al. (2010). Engineering the chloroplast targeted malarial vaccine antigens in *Chlamydomonas* starch granules. *PLoS ONE, 5*, e15424. doi: 10.1371/journal.pone.

Dawson, H. N., Burlingame, R., & Cannons, A. C. (1997). Stable transformation of *Chlorella*: rescue of nitrate reductase-deficient mutants with the nitrate reductase gene. *Current Microbiology, 35*, 356–362.

Debuchy, R., Purton, S., & Rochaix, J.-D. (1989). The arginosuccinate lyase gene of *Chlamydomonas reinhardtii*: an important tool for nuclear transformation and for correlating the genetic and molecular maps of the ARG7 locus. *EMBO Journal, 8*, 2803–2809.

De Clerck, O., Bogaret, K., & Leliaert, F. (2012). Diversity and evolution of algae: primary endosymbiosis. *Advances in Botanical Research, 64*, 55–86.

Desmond, E., & Gribaldo, S. (2009). Phylogenomics of sterol synthesis: insights into the origin, evolution, and diversity of a key eukaryotic feature. *Genome Biology and Evolution, 1*, 364–381.

Dittami, S. M., Riisberg, I., John, U., Orr, R. J. S., Jakobsen, K. S., & Edvardsen, B. (2011). Analysis of expressed sequence tags from the marine microalga *Pseudochattonella farcimen* (Dictyochophyceae). *Protist, 163*, 143–161.

Doetsch, N. A., Favreau, M. R., Kuscuoglu, N., Thompson, M. D., & Hallick, R. B. (2001). Chloroplast transformation in *Euglena gracilis*: splicing of a group III twintron transcribed from a transgenic psbK operon. *Current Genetics, 39*, 49–60.

Doran, P. M. (2006). Foreign protein degradation and instability in plants and plant tissue cultures. *Trends in Biotechnology, 24*, 426–432.

Dove, A. (2002). Uncorking the biomanufacturing bottleneck. *Nature Biotechnology, 20*, 777–779.

Dreesen, I. A. J., Charpin-El Hamri, G., & Fussenegger, M. (2010). Heat-stable oral alga-based vaccine protects mice from *Staphylococcus aureus* infection. *Journal of Biotechnology, 145*, 273–280.

Dunahay, T. G., Jarvis, E. E., & Roessler, P. G. (1995). Genetic transformation of the diatoms *Cyclotella cryptica* and *Navicula saprophila*. *Journal of Phycology, 31*, 1004–1012.

Dunahay, T., Jarvis, E., Dais, S., & Roessler, P. (1996). Manipulation of microalgal lipid production using genetic engineering. *Applied Biochemistry and Biotechnology, 57*(8), 223–231.

Ebenezer, V., Medlin, L., & Ki, J.-S. (2011). Molecular detection, quantification, and diversity evaluation of microalgae. *Marine Biotechnology, 14*(2), 129–142.

Eichler-Stahlberg, A., Weisheit, W., Ruecker, O., & Heitzer, M. (2009). Strategies to facilitate transgene expression in *Chlamydomonas reinhardtii*. *Planta, 229*, 873–883.

Eom, H., Lee, C. G., & Jin, E. (2005). Gene expression profile analysis in astaxanthin-induced *Haematococcus pluvialis* using a cDNA microarray. *Planta, 223*, 1231–1242.

Falciatore, A., Casotti, R., Leblanc, C., Abrescia, C., & Bowler, C. (1999). Transformation of nonselectable reporter genes in marine diatoms. *Marine Biotechnology, 1*, 239–251.

Fan, J., Andre, C., & Xu, C. (2011). A chloroplast pathway for the de novo biosynthesis of triacylglycerol in *Chlamydomonas reinhardtii*. *FEBS Letters, 585*, 1985–1991.

Feng, S. Y., Xue, L. X., Liu, H. T., & Lu, P. J. (2009). Improvement of efficiency of genetic transformation for *Dunaliella salina* by glass beads method. *Molecular Biology Reports, 36*, 1433–1439.

Finazzi, G., Moreau, H., & Bowler, C. (2010). Genomic insights into photosynthesis in eukaryotic phytoplankton. *Trends in Plant Science, 15*, 565–572.

Fischer, H., Robl, I., Sumper, M., & Kroger, N. (1999). Targeting and covalent modification of cell wall and membrane proteins heterologously expressed in the diatom *Cylindrotheca fusiformis* (Bacillariophyceae). *Journal of Phycology, 35*, 113–120.

Fischer, N., & Rochaix, J. D. (2001). The flanking regions of PsaD drive efficient gene expression in the nucleus of the green alga *Chlamydomonas reinhardtii*. *Molecular Genetics and Genomics, 265*, 888–894.

Franklin, S., Ngo, B., Efuet, E., & Mayfield, S. (2002). Development of a GFP reporter gene for *Chlamydomonas reinhardtii* chloroplast. *Plant Journal, 30*, 733–744.

Frommolt, R., Werner, S., Paulsen, H., Goss, R., Wilhelm, C., Zauner, S., et al. (2008). Ancient recruitment by chromists of green algal genes encoding enzymes for carotenoid biosynthesis. *Molecular Biology and Evolution, 25*, 2653–2667.

Fuhrmann, M., Oertel, W., & Hegemann, P. (1999). A synthetic gene coding for the green fluorescent protein (GFP) is a versatile reporter in *Chlamydomonas reinhardtii*. *Plant Journal, 19*, 353–361.

Fukami, K., Nishijima, T., & Ishida, Y. (1997). Stimulative and inhibitory effects of bacteria on the growth of microalgae. *Hydrobiologia, 358*, 185–191.

Gachon, C. M. M., Sime-Ngando, T., Strittmatter, M., Chambouvet, A., & Kim, G. H. (2010). Algal diseases: spotlight on a black box. *Trends in Plant Science, 15*, 633–640.

Gasdaska, J. R., Spencer, D., & Dickey, L. (2003). Advantages of therapeutic protein production in the aquatic plant *Lemna*. *BioProcessing Journal, 2*(2), 49–56.

Geng, D. G., Wang, Y. Q., Wang, P., Li, W. B., & Sun, Y. R. (2003). Stable expression of hepatitis B surface antigen gene in *Dunaliella salina* (Chlorophyta). *Journal of Applied Phycology, 15*, 451–456.

Gobler, C. J., Berry, D. L., Dyhrman, S. T., Wilhelm, S. W., Salamov, A., Lobanov, A. V., et al. (2011). Niche of harmful alga *Aureococcus anophagefferens* revealed through ecogenomics. *Proceedings of the National Academy of Sciences of the United States of America, 108*, 4352–4357.

Gomord, W., Sourrouille, C., Fitchette, A. C., Bardor, M., Pagny, S., Lerouge, P., et al. (2004). Production and glycosylation of plant-made pharmaceuticals: the antibodies as a challenge. *Plant Biotechnology Journal, 2*, 83–100.

Gould, S. B., Waller, R. F., & McFadden, G. I. (2008). Plastid evolution. *Annual Review of Plant Biology, 59*, 491–517.

Gourdon, D., Lin, Q., Oroudjev, E., Hansma, H., Golan, Y., Arad, S., et al. (2008). Adhesion and stable low friction provided by a subnanometer-thick monolayer of a natural polysaccharide. *Langmuir, 24*, 1534–1540.

Gouveia, L., Coutinho, C., Mendonca, E., Batista, A. P., Sousa, I., Bandarra, N. M., et al. (2008). Functional biscuits with PUFA-omega 3 from *Isochrysis galbana*. *Journal of the Science of Food and Agriculture, 88*, 891–896.

Grimsley, N., Thomas, R., Kegel, J., Jacquet, S., Moreau, H., & Desdevises, Y. (2012). Genomics of algal host-virus interactions. *Advances in Botanical Research, 64*, 343–378.

Grobbelaar, J. (2010). Microalgal biomass production: challenges and realities. *Photosynthesis Research, 106*, 135–144.

Guarnieri, M. T., Nag, A., Smolinski, S. L., Darzins, A., Seibert, M., & Pienkos, P. T. (2011). Examination of triacylglycerol biosynthetic pathways via de novo transcriptomic and proteomic analyses in an unsequenced microalga. *PLoS ONE, 6*, e25851. doi: 10.1371/journal.pone.0025851.

Guschina, I. A., & Harwood, J. L. (2006). Lipids and lipid metabolism in eukaryotic algae. *Progress in Lipid Research, 45*, 160–186.

Habib, M. A. B., Huntington, T., & Hasan, M. R. (2008). A review on culture, production and use of *Spirulina* as food for humans and feeds for domestic animals and fish. *FAO Fisheries and Aquaculture Circular*. No. 1034.

Hall, L. M., Taylor, K. B., & Jones, D. D. (1993). Expression of a foreign gene in *Chlamydomonas reinhardtii*. *Gene, 124*, 75–81.

Hallmann, A., Amon, P., Godl, K., Heitzer, M., & Sumper, M. (2007). Algal transgenics and biotechnology. *Transgenic Plant Journal, 1*, 81–98.

Harada, H., & Matsuda, Y. (2005). Promoter analysis of two CO_2-inducible carbonic anhydrases in the marine diatom *Phaeodactylum tricornutum*. *Plant and Cell Physiology, 46*, 88–89.

Hawkins, R. L., & Nakamura, M. (1999). Expression of human growth hormone by the eukaryotic alga, *Chlorella*. *Current Microbiology, 38*, 335–341.

He, D. M., Qian, K. X., Shen, G. F., Zhang, Z. F., Li, Y. N., Su, Z. L., et al. (2007). Recombination and expression of classical swine fever virus (CSFV) structural protein E2 gene in *Chlamydomonas reinhardtii* chroloplasts. *Colloids and Surfaces B-Biointerfaces, 55*, 26–30.

Hempel, F., Bozarth, A. S., Lindenkamp, N., Klingl, A., Zauner, S., Linne, U., et al. (2011). Microalgae as bioreactors for bioplastic production. *Microbial Cell Factories*. doi: 10.1186/1475-2859-10-81.

Hempel, F., Lau, J., Klingl, A., & Maier, U. G. (2011). Algae as protein factories: expression of a human antibody and the respective antigen in the diatom *Phaeodactylum tricornutum*. *PLoS ONE, 6*, e28424. doi: 10.1371/journal.pone.0028424.

Hermsmeier, D., Schulz, R., & Senger, H. (1994). Formation of light-harvesting complexes of photosystem II in *Scenedesmus*. 1. Correlations between amounts of photosynthetic pigments, Lhc messenger RNAs and LHC apoproteins during constitutional dark- and light-dependent Lhc-gene expression. *Planta, 193*, 398–405.

Hirakawa, Y., Kofuji, R., & Ishida, K. (2008). Transient transformation of a chlorarachniophyte alga, *Lotharella amoebiformis* (Chlorarachniophyceae), with uidA and egfp reporter genes. *Journal of Phycology, 44*, 814–820.

Hu, Q., Sommerfeld, M., Jarvis, E., Ghirardi, M., Posewitz, M., Seibert, M., et al. (2008). Microalgal triacylglycerols as feedstocks for biofuel production: perspectives and advances. *Plant Journal, 54*, 621–639.

Huheihel, M., Ishanu, V., Tal, J., & Arad, S. (2002). Activity of *Porphyridium* sp polysaccharide against herpes simplex viruses in vitro and in vivo. *Journal of Biochemical and Biophysical Methods, 50*, 189–200.

Iomini, C., Till, J. E., & Dutcher, S. K. (2009). *Cilia: Model organisms and intraflagellar transport. Methods in cell biology*, Vol. 93. Academic Press. p. 121.

Janssen, M., Tramper, J., Mur, L. R., & Wijffels, R. H. (2003). Enclosed outdoor photobioreactors: light regime, photosynthetic efficiency, scale-up, and future prospects. *Biotechnology and Bioengineering, 81*, 193–210.

Jarvis, E. E., & Brown, L. M. (1991). Transient expression of firefly luciferase in protoplasts of the green alga *Chlorella ellipsoidea*. *Current Genetics, 19*, 317–321.

Karsenti, E., Acinas, S. G., Bork, P., Bowler, C., De Vargas, C., Raes, J., et al. (2011). A holistic approach to marine eco-systems biology. *PLoS Biology, 9*, e1001177. doi: 10.1371/journal.pbio.1001177.

Kathiresan, S., Chandrashekar, A., Ravishankar, G. A., & Sarada, R. (2009). *Agrobacterium*-mediated transformation in the green alga *Haematococcus pluvialis* (Chlorophyceae, Volvocales). *Journal of Phycology, 45*, 642–649.

Kato, T., Kolenic, N., & Pardini, R. S. (2007). Docosahexaenoic acid (DHA), a primary tumor suppressive omega-3 fatty acid, inhibits growth of colorectal cancer independent of p53 mutational status. *Nutrition and Cancer, 58*, 178–187.

Katz, A., Waridel, P., Shevchenko, A., & Pick, U. (2007). Salt-induced changes in the plasma membrane proteome of the halotolerant alga *Dunaliella salina* as revealed by blue native gel electrophoresis and nano-LC-MS/MS analysis. *Molecular and Cellular Proteomics, 6*, 1459–1472.

Khozin-Goldberg, I., & Cohen, Z. (2011). Unraveling algal lipid metabolism: recent advances in gene identification. *Biochimie, 93*, 91–100.

Kilian, O., Benemann, C., Niyogi, K., & Vick, B. (2011). High-efficiency homologous recombination in the oil-producing alga *Nannochloropsis* sp. *Proceedings of the National Academy of Sciences of the United States of America, 108*, 21265–21269.

Kim, D. H., Kim, Y. T., Cho, J. J., Bae, J. H., Hur, S. B., Hwang, I., et al. (2002). Stable integration and functional expression of flounder growth hormone gene in transformed microalga, *Chlorella ellipsoidea*. *Marine Biotechnology, 4*, 63–73.

Kim, J., Lee, W., Kim, B., & Lee, C. (2006). Proteomic analysis of protein expression patterns associated with astaxanthin accumulation by green alga *Haematococcus pluvialis* (Chlorophyceae) under high light stress. *Journal of Microbiology and Biotechnology, 16*, 1222–1228.

Ko, K., Tekoah, Y., Rudd, P. M., Harvey, D. J., Dwek, R. A., Spitsin, S., et al. (2003). Function and glycosylation of plant-derived antiviral monoclonal antibody. *Proceedings of the National Academy of Sciences of the United States of America, 100*, 8013–8018.

Kovar, J. L., Zhang, J., Funke, R. P., & Weeks, D. P. (2002). Molecular analysis of the acetolactate synthase gene of *Chlamydomonas reinhardtii* and development of a genetically engineered gene as a dominant selectable marker for genetic transformation. *Plant Journal, 29*, 109–117.

Kozminski, K. G., Diener, D. R., & Rosenbaum, J. L. (1993). High-level expression of nonacetylatable alpha-tubulin in *Chlamydomonas reinhardtii. Cell Motility and the Cytoskeleton, 25*, 158–170.

Kumar, S., Misquitta, R., Reddy, V., Rao, B., & Rajam, M. (2004). Genetic transformation of the green alga—*Chlamydomonas reinhardtii* by *Agrobacterium tumefaciens. Plant Science, 166*, 731–738.

Lapidot, M., Raveh, D., Sivan, A., Arad, S., & Sapira, M. (2002). Stable chloroplaste transformation of the unicellular red alga *Porphyridium* species. *Plant Physiology, 129*, 7–12.

Lerche, K., & Hallmann, A. (2009). Stable nuclear transformation of *Gonium pectorale. BMC Biotechnology, 9*. doi: 10.1186/1472-6750-9-64.

Levy-Ontman, O., Arad, S. M., Harvey, D. J., Parsons, T. B., Fairbanks, A., & Tekoah, Y. (2011). Unique N-glycan moieties of the 66-kDa cell wall glycoprotein from the red microalga *Porphyridium* sp. *Journal of Biological Chemistry, 286*, 21340–21352.

Li, H., Xue, L., Yana, H., Wang, L., Liu, L., Lu, Y., et al. (2007). The nitrate reductase gene-switch: a system for regulated expression in transformed cells of *Dunaliella salina. Gene, 403*, 132–142.

Li, J., Xue, L. X., Yan, H. X., Liu, H. T., & Liang, J. Y. (2008). Inducible EGFP expression under the control of the nitrate reductase gene promoter in transgenic *Dunaliella salina. Journal of Applied Phycology, 20*, 137–145.

Li, S. S., & Tsai, H. J. (2008). Transgenic microalgae as a non-antibiotic bactericide producer to defend against bacterial pathogen infection in the fish digestive tract. *Fish and Shellfish Immunology, 26*, 316–325.

Li, Y., Han, D., Hu, G., Sommerfeld, M., & Hu, Q. (2010). Inhibition of starch synthesis results in overproduction of lipids in *Chlamydomonas reinhardtii. Biotechnology and Bioengineering, 107*, 258–268.

Li, Y. T., Han, D. X., Hu, G. R., Dauvillee, D., Sommerfeld, M., Ball, S., et al. (2010). Chlamydomonas starchless mutant defective in ADP-glucose pyrophosphorylase hyperaccumulates triacylglycerol. *Metabolic Engineering, 12*, 387–391.

Liska, A. J. (2004). Enhanced photosynthesis and redox energy production contribute to salinity tolerance in *Dunaliella* as revealed by homology-based proteomics. *Plant Physiology, 136*, 2806–2817.

Llamas, A., Igeno, M. I., Galvan, A., & Fernandez, E. (2002). Nitrate signalling on the nitrate reductase gene promoter depends directly on the activity of the nitrate transport systems in *Chlamydomonas. Plant Journal, 30*, 261–271.

Lohr, M., Schwender, J., & Polle, J. E. W. (2012). Isoprenoid biosynthesis in eukaryotic phototrophs: a spotlight on algae. *Plant Science, 185*, 9–22.

Lopez, D., Casero, D., Cokus, S., Merchant, S., & Pellegrini, M. (2011). Algal functional annotation tool: a web-based analysis suite to functionally interpret large gene lists using integrated annotation and expression data. *BMC Bioinformatics, 12*. doi: 10.1186/1471-2105-12-282.

Lorenz, R. T., & Cysewski, G. R. (2000). Commercial potential for *Haematococcus* microalgae as a natural source of astaxanthin. *Trends in Biotechnology, 18*, 160–167.

Lumbreras, V., & Purton, S. (1998). Recent advances in *Chlamydomonas* transgenics. *Protist, 149*, 23–27.

Maheswari, U., Mock, T., Armbrust, E. V., & Bowler, C. (2009). Update of the diatom EST database: a new tool for digital transcriptomics. *Nucleic Acids Research, 37*, 1001–1005.

Manuell, A. L., Beligni, M. V., Elder, J. H., Siefker, D. T., Tran, M., Weber, A., et al. (2007). Robust expression of a bioactive mammalian protein in *Chlamydomonas* chloroplast. *Plant Biotechnology Journal, 5*, 402–412.

Marin-Navarro, J., Manuell, A. L., Wu, J., & Mayfield, S. P. (2007). Chloroplast translation regulation. *Photosynthesis Research, 94*, 359–374.

Matsui, M. S., Muizzuddin, N., Arad, S., & Marenus, K. (2003). Sulfated polysaccharides from red microalgae have antiinflammatory properties in vitro and in vivo. *Applied Biochemistry and Biotechnology, 104*, 13–22.

Matsuo, T., & Ishiura, M. (2011). *Chlamydomonas reinhardtii* as a new model system for studying the molecular basis of the circadian clock. *FEBS Letters, 585*, 1495–1502.

Mayali, X., & Azam, F. (2004). Algicidal bacteria in the sea and their impact on algal blooms. *The Journal of Eukaryotic Microbiology, 51*, 139–144.

Mayfield, S. P., Franklin, S. E., & Lerner, R. A. (2003). Expression and assembly of a fully active antibody in algae. *Proceedings of the National Academy of Sciences of the United States of America, 100*, 438–442.

Mayfield, S., & Schultz, J. (2004). Development of a luciferase reporter gene, luxCt, for *Chlamydomonas reinhardtii* chloroplast. *Plant Journal, 37*, 449–458.

Mayfield, S. P., & Franklin, S. E. (2005). Expression of human antibodies in eukaryotic micro-algae. *Vaccine, 23*, 1828–1832.

Meireles, L., Guedes, A. C., & Malcata, F. X. (2003). Increase of the yields of eicosapentaenoic and docosahexaenoic acids by the microalga *Pavlova lutheri* following random mutagenesis. *Biotechnology and Bioengineering, 81*, 50–55.

Mendoza, H., de la Jara, A., Freijanes, K., Carmona, L., Ramos, A. A., de Sousa Duarte, V., et al. (2008). Characterization of *Dunaliella salina* strains by flow cytometry: a new approach to select carotenoid hyperproducing strains. *Electron Journal of Biotechnology, 11*, 1–13.

Merchant, S. S., Prochnik, S. E., Vallon, O., Harris, E. H., Karpowicz, S. J., Witman, G. B., et al. (2007). The *Chlamydomonas* genome reveals the evolution of key animal and plant functions. *Science, 318*, 245–250.

Merchant, S. S., Kropat, J., Liu, B., Shaw, J., & Warakanont, J. (2011). TAG, You're it! *Chlamydomonas* as a reference organism for understanding algal triacylglycerol accumulation. *Current Opinion in Biotechnology*. doi: 10.1016/j.copbio.2011.12.001.

Miller, R., Wu, G. X., Deshpande, R. R., Vieler, A., Gartner, K., Li, X. B., et al. (2010). Changes in transcript abundance in *Chlamydomonas reinhardtii* following nitrogen deprivation predict diversion of metabolism. *Plant Physiology, 154*, 1737–1752.

Minoda, A., Sakagami, R., Yagisawa, F., Kuroiwa, T., & Tanaka, K. (2004). Improvement of culture conditions and evidence for nuclear transformation by homologous recombination in a red alga, *Cyanidioschyzon merolae* 10D. *Plant and Cell Physiology, 45*, 667–671.

Mitra, A., Higgins, D. W., & Rohe, N. J. (1994). A *Chlorella* virus gene promoter functions as a strong promoter both in plants and bacteria. *Biochemical and Biophysical Research Communications, 204*, 187–194.

Miyagawa, A., Okami, T., Kira, N., Yamaguchi, H., Ohnishi, K., & Adachi, M. (2009). Research note: high efficiency transformation of the diatom *Phaeodactylum tricornutum* with a promoter from the diatom *Cylindrotheca fusiformis*. *Phycological Research, 57*, 142–146.

Miyagawa-Yamaguchi, A., Okami, T., Kira, N., Yamaguchi, H., Ohnishi, K., & Adachi, M. (2011). Stable nuclear transformation of the diatom *Chaetoceros* sp. *Phycological Research, 59*, 113–119.

Mock, T., & Medlin, L. K. (2012). Genomics and genetics of diatoms. *Advances in Botanical Research, 64*, 245–289.

Moellering, E. R., Miller, R., & Benning, C. (2009). Molecular genetics of lipid metabolism in the model green alga *Chlamydomonas reinhardtii*. In H. Wada, & N. Murata (Eds.), *Advances in photosynthesis and respiration. Lipids in photosynthesis: Essential and regulatory functions* (pp. 139–155). Springer.

Moellering, E. R., & Benning, C. (2010). RNA interference silencing of a major lipid droplet protein affects lipid droplet size in *Chlamydomonas reinhardtii. Eukaryotic Cell, 9*, 97–106.

Molina Grima, E., Sanchez Perez, J. A., Garcia Camacho, F., Medina, A. R., Gimenez, A. G., & Alonso, D. L. (1995). The production of polyunsaturated fatty acids by microalgae: from strain selection to product purification. *Process Biochemistry, 30*, 711–719.

Morin, P. J. (2008). Sex as an algal antiviral strategy. *Proceedings of the National Academy of Sciences of the United States of America, 105*, 15639–15640.

Muhlhaus, T., Weiss, J., Hemme, D., Sommer, F., & Schroda, M. (2011). Quantitative shotgun proteomics using a uniform 15N-labeled standard to monitor proteome dynamics in time course experiments reveals new insights into the heat stress response of *Chlamydomonas reinhardtii. Molecular and Cellular Proteomics, 10*. doi: 10.1074/mcp.M110.004739.

Nagasaki, K. (2008). Dinoflagellates, diatoms, and their viruses. *Journal of Microbiology, 46*, 235–243.

Nelson, J., & Lefebvre, P. (1995). Targeted disruption of the NIT8 gene in *Chlamydomonas reinhardtii. Molecular and Cellular Biology, 15*, 5762–5769.

Neupert, J., Karcher, D., & Bock, R. (2009). Generation of *Chlamydomonas* strains that efficiently express nuclear transgenes. *Plant Journal, 57*, 1140–1150.

Nguyen, P., Falcone, D. L., & Graves, M. V. (2009). The A312L 5′-UTR of *Chlorella* virus PBCV-1 is a translational enhancer in *Arabidopsis thaliana. Virus Research, 140*, 138–146.

Nissimov, J. I., Worthy, C. A., Rooks, P., Napier, J. A., Kimmance, S. A., Henn, M. R., et al. (2011). Draft genome sequence of the coccolithovirus *Emiliania huxleyi* virus 203. *Journal of Virology, 85*, 13468–13469.

Niu, Y. F., Zhang, M. H., Xie, W. H., Li, J. N., Gao, Y. F., Yang, W. D., et al. (2011). A new inductible expression system in a transformed green alga, *Clorella vulgaris. Genetic Molecular Research, 10*(4), 3427–3434.

Norsker, N. H., Barbosa, M. J., Vermue, M. H., & Wijffels, R. H. (2011). Microalgal production: a close look at the economics. *Biotechnology Advances, 29*, 24–27.

Norton, T. A., Melkonian, M., & Andersen, R. A. (1996). Algal biodiversity. *Phycologia, 35*, 308–326.

Not, F., Siano, R., Kooistra, W. H. C. F., Simon, N., Vaulot, D., & Probert, I. (2012). Diversity and ecology of eukaryotic marine phytoplankton. *Advances in Botanical Research, 64*, 1–53.

Ohnuma, M., Yokoyama, T., Inouye, T., Sekine, Y., & Tanaka, K. (2008). Polyethylene glycol (PEG)-mediated transient gene expression in a red alga, *Cyanidioschyzon merolae* 10D. *Plant and Cell Physiology, 49*, 117–120.

Ohnuma, M., Misumi, O., Fujiwara, T., Watanabe, S., Tanaka, K., & Kuroiwa, T. (2009). Transient gene suppression in a red alga, *Cyanidioschyzon merolae* 10D. *Protoplasma, 236*, 107–112.

Ohresser, M., Matagne, R. F., & Loppes, R. (1997). Expression of the arylsulphatase reporter gene under the control of the nit1 promoter in *Chlamydomonas reinhardtii. Current Genetics, 31*, 264–271.

Pan, K., Qin, J. J., Li, S., Dai, W. K., Zhu, B. H., Jin, Y. C., et al. (2011). Nuclear monoploidy and asexual propagation of *Nannochloropsis oceanica* (Eustigmatophyceae) as revealed by its genome sequence. *Journal of Phycology, 47*, 1425–1432.

Peled, E., Leu, S., Zarka, A., Weiss, M., Pick, U., Khozin-Goldberg, I., et al. (2011). Isolation of a novel oil globule protein from the green alga *Haematococcus pluvialis* (Chlorophyceae). *Lipids, 46*, 851–861.

Petruccelli, S., Otegui, M. S., Lareu, F., Dinh, O. T., Fitchette, A. C., Circosta, A., et al. (2006). A KDEL-tagged monoclonal antibody is efficiently retained in the endoplasmic reticulum in leaves, but is both partially secreted and sorted to protein storage vacuoles in seeds. *Plant Biotechnology Journal, 4*, 511–527.

Potvin, G., & Zhang, Z. (2010). Strategies for high-level recombinant protein expression in transgenic microalgae: a review. *Biotechnology Advances, 28*, 910–918.

Poulsen, N., & Kroger, N. (2005). A new molecular tool for transgenic diatoms: control of mRNA and protein biosynthesis by an inducible promoter-terminator cassette. *FEBS Journal, 272*, 3413–3423.

Poulsen, N., Chesley, P., & Kroger, N. (2006). Molecular genetic manipulation of the diatom *Thalassiosira pseudonana* (Bacillariophyceae). *Journal of Phycology, 42*, 1059–1065.

Quinn, J. M., & Merchant, S. (1995). 2 Copper-responsive elements associated with the *Chlamydomonas* Cyc6 gene-function as targets for transcriptional activators. *The Plant Cell, 7*, 623–638.

Radakovits, R., Jinkerson, R. E., Darzins, A., & Posewitz, M. C. (2010). Genetic engineering of algae for enhanced biofuel production. *Eukaryotic Cell, 9*, 486–501.

Ramazanov, A., & Ramazanov, Z. (2006). Isolation and characterization of a starchless mutant of *Chlorella pyrenoidosa* STL-PI with a high growth rate, and high protein and polyunsaturated fatty acid content. *Phycological Research, 54*, 255–259.

Rasala, B. A., Muto, M., Lee, P. A., Jager, M., Cardoso, R. M. F., Behnke, C. A., et al. (2010). Production of therapeutic proteins in algae, analysis of expression of seven human proteins in the chloroplast of *Chlamydomonas reinhardtii*. *Plant Biotechnology Journal, 8*, 719–733.

Rasala, B. A., & Mayfield, S. (2011). The microalgae *Chlamydomonas reinhardtii* as a platform for the production of human protein therapeutics. *Bioengineered bugs, 2*, 50–54.

Rasala, B. A., Muto, M., Sullivan, J., & Mayfield, S. P. (2011). Improved heterologous protein expression in the chloroplast of *Chlamydomonas reinhardtii* through promoter and 5′ untranslated region optimization. *Plant Biotechnology Journal, 9*, 674–683.

Raven, J. A., & Falkowski, P. G. (1999). Oceanic sinks for atmospheric CO_2. *Plant, Cell and Environment, 22*, 741–755.

Rismani-Yazdi, H., Haznedaroglu, B. Z., Bibby, K., & Peccia, J. (2011). Transcriptome sequencing and annotation of the microalgae *Dunaliella tertiolecta*: pathway description and gene discovery for production of next-generation biofuels. *BMC Genomics, 12*. doi: 10.1186/1471-2164-12-148.

Rolland, N., Atteia, A., Decottignies, P., Garin, J., Hippler, M., Kreimer, G., et al. (2009). *Chlamydomonas* proteomics. *Current Opinion in Microbiology, 12*, 285–291.

Rouxel, C., Bougaran, G., Doulin-Grouas, S., Dubois, N., & Cadoret, J.-P. (2011). Novel *Isochrysis* sp tahitian clone and uses therefore. EP 11006712.1., EU.

Sakaue, K., Harada, H., & Matsuda, Y. (2008). Development of gene expression system in a marine diatom using viral promoters of a wide variety of origin. *Physiologia Plantarum, 133*, 59–67.

Sasso, S., Pohnert, G., Lohr, M., Mittag, M., & Hertweck, C. (2011). Microalgae in the postgenomic era: a blooming reservoir for new natural products. *FEMS Microbiology*. doi: 10.1111/j.1574–6976.2011.00304.

Sastre, R. R., & Posten, C. (2010). The variety of microalgae applications as a renewable resource. *Chemie Ingenieur Technik, 82*, 1925–1939.

Schiedlmeier, B., Schmitt, R., Muller, W., Kirk, M. M., Gruber, H., Mages, W., et al. (1994). Nuclear transformation of *Volvox-carteri*. *Proceedings of the National Academy of Sciences of the United States of America, 91*, 5080–5084.

Schmidt, F. R. (2004). Recombinant expression systems in the pharmaceutical industry. *Microbiology and Biotechnology, 65*, 363–372.
Schroda, M., Blocker, D., & Beck, C. F. (2000). The HSP70A promoter as a tool for the improved expression of transgenes in *Chlamydomonas*. *Plant Journal, 21*, 121–131.
Schroeder, D. C., Oke, J., Malin, G., & Wilson, W. H. (2002). Coccolithovirus (Phycodnaviridae): characterisation of a new large dsDNA algal virus that infects *Emiliania huxleyi*. *Archives of Virology, 147*, 1685–1698.
Sekar, S., & Chandramohan, M. (2008). Phycobiliproteins as a commodity: trends in applied research, patents and commercialization. *Journal of Applied Phycology, 20*, 113–136.
Shaish, A., Ben-Amotz, A., & Avron, M. (1991). Production and selection of high beta-carotene mutants of *Dunaliella bardawil* (Chlorophyta). *Journal of Phycology, 27*, 652–656.
Sheehan, J., Dunahay, T., Benemann, J., & Roessler, P. (1998). A look back at the U.S. Department of Energy's Aquatic Species program: Biodiesel from algae. In "*US Report NREL/TP-580-24190*" (p. 323). Golden, USA: US Dpt of Energy.
Siaut, M., Heijde, M., Mangogna, M., Montsant, A., Coesel, S., Allen, A., Manfredonia, A., Falciatore, A., & Bowler, C. (2007). Molecular toolbox for studying diatom biology in *Phaeodactylum tricornutum*. *Gene, 406*, 23–35.
Siaut, M., Cuiné, S., Cagnon, C., Fessler, B., Nguyen, M., Carrier, P., et al. (2011). Oil accumulation in the model green alga *Chlamydomonas reinhardtii*: characterization, variability between common laboratory strains and relationship with starch reserves. *BMC Biotechnology, 11*. doi: 10.1186/1472-6750-11-7.
Sizova, I., Fuhrmann, M., & Hegemann, P. (2001). A *Streptomyces rimosus* aphVIII gene coding for a new type phosphotransferase provides stable antibiotic resistance to *Chlamydomonas reinhardtii*. *Gene, 277*, 221–229.
Specht, E., Miyake-Stoner, S., & Mayfield, S. (2010). Micro-algae come of age as a platform for recombinant protein production. *Biotechnology Letters, 32*, 1373–1383.
Spolaore, P., Joannis-Cassan, C., Duran, E., & Isambert, A. (2006). Commercial applications of microalgae. *Journal of Bioscience and Bioengineering, 101*, 87–96.
Steinbrenner, J., & Sandmann, G. (2006). Transformation of the green alga *Haematococcus pluvialis* with a phytoene desaturase for accelerated astaxanthin biosynthesis. *Applied and Environmental Microbiology, 72*, 7477–7484.
Stevens, D. R., Rochaix, J. D., & Purton, S. (1996). The bacterial phleomycin resistance gene ble as a dominant selectable marker in *Chlamydomonas*. *Molecular & General Genetics, 251*, 23–30.
Sun, M., Qian, K. X., Su, N., Chang, H. Y., Liu, J. X., & Chen, G. F. (2003). Foot-and-mouth disease virus VP1 protein fused with cholera toxin B subunit expressed in *Chlamydomonas reinhardtii* chloroplast. *Biotechnology Letters, 25*, 1087–1092.
Sun, Y., Yang, Z. Y., Gao, X. S., Li, Q. Y., Zhang, Q. Q., & Xu, Z. K. (2005). Expression of foreign genes in *Dunaliella* by electroporation. *Molecular Biotechnology, 30*, 185–192.
Surzycki, R., Greenham, K., Kitayama, K., Dibal, F., Wagner, R., Rochaix, J.-D., et al. (2009). Factors effecting expression of vaccines in microalgae. *Biologicals, 37*, 133–138.
Tan, C., Qin, S., Zhang, Q., Jiang, P., & Zhao, F. (2005). Establishment of a micro-particle bombardment transformation system for *Dunaliella salina*. *Journal of Microbiology, 43*, 361–365.
Tang, D., Qiao, S., & Wu, M. (1995). Insertion mutagenesis of *Chlamydomonas reinhardtii* by electroporation and heterologous DNA. *Biochemistry and Molecular Biology International, 36*, 1025–1035.
Ten Lohuis, M. R., & Miller, D. J. (1998). Genetic transformation of dinoflagellates (Amphidinium and Symbiodinium): expression of GUS in microalgae using heterologous promoter constructs. *Plant Journal, 13*, 427–435.

Teng, C., Qin, S., Liu, J., Yu, D., Liang, C., & Tseng, C. (2002). Transient expression of lacZ in bombarded nicellular green alga *Haematococcus pluvialis*. *Journal of Applied Phycology, 14*, 495–500.

Thomas, R., Grimsley, N., Escande, M. L., Subirana, L., Derelle, E., & Moreau, H. (2011). Acquisition and maintenance of resistance to viruses in eukaryotic phytoplankton populations. *Environmental Microbiology, 13*, 1412–1420.

Toulza, E., Blanc-Mathieu, R., Gourbiere, S., & Piganeau, G. (2012). Environmental genomics of microbial algae: power and challenges of metagenomics. *Advances in Botanical Research, 64*, 379–423.

Tran, N., Park, J., Hong, S., & Lee, C. (2009). Proteomics of proteins associated with astaxanthin accumulation in the green algae *Haematococcus lacustris* under the influence of sodium orthovanadate. *Biotechnology Letters, 31*, 1917–1922.

Tran, M., Zhou, B., Pettersson, P. L., Gonzalez, M. J., & Mayfield, S. P. (2009). Synthesis and assembly of a full-length human monoclonal antibody in algal chloroplasts. *Biotechnology and Bioengineering, 104*, 663–673.

Umen, J., & Olson, B. (2012). Genomics of volvocine algae. *Advances in Botanical Research, 64*, 185–243.

Van Etten, J. L., & Dunigan, D. D. (2012). Chloroviruses: not your everyday plant virus. *Trends in Plant Science, 17*, 1–8.

Villalobos, A., Ness, J. E., Gustafsson, C., Minshull, J., & Govindarajan, S. (2006). Gene designer: a synthetic biology tool for constructing artificial DNA segments. *BMC Bioinformatics, 7*. doi: 10.1186/1472-2105-7-285.

Von Dassow, P., Ogata, H., Probert, I., Wincker, P., Da Silva, C., Audic, S., et al. (2009). Transcriptome analysis of functional differentiation between haploid and diploid cells of *Emiliania huxleyi*, a globally significant photosynthetic calcifying cell. *Genome Biology, 10*. doi: 10.1186/gb-2009-10-10-r114.

Walker, T. L., Collet, C., & Purton, S. (2005). Algal transgenics in the genomic ERA. *Journal of Phycology, 41*, 1077–1093.

Wallis, J. G., & Browse, J. (2010). Lipid biochemists salute the genome. *Plant Journal, 61*, 1092–1106.

Wang, C. H., Wang, Y. Y., Su, Q., & Gao, X. R. (2007). Transient expression of the GUS gene in a unicellular marine green alga, *Chlorella* sp MACC/C95, via electroporation. *Biotechnology and Bioprocess Engineering, 12*, 180–183.

Wang, T. Y., Xue, L. X., Hou, W. H., Yang, B. S., Chai, Y. R., Ji, X. A., et al. (2007b). Increased expression of transgene in stably transformed cells of *Dunaliella salina* by matrix attachment regions. *Applied Microbiology and Biotechnology, 76*, 651–657.

Wang, X. F., Brandsma, M., Tremblay, R., Maxwell, D., Jevnikar, A. M., Huner, N., et al. (2008). A novel expression platform for the production of diabetes-associated autoantigen human glutamic acid decarboxylase (hGAD65). *BMC Biotechnology, 8*. doi: 10.1186/1472-6750-8-87.

Wang, Z. T., Ullrich, N., Joo, S., Waffenschmidt, S., & Goodenough, U. (2009). Algal lipid bodies: stress induction, purification, and biochemical characterization in wild-type and starch-less *Chlamydomonas reinhardtii*. *Eukaryotic Cell, 8*(12), 1856–1868.

Watanabe, S., Ohnuma, M., Sato, J., Yoshikawa, H., & Tanaka, K. (2011). Utility of a GFP reporter system in the red alga *Cyanidioschyzon merolae*. *Journal of General and Applied Microbiology, 57*, 69–72.

Weber, A., Oesterhelt, C., Gross, W., Brautigam, A., Imboden, L., Krassovskaya, I., et al. (2004). EST-analysis of the thermo-acidophilic red microalga *Galdieria sulphuraria* reveals potential for lipid A biosynthesis and unveils the pathway of carbon export from rhodoplasts. *Plant Molecular Biology, 55*, 17–32.

Wijffels, R. H., Barbosa, M. J., & Eppink, M. H. M. (2010). Microalgae for the production of bulk chemicals and biofuels. *Biofuels, Bioproducts and Biorefining, 4*, 287–295.

Wilson, W. H., Van Etten, J. L., & Allen, M. J. (2009). The Phycodnaviridae: the story of how tiny giants rule the world. *Current topics in microbiology and immunology, 328*, 1–42.
Winter, J. M., Behnken, S., & Hertweck, C. (2011). Genomics-inspired discovery of natural products. *Current Opinion in Chemical Biology, 15*, 22–31.
Yamano, T., & Fukuzawa, H. (2009). Carbon-concentrating mechanism in a green alga, *Chlamydomonas reinhardtii*, revealed by transcriptome analyses. *Journal of Basic Microbiology, 49*, 42–51.
Yang, Z. Q., Li, Y. N., Chen, F., Li, D., Zhang, Z. F., Liu, Y. X., et al. (2006). Expression of human soluble TRAIL in *Chlamydomonas reinhardtii* chloroplast. *Chinese Science Bulletin, 51*, 1703–1709.
Yu, S., Liu, S., Li, C., & Zhou, Z. (2011). Submesoscale characteristics and transcription of a fatty acid elongase gene from a freshwater green microalgae, *Myrmecia incisa* Reisigl. *Chinese Journal of Oceanology and Limnology, 29*, 87–95.
Zaslavskaia, L., & Lippmeier, J. C. (2000). Transformation of the diatom *Phaeodactylum tricornutum* (Bacillariophyceae) with a variety of selectable marker and reporter genes. *Journal of Phycology, 36*, 379–386.
Zhang, Y. K., Shen, G. F., & Ru, B. G. (2006). Survival of human metallothionein-2 transplastomic *Chlamydomonas reinhardtii* to ultraviolet B exposure. *Acta Biochimica Et Biophysica Sinica, 38*, 187–193.
Zhao, R., Cao, Y., Xu, H., Lv, L., Qiao, D., & Cao, Y. (2011). Analysis of expressed sequence tags from the green alga *Dunaliella salina* (Chlorophyta). *Journal of Phycology, 47*, 1454–1460.

Genomics of Algal Host–Virus Interactions

Nigel H. Grimsley[1,*], Rozenn Thomas*, Jessica U. Kegel[†], Stéphan Jacquet[‡], Hervé Moreau*, and Yves Desdevises*

[*]CNRS, UMR7232, University Pierre et Marie Curie Paris 06, Laboratoire de Biologie Intégrative des Organisms Marins, Observatoire Océanologique, Banyuls-sur-Mer, France
[†]Alfred Wegener Institute for Polar and Marine Research, Am Handelshafen 12, D-27570 Bremerhaven, Germany
[‡]INRA, Stationd'Hydrobiologie Lacustre, 74203 Thonon-les-bains cedex, France
[1]Corresponding author: E-mail: nigel.grimsley@obs-banyuls.fr

Contents

Abstract

Viruses in Earth's aquatic environment outnumber all other forms of life and carry a vast reservoir of genetic information. A large proportion of the characterized viruses infecting eukaryotic algae are large double-stranded DNA viruses, each of their genomes carrying more than a hundred genes, but only a minority of their genes resemble genes with known biological functionalities. Unusual forms of single-stranded DNA and single- and double-stranded RNA viral genomes have been characterized

Advances in Botanical Research, Volume 64
ISSN 0065-2296,
http://dx.doi.org/10.1016/B978-0-12-391499-6.00009-8

over the last 10 years, and the number of novel taxa of viruses being discovered continues to increase. Although viral infections are usually specific to certain host strains in a species, lytic viral infections nevertheless affect a large proportion of algae and have a global impact, for example in the termination of blooms. Resistance to viruses is thus subject to strong selection, but little is known about its mechanism. Lateral gene transfer between host and virus has been shown by comparisons between their complete genomes and must play an important role in coevolution in the microbial world. Recent advances in bioinformatics and the possibility of amplifying complete genomes from single cells promise to revolutionize analyses of viral genomes from environmental samples.

1. INTRODUCTION

When life was born in the oceans, so were viruses. All known life forms are infected, either chronically or lytically, at certain or all stages of their lifetimes, by their specific viruses. When terrestrial life forms evolved, viruses became hitchhikers that were forced to adapt to a drastically different environment. They could no longer diffuse or be carried to another host cell by diffusion or turbulence, so new means for transmission were required. Additionally, terrestrial plant cells have developed a rigid cell wall to resist the reduced osmotic pressure of a freshwater environment, a formidable barrier to viral ingress. Arguably, one of the selective pressures acting in adaptation of marine life to terrestrial conditions may have been to escape viral attack in an environment teeming with viruses that outnumber host populations by an order of magnitude. Nowadays, vegetal viruses are usually carried from plant to plant by sucking or biting insects or less frequently by the mechanical contacts with animals harvesting or moving through vegetation.

In this review, we will turn our attention to viruses of photosynthetic eukaryotes in the euphotic zone of aquatic environments, namely that depth of water that receives enough light for photosynthesis, on average down to about 200 m below the surface in the open sea, and at very variable depths in coastal or freshwater lakes and rivers, because of variable levels of turbidity. The term 'algae' will be used to regroup these organisms, although it has no phylogenetic significance, spanning at least four kingdoms in the tree of life (see Not *et al.* (2012), De Clerck, Bogaret, & Leliaert (2012) and Archibald (2012) in this volume for a review on the diversity of algae). In the context of this volume, we will furthermore consider only viruses whose complete genomic sequences have been analysed, giving clues about their biological functionalities, and apologize for not including important data about partial sequences, individual genes or the growing number of genomes being assembled from metagenomic data (whose host species are not usually known). We will not include much detail about individual viruses but rather refer the reader to more detailed reviews and original research articles. Chloroviruses and other large viruses of protists have been reviewed extensively (Van Etten, 2003; Van Etten and Dunigan, 2011; Yamada, Onimatsu, & Van Etten, 2006), and a comprehensive review of dinoflagellate and diatom viruses is also available (Nagasaki, 2008), though this does not include the most recently discovered viruses. The largest viruses known, such as mimivirus (microbe-mimicking virus), infect non-photosynthetic protists, so we considered them outside of the scope of this botanical journal, although we note that they are in the very diverse family of double-stranded DNA (dsDNA) viruses that includes 'phycodnaviruses'. These giant viruses (giruses) are the subjects of several other reviews (Claverie *et al.*, 2006; Forterre, 2010; Van Etten, 2011; Van Etten *et al.*, 2010).

1.1. What Are Viruses?

Viruses consist of a nucleic acid sequence enclosed within a protein and/or lipid envelope. The simplest viral genomes thus may encode only two biological functionalities – a polymerase to ensure replication of their nucleic acid sequence and a capsid protein (CP) that is produced abundantly and coats the nucleic acid to provide protection in the period when a virus is not within its natural host. The nucleic acid component can be RNA or DNA, single or double stranded, and is a characteristic of the type of virus. Such simple viruses are completely dependent on host cell functionalities (such as protein synthesis), but algal viral genomes can also be very large, encoding hundreds of functionalities (Table 9.1). Terrestrial eukaryotes are dominated largely by only two kingdoms of organisms, animals and plants, but all of five kingdoms among the currently recognized eukaryotic divisions of life (Fig. 9.1) are well represented in aquatic environments. Whereas all the *Plantae* possess plastids, many lineages within the other four kingdoms can harbour photosynthetic plastids. The extent of such endosymbioses varies between kingdoms, most chromalveolates being photosynthetic, and symbiotic associations in the other kingdoms are more or less common depending on the lineage (reviewed in Johnson (2011) and Archibald (2012) in this volume).

1.2. Why Are Algal Viruses Important?

Phytoplankton is responsible for about half of the photosynthetic activity of the planet (Field, Behrenfeld, Randerson, & Falkowski, 1998), the second

Table 9.1 Algal Viruses Whose Complete Genomic DNA or RNA Sequences Are Known

Full name	GenBank Accession	Abbreviation	Source Information	Genome Size, Nucleotides
Acanthamoeba polyphaga mimivirus	NC_014649	APMV	Raoult *et al.* (2004)	1,181,549
Acanthocystis turfacea chlorella virus 1	NC_008724	ATCV-1	Fitzgerald *et al.* (2007)	288,047
Bathycoccus sp. RCC1105 virus BpV1	NC_014765	BpV1	Moreau *et al.* (2010)	198,519
Bathycoccus sp. RCC1105 virus BpV2	HM004430	BpV2	Moreau *et al.* (2010)	187,069
Cafeteria roenbergensis	GU244497	CroV	Fischer *et al.* (2010)	617,453
Chaetoceros lorenzianus DNA virus	NC_015211	ClorDNAV01	Tomaru *et al.* (2011)	5813
Chaetoceros salsugineum DNA virus	NC_007193	CsalDNAV	Nagasaki *et al.* (2005b)	6000
Chaetoceros tenuissimus DNA virus	NC_014748	CtenDNAV06	Shirai *et al.* (2007)	5639
Chaetoceros tenuissimus RNA virus	AB375474	CtenRNAV01	Shirai *et al.* (2008)	9431
Chara australis virus	JF824737	CAV	Gibbs *et al.* (2011)	9065[a]
Ectocarpus siliculosus virus 1	NC_002687	EsV-1	Delaroque *et al.* (2001)	335,593
Emiliania huxleyi virus 84	JF974290	EhV-84	Nissimov *et al.* (2011a)	395,820[a]
E. huxleyi virus 86	NC_007346	EhV-86	Wilson *et al.* (2005)	407,339
E. huxleyi virus 88	JF974310.1	EhV-88	Nissimov *et al.* (2012)	397,298
E. huxleyi virus 163	DQ127552−127818	EhV-163	Allen *et al.* (2006)	400,000[b]
E. huxleyi virus 201	JF974311.1	EhV-201	Nissimov *et al.* (2012)	407,301
E. huxleyi virus 203	JF974291	EhV-203	Nissimov *et al.* (2011b)	400,520[a]
E. huxleyi virus 207	JF974317.1	EhV-207	Nissimov *et al.* (2012)	421,891
E. huxleyi virus 208	JF974318.1	EhV-208	Nissimov *et al.* (2012)	411,003
Feldmannia species virus	NC_011183	FsV-158	Schroeder *et al.* (2009)	154,641
Heterocapsa circularisquama RNA virus	NC_007518	HcRNAV34	Nagasaki *et al.* (2005)	4375
H. circularisquama RNA virus	AB218609	HcRNAV109	Nagasaki *et al.* (2005)	4391

Heterosigma akashiwo RNA virus SOG263	NC_005281	HaRNAV-SOG263	Lang *et al.* (2004)	8587
Megavirus chiliensis	NC_016072	MGVC	Arslan *et al.* (2011)	1,259,197
Micromonas pusilla reovirus	NC_008171–8181	MpRV	Attoui *et al.* (2006)	25,563
Micromonas sp. RCC1109 virus MpV1	NC_014767	MpV1	Moreau *et al.* (2010)	184,095
Ostreococcus lucimarinus virus OlV1	NC_014766	OlV1	Moreau *et al.* (2010)	194,022
Ostreococcus tauri virus 1	NC_013288	OtV1	Weynberg *et al.* (2009)	191,761
O. tauri virus 2	NC_014789	OtV2	Weynberg *et al.* (2011)	184,409
Ostreococcus virus OsV5	NC_010191	OtV5	Derelle *et al.* (2008)	185,373
Paramecium bursaria chlorella virus 1[c]	NC_000852	PBCV-1	Yanai-Balser *et al.* (2010)	330,611
P. bursaria chlorella virus AR158	NC_009899	PBCV-AR158	Fitzgerald, Graves, Li, Feldblyum, *et al.* (2007)	344,691
P. bursaria chlorella virus NY2A	NC_009898	PBCV-NY2A	Fitzgerald, Graves, Li, Feldblyum, *et al.* (2007)	368,683
P. bursaria chlorella virus FR483	NC_008603	PBCV-FR483	Fitzgerald, Graves, Li, Hartigan, *et al.* (2007)	321,240
P. bursaria chlorella virus MT325	DQ491001	PBCV-MT325	Fitzgerald, Graves, Li, Hartigan, *et al.* (2007)	314,335
Rhizosolenia setigera RNA virus	AB243297	RsRNAV	Nagasaki *et al.* (2004)	11,200[b]
Schizochytrium sp. single-stranded RNA virus	NC_007522	SsSRVAV	Takao *et al.* (2006)	9035

Rows on a grey background are examples of viruses from non-photosynthetic protists.
[a]Almost complete.
[b]About 80% of this length complete.
[c]The most recent version of the genome, but this sequence was published in several parts previously.

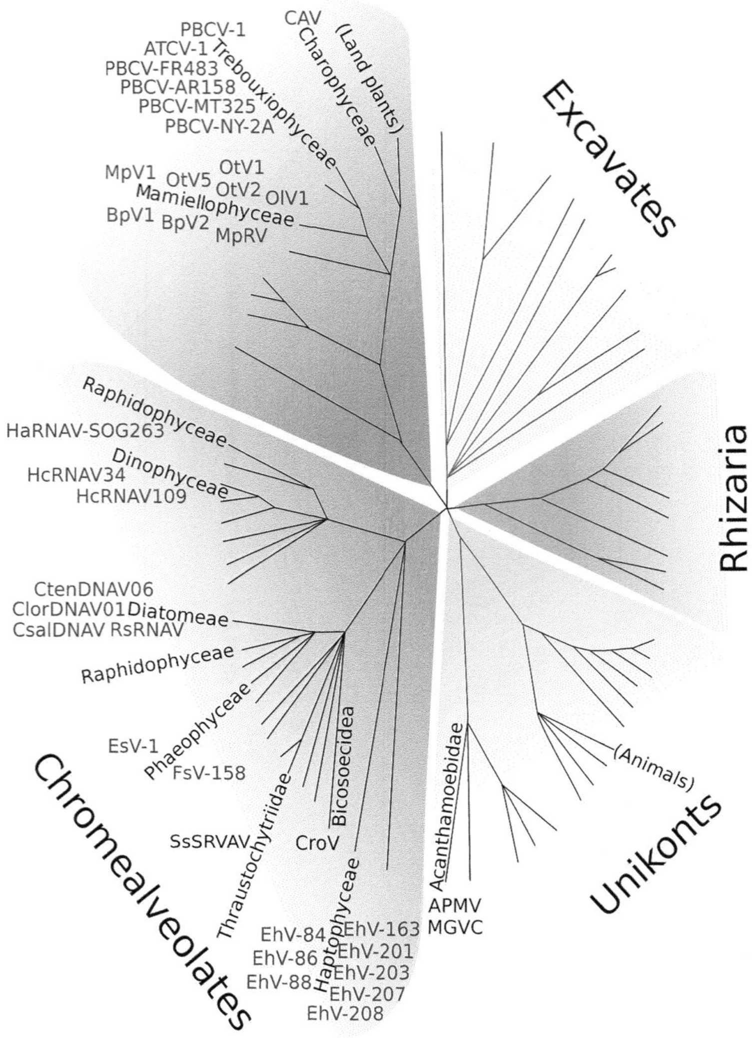

Figure 9.1 ***Algal viruses whose genomes have been sequenced.*** Known genomes of viruses infecting photosynthetic algae (text in blue, please refer to Table 9.1 for the names of hosts and viruses) lie mainly in two of the five eukaryotic kingdoms (coloured backgrounds) of life shown (most Unikonts, grey background, do not carry plastids, so their viruses are not included in this review). Only taxa with viruses mentioned in this review are labelled. Many lineages in Rhizaria, Metazoans (Unikonts) and Excavates can form symbioses with photosynthetic organisms (Johnson *et al.*, 2011). Algae of the Trebouxiophyceae infected by PBCV are usually symbionts of *Paramecium bursaria* (Alveolata) in nature. The positions of land plants and animals are shown for reference (tree simplified from Keeling *et al.*, 2005). See the colour plate.

half being ensured by terrestrial plants. Algae are at the base of the global food web, nourishing all aquatic life. In aquatic environments, while bacteria represent the largest biomass of organisms present, viruses outnumber them by 10 to 1. Algal viruses control blooms and shape the evolution of biodiversity in phytoplankton, yet little is known about their biological functions. In the oceans, viruses are thus the most abundant and diverse biological entities (Fuhrman, 1999; Wommack & Colwell, 2000) and infect all organisms from bacteria to whales (Suttle, 2005). The marine environment contains an estimated 10^{30} virus-like particles (Suttle, 2007). Most of the viruses described to date are species specific, infecting a single host species and sometimes even a single strain within a species. Due to their immobility, viruses depend on passive movement to contact a suitable host (Brussaard, 2004; Weinbauer, 2004). Consequently, the encounter rate between a virus and a host is directly affected by their relative abundances.

Several studies have shown the infection of a wide range of aquatic algae (Van Etten, Lane, & Meints, 1991; Van Etten & Meints, 2003) including bloom-forming marine phytoplankton (Suttle & Chan, 1995, Jacobsen, Bratbak, & Heldal, 1996; Sandaa, Heldal, Castberg, Thyrhaug, & Bratbak, 2001) like *Phaeocystis globosa* (Brussaard *et al.*, 2005), *Heterosigma akashiwo* (Nagasaki *et al.*, 1994a, 1994b; Nagasaki & Yamaguchi, 1997), *Aureococcus anophagefferens* (Gobler *et al.*, 1997, 2004, 2007), *Emiliania huxleyi* (Bratbak *et al.*, 1993) and *Ostreococcus* sp. (Countway & Caron, 2006).

When host organisms are lysed, nutrients are released into the surrounding environment and thus influence biogeochemical and ecological processes (Fuhrman, 1999; Gobler *et al.*, 1997; Sandaa, 2008; Wilhelm & Suttle, 1999). Viral lysis affects the efficiency of the biological pump by increasing or decreasing the relative amount of carbon in exported production (Suttle, 2007). This so-called 'viral shunt' moves material from heterotrophic and phototrophic microorganisms into particulate organic matter and dissolved organic matter (Gobler *et al.*, 1997; Middelboe *et al.*, 1996), which is mostly converted to CO_2 by respiration and photodegradation (Fuhrman, 1999; Suttle, 2005; Weinbauer, 2004; Wilhelm, 1999). Furthermore, in the sea the accelerated sinking rates of virus-infected cells increase the transport of organic molecules from the photic zone to the deep ocean (Lawrence & Suttle, 2004; Lawrence *et al.*, 2002).

Marine microbial virology has mainly concentrated on the infection of marine bacteria regarding abundance, genetic diversity, host specificity and genomics (Sullivan *et al.*, 2006). In comparison to prokaryotic viruses, less is known about viruses that infect marine eukaryotic phytoplankton, although viral proliferation can trim populations or terminate phytoplankton blooms (Wommack & Colwell, 2000; Gobler *et al.*, 2004) and shuttle genetic material (Brown *et al.*, 2007; Rohwer & Thurber, 2009).

Most of the identified eukaryotic phytoplankton viruses are members of the family Phycodnaviridae, a diverse group of large icosahedral viruses with dsDNA genomes ranging from 160 to 560 kb with 100- to 220-nm-sized capsids (Van Etten *et al.*, 2002). Their importance in aquatic environments become clear in several independent studies (Monier, Claverie, & Ogata, 2008; Monier, Larsen, *et al.*, 2008; Short & Short 2008; Short & Suttle, 2002; Fischer, Allen, Wilson, & Suttle, 2010). Members of the Phycodnaviridae are currently grouped into six genera (named after the hosts they infect): *Chlorovirus, Coccolithovirus, Prasinovirus, Prymnesiovirus, Phaeovirus* and *Raphidovirus* (Wilson *et al.*, 2009). Complete genomes have been sequenced from representatives of the *Chlorovirus, Coccolithovirus, Phaeovirus* and *Prasinovirus* genera (Dunigan *et al.*, 2006).

Viruses affect host population dynamics and nutrient flow in aquatic food webs. However, only a small portion of marine viruses has been isolated and described so far, revealing that marine virology is still in its infancy. Each infection has the potential to introduce new genetic information in an organism or progeny virus, thereby driving the evolution of both host and viral assemblages (Suttle, 2007).

Marine viruses have been mainly studied for socio-economic reasons linked to massive and sudden death of microalgae or metazoan host organisms in natural marine environments or in aquacultures (Brussaard, 2004; Nagasaki, 2008). Microalgal bloom is a phenomenon characterized by a rapid increase in population of microscopic unicellular algae. When their pigments discolour the water, blooms are called 'red tide', 'brown tide' or 'green tide' depending on the colour of water. Harmful algal blooms cause large economic damage in fishery, aquaculture, leisure industries and other socio-economic activities in coastal areas. For instance, red tides of raphidophytes and dinoflagellates lead to recurrent serious mass mortality of cultured fishes and bivalves in Japan, Canada, New Zealand and Chile (K. Nagasaki, personal communication). Algal blooms often experience a sudden disintegration and disappearance, and in many cases, giant viruses are the agent infecting and killing bloom-forming algae. Viruses are thus one of the main regulators of the seasonal occurrence and termination of algal blooms. These viruses represent potential anti-algae agents (akin to 'phage therapy') in aquacultures. Algal blooms and associated viruses have been also implicated in other ecological/climatic processes. The cosmopolitan coccolithophore *E. huxleyi* is known for its white blooms covering huge oceanic surfaces (Zondervan, 2007). This photosynthetic microalga contributes to the production of atmospheric dimethyl sulphide, which in turn leads to cloud formation. Coccolithophores drive massive sinking of calcium carbonate into deep oceanic lithosphere, thus contributing also to global carbon fluxes; for instance, Dover's chalk cliffs are largely (80%) made up of beautiful calcium carbonate scales of ancient coccolithophores (ca. 100 million years old) and illustrate the significant geochemical role of coccolithophores. The huge *E. huxleyi* blooms suddenly terminate due to the infection of marine viruses called EhVs (*E. huxleyi* viruses) with a large ~400-kb genome (Bratbak *et al.*, 1993; Wilson *et al.*, 2005b).

Global warming has now become unequivocal, after multiple and serious scientific surveys by international and intergovernmental organizations during the last two decades. The United Nation's Intergovernmental Panel on Climate Change reported significant increases in global average air and ocean temperatures, widespread melting of snow and ice and rising average sea level (the Fourth Assessment Report: http://www.ipcc.ch/pdf/assessment-report/ar4/syr/ar4_syr.pdf). The recent Tara-Arctic expedition (2007–2008) also revealed serious melting of the ice in the most northerly latitudes. Recent studies show that marine viral abundance increases with temperature (see Danovaro *et al.* [2011] for a review), but many other aspects of this complex ecosystem are also affected.

2. WHAT IS KNOWN ABOUT AQUATIC ALGAL VIRUS GENOMICS?

Although interest in aquatic algal viruses is growing (Fig. 9.2), for reasons mentioned above, relatively few algal viruses have been characterized at the level of their genomes. There are several reasons for this paucity of knowledge. The majority of algae have little direct economic importance, and unlike land plants, the large majority of algae are not easily observable, most species being microscopic and distributed around remote regions of the planet. In addition, most algal species are not easily cultured in the laboratory, usually a necessary condition to perform the molecular biological steps required for sequencing the genome. However, given the increased interest in aquatic life arising because of global climate change and the reduction in cost of next-generation sequencing, numerous new algal and viral genomes are now being sequenced.

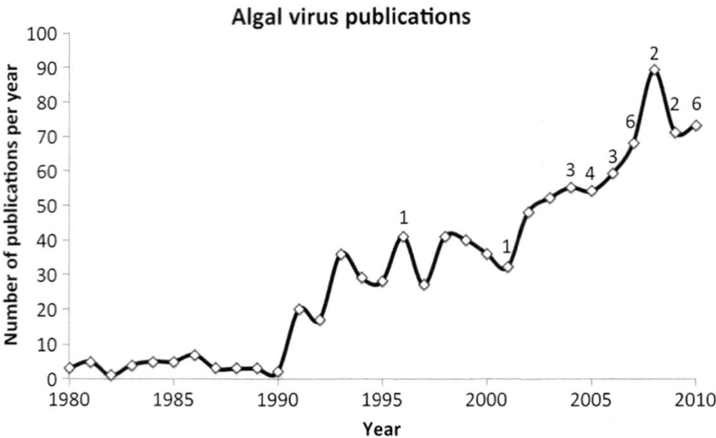

Figure 9.2 The growing interest in algal viruses is witnessed by an increasing number of publications. The curve shows new journal articles appearing each year as seen by an internet keyword search (Web of Knowledge) by combining the key words 'alga' and 'virus' over the 20-year period 1980–2010. Figures above the curve indicate the number of new complete genomes reported for the corresponding year (listed in Table 9.1). In 2011 (data not shown), 99 new articles and 6 complete genomes were published. Four more genomes have been described in the first 3 months of 2012.

2.1. Genomics

A new era is dawning for research on viral genomes with the advent of 'Next Generation Sequencing' (NGS) technologies (Gilbert & Dupont, 2011; Nowrousian *et al.*, 2010; Rodriguez-Brito *et al.*, 2010). Whole genomes of very large viruses can be sequenced more easily at a lower overall cost, and sequence analyses are facilitated by a growing number of bioinformatics programs for assembling and annotating data, permitting prediction of some of the encoded biological functionalities. In addition, the size ranges of viruses in general falls into those classes of organisms that can be collected from diverse environments by filtration (Lauro *et al.*, 2011) or flocculation (John *et al.*, 2011) and the relatively small size of their genomes, much less complex than those of eukaryotes, facilitates interpretation of sequences from metagenomic data or deep sequencing of given polymerase chain reaction-amplified marker genes. NGS technology thus allows the exploration of viral diversity in a range of environments since complete genome data from laboratory-cultured strains can be used to probe and assess the distribution of a species between these environments. The exponentially increasing amount of metagenomic data from diverse environments can thus be exploited. This kind of approach first enabled the distributions of prokaryotic viruses (bacteriophages) to be analysed, revealing an unprecedented abundance and diversity, including astronomic numbers of unknown biological functionalities in these 'viromes' (Bench *et al.*, 2007), but more recently, attention is turning to eukaryotic viruses (e.g. Kristensen, Mushegian, Dolja, & Koonin, 2010; Monier Claverie, J.-M. & Ogata, H. 2008).

In the following sections, we will firstly summarize very briefly some of the history of aquatic algal virus genomes before discussing the evolution of host–virus interactions, finishing with some perspectives about how the field is developing.

2.1.1. dsDNA Viruses (Giruses) Abound in the Aquatic World

The first reports of viruses or viral-like particles in green and brown algae date back as far back as 1958, but further confirmations about their nature, with large particle sizes (100–200 nm) came in the 1970s (reviewed in Brown, 1972; Van Etten *et al.*, 1991). To date, the majority (about two thirds, 23/33 listed in Table 9.1) of algal viruses whose genomes are characterized are phycodnaviruses.

2.1.1.1. Chlorella Viruses

While most species of the unicellular green alga chlorella are free living, certain of them can form symbioses. The freshwater unicellular protozoan *Paramecium bursaria*, or the metazoan *Hydra viridis*, for example, can harbour symbiotic chlorella-like 'zoochlorellae'. In paramecium, each algal cell is enclosed in a perialgal vacuole, and all chlorellae in the host cell are inherited to the progeny, undergoing coordinated division with the host cells, giving a constant population density of several hundred per cell. When such hosts are cultured for some time under suitable conditions without light, the zoochlorellae are released and can be cultured independently on liquid or solidified media. Zoochlorellae in culture are susceptible to lytic attack from phycodnaviruses.

In native freshwater, the titre of PBCV-1 (*P. bursaria* chlorella virus) particles may attain 100,000 plaque-forming units (PFUs) per millilitre but more typically are found to be around 1–100 PFU/mL (Van Etten *et al.*, 1985).

Over the last 30 years, research on PBCV-1 has revealed some fascinating features about the structure and biological functionalities encoded by such large viruses (several reviews are available, Yamada *et al.*, 2006; Van Etten *et al.*, 2010; Van Etten and Dunigan, 2012). Analyses of chlorella virus genomes were pioneered by the assiduous work of J. Van Etten's laboratory, who after beginning work on PBCV-1 in the 1980s sequenced several large regions of the PBCV-1 genome in the 1990s, before publishing an updated corrected version of the complete genome (Kutish *et al.*, 1996; Li *et al.*, 1995, 1997; Lu *et al.*, 1995, 1996; Yanai-Balser *et al.*, 2010). PBCV-1 is a member of the supergroup of viruses known as 'nuclear–cytoplasmic large DNA viruses' (NCLDV; Iyer, Aravind, & Koonin, 2001; Iyer, Balaji, Koonin, & Aravind, 2006) that includes viruses infecting metazoans (such as poxviruses) and viruses infecting algae (phycodnaviruses, see Table 9.1). In contrast to viruses of land plants, phycodnaviruses are really huge. PBCV-1, for example, encodes 365 predicted proteins and 11 transfer RNAs (tRNAs; Yanai-Balser *et al.*, 2011). The molecular structure of PBCV-1 has been examined in detail (Kuznetsov, Gurnon, Van Etten, & McPherson, 2005; Zhang *et al.*, 2011); the virion consists of an icosahedral particle made of glycoproteins containing a membrane-bounded dsDNA genome. After attachment to the wall of its specific host algal cell, the host cell wall is digested and the virion DNA is injected before a lytic infection cycle starts, the infection process thus resembling those of bacteriophages. Several complete genomes of chlorella viruses have now been sequenced and described (Fitzgerald *et al.*, 2010a, 2010b). Biological functionalities encoded by its 330-kb-long genome to govern the host cell during its lytic life cycle include (i) methylation of host histones, (ii) a restriction enzyme/DNA methylation system, (iii) sugar metabolizing enzymes, (iv) channel/transporter proteins, (v) DNA replication enzymes and (vi) polyamine metabolism enzymes, to mention but a few. Several other chlorella virus genomes have now been analysed (ATCV-1, AR158, NY2A, FR483, MT325, see Table 9.1), revealing new gene functionalities and a high genetic diversity within this group.

2.1.1.2. Viruses of Heterotrophic Protists

We mention these giruses here because of their exceptional sizes, remarkable panoplies of biological functionalities and phylogenetic relationship to viruses of algae (Monier, Claverie, & Ogata, 2008; Monier, Larsen, *et al.*, 2008, and see below), but they infect non-photosynthetic unicellular eukaryotes, and we will not review them here. The NCLDV group also includes largest known viruses, whose genome size exceeds those of the smallest bacteria. The first of these, mimivirus, that infects the freshwater amoeba *Acathamoeba polyphaga*, encodes 1018 predicted proteins (Raoult *et al.*, 2004; Renesto *et al.*, 2006; Legendre, Santini, Rico, Abergel, & Claverie, 2011), but an even larger virus of this kind has recently been reported (Arslan, Legendre, Seltzer, Abergel, & Claverie, 2011). In the sea, *Cafeteria roenbergensis* is a common bactivorous flagellate (Fig. 9.1) and is also infected by a girus (Fischer *et al.*, 2010). Infections of both CroV and mimivirus are sometimes accompanied by virophages that depend on girus for growth in the host. Like satellite viruses of high plants, these virophages affect the severity of the girus infection in a host cell.

2.1.1.3. Phaeoviruses

Viruses infecting multicellular brown algae in the order Ectocarpales were recognized over 30 years ago (reviewed in Brown, 1972; Oliveira & Bisalputra, 1978), and the first genome of a virus in this group, EsV-1 (*Ectocarpus siliculosus* virus 1), was analysed in 2001 (Delaroque *et al.*, 2001) and a second complete genome for this group, FsV-158 (*Feldmannia* species virus 158), being published more recently (Schroeder *et al.*, 2009). The life cycles of phaeoviruses are particularly well adapted to those of the brown algae in this group, which undergo an alternation of generations (see Chapter 5, The Ectocarpus Genome Consortium, 2012, and Peters *et al.*, 2008 for further details of the life cycle). Whereas these algae spend most of their lifetimes as sessile filamentous forms, their cells being protected by a cellulose/alginate cell wall, their motile zoospores and gametes are naked cells that can be infected by specific phaeoviruses (Müller, 1991a). Once infected, the virus can be integrated into the host genome and is subsequently inherited in a Mendelian manner (Müller, 1991b). The filamentous plant then developing from the zooid or gamete shows no symptoms until it produces sporangia (fruiting bodies), in which infectious viral particles are produced, these being released under in certain environmental conditions (Müller, 1991a). Filamentous sporophyte plants carrying an integrated phaeovirus thus have reduced fertility and transmit viruses to other plants at propitious times during gamete release. The EsV-1 and FsV-158 genomes are strikingly different in size (336 and 155 kb, see Table 9.1), probably reflecting ancient evolutionary paths since their last common ancestors, providing interesting models for host–virus evolution. This genetic system resembles those of human herpesviruses in some ways. Herpesviruses are likewise large dsDNA viruses that are transmitted vertically to offspring and are integrated in the host genome, remaining latent until their outbreak in certain diseases (e.g. 'Chicken Pox' may reappear as 'Shingles' in later life), but good animal models for studying this disease are lacking (Kennedy, 2002).

2.1.1.4. Coccolithoviruses

Haptophyte algae (see Fig. 9.1) are common and abundant worldwide, some of them forming blooms which may be terminated by viral infections (see above).

In contrast to chloroviruses, phaeoviruses and prasinoviruses (see below), the large *E. huxleyi virus 86* genome (Wilson *et al.*, 2005) carries an RNA polymerase gene, suggesting that at least the virus more directly controls some of its own gene expression. Several other features also set this virus apart from the other phycodnaviruses. *Emiliania huxleyi* viruses are surrounded by a lipid membrane rather than a rigid capsid and enter their host cells via endocytosis (Mackinder *et al.*, 2009). After about 4.5 h, new virions are released by budding from the host cell. In all these characteristics, coccolithoviruses and other NCLDV of non-photosynthetic protists (shaded lines in Table 9.1) more closely resemble animal viruses, such as poxviruses. Remarkably, coccolithoviruses have acquired numerous genes, most likely from their host, that encode the synthesis of complex sphingolipids (Monier *et al.*, 2009). Recent work suggests that syringolipid signalling might play a role in controlling host cell death during infection (Han *et al.*, 2006; Monier *et al.*, 2009; Pagarete, Allen, Wilson, Kimmance, & de Vargas, 2009; reviewed in Michaelson, 2010; Bidle & Vardi, 2011). The life cycle of *E. huxleyi* is known (Laguna, Romo, Read, & Wahlund, 2001), and in nature, the diploid form carrying many coccoliths is far more abundant than its haploid (gametic) form, but both haploid and diploid forms can be grown in culture. Frada, Probert, Allen, Wilson, & de Vargas (2008) showed that only diploid cells were susceptible to viral attack and that this species might escape viral infection by meiosis, producing resistant gametic cells. Eight complete (or nearly complete) genomes are now available for viruses infecting *E. huxleyi* (Allen, Schroeder, Donkin, Crawfurd, & Wilson, 2006; Nissimov *et al.*, 2011a, 2011b, 2012) and a complete host genome is also available for *E. huxleyi* (see the website of the Joint Genome Institute [JGI]: http://www.jgi.doe.gov/).

2.1.1.5. Prasinoviruses

Green algae in the class Mamiellophyceae (formerly Prasinophyceae, see Marin and Melkonian, 2010) are globally distributed in aquatic environments. In coastal regions and marine lagoons, picoplanktonic algae of the order Mamiellales are often dominant, common genera including the genera *Micromonas*, *Ostreococcus* and *Bathycoccus*, the composition of their diversity depending on the environment. In high latitudes, *Micromonas* often prevails (Lovejoy, 2007; Not *et al.*, 2004), whereas *Ostreococcus* is more prevalent in temperate latitudes (Zhu, Massana, Not, Marie, & Vaulot, 2005; Viprey, Guillou, Ferréol, & Vaulot, 2008; Demir-Hilton *et al.*, 2011). These species are among the smallest unicellular organisms known, and complete algal host genomes are available for six species, three *Ostreococcus* (Derelle *et al.*, 2006; Palenik *et al.*, 2007; Grigoriev *et al.*, 2012, see http://genome.jgi-psf.org/OstRCC809_2/OstRCC809_2.home.html), two *Micromonas* (Worden *et al.*, 2009) and one *Bathycoccus* (Moreau *et al.* 2012). Viruses of *Micromonas pusilla* were among the first to be observed in the Phycodnaviridae (Mayer & Taylor, 1979), but the first genome of a *Micromonas* sp. virus became available only recently (Moreau *et al.*, 2010). However, the first sequenced viral genome available in this group was that of OtV5, a virus infecting *Ostreococcus tauri* (Derelle *et al.*, 2008). This virus was chosen first because *O. tauri* is the species of the Mamiellales for which the most physiological data currently exist, including a completely sequenced genome. Eight complete genomes of prasinoviruses are currently available (Fig. 9.1; Derelle *et al.* 2008; Moreau *et al.*, 2010; Weynberg, Allen, Ashelford, Scanlan, & Wilson, 2009; Weynberg, Allen, Gilg, Scanlan, & Wilson, 2011). Perhaps the most surprising finding from comparative genomics within the prasinoviruses is that their genomes show less divergence than those of their host genomes (Moreau *et al.*, 2010, and see below), despite them being mainly species specific (Clerissi *et al.*, 2012), in contrast to classical dogma about the fast evolution of viral genomes. Prasinoviruses were also found to have several genes encoding enzymes for amino acid synthesis not found in other viruses, but it is not clear why these particular pathways have been recruited into the viral genome. Complete genomes are currently available for seven *Prasinovirus* strains (BpV1, BpV2, MpV1, OlV1, OtV1, OtV2 and OtV5, Table 9.1), but this figure will probably double within the next 2 years.

2.1.1.6. Unassigned DNA Viruses – *Chaetoceros salsugineum* Nuclear Inclusion Virus

Chaetoceros is one of the most abundant and widespread genera of diatoms known, with approximately 400 species described (Rines & Theriot, 2003). Temperature, climate, salinity, nutrients and predators are regarded as important factors controlling its abundance and population dynamics. *Chaetoceros salsugineum* nuclear inclusion virus (CsNIV) is a 38-nm icosahedral virus that replicates within the nucleus of *C. salsugineum*. CsNIV has a novel partially dsDNA genome, being a single molecule of covalently closed circular single-stranded DNA (ssDNA; 6005 nucleotides), together with a piece of linear ssDNA (997 nucleotides) that is complementary to a portion of the closed circle (Nagasaki *et al.*, 2005). The putative polymerase shows low but significant similarity that of circoviruses (e.g. beak and feather virus disease of birds). Two other viruses of this kind have now been reported (Table 9.1, CtenDNAV06 and ClorDNAV01).

2.1.2. RNA Viruses

Several kinds of RNA viruses have been found to infect algae. While some of these have been loosely regrouped with previously classified viruses, others have not yet been classified or represent new groups of viruses.

2.1.2.1. Picorna-Like Viruses

Picornaviruses are small positive-strand RNA viruses with icosahedral particles (about 30 nm diameter). In mammals, specific picornaviruses cause diseases such as polio, common colds and foot-and-mouth disease. Their genomes are about 7- to 11-kb long with a long 5' untranslated leader sequence, and they are translated to produce a polyprotein that is proteolytically processed to produce CPs and a replicase (RNA-dependent RNA polymerase or RdRp). Picorna-like viruses are abundant in aquatic environments (Culley *et al.*, 2003; Culley, Lang, & Suttle, 2006; Koonin, Wolf, Nagasaki, & Dolja, 2008). Several species of dinoflagellates and raphidophytes are toxic bloom-forming algae that are ecologically and economically important because they can cause major fish kills. Tai *et al.* (2003) first visualized HaRNAV (*H. akashiwo* RNA virus) as 25-nm diameter particles that can form crystalline lattices in the cytoplasm of its Raphidophyte host *H. akashiwo* during infection, before host cells are lysed. Its positive-strand RNA genome (Lang, Culley, & Suttle, 2004) resembles tomato ringspot virus and certain insect viruses but its overall identity to these is <30% at the amino acid level. Translation probably produces a single polyprotein that is processed to yield three CPs and a replicase. Picorna-like viruses such as RsRNAV (*Rhizosolenia setigera* RNA virus), a 11.2-kb-long single-stranded RNA virus (Nagasaki, K., Tomaru, Y., Katanozaka, N., Shirai, Y., Nishida, K., Itakura, S., *et al.*, 2004), infects *R. setigera,* a diatom. Diatoms (Bacillariophyceae) are among the most widespread organisms on the Earth (Not *et al.* (2012) and Mock and Medlin (2012) in this volume), so it is not surprising that viruses of this kind are found to be abundant in metagenomic data (Culley *et al.*, 2006). Shirai *et al.* (2008) characterized another picorna-like virus, CtenRNAV01 (*Chaetoceros tenuissimus* RNA virus), monophyletic with RsRNAV.

2.1.2.2. dsRNA Viruses

The first double-stranded RNA virus to be sequenced (Attoui, 2006) has a multipartite genome of 11 segments ranging from 0.8 to 5.8 kb, similar to the family Reoviridae and infects the unicellular green alga *M. pusilla* (Brussaard *et al.*, 2004). It is the founder member of the genus *Mimoreovirus*. Phylogenetic analysis using RdRp, the longest predicted protein, revealed only distant relationships to other Reoviruses such as fijivirus (infecting graminaceous plants) and bluetongue virus (infecting cattle).

2.1.2.3. Unassigned RNA Viruses

Two other kinds of single-stranded RNA viruses that have not yet been classified have been discovered. *Heterocapsa circularisquama RNA virus* is a ~4.4-kb virus infecting the dinoflagellate *H. circularisquama*. The virus contains just two open-reading frames (ORFs), one encoding a polyprotein (~3 kb, with two predicted proteins, a protease and an RdRp) and the other encoding a CP (~1 kb). Its closest relatives are higher plant viruses in the Tetravirideae and Luteovirideae (such as beet chlorosis virus and rice yellow mottle virus).

The recently described genome of *Chara australis* virus, whose host is within the Phylum Streptophyta, much more closely related to land plants, has recently become available (Gibbs *et al.*, 2011). Predicted biological functionalites of proteins encoded by 4 of the 12 ORFs identified include RdRp, movement protein (MP), coat/CP and helicase. The closest similarities of the individual proteins were to very divergent groups of viruses (RdRp similarities to benyviruses such as beet necrotic yellow vein virus and helicase similar to pestiviruses such as bovine viral diarrhoea virus) suggesting ancient divergence from known groups. However, the coat protein showed some homology to tobamoviruses (such as tobacco mosaic virus [TMV]), highlighting the relationship of its host with land plants. Indeed, the MP gene also shows homology to the TMV MP group, perhaps not surprising because, as the authors point out, charophyte algae have plasmodesmata-like intercellular connections, and these connections may be used by the virus for systemic spread in the host (Gibbs *et al.*, 2011).

2.2. Transcriptomics

Only a few genomes of algal viruses have been available until recently, so it is not surprising that this field of work is in its infancy. Analyses of the expressed genes of a complete virus permit the activity of the viral coding sequences to be verified *in vivo* in culture. It can further be developed to examine the temporal dynamics of gene expression in a viral life cycle or to examine gene expression in water samples collected from the environment and becomes an even more powerful tool when complete host genomes are also available since the changes in host gene expression in response to viral attacks can also be investigated.

2.2.1. Laboratory-Grown Cultures

Initial analyses of viral expression have usually involved the synthesis of a custom-made set of oligonucleotides representing all the genes of the virus spotted in a microarray on a solid support. RNA is then extracted from infected host cells and is used to make a fluorescently marked complementary DNA library that can be hybridized to the microarray and scanned to assess the level of expression of individual genes by their intensities of fluorescence. Using this technique, expression of 65% of EhV-86 viral genes could be detected in *E. huxleyi* (Wilson *et al.*, 2005), although the biological function of the large majority of the gene products remains unknown.

More recently, using a similar technique, studies aimed at describing the temporal sequence of viral gene expression have been undertaken. Yanai-Balser *et al.* (2010) performed a detailed analysis of infection of PBCV-1 transcription using RNA extracted from seven time points during infection, permitting them to show that 99% of the 365 predicted genes and 11 tRNAs are expressed at some time point during the life cycle, being classed as early (66% of genes before DNA replication) and late (36% after DNA synthesis begins). Many early expressed genes were involved in DNA replication and metabolism, while some of the late expressed genes included proteins known to be virion structural genes (CPs) and others packaged on the virion such as enzymes degrading polysaccharides, necessary for the virus to penetrate a new host cell and enzymes required for injection of its DNA into a new host cell.

A time course of expression of EhV-86 was examined at three time points but this time using RNA expressed sequence tags (ESTs) cloned by production from both virus and hosts. The ESTs were then sequenced and compared to the viral sequence (Wilson *et al.*, 2005; the draft host sequence is available at the JGI website: http://www.jgi.doe.gov/), and grouped in functional categories. In a second set of experiments, oligonucleotide arrays were used to confirm the expression of genes identified by EST analyses from both host and viral genomes and showed that 12% of the assembled ESTs (223 clusters of clones) were similar to the viral genome (Kegel *et al.*, 2007, 2010).

A preliminary transcriptomic analysis of host and viral genes expressed after infection of the unicellular green alga *O. tauri,* comparing healthy and OtV5-infected cells, has been done and used to establish the codon usage preferences in host and virus-infected cells (Michely, S., Toulza, E., Subirana, L., John, U., Cognat, V., Maréchal-Drouard, L., Grimsley, N., Moreau, H., & Piganeau, G., unpublished). Codon usage bias increases with overall gene expression levels, and the tRNAs carried by the virus may help to optimize viral protein translation given the available host tRNAs.

Several questions, such as response to abiotic stress (Dittami *et al.*, 2009) and developmental changes (Peters *et al.*, 2008), are being addressed by transcriptomic analyses of the brown alga *Ectocarpus siliculosus* (The Ectocarpus Genome Consortium, 2012). Since this genome contains an integrated copy of the virus EsV-1, transcriptional analyses also include those viral genes expressed from the latent viral genome. Detailed studies aimed at studying the virus, which would require the study of viral proliferation in the gametophytic stage of the host, have however not yet been attempted.

2.2.2. Environmental Samples

Using a mesocosm (an enclosure of seawater under natural conditions), Pagarete *et al.* (2011) examined the propagation of *E. huxleyi* viruses in a bloom produced after addition of phosphate to enrich natural seawater. The bloom of *E. huxleyi* that resulted, and its subsequent demise by viral lysis, thus occurred with the mixture of strains found at the start of the experiment with the natural populations present in the sampled water (six mesocosms of 11 m^3). Although the transcriptomic probes used to monitor this phenomenon were necessarily limited to those with available ESTs, and the infection could not be synchronous, 'early' and 'late' patterns of gene expression could nevertheless be distinguished during the massive viral lysis observed.

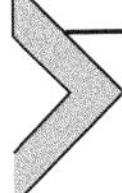

3. HOW DO ALGAE SURVIVE IN THE PRESENCE OF VIRUSES?

While much is known about resistance to viruses in higher plants, different evolutionary paths have been taken in the recruitment of resistance genes and different resistance mechanisms are found (Chapman and Carrington, 2007). We will not review this subject area here but take a more limnological approach to the problem, for which a broader review can be found elsewhere (Thomas, Jacquet, Grimsley, & Moreau, 2012).

Coexistence of phytoplankton hosts and their viruses is a major question to understand how the equilibrium between host and viral populations can be maintained. There are several mechanisms by which a host can avoid interacting with its viruses, including passive or active mechanisms. Among the more passive mechanisms used by the hosts, the 'escape' strategies are well documented in several studies. For example, virus-infected *H. akashiwo* cells sink more rapidly than healthy cells, then moving out of the euphotic zone because of their higher density (Lawrence & Suttle, 2004; Lawrence, Chan, & Suttle, 2001). This may prevent viral infection of conspecifics. This 'altruist' strategy can be explained in terms of evolution for clonal conspecifics. Another example of escape is the narrow host specificity associated with a high host diversity. In this case, the large dilution of infectable hosts reduces their accessibility by viruses (Suttle and Chan, 1994; Suttle, 2005). A third example is the 'Cheshire Cat' strategy (Frada *et al.*, 2008). Here, the haploid phase of the haptophyte *E. huxleyi* does not calcify (i.e. it no longer has hard calcium carbonate scales, but instead has organic surface scales), in contrast to the diploid stage. Frada *et al.* (2008) showed that such 'naked' forms were indeed resistant to EhVs and that exposure of diploid cells to viruses promoted transition to the haploid stage 6 days after infection. These authors suggested that the host's molecules recognized by EhV capsids may be modified or absent in haploid cells, allowing them to become 'invisible' to viruses. Zoochlorellae (such as *P. bursaria* chlorella) are protected from viral attacks (e.g. by PBCV-1) when they are taken up into their host symbiont cells (see above).

More active mechanisms have also been described which can affect each step in the viral cycle, including interaction with the host cell surface, viral DNA entry, viral replication or viral release, as any of these steps might form a barrier to viral propagation.

Adsorption is mediated by receptors used by viruses and which belong to widely different families of proteins, carbohydrates or lipids, often in complex cell surface matrix structures (Baranowski, 2001). Viral receptors are naturally occurring cellular molecules that serve physiological functions for the cell, which have been hijacked by the virus for adsorbing to the cell. Resistance to viral adsorption conferred by host cell mutations could include (i) receptor structure modification, (ii) alteration of receptor accessibility, (iii) decrease in the number of receptors on the cell surface and/or (iv) loss of receptor sites. However, cellular or molecular descriptions of these four processes are rather scarce (Labrie *et al.*, 2010). For eukaryotic phytoplankton, Tarutani, Nagasaki, and Yamaguchi (2006) demonstrated the importance of adsorption in determining viral susceptibility and specificity since viruses of the red tide-forming Raphidophyceae *H. akashiwo* could not adsorb to resistant cells. In another example, Waters and Chan (1982), using a high multiplicity of infection, found that only 50% of the green microalga *M. pusilla* cells lysed after 9 h and a high proportion of viruses did not adsorb while all susceptible cells should have been infected during the first hour. They cloned virus-resistant cells and showed that a mutant virus was able to infect these resistant cells. They suggested that hosts may change composition or conformation of their surface molecules to resist viral pressure, through spontaneous mutation. Another mechanism could be the secretion of extracellular viral inhibitors such as the cell wall sulphated polysaccharide of *Porphyridium* sp., which prevents viral access to cell receptors (Huheihel *et al.*, 2002). Brussaard *et al.* (2007) showed that colonial forms of *Phaeocystis pouchetii* are resistant to viral infection, in contrast to individual cells. Thus, colonial forms, which are surrounded by an 'outer skin', were protected from viral adsorption. In *O. tauri*, two types of resistance were described but in both cases viruses could absorb to the host membrane. However, the mechanism of this immunity to infection remains unknown (Thomas *et al.*, 2011). Overall, the resistance mechanisms of blocking viral entry remain unknown in marine organisms.

Little is known about resistance mechanisms operating at the level of the next step of infection, i.e. the viral replication inside the host cells. The only example described so far for marine phytoplankton is for the dinoflagellate *H. circularisquama* (Tomaru *et al.*, 2009). They transfected the viral RNA genome into resistant cells to see whether this host was permissive or not to viral replication and found that intracellular viral RNA replication was interrupted in virus-resistant cells, but the molecular mechanism responsible is unknown.

The last step of the lytic cycle is the release of viral particles, which in known protist–virus interactions most frequently occurs as a 'burst' from the host cell, at a point when the cell is full of new virus particles. Viral dissemination is thus usually rapid and intense. However, infected host cells can survive and replicate due to lysogeny or chronic infection. Although lysogeny frequently occurs in prokaryotes, a chronic type of infection has only recently come to light in eukaryotic phytoplankton (Thomas *et al.*, 2011). This could be due to an experimental bias since most of the approaches being based on observation of lysis to track viruses. Chronic infections, where viruses adopt a regulated replication within their host and where their expulsion operates by simple diffusion through the cell wall or by budding has been reported in the green picoeukaryote *O. tauri*. In this species, a virus-resistant cell type, named 'resistant producer (R^P)', was able to produce one to three viruses per cell and per day without lysing. Viruses were released through vesicles formed at host membrane (Thomas *et al.*, 2011). R^P cells were so named, rather than chronically infected because they appeared to grow at a similar rate to wild-type cells in parallel cultures. However, competition experiments, mixing R^P and resistant non-producing (immune) cells, did reveal them to have a small reduction in fitness (Thomas *et al.*, 2011). In the *E. huxleyi*–EhV-86 interaction, Pagarete *et al.* (2009) suggested that sphingolipids promote the formation of lipid rafts in the membrane of the alga *E. huxleyi,* which can then become focal points on the membrane for viral budding and release. As the infection progresses, there is a massive increase in sphingolipid requirement and the accumulation of sphingolipids within infected cells triggers virion release through host programmed cell death (PCD). The expression pattern of the sphingolipids pathway in coccolithophorids suggests that sphingolipid biosynthesis driven by the virus is a crucial factor for successful dissemination of viruses. Pagarete *et al.* (2009) speculated that a host-specific sphingolipid could be involved as a bioactive lipid-signalling molecule among host cells, able to trigger meiosis in a fraction of the cell population and allowing them to escape viral infection (Frada *et al.*, 2008).

How can viruses persist in the environment if spontaneous resistance to them occurs frequently? Theory suggests that the coexistence of susceptible and resistant cells may be due to a trade-off between competitive ability and reduced mortality (Lenski, 1988) and that development of viral resistance has physiological costs related to a decrease in fitness of other functionalities (Weinbauer, 2004). Bohannan and Lenski (2000) confirmed this fitness cost associated with the evolution of resistance to viruses. The cost of resistance is often observed to be reduced growth rate. As an example, virus-resistant *P. pouchetii* cells grow 50% more slowly than control cells (Haaber & Middelboe, 2009). Resource uptake might also be reduced by mutations affecting the host cell surface, lowering competitiveness (Bohannan, Travisano, & Lenski, *et al.*, 1999). Susceptible cells allow the virus to persist and the virus in turn prevents the susceptible cell population from competitively excluding the resistant cells, establishing a dynamical equilibrium between hosts and viruses in the environment. The fitness cost of resistance can be eliminated through the reversion of resistance into a susceptible phenotype. When the virus is absent, the resistant organisms are less fit than their susceptible counterparts, increasing the probability of extinction for resistant cells. Variations of fitness associated with virus resistance could then structure microbial communities in aquatic ecosystems.

An adaptation in a virus lineage may change the selection pressure on the host lineage, giving rise to a counter-adaptation. If this occurs reciprocally, an unstable runaway escalation or 'arms race' may result (Comeau & Krisch, 2005; Dawkins & Krebs, 1979) producing strong influences on the dynamics of viruses and their hosts in natural systems (Middelboe *et al.*, 2001). Adaptative responses include viral change to recognize new host receptors, production of proteins by the host that mask the phage receptor and production of extracellular matrices as physical barrier between phages and their receptors (Labrie *et al.*, 2010). Coexistence of hosts and viruses may thus lead to clonal successions of specific host–virus systems over time. In a study of *H. akashiwo,* Tarutani, Nagasaki, and Yamaguchi (2000) observed three periods in which hosts and viruses differed. At first, susceptible cells were dominant and decreased with time due to the viral pressure, then virus-resistant cells dominated, consequently reducing the viral population, and finally, susceptible host cells dominated presumably because a decrease in the abundance of the lytic viruses might allow the growth of susceptible cells.

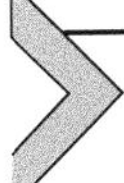

4. DO VIRUSES AND HOSTS SHARE THEIR GENETIC INFORMATION BY LATERAL GENE TRANSFER?

Viruses are frequently seen as 'gene robbers' (Moreira & López-García, 2009), containing a more or less important fraction of genes acquired from an external source, eukaryotic, prokaryotic or viral (see Gogarten & Gogarten, 2009). This might perhaps explain the huge variation in genome sizes of viruses (Monier *et al.*, 2007). However, the amount of transferred genetic material in viral genomes is still subject to intense debate (Monier *et al.*, 2007; Moreira & Brochier-Armanet, 2008), owing to the very high number of genes with no known homologues (ORFans) in viruses. Another explanation of these discrepancies is linked to the different methods used to uncover lateral (or horizontal) gene transfer (LGT): compositional methods can lead to an overestimation of the number of transferred genes (Monier *et al.*, 2007), while phylogenetic approaches may underestimate it (Fischer *et al.,* 2010). In all cases, LGT is acknowledged to be of primary importance in marine viruses, including NCLDV (Iyer *et al.*, 2006, Filée and Chandler, 2010) within which the Phycodnaviridae, algal dsDNA viruses, form a very diverse group (Bidle & Vardi, 2011). Among the organisms that might donate genes to viruses, their hosts are prime candidates because of the intimacy of this relationship and the evolutionary pressure that each partner exerts on the other. Here, we would like to underline a fundamental difference between prokaryotic and eukaryotic host–virus systems. In prokaryotes, viruses (bacteriophages), genetic transduction and the widespread high frequency of homologous recombination facilitates gene flow (see Abedon (2009) for a review). In eukaryotic systems, the genetic material is 'compartmentalized' and species' barriers are usually strict, limiting interspecies genetic flux. Host cells are larger and the virus life cycle usually occurs in specific compartments (nucleus, organelles or cytoplasmic regions), depending on the host–virus interaction in question and on the stage of the viral infection. Only in the last few years, as complete genomes for some eukaryotic hosts and their viruses have become available (Filée, Pouget, & Chandler, 2008), has it been possible to question the presence and amount of LGT between eukaryotic hosts and their viruses. We will focus here on this particular category of LGT, involving algal viruses and their hosts. The amount of gene transferred to viruses from their hosts appears to be low (Monier *et al.,* 2007), whether the number of total LGT is considered to be important or not (most of the 'foreign' genes likely to have been transferred to virus genomes are from bacterial origin). Several studies have suggested the occurrence of gene transfers between phages and their cyanobacterial hosts (see Lindell *et al.*, (2004)), but such transfers between viruses and eukaryotic phototrophic hosts take place too and have only recently started to be documented. Certainly, the most spectacular example suggested for such transfers involves an entire metabolic pathway for sphingolipid biosynthesis in the coccolithophore *E. huxleyi* and its virus EhV (Monier *et al.*, 2009; Pagarete *et al.*, 2009). This pathway seems to be functional in the virus and its host, though its exact role in the infection is still not clear but may involve PCD process. Because of its extent (seven genes are involved) and its functional activity in the virus and its host, this LGT likely reflects the evolutionary arms race within this association (Bidle & Vardi, 2011). Viral sphingolipids are suspected to be involved in virion release (Pagarete *et al.*, 2009), while PCD induced by host sphingolipids could limit the spread of viral infection (Vardi *et al.*, 2009). In *P. bursaria* and its viruses, host—virus LGT are also suspected (Filée *et al.*, 2008). In the Mamiellales–prasinovirus system (Derelle *et al.,* 2008; Moreau *et al.*, 2010), Weynberg *et al.* (2011) have also shown good support for host–virus LGTs, such as the transfer of a heat-shock protein between *Bathycoccus prasinos* and its virus BpV (Moreau *et al.*, 2010) and a gene encoding a phosphate transporter in OtV–2 from its *Ostreococcus* host (Weynberg *et al.*, 2011). However, it is generally difficult to infer whether transfers happened from virus to host or from host to virus. It seems likely that both processes can take place (Monier *et al.*, 2009), making viruses potential shuttles to convey genes from a host to another. This could have some important evolutionary consequences, especially if one viral strain can infect several host species.

5. ARE VIRUSES SPECIFIC TO ONE OR MORE HOST SPECIES, AND HOW ARE THESE PARTNERS EVOLVING TOGETHER?

Most viruses appear to be fairly specific for their host species, though they might infect several host strains within the same species (Clerissi *et al.*, 2012; Nagasaki & Yamaguchi, 1998; Nagasaki *et al.*, 2003; Derelle *et al.*, 2008; Sahlsten, 1998; Sandaa, 2008; see Wommack & Colwell (2000), for a review). However, a problem here is to define the species status of hosts, which in the case of unicellular clonal populations is not usually straightforward because there are few algae for which good genetic data are available. In some cases, algal viruses infect hosts from different geographic regions or from more different phylogenetic clades (e.g. Bellec, Grimsley, & Desdevises, 2010; Zingone *et al.*, 2006). This ability for the same viral clone to infect different host strains can have important consequences in planktonic ecosystems, in their role in the maintenance of genetic diversity in hosts and regarding the possibility to transfer genes between related hosts. The mechanisms for host specificity are still unknown but this subject is being investigated at a molecular level (Tomaru, Shirai, & Nagasaki, 2008), in particular in relation to surface proteins (Nagasaki *et al.*, 2005). This generally high host specificity (at least at the level of a species) suggests the existence of cospeciation patterns between algal viruses and their hosts, but this has never been properly investigated. Cospeciation between viruses and their hosts is often assumed, and information about host divergence is used to infer speciation events in viruses and to estimate their evolutionary rates (Firth *et al.*, 2010). However, this should first be assessed, and previous studies on host–virus evolutionary interactions (Gottschling *et al.*, 2011; Jackson & Charleston, 2004) suggest that a significant cospeciation signal is generally present in the coevolutionary history of viruses and their hosts (but not always, see Nemirov, Leirs, Lundkvist, & Olsson, 2010). However, strict cospeciation is not the rule (e.g. Ramsden, Holmes, & Charleston, 2009) even in the terrestrial ecosystems investigated, where hosts can be separated by physical barriers limiting or precluding contact between host species and then host-switching events (such as in the papillomavirus–host association [Rector *et al.*, 2007]). In the aquatic, and particularly marine, environment, such barriers are more diffuse or absent, and the occurrence of a significant cophylogenetic signal would reflect the close adaptation of viruses to their hosts more than the absence of opportunity to host switch. In the Mamiellales–prasinovirus system, currently available data suggest that cospeciation is not very strict (see Moreau *et al.*, 2010) but that it is probably significant (Bellec, Grimsley, Moreau, & Desdevises, 2009). Further work is needed to clarify this point.

6. RED QUEENS AND WHITE PAWNS – WHICH PARTNER IS EVOLVING THE FASTEST?

Moreau *et al.* (2010) showed from genomic data that the evolutionary divergence between prasinoviruses and their hosts is clearly higher than that between their respective viruses. If significant cospeciation is indeed occurring in this association, this would support the 'White Pawn' hypothesis (see below) that Mamiellales evolve faster than their viruses, a very surprising situation considering current knowledge on virus evolutionary rates (see Duffy & Holmes, 2008). Moreau *et al.* (2010) also hypothesized that one reason for the lower evolutionary divergence of the viruses might be that since viruses must remain dormant (e.g. in the sediment) until they find a suitable host cell, and since host cells must divide frequently to remain viable, then the host genomes would be replicated less frequently than the viral genomes – the so-called 'White Pawn' hypothesis. Providing that coevolution of hosts and viruses has occurred, a topsy–turvy situation could thus arise; in contrast to viruses attacking relatively long-lived animal or multicellular plant hosts, protist viruses are much more likely to attack a host genome that has already undergone numerous replications since its ancestor was infected by the same virus. While there is no concrete evidence that this may be the case, it is known that viral lysates remain infective after many years in the refrigerator and that viruses of toxic dinoflagellates return to infect seasonal algal blooms (Nagasaki *et al.*, 2004). It is difficult to predict what the possible effect of the White Pawn hypothesis might be, especially given the lack of knowledge about effective viral population sizes and mutation rates. However, Ogata and Claverie (2007) found that in contrast to bacterial genomes where ORFs with no predicted biological functionalities (called ORFans) evolve faster than non-ORFans, in large dsDNA viruses, ORFans appear to evolve in a similar way to non-ORFans. This observation thus supports the notion of ancient viral genes but could also be a possible effect of White Pawn evolution.

The high number of ORFans in virus genomes, mentioned above, can be explained in at least two different ways: (1) it reflects the great ancestry of viruses that could have appeared before, and could be at the origin of, the cellular world (Forterre, 2006), and cellular life forms have gained from viruses only a subset of these diverse genes. This would explain why so many genes in viruses have no cellular homologues; and (2) viruses evolve very fast in comparison to their hosts, so that any gene gained from a host quickly accumulates so many substitutions and changes that any homology with the original cellular gene is more or less quickly lost. This hypothesis is compatible with the view that viruses are by-product of cells, mobile genetic elements (Moreira & López-García, 2009). In order to assess the contributions of these two alternatives, that are not necessarily mutually exclusive to the process of evolution, it thus becomes crucial to assess the relative rates of evolution in viruses and their hosts. Ideally, this requires these rates to be independently estimated, for example using fossil or geological calibrations or time-structured data from samples obtained at different dates (Firth *et al.*, 2010). In the absence of such data and as an approximate solution, if a very clear cospeciation pattern exists between viruses and their hosts, temporal information about hosts (which is generally more readily available than for their viruses) can be transposed to corresponding viruses, and relative rates of homologous genes can be compared. To our knowledge, none of these data exist yet for algal viruses, but large dsDNA viruses are frequently assumed to evolve slowly than other kinds of viruses (e.g. Drake, 1991; Gago *et al.*, 2009; Rector *et al.*, 2007). From time-structured data, Firth *et al.* (2010) have suggested that dsDNA viruses may evolve much faster than previously thought, at rates comparable to those of RNA viruses.

7. WHAT IS NEXT?

Historically, the ease of growth of the host organism under laboratory conditions has always played a determinant role for the discovery of new viruses. This was true for bacteriophages and has been important for algal viruses. *Chlorella* and *Emiliana*, for example, are easily grown in liquid media and can be plated out in soft agar. However, culture techniques for the great majority of microbial species are not yet developed (Amann *et al.*, 1990; Cuvelier *et al.*, 2010), and this remains an important hurdle for the analysis of viruses. Until now, it has not usually been possible to produce a sufficient quantity of pure clonal virus particles for sequencing from microbial eukaryotes, except for when the host cells can be cultured in the laboratory.

7.1. Single-Cell Genomics

The technology for sequencing the DNA of individual cells by multiple displacement amplification (MDA; Blanco, Bernad, Esteban, & Salas, 1992, Hellani et al., 2004) is progressing, and this technique promises to yield much information about uncultured and unculturable microbial species. Under carefully controlled laboratory conditions, it is possible to isolate individual microbial cells by flow cytometry and serial dilutions, extract their DNAs and randomly amplify their DNAs. Single bacterial cells were first used to demonstrate the feasibility of this procedure (Martinez-Garcia *et al.*, 2012; Raghunathan *et al.*, 2005; Stepanauskas & Sieracki, 2007). In a similar way, for flow-sorted eukaryotic phytoplankton, MDA has recently permitted whole-genome amplification from a group of cells (Lepere *et al.*, 2011) and genome amplification from single cells (Heywood, Sieracki, Bellows, Poulton, & Stepanauskas, 2011). However, for technical reasons, only partial genomes of the eukaryotic host cells can be amplified at present (M. Sieracki, personal communication). Yoon *et al.* (2011) amplified genomes from single picobiliphyte cells and found that these cells were infected with either large dsDNA viruses or ssDNA viruses. The assembly of new viral genomes from single uncultivated cells thus holds much promise for the future since an appreciable proportion of host cells may be present in a population, and many copies of complete viral are captured within the cell, providing a natural over-representation of these sequences relative to those of the host. In addition, even if the host species is uncultivable, its sequence data should permit its phylogenetic classification to be determined, a cherry on the cake.

7.2. Metagenomics

Using the mimivirus genome sequence as a probe to find sequences in public databases with the BLAST software tool revealed a surprising number of matches, hinting that NCLDV may be common in seawater (Monier, Claverie, & Ogata, 2008; Monier, Larsen, *et al.*, 2008), an observation supported by the recent single-cell genomics analyses (see above). Assembly of complete algal viral genomes from metagenomic data is now a realistic approach to find new viral genomes since new bioinformatics methodologies for assembling complete genomes from paired-end reads promise to revolutionize metagenomics (Iverson *et al.*, 2012, and following chapter on Environmental Genomics, Toulza, Blanc-Mathieu, Gourbiere, & Piganeau, 2012). Given the further development of NGS techniques, sufficient computing capacities, and adequate software programs, the description of the genomes of algae and their viruses in aquatic environments is set to make a quantum leap, but will experimental biology be able to reveal some of the unknown biological functionalities encoded in this bottom line of life on Earth?

ACKNOWLEDGEMENTS

Our sincere thanks are due to members of the GENOPHY team in Banyuls sur Mer for their interest and discussions. Work on prasinoviruses and their hosts in our laboratory has been supported by: the Centre National de Recherche Scientifique, the University of Pierre and Marie Curie Paris 6, the AXA Foundation, the FRB and the Agence National de Recherche (grant numbers PICOVIR ANR-07-BLAN-0210, PICOPOP AAP-FRB2009, TARA-Girus ANR-09-PCS-GENM-218, PHYTADAPT NT09_567009).

REFERENCES

Abedon, S. T. (2009). Phage evolution and ecology. In A. I. Laskin (Ed.), *Advances in applied microbiology*, Vol. 67. San Diego, CA: Elsevier Academic Press Inc. 1–45.

Allen, M., Schroeder, D., Donkin, A., Crawfurd, K., & Wilson, W. (2006). Genome comparison of two Coccolithoviruses. *Virology Journal, 3*, 15.

Amann, R. I., Binder, B. J., Olson, R. J., Chisholm, S. W., Devereux, R., & Stahl, D. A. (1990). Combination of 16S rRNA-targeted oligonucleotide probes with flow cytometry for analyzing mixed microbial populations. *Applied and Environmental Microbiology, 56*, 1919–1925.

Archibald, J. (2012). The evolution of algae by secondary and tertiary endosymbiosis. *Advances in Botanical Research, 64*, 87–118.

Arslan, D., Legendre, M., Seltzer, V., Abergel, C., & Claverie, J.-M. (2011). Distant mimivirus relative with a larger genome highlights the fundamental features of Megaviridae. *Proceedings of the National Academy of Sciences of the United States of America, 108*, 17486–17491.

Attoui, H., Jaafar, F., Belhouchet, M., de Micco, P., de Lamballerie, X., & Brussaard, C. (2006). Micromonas pusilla reovirus: a new member of the family Reoviridae assigned to a novel proposed genus (Mimoreovirus). *Journal of General Virology, 87*, 1375–1383.

Baranowski, E. (2001). Evolution of cell recognition by viruses. *Science, 292*, 1102–1105.

Bellec, L., Grimsley, N., & Desdevises, Y. (2010). Isolation of prasinoviruses of the green unicellular algae *Ostreococcus* spp. on a worldwide geographical scale. *Applied and Environmental Microbiology, 76*, 96–101.

Bellec, L., Grimsley, N., Moreau, H., & Desdevises, Y. (2009). Phylogenetic analysis of new Prasinoviruses (Phycodnaviridae) that infect the green unicellular algae *Ostreococcus, Bathycoccus* and. *Micromonas, Environmental Microbiology Reports, 1*, 114–123.

Benarroch, D., Claverie, J. M., Raoult, D., & Shuman, S. (2006). Characterization of mimivirus DNA topoisomerase IB suggests horizontal gene transfer between eukaryal viruses and bacteria. *Journal of Virology, 80*, 314–321.

Bench, S. R., Hanson, T. E., Williamson, K. E., Ghosh, D., Radosovich, M., Wang, K., et al. (2007). Metagenomic characterization of Chesapeake Bay virioplankton. *Applied and Environmental Microbiology, 73*, 7629–7641.

Bidle, K. D., & Vardi, A. (2011). A chemical arms race at sea mediates algal host-virus interactions. *Current Opinion in Microbiology, 14*, 449–457.

Blanco, L., Bernad, A., Esteban, J. A., & Salas, M. (1992). DNA-independent deoxynucleotidylation of the phi 29 terminal protein by the phi 29 DNA polymerase. *Journal of Biological Chemistry, 267*, 1225–1230.

Bohannan, B. J., & Lenski, R. E. (2000). Linking genetic change to community evolution: insights from studies of bacteria and bacteriophage. *Ecology Letters, 3*, 362–377.

Bohannan, B. J. M., Travisano, M., & Lenski, R. E. (1999). Epistatic interactions can lower the cost of resistance to multiple consumers. *Evolution, 53*, 292–295.

Bratbak, G., Egge, J. K., & Heldal, M. (1993). Viral mortality of the marine alga *Emiliania huxleyi* (Haptophyceae) and termination of algal blooms. *Marine Ecology-Progress Series, 93*, 39–48.

Brown, R. M. J. (1972). Algal viruses. *Advances in Virus Research, 17*, 243–277.

Brown, C. M., Campbell, D. A., & Lawrence, J. E. (2007). Resource dynamics during infection of *Micromonas pusilla* by virus MpV-Sp1. *Environmental Microbiology, 9*, 2720–2727.

Brussaard, C. (2004). Viral control of phytoplankton populations – a review. *Journal of Eukaryotic Microbiology, 51*, 125–138.

Brussaard, C. P. D., Kuipers, B., & Veldhuis, M. J. W. (2005). A mesocosm study of *Phaeocystis globosa* population dynamics – 1. Regulatory role of viruses in bloom. *Harmful Algae, 4*, 859–874.

Brussaard, Corina P. D., Bratbak, G., Baudoux, A.-C., & Ruardij, P. (2007). *Phaeocystis* and its interaction with viruses. *Biogeochemistry, 83*, 201–215.

Chapman, E. J., & Carrington, J. C. (2007). Specialization and evolution of endogenous small RNA pathways. *Nature Reviews Genetics, 8*(11), 884–896.

Claverie, J. M. (2006). Viruses take center stage in cellular evolution. *Genome Biology, 7*(6).

Clerissi, C., Desdevises, Y., & Grimsley, N. H. Prasinoviruses of the marine green alga *Ostreococcus tauri* are mainly species-specific. Journal of Virology, 86, 4611–4619.

Comeau, A., & Krisch, H. (2005). War is peace – dispatches from the bacterial and phage killing fields – Commentary. *Current Opinion in Microbiology, 8*, 488–494.

Countway, P. D., & Caron, D. A. (2006). Abundance and distribution of *Ostreococcus* sp in the San Pedro Channel, California, as revealed by quantitative PCR. *Applied and Environmental Microbiology, 72*, 2496–2506.

Culley, A. I., Lang, A. S., & Suttle, C. A. (2003). High diversity of unknown picorna-like viruses in the sea. *Nature, 424*, 1054–1057.

Culley, A., Lang, A., & Suttle, C. (2006). Metagenomic analysis of coastal RNA virus communities. *Science, 312*, 1795–1798.

Cuvelier, M. L., Allen, A. E., Monier, A., McCrow, J. P., Messié, M., Tringe, S. G., et al. (2010). Targeted metagenomics and ecology of globally important uncultured eukaryotic phytoplankton. *Proceedings of the National Academy of Sciences of the United States of America, 107*, 14679–14684.

Danovaro, R., Corinaldesi, C., Dell'anno, A., Fuhrman, J. A., Middelburg, J. J., & Noble, R. T. (2011). Marine viruses and global climate change. *FEMS Microbiology Reviews, 35*, 993–1034.

Dawkins, R., & Krebs, J. R. (1979). Arms races between and within species. *Proceedings of the Royal Society B: Biological Sciences, 205*, 489–511.

De Clerck, O., Bogaret, K., & Leliaert, F. (2012). Diversity and evolution of algae: primary endosymbiosis. *Advances in Botanical Research, 64*, 55–86.
Delaroque, N., Muller, D., Bothe, G., Pohl, T., Knippers, R., & Boland, W. (2001). The complete DNA sequence of the *Ectocarpus siliculosus* virus EsV-1 genome. *Virology, 287*, 112–132.
Demir-Hilton, E., Sudek, S., Cuvelier, M. L., Gentemann, C. L., Zehr, J. P., & Worden, A. Z. (2011). Global distribution patterns of distinct clades of the photosynthetic picoeukaryote *Ostreococcus*. *ISME Journal, 5*, 1095–1107.
Derelle, E., Ferraz, C., Rombauts, S., Rouze, P., Worden, A., Robbens, S., et al. (2006). Genome analysis of the smallest free-living eukaryote *Ostreococcus tauri* unveils many unique features. *Proceedings of the National Academy of Sciences of the United States of America, 103*, 11647–11652.
Derelle, E., Ferraz, C., Escande, M.-L., Eychenie, S., Cooke, R., Piganeau, G., et al. (2008). Life-cycle and genome of OtV5, a large DNA virus of the pelagic marine unicellular green alga *Ostreococcus tauri*. PLoS ONE*, 3*, 1–13, e2250.
Danovaro, R., Corinaldesi, C., Dell'Anno, A., Fuhrman, J. A., Middelburg, J. J., Noble, R. T., et al. (2011). Marine viruses and global climate change. *FEMS Microbiology Reviews, 35*, 993–1034.
Dittami, S. M., Scornet, D., Petit, J.-L., Ségurens, B., Da Silva, C., Corre, E., et al. (2009). Global expression analysis of the brown alga *Ectocarpus siliculosus (Phaeophyceae)* reveals large-scale reprogramming of the transcriptome in response to abiotic stress. *Genome Biology, 10*, R66.
Drake, J. W. (1991). A constant rate of spontaneous mutation in DNA-based microbes. *Proceedings of the National Academy of Sciences of the United States of America, 88*, 7160.
Duffy, S., & Holmes, E. C. (2008). Phylogenetic evidence for rapid rates of molecular evolution in the single-stranded DNA begomovirus tomato yellow leaf curl virus. *Journal of Virology, 82*, 957–965.
Dunigan, D. D., Fitzgerald, L. A., & Van Etten, J. L. (2006). Phycodnaviruses: a peek at genetic diversity. *Virus Res, 117*, 119–132.
Field, C., Behrenfeld, M., Randerson, J., & Falkowski, P. (1998). Primary production of the biosphere: integrating terrestrial and oceanic components. *Science, 281*, 237–240.
Filée, J., & Chandler, M. (2010). Gene exchange and the origin of giant viruses. *Intervirology, 53*, 354–361.
Filée, J., Pouget, N., & Chandler, M. (2008). Phylogenetic evidence for extensive lateral acquisition of cellular genes by nucleocytoplasmic large DNA viruses. *BMC Evolutionary Biology, 8*, 320.
Firth, C., Kitchen, A., Shapiro, B., Suchard, M. A., Holmes, E. C., & Rambaut, A. (2010). Using time-structured data to estimate evolutionary rates of double-stranded DNA viruses. *Molecular Biology and Evolution, 27*, 2038–2051.
Fischer, M. G., Allen, M. J., Wilson, W. H., & Suttle, C. A. (2010). Giant virus with a remarkable complement of genes infects marine zooplankton. *Proceedings of the National Academy of Sciences of the United States of America, 107*, 19508–19513.
Fitzgerald, L. A., Graves, M. V., Li, X., Feldblyum, T., Nierman, W. C., & Van Etten, J. L. (2007). Sequence and annotation of the 369-kb NY-2A and the 345-kb AR158 viruses that infect Chlorella NC64A. *Virology, 358*, 472–484.
Fitzgerald, L. A., Graves, M. V., Li, X., Hartigan, J., Pfitzner, A. J. P., Hoffart, E., et al. (2007). Sequence and annotation of the 288-kb ATCV-1 virus that infects an endosymbiotic Chlorella strain of the heliozoon *Acanthocystis turfacea*. *Virology, 362*, 350–361.
Fitzgerald, L. A., Zhang, Y., Lewis, G., & Van Etten, J. L. (2009). Characterization of a monothiol glutaredoxin encoded by Chlorella virus PBCV-1. *Virus Genes, 39*, 418–426.

Fitzgerald, L. A., Wu, P. K., Gurnon, J. R., Biffinger, J. C., Ringeisen, B. R., Van Etten, J. L., & (first). (2010). Isolation of the phycodnavirus PBCV-1 by biological laser printing. *Journal of Virological Methods.*

Forterre, P. (2006). Three RNA cells for ribosomal lineages and three DNA viruses to replicate their genomes: a hypothesis for the origin of cellular domain. *Proceedings of the National Academy of Sciences of the United States of America, 103*, 3669–3674.

Forterre, P. (2010). Giant viruses: conflicts in revisiting the virus concept. *Intervirology, 53*(5), 362–378.

Frada, M., Probert, I., Allen, M. J., Wilson, W. H., & de Vargas, C. (2008). The "Cheshire Cat" escape strategy of the coccolithophore *Emiliania huxleyi* in response to viral infection. *Proceedings of the National Academy of Sciences of the United States of America, 105*, 15944–15949.

Fuhrman, J. (1999). Marine viruses and their biogeochemical and ecological effects. *Nature, 399*, 541–548.

Gago, S., Elena, S. F., Flores, R., & Sanjuán, R. (2009). Extremely high mutation rate of a hammerhead viroid. *Science (*New York, N.Y.*), 323*(5919), 1308.

Gibbs, A. J., Torronen, M., Mackenzie, A. M., Wood, J. T., Armstrong, J. S., Kondo, H., et al. (2011). The enigmatic genome of Chara australis virus. *Journal of General Virology, 92*, 2679–2690.

Gilbert, J. A., & Dupont, C. L. (2011). Microbial metagenomics: beyond the genome. *Annual Review of Marine Science, 3*, 347–371.

Gobler, C. J., Deonarine, S., Leigh-Bell, J., Gastrich, M. D., Anderson, O. R., & Wilhelm, S. W. (2004). Ecology of phytoplankton communities dominated by *Aureococcus anophagefferens*: the role of viruses, nutrients, and microzooplankton grazing. *Harmful Algae, 3*, 471–483.

Gobler, Christopher. J., Anderson, O. R., Gastrich, M. D., & Wilhelm, S. W. (2007). Ecological aspects of viral infection and lysis in the harmful brown tide alga. *Aureococcus anophagefferens. Aquatic Microbial Ecology, 47*, 25–36.

Gobler, Christopher J., Hutchins, D. A., Fisher, N. S., Cosper, E. M., & Sanudo-Wilhelmy, S. A. (1997). Release and bioavailability of C, N, P, Se, and Fe following viral lysis of a marine Chrysophyte. *Limnology and Oceanography, 42*, 1492–1504.

Gogarten, M. B., Gogarten, J. P., & Olendzenski, L. (2009). *Horizontal gene transfer: Genomes in flux*. New York: Humana Press.

Gottschling, M., Göker, M., Stamatakis, A., Bininda-Emonds, O. R. P., Nindl, I., & Bravo, I. G. (2011). Quantifying the phylodynamic forces driving papillomavirus evolution. *Molecular Biology and Evolution, 2*, 2101–2113.

Grigoriev, I. V., Nordberg, H., Shabalov, I., Aerts, A., Cantor, M., Goodstein, D., et al. (2012). The genome portal of the Department of Energy Joint Genome Institute. *Nucleic Acids Research, 40*, D26–D32.

Haaber, J., & Middelboe, M. (2009). Viral lysis of *Phaeocystis pouchetii*: implications for algal population dynamics and heterotrophic C, N and P cycling. *The ISME Journal, 3*, 430–441.

Han, G., Gable, K., Yan, L., Allen, M., Wilson, W., Moitra, P., et al. (2006). Expression of a novel marine viral single-chain serine palmitoyl transferase and construction of yeast and mammalian single-chain chimera. *Journal of Biological Chemistry, 281*, 39935–39942.

Heywood, J. L., Sieracki, M. E., Bellows, W., Poulton, N. J., & Stepanauskas, R. (2011). Capturing diversity of marine heterotrophic protists: one cell at a time. *The ISME Journal, 5*, 674–684.

Huheihel, M., Ishanu, V., Tal, J., Arad, S., & (Malis). (2002). Activity of *Porphyridium sp.* polysaccharide against herpes simplex viruses in vitro and in vivo. *Journal of Biochemical and Biophysical Methods, 50*, 189–200.

Iverson, V., Morris, R. M., Frazar, C. D., Berthiaume, C. T., Morales, R. L., & Armbrust, E. V. (2012). Untangling genomes from metagenomes: revealing an uncultured class of marine Euryarchaeota. *Science, 335*, 587–590.

Iyer, L., Aravind, L., & Koonin, E. (2001). Common origin of four diverse families of large eukaryotic DNA viruses. *Journal of Virology, 75*, 11720–11734.

Iyer, L. M., Balaji, S., Koonin, E. V., & Aravind, L. (2006). Evolutionary genomics of nucleo-cytoplasmic large DNA viruses. *Virus Research, 117*, 156–184.

Jackson, A. P., & Charleston, M. A. (2004). A cophylogenetic perspective of RNA-virus evolution. *Molecular Biology and Evolution, 21*, 45–57.

Jacobsen, A., Bratbak, G., & Heldal, M. (1996). Isolation and characterization of a virus infecting Phaeocystis pouchetii (Prymnesiophyceae). *Journal of Phycology, 32*, 923–927.

John, S. G., Mendez, C. B., Deng, L., Poulos, B., Kauffman, A. K. M., Kern, S., et al. (2011). A simple and efficient method for concentration of ocean viruses by chemical flocculation. *Environmental Microbiology Reports, 3*, 195–202.

Johnson, M. D. (2011). The acquisition of phototrophy: adaptive strategies of hosting endosymbionts and organelles. *Photosynthesis Research, 107*, 117–132.

Keeling, P., Burger, G., Durnford, D., Lang, B., Lee, R., Pearlman, R., et al. (2005). The tree of eukaryotes. *Trends in Ecology and Evolution, 20*, 670–676.

Kegel, J., Allen, M. J., Metfies, K., Wilson, W. H., Wolf-Gladrow, D., & Valentin, K. (2007). Pilot study of an EST approach of the coccolithophorid *Emiliania huxleyi* during a virus infection. *Gene, 406*, 209–216.

Kegel, J. U., Blaxter, M., Allen, M. J., Metfies, K., Wilson, W. H., & Valentin, K. (2010). Transcriptional host-virus interaction of *Emiliania huxleyi* (Haptophyceae) and EhV-86 deduced from combined analysis of expressed sequence tags and microarrays. *European Journal of Phycology, 45*, 1–12.

Kennedy, P. G. E. (2002). Varicella-zoster virus latency in human ganglia. *Reviews in Medical Virology, 12*, 327–334.

Koonin, E. V., Wolf, Y. I., Nagasaki, K., & Dolja, V. V. (2008). The big bang of picorna-like virus evolution antedates the radiation of eukaryotic supergroups. *Nature Reviews Microbiology, 6*, 925–939.

Kristensen, D. M., Mushegian, A. R., Dolja, V. V., & Koonin, E. V. (2010). New dimensions of the virus world discovered through metagenomics. *Trends in Microbiology, 18*, 11–19.

Kutish, G., Li, Y., Lu, Z., Furuta, M., Rock, D., & Van Etten, J. (1996). Analysis of 76 kb of the chlorella virus PBCV-1 330-kb genome: map positions 182 to 258. *Virology, 223*, 303–317.

Kuznetsov, Y., Gurnon, J., Van Etten, J., & McPherson, A. (2005). Atomic force microscopy investigation of a chlorella virus, PBCV-1. *Journal of Structural Biology, 149*, 256–263.

Labrie, S., Samson, J., & Moineau, S. (2010). Bacteriophage resistance mechanisms. *Nature Reviews Microbiology, 8*, 317–327.

Laguna, R., Romo, J., Read, B., & Wahlund, T. (2001). Induction of phase variation events in the life cycle of the marine coccolithophorid *Emiliania huxleyi*. *Applied and Environmental Microbiology, 67*, 3824–3831.

Lang, A., Culley, A., & Suttle, C. (2004). Genome sequence and characterization of a virus (HaRNAV) related to picorna-like viruses that infects the marine toxic bloom-forming alga *Heterosigma akashiwo*. *Virology, 320*, 206–217.

Lauro, F. M., DeMaere, M. Z., Yau, S., Brown, M. V., Ng, C., Wilkins, D., et al. (2011). An integrative study of a meromictic lake ecosystem in Antarctica. *The ISME Journal, 5*, 879–895.

Lawrence, J., & Suttle, C. (2004). Effect of viral infection on sinking rates of Heterosigma akashiwo and its implications for bloom termination. *Aquatic Microbial Ecology, 37*, 1–7.

Lawrence, J. E., Chan, A. M., & Suttle, C. A. (2001). A novel virus (HaNIV) causes lysis of the toxic bloom-forming alga Heterosigma Akashiwo (Raphidophyceae). *Journal of Phycology, 37*, 216–222.

Legendre, M., Santini, S., Rico, A., Abergel, C., & Claverie, J.-M. (2011). Breaking the 1000-gene barrier for mimivirus using ultra-deep genome and transcriptome sequencing. *Virology Journal, 8*, 99.

Lenski, R. E. (1988). Experimental studies of pleiotropy and epistasis in Escherichia coli. II. Compensation for maldaptive effects associated with resistance to Virus T4. *Evolution, 42*, 433–440.

Lepere, C., Demura, M., Kawachi, M., Romac, S., Probert, I., & Vaulot, D. (2011). Whole-genome amplification (WGA) of marine photosynthetic eukaryote populations. *FEMS Microbiology Ecology, 76*, 513–523.

Li, Y., Lu, Z., Burbank, D. E., Kutish, G. F., Rock, D. L., & Van Etten, J. L. (1995). Analysis of 43 kb of the chlorella virus PBCV-1 330-kb genome: map positions 45 to 88. *Virology, 212*, 134–150.

Li, Y., Lu, Z., Sun, L., Ropp, S., Kutish, G., Rock, D., et al. (1997). Analysis of 74 kb of DNA located at the right end of the 330-kb chlorella virus PBCV-1 genome. *Virology, 237*, 360–377.

Lindell, D., Sullivan, M. B., Johnson, Z. I., Tolonen, A. C., Rohwer, F., & Chisholm, S. W. (2004). Transfer of photosynthesis genes to and from Prochlorococcus viruses. *Proceedings of the National Academy of Sciences of the United States of America, 101*, 11013–11018.

Lovejoy, C. (2007). Distribution, phylogeny, and growth of cold-adapted picoprasinophytes in arctic seas. *Journal of Phycology, 43*, 78–89.

Lu, Z., Li, Y., Que, Q., Kutish, G., Rock, D., & Van Etten, J. (1996). Analysis of 94 kb of the chlorella virus PBCV-1 330-kb genome: map positions 88 to 182. *Virology, 216*, 102–123.

Lu, Z., Li, Y., Zhang, Y., Kutish, G., Rock, D., & Van Etten, J. (1995). Analysis of 45 kb of DNA located at the left end of the chlorella virus PBCV-1 genome. *Virology, 206*, 339–352.

Marin, B., & Melkonian, M. (2010). Molecular Phylogeny and Classification of the Mamiellophyceae class. nov (Chlorophyta) based on Sequence Comparisons of the Nuclear- and Plastid-encoded rRNA Operons. *Protist, 161*, 304–336.

Mackinder, L. C. M., Worthy, C. A., Biggi, G., Hall, M., Ryan, K. P., Varsani, A., et al. (2009). A unicellular algal virus, *Emiliania huxleyi* virus 86, exploits an animal-like infection strategy. *Journal of General Virology, 90*, 2306–2316.

Martinez-Garcia, M., Swan, B. K., Poulton, N. J., Gomez, M. L., Masland, D., Sieracki, M. E., et al. (2012). High-throughput single-cell sequencing identifies photoheterotrophs and chemoautotrophs in freshwater bacterioplankton. *The ISME Journal, 6*, 113–123.

Mayer, J., & Taylor, F. (1979). Virus which lyses the marine nanoflagellate *Micromonas pusilla*. *Nature, 281*, 299–301.

Michaelson, L. V., Dunn, T. M., & Napier, J. A. (2010). Viral trans-dominant manipulation of algal sphingolipids. *Trends in Plant Science, 15*, 651–655.

Middelboe, M., Jorgensen, N. O. G., & Kroer, N. (1996). Effects of viruses on nutrient turnover and growth efficiency of noninfected marine bacterioplankton. *Applied and Environmental Microbiology, 62*, 1991–1997.

Middelboe, M., Hagstrom, A., Blackburn, N., Sinn, B., Fischer, U., Borch, N., et al. (2001). Effects of bacteriophages on the population dynamics of four strains of pelagic marine bacteria. *Microbial Ecology, 42*, 395–406.

Mizumoto, H., Tomaru, Y., Takao, Y., Shirai, Y., & Nagasaki, K. (2007). Intraspecies Host Specificity of a Single-Stranded RNA Virus Infecting a Marine Photosynthetic Protist Is Determined at the Early Steps of Infection. *Journal of Virology, 81*, 1372–1378.

Mock, T., & Medlin, L. K. (2012). Genomics and genetics of diatoms. *Advances in Botanical Research, 64*, 245–284.

Monier, A., Claverie, J., & Ogata, H. (2007). Horizontal gene transfer and nucleotide compositional anomaly in large DNA viruses. *BMC Genomics, 8*, 456.

Monier, A., Claverie, J.-M., & Ogata, H. (2008). Taxonomic distribution of large DNA viruses in the sea. *Genome Biology, 9*, R106.

Monier, A., Larsen, J. B., Sandaa, R.-A., Bratbak, G., Claverie, J.-M., & Ogata, H. (2008). Marine mimivirus relatives are probably large algal viruses. *Virology Journal, 5*, 12.

Monier, A., Pagarete, A., de Vargas, C., Allen, M., Read, B., Claverie, J., et al. (2009). Horizontal gene transfer of an entire metabolic pathway between a eukaryotic alga and its DNA virus. *Genome Research, 19*, 1441–1449.

Moreau, H., Piganeau, G., Desdevises, Y., Cooke, R., Derelle, E., & Grimsley, N. (2010). Marine prasinovirus genomes show low evolutionary divergence and acquisition of protein metabolism genes by horizontal gene transfer. *Journal of Virology, 84*, 12555–12563.

Moreau, M., Verhelst, B., Couloux, A., Derelle, E., Rombauts, S., Grimsley, N., et al. (2012). Gene functionalities and genome structure in Bathycoccus prasinos reflect cellular specializations at the base of the green lineage. *Genome Biology, 13*(8), R74.

Moreira, D., & Brochier-Armanet, C. (2008). Giant viruses, giant chimeras: The multiple evolutionary histories of Mimivirus genes. *BMC Evolutionary Biology, 8*, 12.

Moreira, D., & López-García, P. (2009). Ten reasons to exclude viruses from the tree of life. *Nature Reviews in Microbiology, 7*, 306–311.

Müller, D. G. (1991a). Marine virioplankton produced by infected Ectocarpus siliculosus (*Phaeophyceae*). *Marine Ecology Progress Series, 76*, 101–102.

Müller, D. G. (1991b). Mendelian Segregation of a Virus Genome during Host Meiosis in the Marine Brown Alga Ectocarpus siliculosus. *Journal of Plant Physiology, 137*, 739–743.

Nagasaki, K. (2008). Dinoflagellates, diatoms, and their viruses. *The Journal of Microbiology, 46*(3), 235–243.

Nagasaki, K., Ando, M., Itakura, S., Imai, I., & Ishida, Y. (1994). Viral mortality in the final stage of *Heterosigma akashiwo* (Raphidophyceae) red tide. *Journal of Plankton Research, 16*, 1595–1599.

Nagasaki, K., Ando, M., Imai, I., Itakura, S., & Ishida, Y. (1994). Virus-like particles in *Heterosigma akashiwo* (Raphidophyceae): a possible red tide disintegration mechanism. *Marine Biology, 119*, 307–312.

Nagasaki, K., Shirai, Y., Takao, Y., Mizumoto, H., Nishida, K., & Tomaru, Y. (2005). Comparison of Genome Sequences of Single-Stranded RNA viruses infecting the bivalve-killing dinoflagellate *Heterocapsa circularisquama. Applied and Environmental Microbiology, 71*, 8888–8894.

Nagasaki, K., Tomaru, Y., Katanozaka, N., Shirai, Y., Nishida, K., Itakura, S., et al. (2004). Isolation and characterization of a novel single-stranded RNA virus infecting the bloom-forming diatom *Rhizosolenia setigera. Applied and Environmental Microbiology, 70*, 704–711.

Nagasaki, K., Tomaru, Y., Nakanishi, K., Hata, N., Katanozaka, N., & Yamaguchi, M. (2004). Dynamics of Heterocapsa circularisquama (Dinophyceae) and its viruses in Ago Bay, Japan. *Aquatic Microbial Ecology, 34*, 219–226.

Nagasaki, K., Tomaru, Y., Tarutani, K., Katanozaka, N., Yamanaka, S., Tanabe, H., et al. (2003). Growth characteristics and intraspecies host specificity of a large virus infecting the dinoflagellate *Heterocapsa circulatisquama. Applied and Environmental Microbiology, 69*, 2580–2586.

Nagasaki, K., Tomaru, Y., Nakanishi, K., Hata, N., Katanozaka, N., & Yamaguchi, M. (2004). Dynamics of Heterocapsa circularisquama (Dinophyceae) and its viruses in Ago Bay, Japan. *Aquatic Microbial Ecology, 34*, 219–226.

Nagasaki, K., Tomaru, Y., Takao, Y., Nishida, K., Shirai, Y., Suzuki, H., & Nagumo, T. (2005). Previously unknown virus infects marine diatom. *Applied and Environmental Microbiology, 71*, 3528–3535.

Nagasaki, K., & Yamaguchi, M. (1997). Isolation of a virus infectious to the harmful bloom causing microalga Heterosigma akashiwo (Raphidophyceae). *Aquatic Microbial Ecology, 13*, 135–140.

Nagasaki, K., & Yamaguchi, M. (1998). Intra-species host specificity of HaV (Heterosigma akashiwo virus) clones. *Aquatic Microbial Ecology, 14*, 109–112.

Nemirov, K., Leirs, H., Lundkvist, A., & Olsson, G. E. (2010). Puumala hantavirus and *Myodes glareolus* in northern Europe: no evidence of co-divergence between genetic lineages of virus and host RID B-8197-2008. *Journal of General Virology, 91*, 1262–1274.

Nissimov, J. I., Worthy, C. A., Rooks, P., Napier, J. A., Kimmance, S. A., Henn, M. R., et al. (2011a). Draft genome sequence of the coccolithovirus EhV-84. *Standards in Genomic Sciences, 5*, 1–11.

Nissimov, J. I., Worthy, C. A., Rooks, P., Napier, J. A., Kimmance, S. A., Henn, M. R., et al. (2011b). Draft genome sequence of the coccolithovirus *Emiliania huxleyi* virus 203. *Journal of Virology, 85*, 13468–13469.

Nissimov, J. I., Worthy, C. A., Rooks, P., Napier, J. A., Kimmance, S. A., Henn, M. R., Ogata, H., et al. (2012). Draft Genome Sequence of Four Coccolithoviruses: *Emiliania huxleyi* Virus EhV-88, EhV-201, EhV-207, and EhV-208. *Journal of Virology, 86*, 2896–2897.

Not, F., Latasa, M., Marie, D., Cariou, T., Vaulot, D., & Simon, N. (2004). A single species, *Micromonas pusilla* (Prasinophyceae), dominates the eukaryotic picoplankton in the western English channel. *Applied and Environmental Microbiology, 70*, 4064–4072.

Not, F., Siano, R., Kooistra, W. H. C. F., Simon, N., Vaulot, D., & Probert, I. (2012). Diversity and ecology of eukaryotic marine phytoplankton. *Advances in Botanical Research, 64*, 1–53.

Nowrousian, M. (2010). Next-generation sequencing techniques for eukaryotic microorganisms: sequencing-based solutions to biological problems. *Eukaryotic Cell, 9*, 1300–1310.

Oliveira, L., & Bisalputra, T. (1978). A Virus Infection in the Brown Alga *Sorocarpus uvaeformis* (Lyngbye) Pringsheim (*Phaeophyta, Ectocarpales*). *Annals of Botany, 42*, 439–445.

Ogata, H., & Claverie, J.-M. (2007). Unique genes in giant viruses: regular substitution pattern and anomalously short size. *Genome Research, 17*, 1353–1361.

Ortmann, A., Lawrence, J., & Suttle, C. (2002). Lysogeny and lytic viral production during a bloom of the cyanobacterium. *Synechococcus spp. Microbial Ecology, 43*, 225–231.

Pagarete, A., Allen, M. J., Wilson, W. H., Kimmance, S. A., & de Vargas, C. (2009). Host-virus shift of the sphingolipid pathway along an *Emiliania huxleyi* bloom: survival of the fattest. *Environmental Microbiology, 11*, 2840–2848.

Pagarete, A., Corguillé, G., Tiwari, B., Ogata, H., Vargas, C., Wilson, W. H., et al. (2011). Unveiling the transcriptional features associated with coccolithovirus infection of natural *Emiliania huxleyi* blooms. *FEMS Microbiology Ecology, 78*, 555–564.

Palenik, B., Grimwood, J., Aerts, A., Rouze, P., Salamov, A., Putnam, N., et al. (2007). The tiny eukaryote Ostreococcus provides genomic insights into the paradox of plankton speciation. *Proceedings of the National Academy of Sciences of the United States of America, 104*, 7705–7710.

Peters, A. F., Scornet, D., Ratin, M., Charrier, B., Monnier, A., Merrien, Y., et al. (2008). Life-cycle-generation-specific developmental processes are modified in the immediate upright mutant of the brown alga *Ectocarpus siliculosus*. *Development, 135*, 1503–1512.

Raghunathan, A., Ferguson, H. R., Bornarth, C. J., Song, W., Driscoll, M., & Lasken, R. S. (2005). Genomic DNA Amplification from a Single Bacterium. *Applied and Environmental Microbiology, 71*, 3342–3347.

Ramsden, C., Holmes, E. C., & Charleston, M. A. (2009). Hantavirus evolution in relation to its rodent and insectivore hosts: no evidence for codivergence. *Molecular Biology and Evolution, 26*, 143–153.

Raoult, D., Audic, S., Robert, C., Abergel, C., Renesto, P., Ogata, H., et al. (2004). The 1.2-megabase genome sequence of mimivirus. *Science, 306*, 1344–1350.

Rector, A., Lemey, P., Tachezy, R., Mostmans, S., Ghim, S.-J., Van Doorslaer, K., et al. (2007). Ancient papillomavirus-host co-speciation in Felidae. *Genome Biology, 8*, R57.

Renesto, P., Abergel, C., Decloquement, P., Moinier, D., Azza, S., Ogata, H., et al. (2006). Mimivirus giant particles incorporate a large fraction of anonymous and unique gene products. *Journal of Virology, 80*, 11678–11685.

Rines, J. E. B., & Theriot, E. C. (2003). Systematics of *Chaetocerotaceae (Bacillariophyceae)*. I. A phylogenetic analysis of the family. *Phycological Research, 51*, 83–98.

Rodriguez-Brito, B., Li, L., Wegley, L., Furlan, M., Angly, F., Breitbart, M., et al. (2010). Viral and microbial community dynamics in four aquatic environments. *The ISME Journal, 4*, 739–751.

Sahlsten, E. (1998). Seasonal abundance in Skagerrak-Kattegat coastal waters and host specificity of viruses infecting the marine photosynthetic flagellate *Micromonas pusilla*. *Aquatic Microbial Ecology, 16*, 103–108.

Sandaa, R. A., Heldal, M., Castberg, T., Thyrhaug, R., & Bratbak, G. (2001). Isolation and characterization of two viruses with large genome size infecting *Chrysochromulina ericina* (*Prymnesiophyceae*) and *Pyramimonas orientalis* (*Prasinophyceae*). *Virology, 290*, 272–280.

Sandaa, R.-A. (2008). *Chrysochromulina ericina* burden or benefit? Virus-host interactions in the marine environment. *Research in Microbiology, 159*, 374–381.

Schroeder, D. C., Oke, J., Hall, M., Malin, G., & Wilson, W. H. (2003). Virus succession observed during an Emiliania huxleyi bloom. *Applied and Environmental Microbiology, 69*(5), 2484–2490.

Schroeder, D. C., Park, Y., Yoon, H. M., Lee, Y. S., Kang, W., Meints, R. H., et al. (2009). Genomic analysis of the smallest giant virus – *Feldmannia sp* virus 158. *Virology, 384*, 223–232.

Shirai, Y., Tomaru, Y., Takao, Y., Suzuki, H., Nagumo, T., & Nagasaki, K. (2008). Isolation and characterization of a single-stranded RNA virus infecting the marine planktonic diatom *Chaetoceros tenuissimus* Meunier. *Applied and Environmental Microbiology, 74*, 4022–4027.

Short, S. M., & Suttle, C. A. (2002). Sequence analysis of marine virus communities reveals that groups of related algal viruses are widely distributed in nature. *Applied and Environmental Microbiology, 68*, 1290–1296.

Stepanauskas, R., & Sieracki, M. E. (2007). Matching phylogeny and metabolism in the uncultured marine bacteria, one cell at a time. *Proceedings of the National Academy of Sciences of the United States of America, 104*, 9052–9057.

Suttle, C. A., & Chan, A. M. (1994). Dynamics and distribution of cyanophages and their effect on marine *Synechococcus spp*. *Applied and Environmental Microbiology, 60*, 3167–3174.

Suttle, C. (2005). Viruses in the sea. *Nature, 437*, 356–361.

Suttle, C. (2007). Marine viruses – major players in the global ecosystem. *Nature Reviews in Microbiology, 5*, 801–812.

Suttle, C., & Chan, A. (1995). Viruses infecting the marine Prymnesiophyte Chrysochromulina spp.: isolation, preliminary characterization and natural abundance. *Marine Ecology Progress Series, 118*, 275–282.

Tai, V., Lawrence, J., Lang, A., Chan, A., Culley, A., & Suttle, C. (2003). Characterization of HaRNAV, a single-stranded RNA virus causing lysis of *Heterosigma akashiwo* (Raphidophyceae) RID C-3150-2008. *Journal of Phycology, 39*, 343–352.

Takao, Y., Nagasaki, K., & Honda, D. (2007). Squashed ball-like dsDNA virus infecting a marine fungoid protist *Sicyoidochytrium minutum* (Thraustochytriaceae, Labyrinthulomycetes). *Aquatic Microbial Ecology, 49*, 101–108.

Tarutani, K., Nagasaki, K., & Yamaguchi, M. (2000). Viral impacts on total abundance and clonal composition of the harmful bloom-forming phytoplankton Heterosigma akashiwo. *Applied and Environmental Microbiology, 66*, 4916–4920.

Tarutani, K., Nagasaki, K., & Yamaguchi, M. (2006). Virus adsorption process determines virus susceptibility in *Heterosigma akashiwo* (*Raphidophyceae*). *Aquatic Microbial Ecology, 42*, 209–213.

The Ectocarpus Genome Consortium. (2012). The Ectocarpus genome and brown algal genomics. *Advances in Botanical Research, 64*, 141–184.

Thomas, R., Grimsley, N., Escande, M.-L., Subirana, L., Derelle, E., & Moreau, H. (2011). Acquisition and maintenance of resistance to viruses in eukaryotic phytoplankton populations. *Environmental Microbiology, 13*, 1412–1420.

Thomas, R., Jacquet, S., Grimsley, N., & Moreau, H. (2012). Strategies and mechanisms of viral resistance in aquatic microorganisms. *Advances in Oceanography and Limnology, 3*, 1–15.

Tomaru, Y., Shirai, Y., & Nagasaki, K. (2008). Ecology, physiology and genetics of a phycodnavirus infecting the noxious bloom-forming raphidophyte *Heterosigma akashiwo*. *Fisheries Science, 74*, 701–711.

Tomaru, Y., Shirai, Y., Suzuki, H., Nagumo, T., & Nagasaki, K. (2008). Isolation and characterization of a new single-stranded DNA virus infecting the cosmopolitan marine diatom *Chaetoceros dehilis*. *Aquatic Microbial Ecology, 50*, 103–112.

Tomaru, Yuji, Mizumoto, H., & Nagasaki, K. (2009). Virus resistance in the toxic bloom-forming dinoflagellate *Heterocapsa circularisquama* to single-stranded RNA virus infection. *Environmental Microbiology, 11*, 2915–2923.

Tomaru, Y., Takao, Y., Suzuki, H., Nagumo, T., Koike, K., & Nagasaki, K. (2011). Isolation and characterization of a single-stranded DNA virus infecting *Chaetoceros lorenzianus* Grunow. *Applied and Environmental Microbiology, 77*, 5285–5293.

Toulza, E., Blanc-Mathieu, R., Gourbiere, S., & Piganeau, G. (2012). Environmental genomics of microbial algae: power and challenges of metagenomics. *Advances in Botanical Research, 64*, 383–427.

Van Etten, J. L., Van Etten, C. H., Johnson, J. K., & Burbank, D. E. (1985). A survey for viruses from fresh water that infect a eucaryotic chlorella-like green alga. *Applied and Environmental Microbiology, 49*, 1326–1328.

Van Etten, J. (2003). Unusual life style of giant chlorella viruses. *Annual Review of Genetics, 37*, 153–195.

Van Etten, J., Lane, L., & Meints, R. (1991). Viruses and virus like particles of eukaryotic algae. *Microbiological Review, 55*, 586–620.

Van, Etten, James, L., Lane, L. C., & Dunigan, D. D. (2010). DNA viruses: the really big ones (giruses). *Annual Review of Microbiology, 64*, 83–99.

Van, Etten, James, L., & Dunigan, D. D. (2012). Chloroviruses: not your everyday plant virus. *Trends in Plant Science, 17*, 1–8.

Van Etten, J. L., & Dunigan, D. D. (2011). Chloroviruses: not your everyday plant virus. *Trends in Plant Science, 17*, 1–8.

Vardi, A., Van Mooy, B. A. S., Fredricks, H. F., Popendorf, K. J., Ossolinski, J. E., Haramaty, L., et al. (2009). Viral glycosphingolipids induce lytic infection and cell death in marine phytoplankton. *Science, 326*, 861–865.

Viprey, M., Guillou, L., Ferréol, M., & Vaulot, D. (2008). Wide genetic diversity of picoplanktonic green algae (Chloroplastida) in the Mediterranean Sea uncovered by a phylum-biased PCR approach. *Environmental Microbiology, 10*, 1804–1822.

Waters, R. E., & Chan, A. T. (1982). Micromonas pusilla Virus - the Virus Growth-Cycle and associated physiological events within the host-cells - host range mutation. *Journal of General Virology, 63*, 199–206.
Weinbauer, M. (2004). Ecology of prokaryotic viruses. *FEMS Microbiology Review, 28*, 127–181.
Weynberg, K. D., Allen, M. J., Ashelford, K., Scanlan, D. J., & Wilson, W. H. (2009). From small hosts come big viruses: the complete genome of a second Ostreococcus tauri virus, OtV-1. *Environmental Microbiology, 11*, 2821–2839.
Weynberg, K. D., Allen, M. J., Gilg, I. C., Scanlan, D. J., & Wilson, W. H. (2011). Genome sequence of Ostreococcus tauri virus OtV-2 enlightens the role of picoeukaryote niche separation in the ocean. *Journal of Virology, 85*, 4520–4529.
Wilhelm, S. W., & Suttle, C. A. (1999). Viruses and nutrient cycles in the sea. *Bioscience, 49*, 781–788.
Wilson, W. H., Schroeder, D. C., Allen, M. J., Holden, M. T. G., Parkhill, J., Barrell, B. G., et al. (2005). Complete genome sequence and lytic phase transcription profile of a coccolithovirus. *Science, 309*, 1090–1092.
Wilson, W. H., Dale, A. L., Davy, J. E., & Davy, S. K. (2005). An enemy within? Observations of virus-like particles in reef corals. *Coral Reefs, 24*, 145–148.
Wilson, W. H., Van Etten, J. L., & Allen, M. J. (2009). The Phycodnaviridae: the story of how tiny giants rule the world. *Current Topics in Microbiology and Immunology, 328*, 1–42.
Wommack, K., & Colwell, R. (2000). Virioplankton: viruses in aquatic ecosystems. *Microbiology and Molecular Biology Reviews, 64*, 69–114.
Worden, A. Z., Lee, J.-H., Mock, T., Rouzé, P., Simmons, M. P., Aerts, A. L., et al. (2009). Green evolution and dynamic adaptations revealed by genomes of the marine picoeukaryotes Micromonas. *Science, 324*, 268–272.
Yamada, T., Onimatsu, H., & Van Etten, J. (2006). Chlorella viruses. *Advances in Virus Research, 66*, 293–336.
Yanai-Balser, G. M., Duncan, G. A., Eudy, J. D., Wang, D., Li, X., Agarkova, I. V., et al. (2010). Microarray analysis of *Paramecium bursaria* chlorella virus 1 transcription. *Journal of Virology, 84*, 532–542.
Yoon, H. S., Price, D. C., Stepanauskas, R., Rajah, V. D., Sieracki, M. E., Wilson, W. H., Yang, E. C., et al. (2011). Single-Cell Genomics Reveals Organismal Interactions in Uncultivated Marine Protists. *Science* (New York, N.Y.*), 332*, 714–717.
Zhang, X., Xiang, Y., Dunigan, D. D., Klose, T., Chipman, P. R., Van Etten, J. L., et al. (2011). Three-dimensional structure and function of the *Paramecium bursaria* chlorella virus capsid. *Proceedings of the National Academy of Sciences of the United States of America, 108*, 14837–14842.
Zhu, F., Massana, R., Not, F., Marie, D., & Vaulot, D. (2005). Mapping of picoeucaryotes in marine ecosystems with quantitative PCR of the 18S rRNA gene. *FEMS Microbiology Ecology, 52*, 79–92.
Zingone, A., Natale, F., Biffali, E., Borra, M., Forlani, G., & Sarno, D. (2006). Diversity in morphology, infectivity, molecular characteristics and induced host resistance between two viruses infecting *Micromonas pusilla*. *Aquatic Microbial Ecology, 45*, 1–14.
Zondervan, I. (2007). The effects of light, macronutrients, trace metals and CO2 on the production of calcium carbonate and organic carbon in coccolithophores—A review. *Deep Sea Research Part II: Topical Studies in Oceanography, 54*, 521–537.

CHAPTER TEN

Environmental and Evolutionary Genomics of Microbial Algae: Power and Challenges of Metagenomics

Eve Toulza[1,*,†], Romain Blanc-Mathieu[*,†], Sébastien Gourbière[‡], and Gwenael Piganeau[*,†,‡]

[*]UPMC Univ Paris 06, UMR 7232, Observatoire Océanologique, Avenue du Fontaulé, BP44, 66651 Banyuls-sur-Mer, France

[†]CNRS, UMR 7232, Observatoire Océanologique, Avenue du Fontaulé, BP44, 66651, Banyuls-sur-Mer, France

[‡]UMR 5244 CNRS-UPVD, Ecologie et Evolution des Interactions, Université de Perpignan via Domitia, 66860 Perpignan, France

[1]Corresponding author: E-mail: eve.toulza@obs-banyuls.fr

Contents

Advances in Botanical Research, Volume 64
ISSN 0065-2296,
http://dx.doi.org/10.1016/B978-0-12-391499-6.00010-4

Abstract

Metagenomics is the study of the DNA content of a community of microorganisms. With the advent of next-generation sequencing technologies, more and more metagenomes from various environments are being produced. This DNA sequence profusion has revolutionized microbiology, where many cellular-based molecular approaches are hampered by cultivation difficulties. Most metagenomes correspond to sequence data from viral and bacterial communities, though allowing many new groups to be identified, though recent technological advances are now extending the approach to microbial eukaryotes. Most metagenomic studies address the fundamental issue of species richness: the number of species and their abundance distribution across environments. However, there is much more at stake than species accountancy; there is functional gene diversity, and there is also the opportunity to test current hypotheses about ecological and evolutionary processes behind community structures and dynamics. In this chapter, we aim to review some of the methodological and conceptual challenges brought by metagenomics and point to the perspectives opened to better understand the microbial algae.

1. INTRODUCTION

Metagenomes are defined as the sum of the genomes living in an environment (Handelsman, Rondon, Brady, Clardy, & Goodman, 1998; Riesenfeld, Schloss, & Handelsman, 2004) and thus set the stage for the assessment of the direct relationship between genes and environment. Traditionally, biologists have not investigated this relationship; rather, they have dissected the links between the genotype and the phenotype (molecular and population genetics and genomics) and the link between the phenotype and the environment (quantitative genetics, ecology and life history traits).

Practically, a metagenome is a sum of short DNA sequences, called reads, which are 70- to 800-bp random fragments of the sampled genomes, eventually assembled into larger contigs. These new data sets are conceptually challenging because several adjustments need to be made to infer gene and species frequencies from the number of reads in a metagenome, rather than the number of individuals sampled from a population or community. Information about the phenotype of the most abundant species may be indirectly inferred from the functional gene repertoire of longer read assemblies (i.e. partial chromosomes), but many reads or small assemblies have to be considered in the absence of any relationship to other reads or genomes. For example, the pilot shotgun metagenomic study of the Sargasso Sea (Venter *et al.*, 2004) provided 1.6×10^6 reads, of which 16% could not be assembled with any other read.

This DNA sequence bounty has revolutionized microbiology, where cell-based approaches are hampered by cultivation techniques. Metagenomics has revealed many new species through their genes. Although, at present, the use of metagenomics is largely restricted to the study of viruses and prokaryotes, recent technological advances are now extending the approach to microbial eukaryotes.

Most metagenomic studies address the fundamental issue of species richness: the number of species and their abundance distribution. There is much more at stake than accountancy; first there is functional diversity – the metabolic potential of a community through the description of gene functions. Second, there is the possibility of testing current hypotheses about ecological and evolutionary processes shaping community structures and dynamics. In this chapter, we aim to review some of the methodological and conceptual challenges of metagenomes and point to the opened perspectives to better understand the microbial algae.

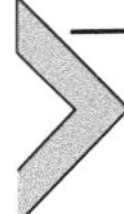

2. HOW DOES ONE SORT THE METAGENOME SEQUENCES INTO TAXONOMIC OR FUNCTIONAL GROUPS?

2.1. The Many 'Ways and Reads' of Metagenomics

There are different methods for each of the steps to produce a metagenome: cell sampling, DNA preparation (random shotgun or barcoding gene amplification), cloning or direct sequencing and sequencing method (Sanger, 454 pyrosequencing, Illumina or SOLiD; single reads or paired ends [PEs]) that can be used to produce a metagenome (Fig. 10.1). This diversity is important to consider when doing comparative metagenomics, i.e. comparison of metagenome gene content across environment.

We here shortly describe these three steps. The important initial step is how you isolate the cells from the environment, either by size fractionation (using filters with different porosities) or by flow cytometry. In the case of cell sorting, the extracted DNA has to be amplified by whole-genome amplification (WGA) due to the small amount of material available (Yilmaz & Singh, 2011). Another way to look at a sample is to sequence the whole DNA content by shotgun metagenomics, i.e. the random sequencing of the whole DNA in the community or target copies of a often ubiquitous gene (e.g. a 'barcoding' gene such as the 16S ribosomal DNA [rDNA] for

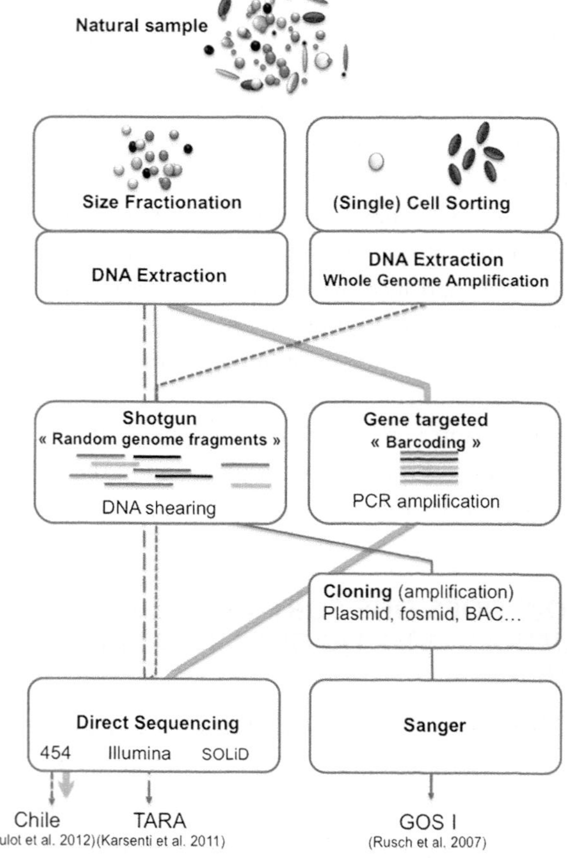

Figure 10.1 Overview of the different types of metagenomes. For colour version of this figure, the reader is referred to the online version of this book.

prokaryotes or 18S rDNA for eukaryotes). In both cases, DNA fragments can be either directly sequenced using next-generation sequencing technologies (NGS) or cloned in a vector for amplification and sequenced subsequently using either NGS or Sanger methods. The size of the generated sequences or 'reads' depends on the technology used, from 50 nucleotides (SOLiD) to about 800 (Sanger). Paired-end sequencing is the sequencing of the two extremities of the same DNA fragment that will enable to join the two (not necessarily overlapping) reads to be joined in the same sequence assembly (see Fullwood, Wei, Liu, & Ruan, 2009, for a review on PE sequencing).

Metagenomics has been applied to a variety of ecosystems, from soil to the human or termite gut to acid mine drainages (reviewed in the Genomes OnLine Database [GOLD, Pagani *et al.*, 2012]). The largest metagenomic study to date is the Global Ocean Sampling (GOS) expedition, which aims to explore ocean microbial diversity across the world's oceans (Rusch *et al.*, 2007). The thin solid line in Fig. 10.1 presents the methodological steps used for the GOS project. In short, seawater samples are filtered through two or more filters, but mostly recovered from the 0.1- to 0.8-μm fraction of organism size (0.8–3 μm for a small number of sites). Extracted DNA is cloned into plasmids (2-kb insert size) or fosmids (40-kb insert size) and sequenced at both ends using Sanger (average of 820 bp for each sequence). Recently, another pan-oceanographic campaign, Tara Oceans, has been launched (Karsenti *et al.*, 2011) to explore ocean's biodiversity, sampling all sizes from viruses to zooplanktonic metazoans, and collecting at both surface, mesopelagic and deep chlorophyll maximum (DCM) depths. Extracted DNA is used for gene-targeted diversity (18S, 28S, 16S and Cox1) as well as shotgun metagenomics. Analysis of transcriptional activity, 'metatranscriptomics', is also performed using the same strategy on cDNAs (obtained from both messenger and total RNA). The methodological steps used for Tara Oceans are presented by a grey large dotted line in Fig. 10.1. Because the sequence data from this study also target the eukaryotic fraction, Tara is expected to provide many new sequences of marine planktonic algae. Also presented in Fig. 10.1 (small dotted line) are metagenomics data obtained from sorted cells targeting picoeukaryotic algae (Vaulot *et al.*, 2012, see below Section 10.3.2).

The metagenomic sequences have their own 'env' section in GenBank, and metagenomic-dedicated portals, like CAMERA or GOLD, provide a centralized platform for metagenomic projects and data deposition (Pagani *et al.*, 2012; Sun *et al.*, 2011).

Taxonomic assignation of shotgun metagenomic sequences is a challenging task because of the highly fragmented nature of the sequences, and the unbalanced set of reference genomes. Bioinformatics analysis has become the main bottleneck for metagenomic projects. Annotation is a time-consuming task requiring comprehensive bioinformatics skills and their application to fragmented metagenomic sequences has even fostered the development of bioinformatics university training courses specifically targeted at undergraduate university students (e.g. Hingamp *et al.*, 2008).

The primary objectives when analysis metagenomes are to sort the reads and to assess 'who is in there?' and 'what are they doing?'. We will focus below on analysis of shotgun metagenomic sequences.

2.2. Sorting Reads for Taxonomic Annotation

Community composition can be inferred by two types of methods, similarity based or composition based. We cite here only a small number of existing programs for each type of method, as new tools are becoming available regularly in this fast-evolving field.

2.2.1. Similarity-Based Methods Rely on Sequences We Already Know

Similarity-based searches typically rely on a first comparison step with a reference database using algorithm such as BLAST (Altschul, Gish, Miller, Myers, & Lipman, 1990). The results are then analysed to infer taxonomic affiliation using either taxonomy of homologues or phylogenetic mapping of marker genes. Genes encoding ribosomal RNA are widely used for such purposes (Pace, 1997). These sequences are ubiquitous in all cells, as a result of their essential function in the translation machinery, and such genes therefore have the best taxonomic representation in the databases. Relying on a handful of genes may sound disappointing, but these marker or 'barcoding' genes present both well-conserved motifs and rapidly evolving regions that allow assignation at different taxonomic levels from phylum to species level. Furthermore, these genes are the best-known genes across all cellular organisms and enable the classification of any sequence into known taxonomic clades. A number of tools are focused on 16S or 18S rDNA analysis like Qiime (Caporaso *et al.*, 2010) or Mothur (Schloss *et al.*, 2009). Ribosomal barcoding then use classical methods to classify sequences into a taxonomic group, such as clustering into molecular operational taxonomic units (OTUs).

Several tools rely directly on similarity-based methods. To cite only a few, MG-RAST (Meyer *et al.*, 2008) uses best hit against a comprehensive protein database, whereas MEGAN (Huson, Auch, Qi, & Schuster, 2007) produces taxonomic profiling of metagenomic input from BLAST search against the NCBI non-redundant database using the 'lowest common ancestor' from the NCBI taxonomy node (http://www.ncbi.nlm.nih.gov/taxonomy). CARMA3 (Gerlach & Stoye, 2011) can handle results from BLAST and also from more sophisticated homology searches using hidden Markov models.

Many resources like MG-RAST (Meyer *et al.*, 2008) or Galaxy (Kosakovsky Pond *et al.*, 2009) enable analyses beyond annotation, such as functional pathway identification, as well as the comparison of several metagenomes. These workflows can be used with preloaded publicly available metagenomes, as well as on original data; a quota limit of 250 Gb for registered users is available in Galaxy, and the latter can also be downloaded and installed locally.

Phylogenetic mapping is another fast and accurate method for taxonomic classification of anonymous sequences such as metagenomic reads (Von Mering *et al.*, 2007). Phylogenetic mapping requires a reference alignment and a corresponding reference phylogenetic tree. The positions of the query sequences are then examined using phylogenetic tree reconstruction algorithms. This procedure has been used for a set of 31 protein-coding marker genes, which are present in single copy in most genomes (Ciccarelli *et al.*, 2006). Several dedicated programs able to handle very large data sets are available, among others Treephyler (Schreiber, Gumrich, Daniel, & Meinicke, 2010), pplacer (Matsen, Kodner, & Armbrust, 2010), or evolutionary placement algorithm (EPA; Berger, Krompass, & Stamatakis, 2011).

2.2.2. K-mer Composition and Mixture Modelling

The second family of methods relies on nucleotide composition and k-mer frequencies. These are machine-learning methods that use statistical signatures of reads to infer taxonomy. They often use reference genomes to precalculate oligomer frequencies that correspond to known taxonomic groups. For example, TETRA provides a statistical framework for tetranucleotide distribution analysis and discriminating sequences of common taxonomic origin (Teeling, Waldmann, Lombardot, Bauer, & Glockner, 2004). PhyloPythia has been developed to characterize composition of more complex communities by using multidimensional space of variable oligomer frequencies (Mchardy, Martin, Tsirigos, Hugenholtz, & Rigoutsos, 2007).

Moreover, it allows machine learning on sample-specific clades using fragments carrying marker genes. Phymm uses interpolated Markov models to classify oligonucleotides of variable length. It is trained on 539 complete, curated genomes, and allows sequence binning from reads as short as 100 base pairs (Brady & Salzberg, 2009). The latter are best used in combination with similarity-based methods.

Similarity-based programs are biased by database under- and over-representation and their accuracy dramatically increases with sequence length, so that short reads represent a significant challenge (Wommack, Bhavsar, & Ravel, 2008). Composition-based methods are also less accurate for shorter reads. A new class of algorithms use mixture modelling of overall oligonucleotide composition instead of single-sequence analysis (Meinicke, Asshauer, & Lingner, 2011). Unlike other classification approaches, the accuracy of this method is independent of sequence length.

2.3. Sorting Reads for Functional Annotation

Bioprospecting, i.e. experimental function-based discovery of new biological activities from microbial communities, has been developed further with the advent of metagenomes, cloning random DNA fragments and screening them for biological activities (reviewed in Kennedy *et al.*, 2011 and Cadoret, Garnier, & Saint-Jean, 2012 in this volume). Sequence data mining from shotgun metagenomics has fostered the development of bioinformatics tools to discover new genes from these highly fragmented sequences.

Tools used for taxonomic assignation from protein-coding metagenomic sequences (i.e. similarity based) can also be used for functional classification of sequences. Pairwise amino acid homology searches like BLASTX are routinely used to identify gene functions in metagenomes (implemented, e.g. in the MG-RAST metagenome analysis pipeline). However, there are several cases of gene families where there is no relation between protein similarity and function (e.g. ABC transporters share sequence similarity because of their ATP-binding domains but may share little or no substrate specificity (Davidson, Dassa, Orelle, & Chen, 2008)). Moreover, as with taxonomic assignation, functional analysis using a best BLAST hit, although computationally efficient, is not very accurate and strongly depends on the database used for the search and its annotation.

To give a gene-centric view of the diversity in a natural sample (the so-called functional diversity), previous gene prediction methods have to be adapted because of the fragmentary nature of sequences (sometimes of poor quality with low depth) and their multiple taxonomic origins that prevent the use of species-trained algorithms. In this context, assembly procedures are crucial both to reduce computational effort and to maximize sequence size. A new version of the Glimmer gene prediction software has been developed for *ab initio* gene prediction from metagenomic data (Kelley, Liu, Delcher, Pop, & Salzberg, 2012). It uses unsupervised clustering of sequences to generate clusters before gene prediction. This version can also deal with sequencing errors (substitutions as well as insertions/deletions) by passing through stop codons or detecting frameshifts. MetaGene (Noguchi, Park, & Takagi, 2006) and FragGeneScan (Rho, Tang, & Ye, 2010) are other examples of metagenome-dedicated gene prediction tools.

Another strategy to identify protein-coding sequences in metagenomes is to search for orthologous groups. HMMER (Finn, Clements, & Eddy, 2011) uses probabilistic methods after protein sequence profile searches against a database of Pfam protein domains (Bateman *et al.*, 2002). This approach has recently been used to analyse genes involved in iron uptake in bacterial genomes and metagenomes (Hopkinson & Barbeau, 2011).

From the GOS data set (at that time the pilot study [Venter *et al.*, 2004] and the first route [Rusch *et al.*, 2007]), more than 6 million proteins were predicted using sequence similarity clustering of predicted open-reading frames (Yooseph *et al.*, 2007). At the time of the study, these environmental protein sequences corresponded to a doubling of the proteins in the databases, thus substantially expanding the universe of protein families. Due to their larger genomes, lower gene densities and large protein families (see Fig. 10.3, Section 10.5 for paralogy and orthology definitions), these methods are less performant on eukaryotic sequences. OrthoMCL has been specially developed for eukaryotic data (Li, Stoeckert, & Roos, 2003).

2.4. Transforming Read Counts into Species Diversity and Richness Estimates

Transforming the observed number of reads assigned to a taxonomic or functional group into species richness (the number of different species in a sample) or abundance (relative representation of species) is not a trivial issue, as two types of variation must be taken into account.

First, the number of different operational taxonomic units (OTUs) may be biased at different taxonomic levels. Some taxonomical groups like GC-rich haptophytes (Liu *et al.*, 2009) may be omitted because of polymerase chain reaction (PCR) amplification bias. Some barcoding genes, like the 18S rDNA, 'lump' several species together, and this is expected to be more the case for unicellular eukaryotes with large effective population sizes (Piganeau, Eyre-Walker, Grimsley, & Moreau, 2011). The opposite bias has also been reported: *Elphidium macellum*, a Foraminifera, contains up to five different copies of the 18S rDNA gene within the same genome (Pillet, Fontaine, & Pawlowski, 2012), leading to an overestimation of species richness.

Second, the high sequencing error rate of some sequencing technologies is also an issue, especially pyrosequencing, as this may lead to overestimation of diversity when clustering OTUs at level of 1% dissimilarity (Kunin, Engelbrektson, Ochman, & Hugenholtz, 2010). It has been estimated that even with a sequencing error rate below 0.5% (454 rate is about 1%, see Section 10.5, Table 2), the standard method used for OTU clustering significantly increases richness estimates (Huse, Welch, Morrison, & Sogin, 2010).

Analysis of additional marker genes like the 16S plastid rDNA of sorted cells from the South Pacific suggest that a single-gene marker, even one as widely used as 18S rDNA, provides a biased view of eukaryotic communities and that the use of several markers is necessary to obtain a more accurate estimate of diversity (Shi, Lepere, Scanlan, & Vaulot, 2011).

Once the reads have been assigned to a taxonomic group, how do their counts correlate to species richness, which is the number of different species, and their abundance?

This depends on the species abundance distribution: if the community is composed of a few very abundant species and a large number of rare species, the abundance of the rare species may be hard to estimate. In contrast, in a low-complexity community with *n* equally abundant species, the read numbers will enable a direct estimate of the abundance of species. Even considering unbiased sequences containing no errors, the coverage of 90% of the diversity (based on the standard 97% 16S identity) was thought to require a fivefold to tens of thousands of times higher sequencing depth than that currently achieved (Quince, Curtis, & Sloan, 2008). Ultra-deep sequencing allowed by Illumina platform can overcome these difficulties and greatly improve understanding of species richness (Caporaso *et al.*, 2012; Lecroq *et al.*, 2011).

Genome size is a confounding factor in assessing read frequencies in metagenomes because the probability of observing a read from a random-genome sample is proportional to the size of the genome. This effect is expected to be particularly important for eukaryotes; read counts should therefore be normalized by genome sizes (Beszteri, Temperton, Frickenhaus, & Giovannoni, 2010). Thus, the proportion of prokaryotic versus eukaryotic or viral sequences can have a confounding effect on relative abundances between communities (Delmont *et al.*, 2011).

In conclusion, assessment of gene frequencies in a community and its conversion into species abundancies need to be undertaken using a rigorous statistical framework, especially so when doing comparative metagenomics (Temperton *et al.*, 2009; Øvreas and Curtis, 2011).

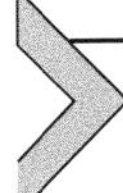

3. RECENT ADVANCES IN ALGAL METAGENOMICS

3.1. Gene-Targeted 'Barcoding' Diversity Estimates

3.1.1. 18S rDNA Insights into Diversity

Most studies on microbial algal diversity rely on 18S rDNA sequencing (Not *et al.* Chapter 1 of this volume). Here, we give a short overview of some milestones in molecular surveys from environmental samples, especially those that have allowed the identification of new algal groups.

Pioneering analyses of picoeukaryote diversity in the Equatorial Pacific from just 35 sequences highlighted the astounding diversity of these microorganisms (Moon-van der Staay, De Wachter, & Vaulot, 2001). Most 18S rDNA sequences were unknown and affiliated to important marine phyla including prasinophytes, haptophytes, dinoflagellates, stramenopiles, choanoflagellates and acantharians. This study also led to the description of a novel lineage, closely related to dinoflagellates. In a separate study of 225 clones from North Atlantic, Southern Ocean and Mediterranean Sea surface waters (Diez, Pedros-Alio, & Massana, 2001), 126 were affiliated to algal classes, especially the Prasinophyceae, the Prymnesiophyceae, the Bacillariophyceae and the Dinophyceae. Prasinophytes and novel stramenopiles were very abundant in all libraries. A new picoplanktonic algal group, the picobiliphytes, was identified from environmental 18S data (Not *et al.*, 2007, 2012).

In the North Sea, prasinophytes dominated the photosynthetic group at 40%, but other groups, such as bolidomonads and chrysophytes, were also present (Medlin, Metfies, Mehl, Wiltshire, & Valentin, 2006). Novel groups were found in the cryptomonads and in the dinoflagellates, as well as a new algal group, sister to the cryptophyte and the glaucocystophytes. In the Arctic Ocean, a new group within the photosynthetic stramenopiles has been identified besides new representatives from five of eight major marine eukaryotic lineages (Lovejoy, Massana, & Pedros-Alio, 2006). Using a similarity threshold of 98% identity on a region of the 18S rDNA gene to bin sequences into the same OTU, 42% of the arctic sequences were not similar to any publicly available sequence, highlighting the unknown diversity of these organisms.

Because of the potential representation bias induced by universal 18S primers, clone libraries made with specific primers targeting Haptophyta 18S have been produced from subpolar and subtropical oceanic waters (Liu *et al.*, 2009). This analysis revealed extreme diversity, especially for non-calcifying haptophytes. Another limitation of using 18S barcoding to target algae is that many eukaryotic groups contain both heterotrophic and phototrophic organisms (e.g. Not *et al.*, 2012 in this volume). A possible workaround is to target directly photosynthetic cells. Clone libraries from flow cytometry-sorted cells targeting photosynthetic picoeukaryotes in the South East Pacific showed that despite numerous prasinophyte reference genomes, most sequences belong to uncultured organisms (Shi, Marie, Jardillier, Scanlan, & Vaulot, 2009).

3.1.2. Other Barcoding Genes

Sequences of the mitochondrial encoded cytochrome-c oxidase subunit 1 (*cox1*, also known as cyclooxygenase I) from 336 cultures of dinoflagellates were used to investigate three marine environments (Northeast Pacific, Northwest Atlantic and Caribbean), revealing previously underestimated diversity (Stern *et al.*, 2010). Compared to *rbcL* (encoding the large Rubisco subunit), 18S and ribosomal internal transcribed spacer, *cox1* performs better in distinguishing cryptic species within diatom morphospecies (Evans, Wortley, & Mann, 2007). Cryptic species of the Mamiellophyceae *Micromonas pusilla* have been identified using several nuclear (β-tubulin), mitochondrial (cox1), and chloroplast (rbcL) genes (Slapeta, Lopez-Garcia, & Moreira, 2006).

Diversity of eukaryotic picophytoplankton in the South East Pacific Ocean was revealed using flow cytometry cell sorting and plastid 16S libraries (Shi *et al.*, 2011). Contrary to 18S analysis, it showed that Chlorophyta, Chrysophyceae (mostly in surface), Pelagophyceae (mostly at the DCM), Dictyochophyceae and Haptophyta were prevalent contributors to small photosynthetic eukaryote communities. This work also identified a new clade of Prasinophyceae, possibly specific to hyper-oligotrophic environments.

3.2. *De Novo* Genome Assemblies Studies

There are now many examples of prokaryotic or viral genome assemblies from metagenomic data. Reads from the GOS were assembled as 'an ensemble view' of 24 scaffolds assigned to the cyanobacteria *Prochlorococcus*. Even though these assemblies are not equivalent to draft genome assemblies, they nevertheless revealed two uncharacterized co-occuring *Prochlorococcus* clades (Rusch *et al.*, 2010).

There are minimal sequencing requirements to assemble a bacterial genome from metagenomic data in complex environments (Luo, Tsementzi, Kyrpides, & Konstantinidis, 2011). Coverage of at least 20× for a single species is necessary to avoid chimeras in the assembly. Recently, a complete *de novo* Euryarchaeota genome has been assembled with 100-fold coverage, despite the fact that it represented only less than 2% of the metagenome sequences (Iverson *et al.*, 2012). This was achieved by producing 58 Gbp of 50-nucleotide mate pairs with SOLiD sequencing technology on two samples of Puget Sound (Washington, DC).

Considering both the proportion of eukaryotic organisms in the seawater, and their larger genome sizes, the sequencing effort required to recover sufficient sequence information to assemble a whole algal genome from a metagenome is impractical for the moment, unless closely related reference genomes are available (Piganeau & Moreau, 2007). Flow cytometry sorting followed by WGA is an alternative to the sequencing of the eukaryotic community. Cells can be sorted from natural samples on the basis of their size and pigment autofluorescence. Several methods of DNA amplification from environmental samples have been developed that allow sequencing from small number of cells or even single cells. For example, multiple displacement amplification (MDA) is a highly efficient and popular method for WGA based on the strand-displacing DNA polymerase of bacteriophage phi29 DNA (Dean, Nelson, Giesler, & Lasken, 2001). However, this method can be biased towards the amplification of AT-rich sequences (Yilmaz, Allgaier, & Hugenholtz, 2010) and may produce chimeric sequences (Lasken & Stockwell, 2007). Nextera is a new amplification method based on transposase fragmentation that circumvents amplification biases observed with MDA and is particularly well suited for 454 pyrosequencing (Marine *et al.*, 2011).

A few recent studies have targeted eukaryotic algae, especially picophytoeukaryotes (Table 10.1). Prymnesiophytes (also known as haptophytes), although abundant primary producers in the seawater, remain poorly understood. Uncultured picoprymnesiophyte ecology has been addressed using flow cytometry sorting followed by MDA amplification from a range of tropical to subpolar samples (Cuvelier *et al.*, 2010). This study revealed a mixed-lineage nuclear gene repertoire, apparently distinct from plastid evolutionary history, which may have favoured niche differentiation, thus contributing to their global evolutionary success. Two

Table 10.1 Targeted Metagenomic Analyses of Marine Algae

Sample size	Method	Number of Reads	Nb of Contigs	Maximum Contig Size	Remarks and Reference
Eight samples, ~300 cells/ sort	MDA - Shotgun 454 (200 bp) - Cloning (16S/ 18S) and Sanger sequencing	10^6 (454)	89,375 (27 Mb)	21 kb	1015 Scaffolds corresponding to 2.5 Mb of plastid genomes and 2 Mb of nuclear genomes of uncultured prymnesiophyte (Cuvelier *et al.*, 2010)
Two samples ~104,000 Cells ~233,000 Cells	MDA - 454 (420 bp)	7×10^5	23,187 (23 Mb) 34,839 (35 Mb)	35 kb 43 kb	11 and 19 Mb of non-redundant *Bathycoccus* genome sequences with different genotypes (Vaulot *et al.*, 2012)
Three cells	MDA - 3 Cells: 454 (370 bp) - 2 Cells: 454 + Illumina PEs	7 10^5 Reads (454) 10^9 Reads (Illumina)	454: - 123 (0.2 Mb) - 472 (0.8 Mb) - 268 (0.4 Mb) 454 and Illumina: - 73,286 (28 Mb) - 74,660 (29 Mb)	N/A	Picobiliphytes with no plastidial sequences The genome sequence of one nanovirus (1.8 kb) (Yoon *et al.*, 2011)

N/A: not applicable.

samples from the Chile upwelling have been subjected to analysis of the picophytoeukaryotic fraction by flow cytometry sorting, followed by WGA and 454 pyrosequencing (Vaulot *et al.*, 2012). Communities were dominated by Mamiellophyceae, in particular the genus *Bathycoccus* with half of the reads sharing 96% identity with the reference genome of the Mediterranean *Bathycoccus* strain RCC1105 (Moreau *et al.*, 2012). Mapping of the reads affiliated to *Bathycoccus* to the reference genome allowed the discovery of several distinct genetic strains within the sample.

Flow cytometry followed by genome amplification is powerful enough to be used at the single-cell level. Libraries of single amplified genomes revealed higher diversity of heterotrophic protists than clone libraries of the PCR-amplified 18S rRNA gene from the same coastal water sample (Heywood, Sieracki, Bellows, Poulton, & Stepanauskas, 2011). This technique is also very useful as a means of studying algae. A new variant of the haptophyte alga *Emiliania huxleyi* virus has been sequenced from a single cultured infected cell (Martinez Martinez, Poulton, Stepanauskas, Sieracki, & Wilson, 2011). In order to obtain new insights into plastid acquisition of marine picobiliphytes, a previously unknown group of pigmented eukaryotes, single cells have been isolated from the Gulf of Maine (Yoon *et al.*, 2011). Following 18S-based identification, three individuals lacking chlorophyll fluorescence were sequenced. Genome data revealed no traces of plastid-originating genes and argue against a photosynthetic ability in these picobiliphytes. These three cells each gave a draft genome assembly of at least 50% of the genome. One cell was infected by a nanovirus (a widespread single-stranded DNA virus) whose sequence has been completely assembled. More generally, this study highlights the capability of single-cell sorting and genome amplification to obtain substantial amounts of genome data and its potential to reveal ecological lifestyles of uncharacterized, rare and uncultured microorganisms.

Multiple and single-cell sorting followed by genome amplification is a promising solution for eukaryote environmental genomics. It has already been shown that WGA is suitable for photosynthetic eukaryotes and the first studies led to the production of large amounts of genomic data (Lepere *et al.*, 2011). In the near future, we anticipate that the first complete eukaryotic algae genomes from natural samples may be released.

3.3. Perspectives on Microbial Algae Transcriptomics

To date, the use of metagenomic techniques in eukaryotes has largely been restricted to taxonomic classification based on 18S barcoding. Beyond taxonomic diversity, metabolic capabilities and gene expression levels are expected to shed light on key functional groups and pathways that are active at the community level.

Metatranscriptomics (the sequencing of community transcripts) will prove an important tool in characterizing the metabolic activity of a natural microbial assemblage (Helbling, Ackermann, Fenner, Kohler, & Johnson, 2012). The exploration of the functional dynamics in the bacterioplankton has begun, for example with the comparison of gene expression between day and night in the North Pacific subtropical gyre (Poretsky *et al.*, 2009) and with analysis of metatranscriptomes in a permanent oxygen minimum zone in the Eastern Tropical South Pacific (Stewart, Ulloa, & Delong, 2012) and in the course of a microcosm phytoplankton bloom (Rinta-Kanto, Sun, Sharma, Kiene, & Moran, 2012).

In eukaryotes, metatranscriptomics has the potential to directly target messenger RNAs, because they are polyadenylated, whereas prokaryotic are generally not. It can thus overcome problems caused by non-coding regions and low gene densities that are expected to make gene discovery in eukaryote metagenomes harder. A first small-scale metatranscriptomic analysis of freshwater dinoflagellate algae aimed to describe expressed genes as a proof of principle (Lin, Zhang, Zhuang, Tran, & Gill, 2010). In this lineage (with no sequenced reference genome), the presence of a specific spliced leader in messenger RNA allows moreover selective sequencing.

The exploration of eukaryotic transcriptomes from natural samples will be a promising way to extract information from these organisms and investigate differential gene expression associated with environmental variations in the context of climate change.

4. ENVIRONMENTAL GENOMICS : LINKING TAXONOMIC/FUNCTIONAL DIVERSITY TO THE ENVIRONMENT

4.1. How Can OTUs Inform Us about Species Biogeography?

Biogeography refers to the distribution patterns of species, whereas phylogeography refers to the historical processes responsible for this geographical distribution. In this review, we discuss how metagenomics addresses microbial biogeography. Metagenome mining cannot pretend to solve biogeography at the species level but can provide insights about the geographical distribution of organisms defined as OTUs.

Understanding an organism's biogeography is crucial to understand its ecological role. The question at the individual species level is whether it is a cosmopolitan (globally distributed) or endemic (niche adapted) species. It is still debated whether microorganisms are globally dispersed or whether they have distinct phylogeographies, in particular in aquatic environments. In the oceans, because of an assumed high dispersal rate, the neutral hypothesis postulates that free-living microbial species have a worldwide distribution with large population sizes (Finlay, 2002; Finlay & Fenchel, 2004). This theory mirrors the 'everything is everywhere, but the environment selects' principle (see O'malley [2007] for a historical review). For example, thermophilic bacteria, unless metabolically inactive (endospores), can be dispersed over large oceanic areas including cold Arctic environments (Hubert *et al.*, 2009). This view has been challenged by experimental evidence in aquatic environments, e.g. for diatom morphospecies for which regional-scale genetic variability can be detected (Telford, Vandvik, & Birks, 2006).

The alternative hypothesis to the ubiquitous dispersal is that species may have adapted to a specific environment where they have become endemic. Patterns of marine bacterioplankton diversity have been described using 16S rDNA libraries (Pommier *et al.*, 2007) and by examining the GOS data set (Biers, Sun, & Howard, 2009). Ribosomal DNA analyses at a global scale show that OTU richness is negatively correlated with latitude and positively correlated with temperature (Fuhrman *et al.*, 2008), consistent with the biogeography pattern of macrospecies.

For eukaryotic microorganisms, trends in biogeographical distributions have been analysed at smaller scales using mainly conventional approaches. Integration of numerous studies about free-living microorganisms from every ecosystem supports the existence of biogeographic patterns (Martiny *et al.*, 2006). Recently, however, a global distribution of fossil diatoms was reported, supporting a dispersal-dominated model (Cermeno & Falkowski, 2009). These apparently paradoxical findings may be reconciled by the 'rare biosphere' concept where a small number of species dominates a given environment at a given time, with a huge diversity of low-abundance species (Sogin *et al.*, 2006).

In addition, in the microbial world, both morphospecies and barcoding gene sequences can refer to many different cryptic species (see also Not *et al.*, 2012 in this volume). Metagenomics will be crucial in addressing these aspects of algae diversity and biogeography as it permits a direct assessment of community composition at different taxonomic levels based on sequence information. Comparison of species richness among environments and distribution patterns in a broad range of latitudes and conditions is expected to give a comprehensive view of how planktonic microbes reach, develop and adapt to an environment. This has been addressed for prokaryotic communities with the GOS as the largest metagenomic data set, where functional richness and diversity are linked to primary production and show a latitudinal gradient (Raes, Korbel, Lercher, Von Mering, & Bork, 2011). Circumglobal expeditions such as Tara Oceans which sample large size plankton fraction are poised to extend such analyses to eukaryotes (Karsenti *et al.*, 2011).

4.2. How Can Reads Provide Evidence for Adaptation to the Abiotic Environment?

One of the big challenges of ecology today is to link taxonomic diversity and ecosystem functioning to understand the interplay between communities and their environment. To this end, meaningful information has to be extracted from large amounts of sequence data in a statistically rigorous way. Multivariate analyses for quantitative ecology rely on gene or species relative abundances (i.e. frequencies). The first issue when analysing correlations between these frequencies and environmental factors is thus how these frequencies are estimated (see Section 10.2.4 below).

Trophic strategies and microbial lifestyles have been predicted from metagenomics data using models of genomic strategies (Lauro *et al.*, 2009). The method that uses clustering methods of self-organizing maps based on clusters of orthologous groups may be easily applicable to eukaryotic sequences. Multivariate analyses have been recently adapted to the analysis of metagenomic sequences in relation with environmental variables. Gianoulis *et al.* explored the GOS data set using 'metabolic footprinting' (Gianoulis *et al.*, 2009). They assessed relationships within and between gene content (metabolic pathways) and multiple environmental features using approaches derived from canonical correlation analysis. They established covariations between metabolic pathways and environmental gradients. They identified, for instance, a link between energy-conversion strategies and environmental constraints such as temperature. Using clustering as well as a new method of canonical correspondence analysis visualization based on network representation, the same group analysed membrane protein families' variation in the same data set (Patel *et al.*, 2010). Interestingly, they discovered a strong negative correlation between phosphate concentration in the seawater and transporter affinity for phosphate. Finally, in a effort to deduce ecological descriptors from sequence data, available geographical, meteorological and geophysicochemical data were confronted to the same GOS metagenomes (Raes *et al.*, 2011). This study confirmed a link between gene repertoire and environmental data (e.g. the abundance of genes encoding the photosynthetic machinery is positively correlated to temperature and hours of sunlight) as well as ecosystem processes (e.g. sulphur-related processes are negatively correlated to temperature).

Nevertheless, one must keep in mind that the biological interpretation of nutrient composition in the environment may be controversial. Geochemical measures may not be directly correlated to availability of elements for microorganisms, as is the case, for example, for iron, a limiting nutrient in large oceanic regions. Its chemistry, and hence bioavailability, is modified by ocean acidification (Shi, Xu, Hopkinson, & Morel, 2011), whereas saccharides enhance its bioavailability for eukaryotic algae (Hassler, Schoemann, Nichols, Butler, & Boyd, 2011).

Conversely, genetic markers can be used as natural biomarkers for element limitation in the environment as is the case for the iron storage *bacterioferritin* gene, which is positively correlated with iron concentration in the seawater (Toulza, Tagliabue, Blain, & Piganeau, 2012). In diatoms, *ferritin* has been acquired by horizontal gene transfer and is a genomic hallmark of low-iron concentration in the environment (Marchetti *et al.*, 2009). At the community scale, there is a statistically significant relationship between iron metabolism pathways and predicted iron concentrations, highlighting genomic strategies to face environmental iron limitation such as siderophore uptake (Toulza, Tagliabue, Blain, & Piganeau, 2012).

4.3. How Can Metagenome Sequences Be Informative about Associations and Interactions within the Community?

4.3.1. Host–Associate Interactions

Community assemblages in the aquatic environment rely on complex interactions, trophic, parasitic and symbiotic, among eukaryotic algae, viruses and bacteria (and higher trophic levels). In the context of host–virus interactions, there is some hope that these questions might be answered using environmental genomics (see also Grimsley *et al.* 2012, this volume). A recent Antarctic lake metaproteogenomic analysis (i.e. analysis of both genomic and proteomic data) highlighted the dynamics between Prasinophyceae, their viruses and associated virophage (a virus of a virus) (Yau *et al.*, 2011). The complete genome of the virophage and the nearly complete genome of the phycodnavirus could be assembled from this metagenome. Using single-cell genomics, Yoon *et al.* assembled a nanovirus genome from an incidentally infected picobiliphyte cell (Yoon *et al.*, 2011, see also Section 10.3.2). More recently, a method for detection and sorting of individual infected cells of the microalga *E. huxleyi* has been developed (Martinez Martinez *et al.*, 2011). Single-cell sorting approaches will greatly help to disentangle virus/host or more generally symbiotic interactions at the cellular level. Metagenome-based co-occurrence can also shed light on tight host association with parasite or symbiont but needs very large and uniformly sampled sites. Co-occurrence networks can be used to define interlineage associations in ecosystems (Chaffron, Rehrauer, Pernthaler, & Von Mering, 2010).

4.3.2. Community Interactions

The interactions between species, especially prokaryotes, in ecosystem functioning have been well documented in the context of biogeochemical cycles. Microbial organisms compete for nutrient resources and may act at different, successive levels of the same biogeochemical cycles.

In the oceans, bacteria and eukaryotic algae interact for nutrient uptake. For example, dimethylsulphoniopropionate (DMSP) is a metabolite produced by marine phytoplankton, which is catabolised by heterotrophic bacteria and accounts for up to 10% of carbon fixation (Reisch *et al.*, 2011). The analysis of surface ocean metagenomes led to the identification of new sequences of genes encoding enzymes involved in DMSP assimilation, thus highlighting the importance of this pathway in marine bacteria. This study provided new insights into the acquisition of reduced carbon and sulphur by surface ocean heterotrophs (Reisch *et al.*, 2011).

This topic is also particularly relevant in the context of global climate change, as dimethyl sulphide, which is released upon bacterial catabolism of DMSP, impacts on sulphur flux into the atmosphere. This is one example of what metagenomics can bring to the understanding of microbial interactions in ecosystem functioning. Ecosystem modelling can also be used to explore relations between organisms and environmental conditions (Follows, Dutkiewicz, Grant, & Chisholm, 2007).

4.4. Metagenomics and Evolutionary Stable Strategies

Evolutionary game theory (EGT) has been widely and successfully applied to understand the evolution of behavioural and life history traits in the context of basic (Dieckmann, Doebeli, Metz, & Tautz, 2004; Maynard Smith, 1982) or applied research (Ferrière, Dieckmann, & Couvet, 2004). One of the most studied evolutionary game, which has originally motivated the development of the EGT, is the evolution of cooperative or defective animal behaviour. The underlying concepts and modelling approach can readily be expanded to look at interesting and important evolutionary issues faced by marine microorganisms in their environment, which can be further confronted to metagenome data analysis.

We will consider a population of individuals whose survival is limited by a key resource, for example iron in the open ocean, and a set of strategies that individuals can adopt to extract such resource from the environment. In traditional studies of life history evolution, strategies are measured at the individual level. The set of observed strategies is then assumed to correspond to a set of different genotypes in a one-to-one relationship since the actual genetic determinism of life history traits being considered is usually unknown. Conversely, metagenomes and metatranscriptomes provide measures of the genes, corresponding to different strategies, as well measures of their level of expression, though they do not allow identifying the strategies played by every individual. These two fields thus provide complementary pictures of the individual-strategy–genotype relationship. Interestingly, they can both benefit of the EGT as the central elements of this theory are strategies. We will illustrate the potential contribution of the EGT by taking the example of the well-studied bacterial strategies to produce and/or use siderophores to uptake iron from the ocean environment, thought this framework applies to any molecule excreted in the environment to increase the fitness of the cell. Possible examples for algae are carbon-rich molecules implied in algal–bacterial symbiosis, antibotics or fongicides.

When the concentration of iron in the marine environment is low, some microorganisms produce a 'public good'; siderophores that efficiently fix iron and subsequently facilitate its uptake. Clearly, this strategy pays off only if the net balance between the metabolic cost of producing siderophores and the energetic gain is positive. Now, because siderophores are released in the environment, they are made publicly available. Potentially, such 'public goods' can thus be used by cheaters not paying the cost of producing them. This cheating strategy has obvious advantages, but it critically relies on the presence of individuals producing siderophores in the environment. This raises two typical evolutionary questions. First, will these two strategies coexist in the environment, or will one of them persist on its own? Second, what are the conditions which allow such polymorphic and monomorphic states to evolve?

Following the standard literature, we refer to these two strategies as 'cooperative' (C) and 'defective' (D). We first assume that, in absence of defectors, siderophores produced by cooperators provide them with an additive gain, λ_C. When defectors are present in the environment, this gain drops because a proportion of the siderophores they produce are now used by defectors. Denoting c such a proportion, the simplest way to describe the gain of cooperators in the presence of defectors is $(1-c)\lambda_C$. Clearly, defectors then benefit from the presence of cooperators, and their gain can be assumed to be proportional to c, and to their efficiency λ_D in turning additional iron into growth. For simplicity, we will thus note $c\lambda_D$, the gain of defectors in the presence of cooperators. The 'pay-off' matrix (Maynard Smith, 1982) for this evolutionary game then simply reads:

$$P = \begin{pmatrix} E(C,C) = \lambda_C & E(C,D) = \lambda_C(1-c) \\ E(D,C) = c\lambda_D & E(D,D) = 0 \end{pmatrix}$$

where $E(X, Y)$ refers to the expected gain of an individual adopting a strategy X in a population where all individuals play a strategy Y.

From this matrix, one can first determine whether or not strategies C and D are evolutionary stable strategies (ESS). An ESS is a strategy that once adopted by all members of a population cannot be invaded by any alternative mutant strategy (M) through selective processes. Standard results of EGT state that a resident strategy R is an ESS if $E(R, R) > E(M, R)$. The defective strategy (D) will thus never be an ESS since $E(D,D) - E(C,D) = \lambda_C(1-c) > 0$ (λ and c being positive). On the contrary, the cooperative strategy (C) can be an ESS if $E(C,C) - E(D,C) = \lambda_C - c\lambda_D > 0$, which is equivalent to $\lambda_C/\lambda_D > c$. If this condition is verified, one should observe only cooperative individuals in the environment. One can then start specifying the definition of λ_C and λ_D, to make this condition more informative. Essentially, λ_C can be seen as the product between (i) the quantity of siderophores produced (q) and (ii) the ratio between the increase in replication the siderophores can allow for ($b > 0$) and the cost of their production ($d > 1$): $\lambda_C = qb/d$. With a similar reasoning, one can set $\lambda_C = qb$, and the condition for C to be an ESS then simply reads $dc < 1$ (Fig. 10.2). This condition is biologically appealing; the strategy of producing siderophores (C) will resist invasion by defectors (D) if the cost of siderophores production and/or the proportion of those siderophores used by defectors are low. It also provides us with a clear-cut quantitative prediction about the conditions where one should expect a monomorphic population of cooperators. One might wonder what the predicted evolutionary outcome is, when none of the two strategies is an ESS. Given that C can invade D and D can invade C, one shall then now look for a polymorphic state to be evolutionary stable, whereby strategies C and D appear in proportion p and $1-p$ in the environment. Standard methods of

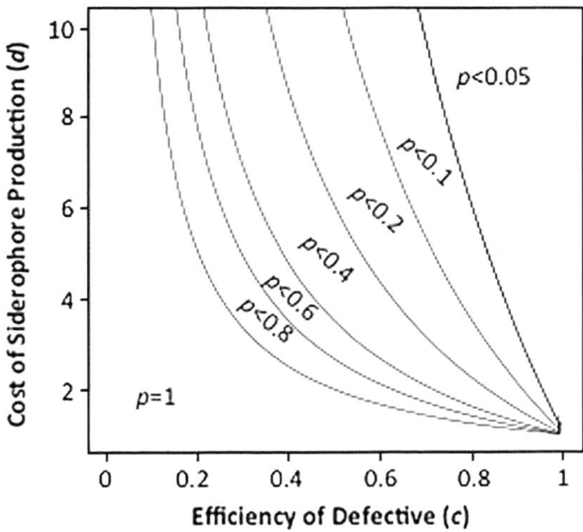

Figure 10.2 Expected outcomes of the cooperative–defective evolutionary game. As long as $cd < 1$, the cooperative strategy ($p = 1$) is an ESS, and one should not observe any defective individuals in the population. Once this condition no longer holds, the evolutionary stable frequency of cooperative individual in the population (p) progressively decreases.

EGT provide us with the possibility to predict such proportions. Following basic criteria (Maynard Smith, 1982), one can show that the expected proportion of cooperators in the environment is $p = (1 - c)\lambda_C / c(\lambda_D - \lambda_C)$ and using the above specification of λ_C and λ_D, this can be expressed in a simple way: $p = (1 - c)/c(d - c)$. This result is again biologically very consistent as the proportion of cooperators in the environment decreases with the cost of siderophores production and/or the proportion of those siderophores used by defectors. Finally, this simple quantitative prediction nicely combines with the previous condition to provide us with a full picture of the expected outcomes of the evolutionary game (Fig. 10.2).

Predictions on the persistence (or not) of strategies can be compared to the measure of the presence (or absence) of the corresponding genes in metagenomes. Recent analysis of gene frequencies of the GOS metagenomes for siderophore synthesis genes (cooperative genome strategy) versus siderophore uptake genes (cooperative + defective strategy) suggests that the frequency of the cooperative genes in the community is very low, indicating that siderophores synthesis is not an ESS in most environments, which in turns suggests that the cost of siderophores production and/or the use of these siderophores by defectors is too important. However, future metagenomes with higher sequence coverage may enable a more precise estimation of gene prevalence in the community and possibly allow for indirect estimates of the costs and benefits of producing and using siderophores in different assemblies.

A strong limitation of the use of EGT in the context of life history evolution is that when co-existence of strategies is expected (if $dc > 1$ in the above model), such diversity could results from a genetic polymorphism of pure strategies (individuals always cooperating or always defecting) or the existence of a 'mixed strategy' whereby an individual express either of the two 'pure' strategies with probability p and $1 - p$ (Gourbière & Menu, 2009). Investigating which of these two alternative is the most likely to explain the situation observed in the field require developing eco-genetic models based on data accumulated through long-term field studies (Gourbière & Menu, 2009), and the answer is likely to depend upon the specific evolutionary game at hands. Shotgun metagenomic and metatranscriptomic are likely to provide novel insights into the genetic basis of phenotypic strategies and help resolving such a longstanding difficulty. Pure strategies should be reflected by gene presence and thus gene frequencies within metagenomes. On the other hand, mixed strategies should be reflected by RNA presence and thus gene transcript frequencies in the environment. Combining typical measure of individual life history traits with metagenomes and metatranscriptomes is an exciting and promising avenue, and the EGT provides an ideal framework for such integrative studies to develop.

5. INTRASPECIFIC DIVERSITY: TOWARDS POPULATION METAGENOMICS?

5.1. Why Should We Care about Intraspecific Polymorphism?

Population genomics is defined as the study of genomic regions from different individuals to better understand the roles of evolutionary processes such as mutation, random genetic drift, gene flow and natural selection in the diversity pattern within a genome and between genomes of a population. Population genomics approaches can also provide information about the life cycle and can be informative about the prevalence of sexual reproduction as shown for the green alga *Ostreococcus tauri* (Grimsley, Pequin, Bachy, Moreau, & Piganeau, 2010).

What population genomics can offer for microbial eukaryotes (and any species in general) is the estimation of their intraspecific diversity, polymorphism, to provide information on both their evolutionary history and their potential to adapt because standing genetic diversity is the prerequisite for selection of fitter individuals in future generations. Given the ecological role of microbial eukaryotes in biological cycles and ecosystem structure (Not *et al.,* 2012, Chapter 1 of this volume), a better knowledge of their ability to adapt to environmental changes is crucial.

Polymorphism can be defined as the percent of non-identical nucleotides of the same genomic region within a population of individuals. Two main processes influencing genetic diversity are (i) molecular processes engineering the DNA, like mutation and recombination and (ii) population processes, like selection, genetic drift or demography. Under the assumption of neutrality, the expected genetic diversity, θ, is a function of two parameters: the effective population size, N_e, and the mutation rate, μ, $\theta = 4N_e\mu$ (Hartl and Clark, 1997; Kimura, 1969). Roughly, N_e corresponds to the average number of individuals contributing to the next generation at each generation over a given period of time; it is not the census population size.

At any location along the genome, departures from the expected value of θ can reflect the action of selection or genetic drift at this locus (Hartl & Clark, 1997). A reduction of θ in a given region can be interpreted as the result of either positive (hitchhiking) (Smith & Haigh, 1974) or negative selection (background selection, Charlesworth, Morgan, & Charlesworth, 1993). The relative importance of selection and genetic drift is affected by the effective population size. Populations with small N_e are more affected by genetic drift. These random changes in allele frequencies can lead to fixation of deleterious alleles and to elimination of advantageous ones by chance alone. The selection is therefore more efficient in population with large N_e. Demographic processes (reflected by N_e) thus have a real potential to affect the adaptive potential of a population.

In the extreme case of an endangered population, there is a N_e cut-off value under which the genetic load (accumulation of deleterious mutations) increases quickly as selection becomes a negligible evolutionary pressure in comparison to drift. This is the minimum viable population (MVP) size. When populations pass below the MVP, accumulation of deleterious mutation leads to a reduction of demographic parameters eventually leading to a decrease in N_e (Lynch, Conery, & Burger, 1995). As the population size decreases, impact of the genetic drift becomes stronger and increases the accumulative rate of deleterious mutations; this has been named the extinction vortex (Shaffer, 1981).

One can ask whether it is really relevant to worry about species extinction in the marine eukaryotic microbial world because their population sizes are usually thought to be large, on account of their immense habitat. However, very little is known about effective population size and dynamics of these organisms.

So far, most algal populations genetics studies have been performed using a handful of microsatellites. Microsatellites, also called simple sequence repeats, are tandem repeated motifs of one to six bases (Zane, Bargelloni, & Patarnello, 2002). Their high degree of length polymorphism makes them suitable markers for population genetics. Microsatellites revealed a high standing genetic diversity in the dinoflagellate *Alexandrium minutum* (Casabianca *et al.*, 2012), the Rhodophyta *Chondrus crispus* (Krueger-Hadfield, Collen, Daguin-Thiebaut, & Valero, 2011), the diatoms *Ditylum brightwelli* (Rynearson and Armbrust, 2000, 2005; Rynearson, Lin, & Armbrust, 2009; Rynearson, Newton, & Armbrust, 2006), *Pseudo-nitzschia punges* (Casteleyn *et al.*, 2010) and *Pseudo-nitzschia multiseries*. For the well-studied diatom *D. brightwelli*, this diversity remains high during a bloom (Rynearson & Armbrust, 2005) and could be partitioned in time and space and linked to environmental variables such as solar irradiance and silicic acid concentration (Rynearson *et al.*, 2006, 2009).

Microsatellite analyses further suggest that gene flow is limited among structured populations of *D. brightwelli* (Rynearson *et al.*, 2009), *P. punges* (Casteleyn *et al.*, 2010) and *C. crispus* (Krueger-Hadfield *et al.*, 2011).

However, designing microsatellite experiments is a labour-intensive task (Zane *et al.*, 2002) and population genetics inferences are made on few markers. Doing population genetics at the genome level would allow a much finer resolution of the extent of genetic diversity. In populations with effective population size that is large enough to avoid drift and with a strong selection strength relative to effective population size, it will also allow investigation of the molecular basis of adaptation by looking for loci showing signatures of selection (Luikart, England, Tallmon, Jordan, & Taberlet, 2003). By randomly sequencing any part of any genome in a given community, a metagenomic approach theoretically provides the raw material for large-scale algae population genomics. However, extracting a population data set for eukaryotic algae from a metagenome remains a significant challenge.

5.2. Challenges of Population Metagenomics

5.2.1. Sampling Issues

In traditional population genetics, the proportion of any allele is directly estimated from the sample of this population (e.g. in Fig. 10.3A, the frequency of allele 1 for gene A can be estimated as $1/4 = 0.25$). Each

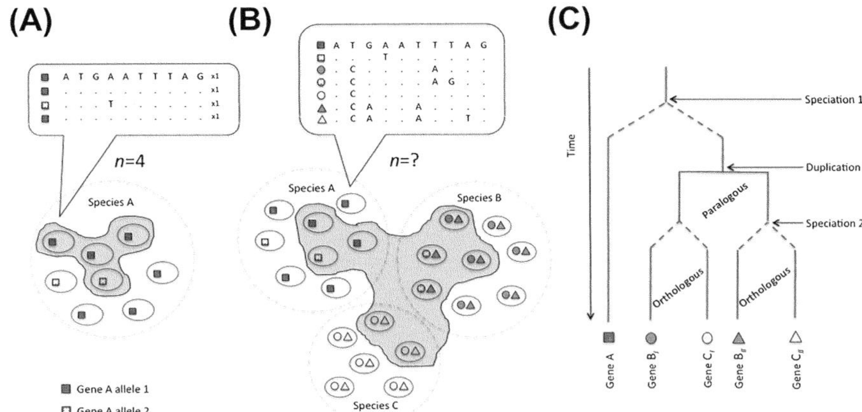

Figure 10.3 Sample size and homologous relationships in metagenomes. Species are delimited by dash–dotted circles. Genes are represented with circles or triangles and possible alleles are marked with a white cross. In A and B, individuals are represented by ovals, the sampling area is grey, *n* is the sample size. (A) In a traditional population genomics, sample alleles frequencies are directly estimated from the data set and inferences of population genetics parameters are based on this estimation. (B) In metagenome allele frequencies estimation is hampered by unknown sample size and unknown homology relationships within gene families'. (C) Homology relationship (orthology and paralogy) for five genes in three species related by speciation and duplication events. For colour version of this figure, the reader is referred to the online version of this book.

nucleotide of a sequence can potentially be polymorphic in the population and therefore the sequencing of one single-genomic region will provide several alleles to estimate θ. In a metagenome, things are complicated by two additional unknown variables; first, the sample size of species A, due to the random nature of the sequencing (Fig. 10.3B), and second, the interpretation of variants as true polymorphisms as opposed to orthologous (same genes in different species) or paralogous (different genes in same or different species) relationships (Fig. 10.3C). On Fig. 10.3C, we can see that genes A, B1 and C1 or B2 and C2 are orthologous because they have evolved from a common ancestral gene by speciation, whereas genes B1 and B2 or C1 and C2 are paralogous because they originated after a duplication event within a genome. Reads resulting from the sequencing of these homologous regions should be discarded from the analysis because their potential nucleotide differences with the allele of interest are due to divergence relationship and do not reflect polymorphism within species. Depending on how the metagenome has been obtained (Fig. 10.1), the frequency of reads at a given genomic location may change greatly (e.g. genome amplification), which has also to be taken into account when calculating allele frequencies.

5.2.2. Sequencing Issues

Another current limit is that we cannot estimate the level of polymorphism in a population where it is lower than the error rate of the sequencing technology used. Recent estimate of the error rate for Roche 454 GS-FLX pyrosequencing technology is around 1% (Gilles *et al.*, 2011). This error rate is much lower for Illumina GA: 0.168 % (Minoche, Dohm, & Himmelbauer, 2011) or for Sanger sequencing: 0.1% (Table 10.2). Furthermore, the type of sequencing errors is not the same from one technology to another. Errors are biased towards single-nucleotide changes for Illumina GA, whereas insertion/deletions are more prevalent for Roche 454, especially around homopolymeric regions. A high read coverage of

Table 10.2 Sequencing Technologies Associated Features

Technology	Read Length (bp)	Maximum Distance Between Mate Pairs (kbp)	Coverage	Error Rate (%)
ABI 3730xL Sanger	800	200 (BAC, Bacterial Artificial Chromosomes)	Low	0.1
Roche 454 GS-FLX+	600	20	Low	1.07
Illumina PhiX GAIIx	150	10	High	0.168

a site may help to correct for errors, this is particularly true for sequencing technologies with high sequence coverage like Illumina (Table 10.2).

Another sequencing issue is the linkage information: the degree to which two sites are linked to each other in a genome. Linkage can be used to estimate the recombination rate across a genome. Along with mutation, recombination is the other important source of genetic variation because it accelerates the rate of formation of beneficial gene combinations (Hartl and Clark, 1997). Knowledge of recombination rates therefore also supplies information about the adaptive potential of a population. However, to measure linkage from metagenomic reads, we would need to know to which individual each read belongs. Such information is usually unavailable and linkage information is thus dictated by read length and insert size between two PE reads of the sequencing experiment. In the best case, we would recover linkage information from 800-bp PE reads distant of 200 kbp from each other (Table 10.2).

5.2.3. Solutions

These three mains drawbacks can be overcome in two opposite ways. The first consists in the integration of the sample size, homology relationships and error rates as three unknown values (in addition to other parameters such as θ, N_e, u) into a statistical framework and work back the parameters. The second would be to reduce the complexity of the data set to produce a traditional population genetic-like data set. The solution may lie somewhere between these two extremes. As an example, Johnson and Slatkin overcame these problems by assuming that the species of interest are highly over-represented with low species diversity, so that each read in the sample is supposed to belong to a different individual of the same species (no orthology problem) (Johnson & Slatkin, 2006, 2009). Therefore, the number of reads at a given site of the genome is taken as a proxy to the sample size and inferences are based on this approximation. They further assume no paralogous genes and eventually integrate the error rate for each base of each read as a parameter in their model. However, for a species-rich environment where several closely related species coexist, these assumptions do not hold.

One technological solution to the species richness problem would consist in enriching the sample in cells of interest using flow cytometry.

The paralogy problem could be overcome by using single-copy genes (Ciccarelli *et al.*, 2006), but this would greatly reduce the size of the data set.

The orthology problem could be solved performing the analysis on highly polymorphic markers (intergenic regions) so that any alignment of the targeted region with another species would be very unlikely. Using available genome data from metagenomes (Piganeau & Moreau, 2007) and genome projects (Piganeau, Vandepoele, Gourbiere, Van De Peer, & Moreau, 2009), it has for instance been shown that intergenic sequences are good markers for polymorphism analysis in the marine algae *Ostreococcus*.

Metagenomics is a promising approach for population genomics of eukaryotic marine microorganisms because it allows the direct analysis of the on-going genetic diversity from a sample. However, population genomics inferences from metagenomic data set are still hampered by the random nature of a metagenome. Development of new statistical frameworks together with technological improvement should allow population genomics inferences to be made from these data sets in the near future. These studies will provide a better understanding of these ecologically important organisms.

6. EVOLUTIONARY (META)GENOMICS: METAGENOMIC INSIGHTS INTO GENOME ARCHITECTURE

We will here only discuss two well-studied facets of genome architecture: genome base composition and genome size. There is ample evidence that parasitic or symbiotic lifestyles affect genome architecture (e.g. Andersson *et al.*, 2002; Tamas, Klasson, Sandstrom, & Andersson, 2001), for these microorganism for which the environment is its host. The evolution of the genomes of the chloroplast and the nucleomorph from the genomes of their free-living ancestors has been addressed in previous chapters (Archibald, 2012; De Clerck, Bogaret, & Leliaert, 2012). We will focus here on the expected and observed relationships between genomes and their non-biotic environment.

Several processes have been proposed to account for variation in base composition or genome size, and we may categorize them either as mutational processes or as the consequence of selection. We will shortly review these two processes and discuss how they may be linked to the environment.

6.1. Base Composition

6.1.1. Mutation Biases: Molecular Processes Biasing Base Composition

The issue about the origin of the variability in AT versus GC base composition between double-stranded DNA genomes traces back to early observations in 1962 (Sueoka, 1962) and within-genome variation was discovered later in human (Bernardi *et al.*, 1985). To illustrate these two levels of variation, we have presented the average GC composition of 17 nuclear and 22

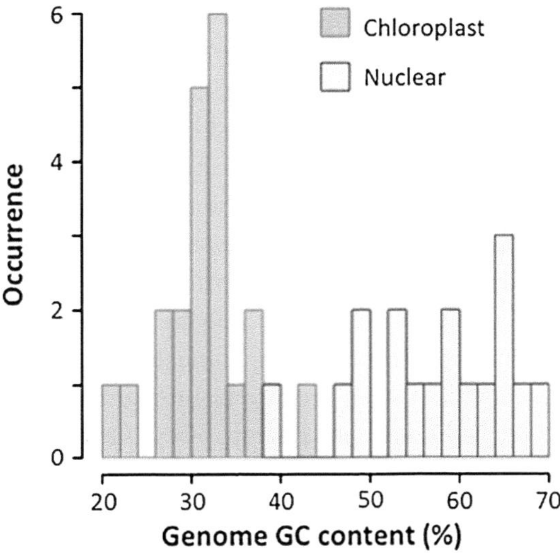

Figure 10.4 Average genomic GC content variation between algae. For colour version of this figure, the reader is referred to the online version of this book.

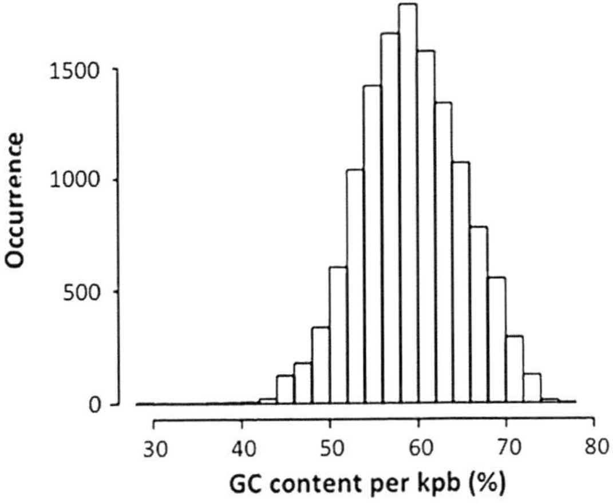

Figure 10.5 Within-genome GC variation in *Ostreococcus tauri*.

chloroplastic genomes of algae (Fig. 10.4) and the within-genome variation of GC content in *O. tauri* (Fig. 10.5). The between-species GC for nuclear and chloroplastic genomes ranges from 38% to 70% and 21% to 42% GC content, respectively. The within-genome variation of GC content within *O. tauri* ranges from 30% to 77%. Between- and within-base composition variation may be the consequence of the same mechanisms that apply differently on different genomic regions for the within-genome heterogeneity.

A mutation bias is defined as a mechanism favouring GC towards AT changes over AT towards GC changes (or the reverse), so that the average GC content is not 50%. This bias may result from the replication errors of the DNA polymerase that may be triggered by a bias in the nucleotide pool available in the cell during replication (Wolfe, Sharp, & Li, 1989). Another molecular mechanism inducing biased base composition is repeat-induced biased point mutations reported in some yeasts (Cambareri, Singer, & Selker, 1991), where G and C nucleotides are replaced by A and T nucleotides in repeated regions. Mutation biases may also be the consequence of biased mismatch repair during recombination, a mechanism called gene conversion: an A–C mismatch between the two homologous DNA strands may be repaired as A–T or G–C with equal probability. However, several lines of evidence suggest that the repair mechanism is biased towards G–C repair in many eukaryotes. As a consequence of GC-biased gene conversion (GC-BGC), GC composition is positively correlated to recombination rates in many eukaryotes (Duret & Galtier, 2009, for a recent review).

Each of these mutation processes leaves different fingerprints on genomic features, i.e. the relationship between recombination rates and GC composition is as expected under GC-BGC, whereas the relationship between GC content and replication timing is as expected under biased mutation rates due to replication. The tests of these predictions in different eukaryotes provide some evidence that biased gene conversion is the main cause of GC variation within mammalian genomes (Duret & Galtier, 2009), whereas repeat-induced mutation is the cause of GC variation along the genome of the fungi *Leptosphaeria maculans* (Rouxel *et al.*, 2011).

However, while we now have a good knowledge of the molecular mechanisms involved in biased base composition, we still have little insights about the evolution of these processes and whether they have evolved as a consequence of selection on base composition or as a neutral by-product of evolution.

6.1.2. Selection: The Metabolic Cost Hypothesis

Direct selection on each nucleotide is theoretically impractical because of the high selection coefficients required to maintain base composition away from the mutational equilibrium (Piganeau, Westrelin, Tourancheau, & Gautier, 2001). Therefore, GC composition may rather be under indirect selection, that is selection of an enzyme biasing GC composition.

We have to think about the advantage that a higher (or lower) GC composition would confer to the cell. A selective advantage for high GC composition was originally proposed, based on the higher thermostability of GC-rich genomes as a result of the higher number of hydrogen bonds between G and C bases (Bernardi & Bernardi, 1986). However, the evidence of GC content variations in cold-blooded vertebrates (Hughes, Zelus, & Mouchiroud, 1999) and the absence of a correlation between genomic GC contents and habitat temperature in bacteria (Galtier & Lobry, 1997) argues against the thermostability hypothesis as a selective force for higher GC content.

The fitness consequences of a biased GC composition may instead be understood in terms of the relationship between the availability of key elements in the environment and the distribution of base composition within the community.

The building blocks of life, amino acids and nucleotides, contain different amount of key elements like nitrogen, N, carbon, C, sulphur, S, or phosphate, P (Fig. 10.6). Environmental differences in these key elements may thus trigger nucleotide and/or amino acid composition (Elser, Acquisti, & Kumar, 2011). Several case studies suggest that there is indeed selection for lower sulphur content in sulphur-depleted organisms (Mazel & Marliere, 1989) and lower nitrogen content for organisms living under nitrogen limiting conditions, like plants (Acquisti, Elser, & Kumar, 2009). In yeast, it has been shown that gene expression of sulphur

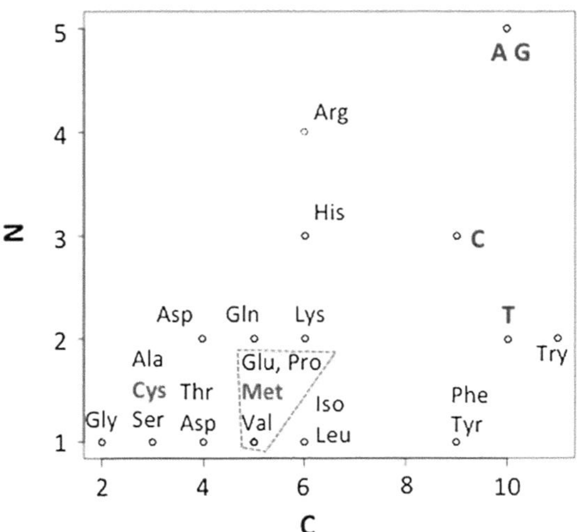

Figure 10.6 Carbon and nitrogen chemical composition of amino acids and nucleotides, sulfur containing amino acids in grey. For colour version of this figure, the reader is referred to the online version of this book.

containing genes is downregulated as a consequence of sulphur depletion (Fauchon et al., 2002).

Using the same line of reasoning for DNA base composition, the metabolic cost of synthesis of the four nucleotides differ: A nucleotides cost one ATP molecule less to synthesize than G, leading Rocha & Danchin (2002) to propose that this may cause AT richer genomes to be favoured in environments where competition for metabolic resources is strong.

6.1.3. Linking Base Composition to the Environment

Metagenome data enable us to test whether there is a relationship between the bioavailability of key elements in the environment and the base composition of the community. It is necessary to stress here that microorganisms take up molecules, not elements, so that the different metabolic pathways at hand may complicate the relationship between measurable environmental concentrations and bioavailability.

Also, a base composition–environment relationship can only evolve if the environment is stable on an evolutionary timescale, so that microorganisms had time to evolve to adapt to these environmental conditions. These analyses are therefore not only relevant to explain the evolution of base composition but also provide information about the stability of the environment on an evolutionary timescale.

The way to test whether base composition varies between environments is to compare the DNA content of orthologous genes between different environments and correct for phylogenetic inertia. Phylogenetic independence is essential to make sure that the difference you may observe between environments is not the consequence of shared ancestry within each environment. Foerstner, Von Mering, Hooper, & Bork (2005) performed such an analysis and found that bacterial sequences from the ocean surface had a significantly lower GC base composition than orthologous sequences from soil metagenomes. They suggest that this is the hallmark of the adaptation of bacteria to oligotrophic environments, where it pays to live with as low metabolic cost as possible. This GC base composition shift has not been yet reported in the eukaryotic fraction of these metagenomes, which represent 4–8% of the sequences (Piganeau, Desdevises, Derelle, & Moreau, 2008). Available nuclear genome sequences of algae suggest a high average GC content (average of data available in Fig. 4 is 57% of GC).

The effect of the bioavailability of key elements like N, P and S is expected to be more dramatic on RNA composition because RNA make up for most of the nucleic acid biomass of the cell (Leick, 1968). The future development of metatranscriptomes on a global scale (Karsenti *et al.*, 2011) will enable us to further test this hypothesis on the most transcribed genes of the community.

6.2. Genome Size

6.2.1. Current Hypothesis about Genome Size Variations

Genome size can be inferred from diverse techniques like Feulgen densitometry, flow cytometry, electrophoretic methods, or complete genome sequencing (Gregory et al., 2007). Available data on nuclear genome sizes in 255 algae (data from the Eukaryotic genome size databases, Gregory *et al.*, 2007) completed with dinoflagellate genome size estimates (Lajeunesse, Lambert, Andersen, Coffroth, & Galbraith, 2005) suggest a four-order magnitude variation in genome size, the *Dinophycean* algae having 10,000-fold larger genomes than the green alga *Ostreococcus* (Fig. 10.7).

Early analysis of genome content refuted the claims that genome size was a proxy for the complexity of organisms in eukaryotes. A small amount of genome size variation is due to gene number variation (as a consequence of duplications), whereas most variation in genome sizes is the consequence of the expansion of non-coding elements, 'selfish' DNA, within genomes (see Chapter 2 in Lynch, 2007, for a review). This 'selfish' DNA is a burden for the genome; therefore, it can only spread in species with small effective population sizes, where drift, the random process that may bring even deleterious mutations to fixation, prevails. This effective population size hypothesis suggests that genome obesity in dinoflagellates is the consequence

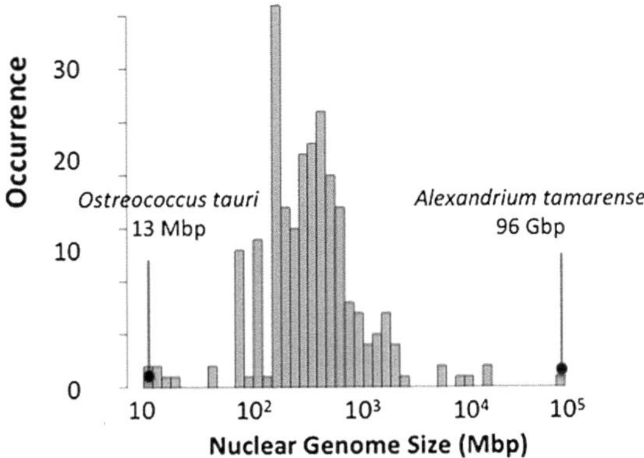

Figure 10.7 Genome size variations within algae.

of their small effective population sizes, whereas the slim genomes of *Ostreococcus* alga are the consequence of huge effective population sizes.

However, careful statistical analyses of current available data after correction for phylogenetic inertia do not support the hypothesis of a mechanistic connection between effective population size and genome sizes (Whitney, Boussau, Baack, & Garland, 2011; Whitney & Garland, 2010).

Metagenome analysis may shed light on the mechanisms underlying genome size evolution, first by providing the geographic range of large versus streamlined algal species (testing the 'large N_e, large range, small genome' hypothesis) and second by providing the link between environmental parameters like phosphorus and nitrate bioavailability on genome size (testing the 'genome size is shaped by environmental factors' hypothesis).

6.2.2. Estimating Genome Sizes and Selfish DNA Content

When harvesting functional and taxonomic diversity for the majority of reads in a metagenome is problematic (see Section 10.2), how might a metagenome be informative about the genome sizes of the community?

Raes, Korbel, Lercher, Von Mering, and Bork (2007) devised a method to estimate the average genome size from a metagenome, called effective genome size, from two simple observations: (i) there are genes in single copy in all cellular organisms and (ii) the probability of observing these genes in a metagenome is inversely correlated to the genome size. Obviously (ii) is only true for metagenomes obtained from random sequencing of the community (see Fig. 10.1 for amplification bias that may affect read frequencies).

Using available data from 154 bacterial and archaeal genomes, and randomly sampling to generate a metagenome, they empirically inferred the relationship between the average genome size of the community, G, as a function of the read lengths, L, and the frequency of the single-copy genes in the metagenome, F, as:

$$\overline{G} = \frac{a + bL^{-c}}{F}$$

To avoid biases due to viral sequences, which would decrease F and thus lead to an overestimation of the average genome size, F can be estimated for the targeted community (e.g. bacterial or eukaryotic), by estimating the metagenome size as the sum of bacterial or eukaryotic hits to a specific database.

The effective genome size gives the average genome size for the sampled microorganisms, while the genome size distribution of the sampled species (not the more abundant microorganisms) may be more informative to detect an environmental effect on genome size. To infer the genome size of a species, we need to obtain the information about the number of reads from this species. This can be inferred for closely related species to already sequenced genomes, for which we already have the genome size (Angly *et al.*, 2009; Frank & Sorensen, 2011).

Screening 187 metagenomes, the most abundant genes, estimated by the proportion of metagenomes that contain at least one representative and the number of copies within metagenomes, encode a transposition activity (i.e. they copy and insert themselves in DNA) (Aziz, Breitbart, & Edwards, 2010). These genes with a transposition activity are well-known selfish elements, which propagate within the genome at a cost for the host genome, except in rare cases where they have been recruited for a cellular function. The first 'meaningful' genes in their list are transporters and ATP-binding proteins, followed by replication-associated proteins and photosystem II protein PsbC.

7. CONCLUDING REMARK

The use of metagenomics and metatranscriptomics approaches to investigate microbial eukaryotes is still in its infancy. This new type of sequence data will continue to foster many methodological and conceptual developments in the future to understand the processes responsible for communities' genome diversity and evolution.

REFERENCES

Acquisti, C., Elser, J. J., & Kumar, S. (2009). Ecological nitrogen limitation shapes the DNA composition of plant genomes. *Molecular Biology and Evolution, 26*, 953–956.

Altschul, S. F., Gish, W., Miller, W., Myers, E. W., & Lipman, D. J. (1990). Basic local alignment search tool. *Journal of Molecular Biology, 215*, 403–410.

Andersson, S. G., Alsmark, C., Canback, B., Davids, W., Frank, C., Karlberg, O., et al. (2002). Comparative genomics of microbial pathogens and symbionts. *Bioinformatics, 18*(Suppl. 2), S17.

Angly, F. E., Willner, D., Prieto-Davo, A., Edwards, R. A., Schmieder, R., Vega-Thurber, R., et al. (2009). The GAAS metagenomic tool and its estimations of viral and microbial average genome size in four major biomes. *PLoS Computional Biology, 5*, e1000593.

Archibald, J. (2012). The evolution of algae by secondary and tertiary endosymbiosis. *Advances in Botanical Research, 64*, 87–118.

Aziz, R. K., Breitbart, M., & Edwards, R. A. (2010). Transposases are the most abundant, most ubiquitous genes in nature. *Nucleic Acids Research, 38*, 4207–4217.

Bateman, A., Birney, E., Cerruti, L., Durbin, R., Etwiller, L., Eddy, S. R., et al. (2002). The Pfam protein families database. *Nucleic Acids Research, 30*, 276–280.

Berger, S. A., Krompass, D., & Stamatakis, A. (2011). Performance, accuracy, and Web server for evolutionary placement of short sequence reads under maximum likelihood. *Systematic Biology, 60*, 291–302.

Bernardi, G., & Bernardi, G. (1986). Compositional constraints and genome evolution. *Journal of Molecular Evolution, 24*, 1–11.

Bernardi, G., Olofsson, B., Filipski, J., Zerial, M., Salinas, J., Cuny, G., et al. (1985). The mosaic genome of warm-blooded vertebrates. *Science, 228*, 953–958.

Beszteri, B., Temperton, B., Frickenhaus, S., & Giovannoni, S. J. (2010). Average genome size: a potential source of bias in comparative metagenomics. *The ISME Journal, 4*, 1075–1077.

Biers, E. J., Sun, S., & Howard, E. C. (2009). Prokaryotic genomes and diversity in surface ocean waters: interrogating the global ocean sampling metagenome. *Applied and Environmental Microbiology, 75*, 2221–2229.

Brady, A., & Salzberg, S. L. (2009). Phymm and PhymmBL: metagenomic phylogenetic classification with interpolated Markov models. *Nature Methods, 6*, 673–676.

Cadoret, J., Garnier, M., & Saint-Jean, B. (2012). Microalgae, functional genomics and biotechnology. *Advances in Botanical Research, 64*, 285–341.

Cambareri, E. B., Singer, M. J., & Selker, E. U. (1991). Recurrence of repeat-induced point mutation (Rip) in Neurospora-Crassa. *Genetics, 127*, 699–710.

Caporaso, J. G., Kuczynski, J., Stombaugh, J., Bittinger, K., Bushman, F. D., Costello, E. K., et al. (2010). QIIME allows analysis of high-throughput community sequencing data. *Nature Methods, 7*, 335–336.

Caporaso, J. G., Lauber, C. L., Walters, W. A., Berg-Lyons, D., Huntley, J., Fierer, N., et al. (2012). Ultra-high-throughput microbial community analysis on the Illumina HiSeq and MiSeq platforms. *The ISME Journal, 6*, 1621–1624.

Casabianca, S., Penna, A., Pecchioli, E., Jordi, A., Basterretxea, G., & Vernesi, C. (2012). Population genetic structure and connectivity of the harmful dinoflagellate Alexandrium minutum in the Mediterranean Sea. *Proceedings of the Royal Society B: Biological Sciences, 279*, 129–138.

Casteleyn, G., Leliaert, F., Backeljau, T., Debeer, A. E., Kotaki, Y., Rhodes, L., & et al.. (2010). Limits to gene flow in a cosmopolitan marine planktonic diatom. *Proceedings of the National Academy of Sciences of the United States of America, 107*, 12952–12957.

Cermeno, P., & Falkowski, P. G. (2009). Controls on diatom biogeography in the ocean. *Science, 325*, 1539–1541.

Chaffron, S., Rehrauer, H., Pernthaler, J., & Von Mering, C. (2010). A global network of coexisting microbes from environmental and whole-genome sequence data. *Genome Research, 20*, 947–959.

Charlesworth, B., Morgan, M. T., & Charlesworth, D. (1993). The effect of deleterious mutations on neutral molecular variation. *Genetics, 134*, 1289–1303.

Ciccarelli, F. D., Doerks, T., Von Mering, C., Creevey, C. J., Snel, B., & Bork, P. (2006). Toward automatic reconstruction of a highly resolved tree of life. *Science, 311*, 1283–1287.

Cuvelier, M. L., Allen, A. E., Monier, A., Mccrow, J. P., Messie, M., Tringe, S. G., et al. (2010). Targeted metagenomics and ecology of globally important uncultured eukaryotic phytoplankton. *Proceedings of the National Academy of Sciences of the United States of America, 107*, 14679–14684.

Davidson, A. L., Dassa, E., Orelle, C., & Chen, J. (2008). Structure, function, and evolution of bacterial ATP-binding cassette systems. *Microbiology and Molecular Biology Reviews, 72*, 317–364.
De Clerck, O., Bogaret, K., & Leliaert, F. (2012). Diversity and evolution of algae: primary endosymbiosis. *Advances in Botanical Research, 64*, 55–86.
Dean, F. B., Nelson, J. R., Giesler, T. L., & Lasken, R. S. (2001). Rapid amplification of plasmid and phage DNA using Phi 29 DNA polymerase and multiply-primed rolling circle amplification. *Genome Research, 11*, 1095–1099.
Delmont, T. O., Malandain, C., Prestat, E., Larose, C., Monier, J. M., Simonet, P., et al. (2011). Metagenomic mining for microbiologists. *The ISME Journal, 5*, 1836–1843.
Dieckmann, U., Doebeli, M., Metz, J. A. J., & Tautz, D. (2004). *Adaptive speciation*. Cambridge, UK: Cambridge University Press.
Diez, B., Pedros-Alio, C., & Massana, R. (2001). Study of genetic diversity of eukaryotic picoplankton in different oceanic regions by small-subunit rRNA gene cloning and sequencing. *Applied and Environmental Microbiology, 67*, 2932–2941.
Duret, L., & Galtier, N. (2009). Biased gene conversion and the evolution of mammalian genomic landscapes. *Annual Review of Genomics and Human Genetics, 10*, 285–311.
Elser, J. J., Acquisti, C., & Kumar, S. (2011). Stoichiogenomics: the evolutionary ecology of macromolecular elemental composition. *Trends in Ecology and Evolution, 26*, 38–44.
Evans, K. M., Bates, S. S., Medlin, L. K., & Hayes, P. K. (2004). Microsatellite marker development and genetic variation in the toxic marine diatom Pseudo-nitzschia multiseries (Bacillariophyceae). *Journal of Phycology, 40*, 911–920.
Evans, K. M., Wortley, A. H., & Mann, D. G. (2007). An assessment of potential diatom "barcode" genes (cox1, rbcL, 18S and ITS rDNA) and their effectiveness in determining relationships in Sellaphora (Bacillariophyta). *Protist, 158*, 349–364.
Fauchon, M., Lagniel, G., Aude, J. C., Lombardia, L., Soularue, P., Petat, C., et al. (2002). Sulfur sparing in the yeast proteome in response to sulfur demand. *Molecular Cell, 9*, 713–723.
Ferrière, R., Dieckmann, U., & Couvet, D. (2004). *Evolutionary conservation biology*. Cambridge, UK: Cambridge University Press.
Finlay, B. J. (2002). Global dispersal of free-living microbial eukaryote species. *Science, 296*, 1061–1063.
Finlay, B. J., & Fenchel, T. (2004). Cosmopolitan metapopulations of free-living microbial eukaryotes. *Protist, 155*, 237–244.
Finn, R. D., Clements, J., & Eddy, S. R. (2011). HMMER web server: interactive sequence similarity searching. *Nucleic Acids Research, 39*, W29–W37.
Foerstner, K. U., Von Mering, C., Hooper, S. D., & Bork, P. (2005). Environments shape the nucleotide composition of genomes. *EMBO Reports, 6*, 1208–1213.
Follows, M. J., Dutkiewicz, S., Grant, S., & Chisholm, S. W. (2007). Emergent biogeography of microbial communities in a model ocean. *Science, 315*, 1843–1846.
Frank, J. A., & Sorensen, S. J. (2011). Quantitative metagenomic analyses based on average genome size normalization. *Applied and Environmental Microbiology, 77*, 2513–2521.
Fuhrman, J. A., Steele, J. A., Hewson, I., Schwalbach, M. S., Brown, M. V., Green, J. L., et al. (2008). A latitudinal diversity gradient in planktonic marine bacteria. *Proceedings of the National Academy of Sciences of the United States of America, 105*, 7774–7778.
Fullwood, M. J., Wei, C. L., Liu, E. T., & Ruan, Y. (2009). Next-generation DNA sequencing of paired-end tags (PET) for transcriptome and genome analyses. *Genome Research, 19*, 521–532.
Galtier, N., & Lobry, J. R. (1997). Relationships between genomic G+C content, RNA secondary structures, and optimal growth temperature in prokaryotes. *Journal of Molecular Evolution, 44*, 632–636.

Gerlach, W., & Stoye, J. (2011). Taxonomic classification of metagenomic shotgun sequences with CARMA3. *Nucleic Acids Research, 39*, e91.
Gianoulis, T. A., Raes, J., Patel, P. V., Bjornson, R., Korbel, J. O., Letunic, I., et al. (2009). Quantifying environmental adaptation of metabolic pathways in metagenomics. *Proceedings of the National Academy of Sciences of the United States of America, 106*, 1374–1379.
Gilles, A., Meglecz, E., Pech, N., Ferreira, S., Malausa, T., & Martin, J. F. (2011). Accuracy and quality assessment of 454 GS-FLX Titanium pyrosequencing. *BMC Genomics, 12*, 245.
Gourbière, S., & Menu, F. (2009). Adaptive dynamics of dormancy duration variability: evolutionary trade-off and priority effect lead to sub-optimal adaptation. *Evolution, 63*, 1879–1892.
Gregory, T. R., Nicol, J. A., Tamm, H., Kullman, B., Kullman, K., Leitch, I. J., et al. (2007). Eukaryotic genome size databases. *Nucleic Acids Research, 35*, D332–D338.
Grimsley, N., Pequin, B., Bachy, C., Moreau, H., & Piganeau, G. (2010). Cryptic sex in the smallest eukaryotic marine green alga. *Molecular Biology and Evolution, 27*, 47–54.
Grimsley, N., Thomas, R., Kegel, J., Jacquet, S., Moreau, H., & Desdevises, Y. (2012). Genomics of algal host-virus interactions. *Advances in Botanical Research, 64*, 343–378.
Handelsman, J., Rondon, M. R., Brady, S. F., Clardy, J., & Goodman, R. M. (1998). Molecular biological access to the chemistry of unknown soil microbes: a new frontier for natural products. *Chemistry & Biology, 5*, R245–R249.
Hartl, D. L., & Clark, A. G. (1997). *Principles of population genetics*. Sunderlands, MA: Sinauer Associates.
Hassler, C. S., Schoemann, V., Nichols, C. M., Butler, E. C., & Boyd, P. W. (2011). Saccharides enhance iron bioavailability to Southern Ocean phytoplankton. *Proceedings of the National Academy of Sciences of the United States of America, 108*, 1076–1081.
Helbling, D. E., Ackermann, M., Fenner, K., Kohler, H. P., & Johnson, D. R. (2012). The activity level of a microbial community function can be predicted from its metatranscriptome. *The ISME Journal, 6*, 902–904.
Heywood, J. L., Sieracki, M. E., Bellows, W., Poulton, N. J., & Stepanauskas, R. (2011). Capturing diversity of marine heterotrophic protists: one cell at a time. *The ISME Journal, 5*, 674–684.
Hingamp, P., Brochier, C., Talla, E., Gautheret, D., Thieffry, D., & Herrmann, C. (2008). Metagenome annotation using a distributed grid of undergraduate students. *PLoS Biology, 6*, e296.
Hopkinson, B. M., & Barbeau, K. A. (2011). Iron transporters in marine prokaryotic genomes and metagenomes. *Environmental Microbiology, 14*, 114–128.
Hubert, C., Loy, A., Nickel, M., Arnosti, C., Baranyi, C., Bruchert, V., et al. (2009). A constant flux of diverse thermophilic bacteria into the cold Arctic seabed. *Science, 325*, 1541–1544.
Hughes, S., Zelus, D., & Mouchiroud, D. (1999). Warm-blooded isochore structure in Nile crocodile and turtle. *Molecular Biology and Evolution, 16*, 1521–1527.
Huse, S. M., Welch, D. M., Morrison, H. G., & Sogin, M. L. (2010). Ironing out the wrinkles in the rare biosphere through improved OTU clustering. *Environmental Microbiology, 12*, 1889–1898.
Huson, D. H., Auch, A. F., Qi, J., & Schuster, S. C. (2007). MEGAN analysis of metagenomic data. *Genome Research, 17*, 377–386.
Iverson, V., Morris, R. M., Frazar, C. D., Berthiaume, C. T., Morales, R. L., & Armbrust, E. V. (2012). Untangling genomes from metagenomes: revealing an uncultured class of marine Euryarchaeota. *Science, 335*, 587–590.
Johnson, P. L. F., & Slatkin, M. (2006). Inference of population genetic parameters in metagenomics: a clean look at messy data. *Genome Research, 16*, 1320–1327.
Johnson, P. L. F., & Slatkin, M. (2009). Inference of microbial recombination rates from metagenomic data. *PLoS Genetics, 5*. 5, e1000674.

Karsenti, E., Acinas, S. G., Bork, P., Bowler, C., De Vargas, C., Raes, J., et al. (2011). A holistic approach to marine eco-systems biology. *PLoS Biology, 9*, e1001177.

Kelley, D. R., Liu, B., Delcher, A. L., Pop, M., & Salzberg, S. L. (2012). Gene prediction with Glimmer for metagenomic sequences augmented by classification and clustering. *Nucleic Acids Research, 40*, e9.

Kennedy, J., O'leary, N. D., Kiran, G. S., Morrissey, J. P., O'gara, F., Selvin, J., et al. (2011). Functional metagenomic strategies for the discovery of novel enzymes and biosurfactants with biotechnological applications from marine ecosystems. *Journal of Applied Microbiology, 111*, 787–799.

Kimura, M. (1969). The number of heterozygous nucleotide sites maintained in a finite population due to steady flux of mutations. *Genetics, 61*, 893–903.

Kosakovsky Pond, S., Wadhawan, S., Chiaromonte, F., Ananda, G., Chung, W. Y., Taylor, J., et al. (2009). Windshield splatter analysis with the Galaxy metagenomic pipeline. *Genome Research, 19*, 2144–2153.

Krueger-Hadfield, S. A., Collen, J., Daguin-Thiebaut, C., & Valero, M. (2011). Genetic population structure and mating system in Chondrus Crispus (Rhodophyta). *Journal of Phycology, 47*, 440–450.

Kunin, V., Engelbrektson, A., Ochman, H., & Hugenholtz, P. (2010). Wrinkles in the rare biosphere: pyrosequencing errors can lead to artificial inflation of diversity estimates. *Environmental Microbiology, 12*, 118–123.

Lajeunesse, T. C., Lambert, G., Andersen, R. A., Coffroth, M. A., & Galbraith, D. W. (2005). Symbiodinium (Pyrrhophyta) genome sizes (DNA content) are smallest among dinoflagellates. *Journal of Phycology, 41*, 880–886.

Lasken, R. S., & Stockwell, T. B. (2007). Mechanism of chimera formation during the multiple displacement amplification reaction. *BMC Biotechnology, 7*, 19.

Lauro, F. M., Mcdougald, D., Thomas, T., Williams, T. J., Egan, S., Rice, S., et al. (2009). The genomic basis of trophic strategy in marine bacteria. *Proceedings of the National Academy of Sciences of the United States of America, 106*, 15527–15533.

Lecroq, B., Lejzerowicz, F., Bachar, D., Christen, R., Esling, P., Baerlocher, L., et al. (2011). Ultra-deep sequencing of foraminiferal microbarcodes unveils hidden richness of early monothalamous lineages in deep-sea sediments. *Proceedings of the National Academy of Sciences of the United States of America, 108*, 13177–13182.

Leick, V. (1968). Ratios between contents of DNA RNA and protein in different microorganisms as a function of maximal growth rate. *Nature, 217*, 1153–1155.

Lepere, C., Demura, M., Kawachi, M., Romac, S., Probert, I., & Vaulot, D. (2011). Whole-genome amplification (WGA) of marine photosynthetic eukaryote populations. *FEMS Microbiology Ecology, 76*, 513–523.

Li, L., Stoeckert, C. J., & Roos, D. S. (2003). OrthoMCL: identification of ortholog groups for eukaryotic genomes. *Genome Research, 13*, 2178–2189.

Lin, S., Zhang, H., Zhuang, Y., Tran, B., & Gill, J. (2010). Spliced leader-based metatranscriptomic analyses lead to recognition of hidden genomic features in dinoflagellates. *Proceedings of the National Academy of Sciences of the United States of America, 107*, 20033–20038.

Liu, H., Probert, I., Uitz, J., Claustre, H., Aris-Brosou, S., Frada, M., et al. (2009). Extreme diversity in noncalcifying haptophytes explains a major pigment paradox in open oceans. *Proceedings of the National Academy of Sciences of the United States of America, 106*, 12803–12808.

Lovejoy, C., Massana, R., & Pedros-Alio, C. (2006). Diversity and distribution of marine microbial eukaryotes in the Arctic Ocean and adjacent seas. *Applied and Environmental Microbiology, 72*, 3085–3095.

Luikart, G., England, P. R., Tallmon, D., Jordan, S., & Taberlet, P. (2003). The power and promise of population genomics: from genotyping to genome typing. *Nature Reviews Genetics, 4*, 981–994.

Luo, C., Tsementzi, D., Kyrpides, N. C., & Konstantinidis, K. T. (2011). Individual genome assembly from complex community short-read metagenomic datasets. *The ISME Journal, 6*, 898–901.

Lynch, M. (2007). *The origins of genome architecture*. Sunderland, MA: Sinauer Associates.

Lynch, M., Conery, J., & Burger, R. (1995). Mutational meltdowns in sexual populations. *Evolution, 49*, 1067–1080.

Marchetti, A., Parker, M. S., Moccia, L. P., Lin, E. O., Arrieta, A. L., Ribalet, F., & et al.. (2009). Ferritin is used for iron storage in bloom-forming marine pennate diatoms. *Nature, 457*, 467–470.

Marine, R., Polson, S. W., Ravel, J., Hatfull, G., Russell, D., Sullivan, M., et al. (2011). Evaluation of a transposase protocol for rapid generation of shotgun high-throughput sequencing libraries from nanogram quantities of DNA. *Applied and Environmental Microbiology, 77*, 8071–8079.

Martinez Martinez, J., Poulton, N. J., Stepanauskas, R., Sieracki, M. E., & Wilson, W. H. (2011). Targeted sorting of single virus-infected cells of the coccolithophore Emiliania huxleyi. *PLoS One, 6*, e22520.

Martiny, J. B., Bohannan, B. J., Brown, J. H., Colwell, R. K., Fuhrman, J. A., Green, J. L., et al. (2006). Microbial biogeography: putting microorganisms on the map. *Nature Reviews Microbiology, 4*, 102–112.

Matsen, F. A., Kodner, R. B., & Armbrust, E. V. (2010). pplacer: linear time maximum-likelihood and Bayesian phylogenetic placement of sequences onto a fixed reference tree. *BMC Bioinformatics, 11*, 538.

Maynard Smith, J. (1982). *Evolution and the theory of games*. Cambridge, UK: Cambridge University Press.

Mazel, D., & Marliere, P. (1989). Adaptive eradication of methionine and cysteine from cyanobacterial light-harvesting proteins. *Nature, 341*, 245–248.

Mchardy, A. C., Martin, H. G., Tsirigos, A., Hugenholtz, P., & Rigoutsos, I. (2007). Accurate phylogenetic classification of variable-length DNA fragments. *Nature Methods, 4*, 63–72.

Medlin, L. K., Metfies, K., Mehl, H., Wiltshire, K., & Valentin, K. (2006). Picoeukaryotic plankton diversity at the Helgoland time series site as assessed by three molecular methods. *Microbial Ecology, 52*, 53–71.

Meinicke, P., Asshauer, K. P., & Lingner, T. (2011). Mixture models for analysis of the taxonomic composition of metagenomes. *Bioinformatics, 27*, 1618–1624.

Meyer, F., Paarmann, D., D'souza, M., Olson, R., Glass, E. M., Kubal, M., et al. (2008). The metagenomics RAST server – a public resource for the automatic phylogenetic and functional analysis of metagenomes. *BMC Bioinformatics, 9*, 386.

Minoche, A. E., Dohm, J. C., & Himmelbauer, H. (2011). Evaluation of genomic high-throughput sequencing data generated on Illumina HiSeq and Genome Analyzer systems. *Genome Biology, 12*, R112.

Moon-van der staay, S. Y., De Wachter, R., & Vaulot, D. (2001). Oceanic 18S rDNA sequences from picoplankton reveal unsuspected eukaryotic diversity. *Nature, 409*, 607–610.

Moreau, H., Verhelst, B., Couloux, A., Derelle, E., Rombauts, S., Grimsley, N., et al. (2012). Gene functionalities and genome structure in Bathycoccus prasinos reflect cellular specializations at the base of the green lineage. *Genome Biology, 13*(8), R74.

Noguchi, H., Park, J., & Takagi, T. (2006). MetaGene: prokaryotic gene finding from environmental genome shotgun sequences. *Nucleic Acids Research, 34*, 5623–5630.

Not, F., Siano, R., Kooistra, W. H. C. F., Simon, N., Vaulot, D., & Probert, I. (2012). Diversity and Ecology of eukaryotic marine phytoplankton. *Advances in Botanical Research, 64*, 1–53.

Not, F., Valentin, K., Romari, K., Lovejoy, C., Massana, R., Tobe, K., et al. (2007). Picobiliphytes: a marine picoplanktonic algal group with unknown affinities to other eukaryotes. *Science, 315*, 253–255.

O'malley, M. A. (2007). The nineteenth century roots of 'everything is everywhere'. *Nature Reviews Microbiology, 5*, 647–651.

Pace, N. R. (1997). A molecular view of microbial diversity and the biosphere. *Science, 276*, 734–740.

Pagani, I., Liolios, K., Jansson, J., Chen, I. M., Smirnova, T., Nosrat, B., et al. (2012). The Genomes OnLine Database (GOLD) v.4: status of genomic and metagenomic projects and their associated metadata. *Nucleic Acids Research, 40*, D571–D579.

Patel, P. V., Gianoulis, T. A., Bjornson, R. D., Yip, K. Y., Engelman, D. M., & Gerstein, M. B. (2010). Analysis of membrane proteins in metagenomics: networks of correlated environmental features and protein families. *Genome Research, 20*, 960–971.

Piganeau, G., Desdevises, Y., Derelle, E., & Moreau, H. (2008). Picoeukaryotic sequences in the Sargasso sea metagenome. *Genome Biology, 9*, R5.

Piganeau, G., Eyre-Walker, A., Grimsley, N., & Moreau, H. (2011). How and why DNA barcodes underestimate the diversity of microbial eukaryotes. *PLoS One, 6*, e16342.

Piganeau, G., & Moreau, H. (2007). Screening the Sargasso Sea metagenome for data to investigate genome evolution in Ostreococcus (Prasinophyceae, Chlorophyta). *Gene, 406*, 184–190.

Piganeau, G., Vandepoele, K., Gourbiere, S., Van De Peer, Y., & Moreau, H. (2009). Unravelling cis-regulatory elements in the genome of the smallest photosynthetic eukaryote: phylogenetic footprinting in Ostreococcus. *Journal of Molecular Evolution, 69*, 249–259.

Piganeau, G., Westrelin, R., Tourancheau, B., & Gautier, C. (2001). Multiplicative versus additive selection in relation to genome evolution: a simulation study. *Genetics Research, 78*, 171–175.

Pillet, L., Fontaine, D., & Pawlowski, J. (2012). Intra-genomic ribosomal RNA polymorphism and morphological variation in Elphidium macellum suggests inter-specific hybridization in foraminifera. *PLoS One, 7*, e32373.

Pommier, T., Canback, B., Riemann, L., Bostrom, K. H., Simu, K., Lundberg, P., & et al.. (2007). Global patterns of diversity and community structure in marine bacterioplankton. *Molecular Ecology, 16*, 867–880.

Poretsky, R. S., Hewson, I., Sun, S., Allen, A. E., Zehr, J. P., & Moran, M. A. (2009). Comparative day/night metatranscriptomic analysis of microbial communities in the North Pacific subtropical gyre. *Environmental Microbiology, 11*, 1358–1375.

Quince, C., Curtis, T. P., & Sloan, W. T. (2008). The rational exploration of microbial diversity. *The ISME Journal, 2*, 997–1006.

Raes, J., Korbel, J. O., Lercher, M. J., Von Mering, C., & Bork, P. (2007). Prediction of effective genome size in metagenomic samples. *Genome Biology, 8*, R10.

Raes, J., Letunic, I., Yamada, T., Jensen, L. J., & Bork, P. (2011). Toward molecular trait-based ecology through integration of biogeochemical, geographical and metagenomic data. *Molecular Systems Biology, 7*, 473.

Reisch, C. R., Stoudemayer, M. J., Varaljay, V. A., Amster, I. J., Moran, M. A., & Whitman, W. B. (2011). Novel pathway for assimilation of dimethylsulphoniopropionate widespread in marine bacteria. *Nature, 473*, 208–211.

Rho, M., Tang, H., & Ye, Y. (2010). FragGeneScan: predicting genes in short and error-prone reads. *Nucleic Acids Research, 38*, e191.

Riesenfeld, C. S., Schloss, P. D., & Handelsman, J. (2004). Metagenomics: genomic analysis of microbial communities. *Annual Review of Genetics, 38*, 525–552.

Rinta-Kanto, J. M., Sun, S., Sharma, S., Kiene, R. P., & Moran, M. A. (2012). Bacterial community transcription patterns during a marine phytoplankton bloom. *Environmental Microbiology, 14*, 228–239.

Rocha, E., & Danchin, A. (2002). Base composition bias might result from competition for metabolic resources. *Trends in Genetics, 18*, 291–294.

Rouxel, T., Grandaubert, J., Hane, J. K., Hoede, C., Van De Wouw, A. P., Couloux, A., et al. (2011). Effector diversification within compartments of the Leptosphaeria maculans genome affected by repeat-induced point mutations. *Nature Communications, 2*(202).

Rusch, D. B., Halpern, A. L., Sutton, G., Heidelberg, K. B., Williamson, S., Yooseph, S., et al. (2007). The Sorcerer II Global Ocean Sampling expedition: northwest Atlantic through eastern tropical Pacific. *PLoS Biology, 5*, e77.

Rusch, D. B., Martiny, A. C., Dupont, C. L., Halpern, A. L., & Venter, J. C. (2010). Characterization of Prochlorococcus clades from iron-depleted oceanic regions. *Proceedings of the National Academy of Sciences of the United States of America, 107*, 16184–16189.

Rynearson, T. A., & Armbrust, E. V. (2000). DNA fingerprinting reveals extensive genetic diversity in a field population of the centric diatom Ditylum brightwellii. *Limnology and Oceanography, 45*, 1329–1340.

Rynearson, T. A., & Armbrust, E. V. (2005). Maintenance of clonal diversity during a spring bloom of the centric diatom Ditylum brightwellii. *Molecular Ecology, 14*, 1631–1640.

Rynearson, T. A., Lin, E. O., & Armbrust, E. V. (2009). Metapopulation structure in the planktonic diatom Ditylum brightwellii (Bacillariophyceae). *Protist, 160*, 111–121.

Rynearson, T. A., Newton, J. A., & Armbrust, E. V. (2006). Spring bloom development, genetic variation, and population succession in the planktonic diatom Ditylum brightwellii. *Limnology and Oceanography, 51*, 1249–1261.

Schloss, P. D., Westcott, S. L., Ryabin, T., Hall, J. R., Hartmann, M., Hollister, E. B., et al. (2009). Introducing mothur: open-source, platform-independent, community-supported software for describing and comparing microbial communities. *Applied and Environmental Microbiology, 75*, 7537–7541.

Schreiber, F., Gumrich, P., Daniel, R., & Meinicke, P. (2010). Treephyler: fast taxonomic profiling of metagenomes. *Bioinformatics, 26*, 960–961.

Shaffer, M. L. (1981). Minimum population sizes for species conservation. *Bioscience, 31*, 131–134.

Shi, D., Xu, Y., Hopkinson, B. M., & Morel, F. M. M. (2011). Effect of ocean acidification on iron availability to marine phytoplankton. *Science, 327*, 676–679.

Shi, X. L., Lepere, C., Scanlan, D. J., & Vaulot, D. (2011). Plastid 16S rRNA gene diversity among eukaryotic picophytoplankton sorted by flow cytometry from the South Pacific Ocean. *PLoS One, 6*, e18979.

Shi, X. L., Marie, D., Jardillier, L., Scanlan, D. J., & Vaulot, D. (2009). Groups without cultured representatives dominate eukaryotic picophytoplankton in the oligotrophic South East Pacific Ocean. *PLoS One, 4*, e7657.

Slapeta, J., Lopez-Garcia, P., & Moreira, D. (2006). Global dispersal and ancient cryptic species in the smallest marine eukaryotes. *Molecular Biology and Evolution, 23*, 23–29.

Smith, J. M., & Haigh, J. (1974). Hitch-hiking effect of a favorable gene. *Genetical Research, 23*, 23–35.

Sogin, M. L., Morrison, H. G., Huber, J. A., Mark Welch, D., Huse, S. M., Neal, P. R., et al. (2006). Microbial diversity in the deep sea and the underexplored "rare biosphere.". *Proceedings of the National Academy of Sciences of the United States of America, 103*, 12115–12120.

Stern, R. F., Horak, A., Andrew, R. L., Coffroth, M. A., Andersen, R. A., Kupper, F. C., et al. (2010). Environmental barcoding reveals massive dinoflagellate diversity in marine environments. *PLoS One, 5*, e13991.

Stewart, F. J., Ulloa, O., & Delong, E. F. (2012). Microbial metatranscriptomics in a permanent marine oxygen minimum zone. *Environmental Microbiology, 14*, 23–40.

Sueoka, N. (1962). On genetic basis of variation and heterogeneity of DNA base composition. *Proceedings of the National Academy of Sciences of the United States of America, 48*, 582–592.

Sun, S., Chen, J., Li, W., Altintas, I., Lin, A., Peltier, S., et al. (2011). Community cyberinfrastructure for advanced microbial ecology research and analysis: the CAMERA resource. *Nucleic Acids Research, 39*, D546–D551.

Tamas, I., Klasson, L. M., Sandstrom, J. P., & Andersson, S. G. (2001). Mutualists and parasites: how to paint yourself into a (metabolic) corner. *FEBS Letters, 498*, 135–139.

Teeling, H., Waldmann, J., Lombardot, T., Bauer, M., & Glockner, F. O. (2004). TETRA: a web-service and a stand-alone program for the analysis and comparison of tetranucleotide usage patterns in DNA sequences. *BMC Bioinformatics, 5*, 163.

Telford, R. J., Vandvik, V., & Birks, H. J. (2006). Dispersal limitations matter for microbial morphospecies. *Science, 312*, 1015.

Toulza, E., Tagliabue, A., Blain, S., & Piganeau, G. (2012). Analysis of the global ocean sampling (GOS) project for trends in iron uptake by surface ocean microbes. *PLoS One, 7*, e30931.

Vaulot, D., Lepere, C., Toulza, E., De La Iglesia, R., Poulain, J., & Gaboyer, F. (2012). Metagenomes of the picoalga Bathycoccus from the Chile coastal upwelling. *PLoS One, 7*, e39648.

Venter, J. C., Remington, K., Heidelberg, J. F., Halpern, A. L., Rusch, D., Eisen, J. A., et al. (2004). Environmental genome shotgun sequencing of the Sargasso Sea. *Science, 304*, 66–74.

Von Mering, C., Hugenholtz, P., Raes, J., Tringe, S. G., Doerks, T., Jensen, L. J., et al. (2007). Quantitative phylogenetic assessment of microbial communities in diverse environments. *Science, 315*, 1126–1130.

Whitney, K. D., Boussau, B., Baack, E. J., & Garland, T. (2011). Drift and genome complexity revisited. *PLoS Genetics, 7*, e1002092.

Whitney, K. D., & Garland, T. (2010). Did genetic drift drive increases in genome complexity? *PLoS Genetics, 6*, e1001080.

Wolfe, K. H., Sharp, P. M., & Li, W. H. (1989). Mutation-rates differ among regions of the mammalian genome. *Nature, 337*, 283–285.

Wommack, K. E., Bhavsar, J., & Ravel, J. (2008). Metagenomics: read length matters. *Applied and Environmental Microbiology, 74*, 1453–1463.

Yau, S., Lauro, F. M., Demaere, M. Z., Brown, M. V., Thomas, T., Raftery, M. J., et al. (2011). Virophage control of antarctic algal host-virus dynamics. *Proceedings of the National Academy of Sciences of the United States of America, 108*, 6163–6168.

Yilmaz, S., Allgaier, M., & Hugenholtz, P. (2010). Multiple displacement amplification compromises quantitative analysis of metagenomes. *Nature Methods, 7*, 943–944.

Yilmaz, S., & Singh, A. K. (2011). Single cell genome sequencing. *Current Opinion in Biotechnology, 23*, 437–443.

Yoon, H. S., Price, D. C., Stepanauskas, R., Rajah, V. D., Sieracki, M. E., Wilson, W. H., et al. (2011). Single-cell genomics reveals organismal interactions in uncultivated marine protists. *Science, 332*, 714–717.

Yooseph, S., Sutton, G., Rusch, D. B., Halpern, A. L., Williamson, S. J., Remington, K., et al. (2007). The Sorcerer II global ocean sampling expedition: expanding the universe of protein families. *PLoS Biology, 5*, e16.

Zane, L., Bargelloni, L., & Patarnello, T. (2002). Strategies for microsatellite isolation: a review. *Molecular Ecology, 11*, 1–16.

AUTHOR INDEX

C

D

E

F

G

H

I

J

L

M

N

Q

R

S

T

U

V

X

Y

Z

SUBJECT INDEX

Note: Page numbers with "f" denote figures; "t" tables.

C

O

P

R

S

W

X

Y

Z

COLOR PLATES

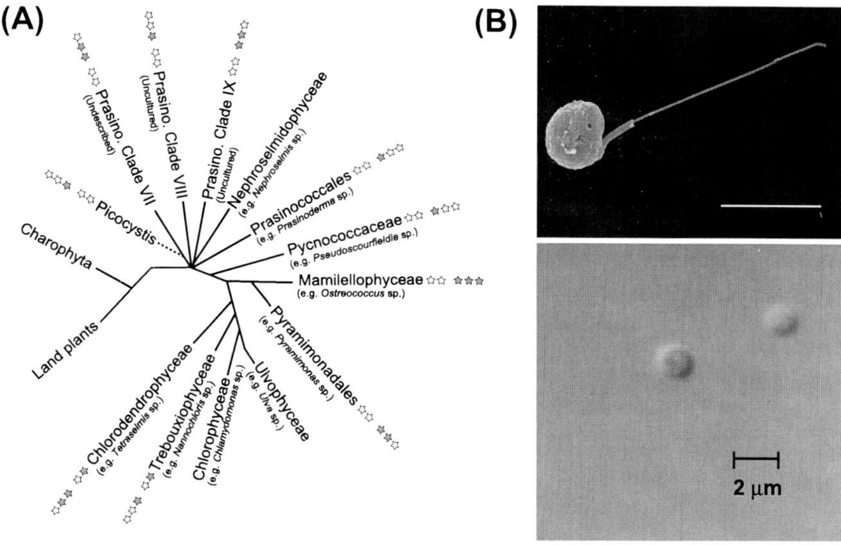

Figure 1.2 A) Schematic phylogenetic tree of the green algae and land plants lineages showing the relationships among major phytoplanktonic taxa and an estimation of their ecological significance. Typical representative of each lineage is indicated in brackets. The overall ecological significance (illustrated by a five-star ranking) is subjective and has been established based on parameters such as abundance, distribution, bloom formation, trophic strategies, toxicity, etc. (color code is 1 blue star = having freshwater members, 1 red star = important toxic or harmful species, 1–3 green stars range = other relevant ecological parameters, no stars means multi-cellular or no marine species). (B) Illustration of two important prasinophytes belonging to the Mamiellophyceae. Top: scanning electron microscopy of the common and abundant *Micromonas* sp. (E. Foulon). Bottom: the smallest photosynthetic eukaryote *Ostreococcus* sp. (D. Vaulot).

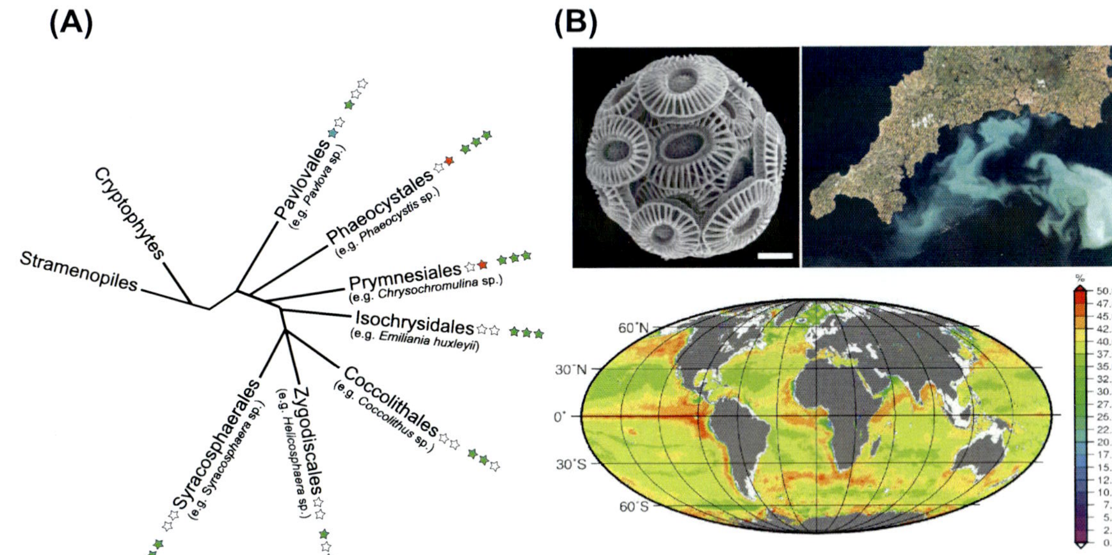

Figure 1.3 A) Legend as Figure 2A but for Haptophyta. (B) Scanning electronic microscopy illustration of the coccolithophore *Emiliania huxleyi*, scale bar 1 μm (top left), satellite image showing a coccolithophore bloom off south-western England (image source: http://ina.tmsoc.org/galleries/photodujour/source/cornwall-bloom_ehux.htm) (top right) and Haptophyta pigment concentration estimates across the world oceans. *(adapted from Liu* et al.*, 2009)*

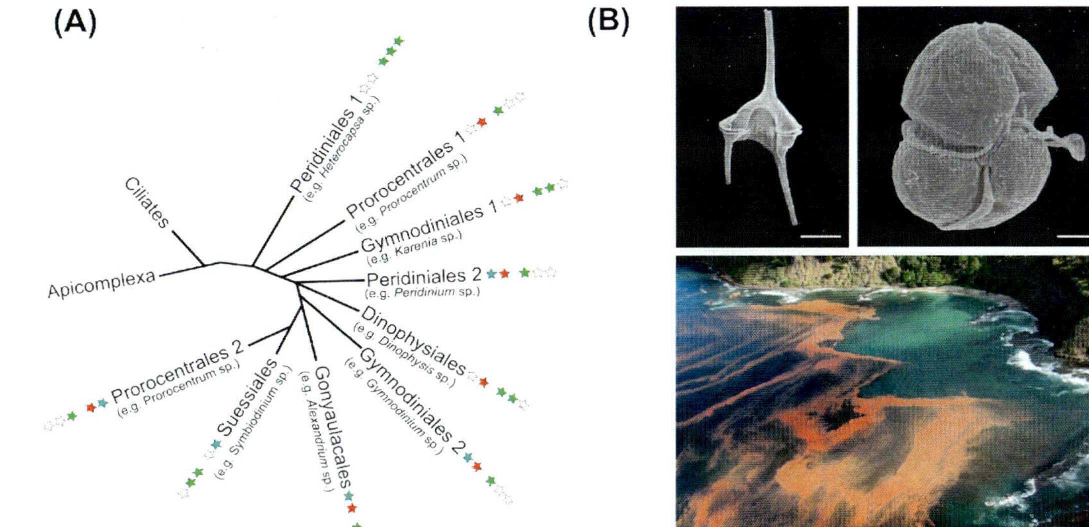

Figure 1.4 A) Legend as Figure 2A, but for dinoflagellates. (B). SEM micrographs of *Neoceratium candelabrum*, a non-toxic thecate micro-dinoflagellate (scale bar = 30 μm), a species of the Gonyaulacales lineage (top left) and *Karlodinium veneficum*, belonging to lineage Gymnodiniales 1, an athecate nano-dinoflagellate (scale bar 2 μm) that is toxic for a range of marine invertebrates and fish (top right). Red discoloration caused by the dinoflagellate *Noctiluca scintillans* off Waiheke Island (New Zealand) (http://www.TeAra.govt.nz/en/plankton/1/4) (bottom).

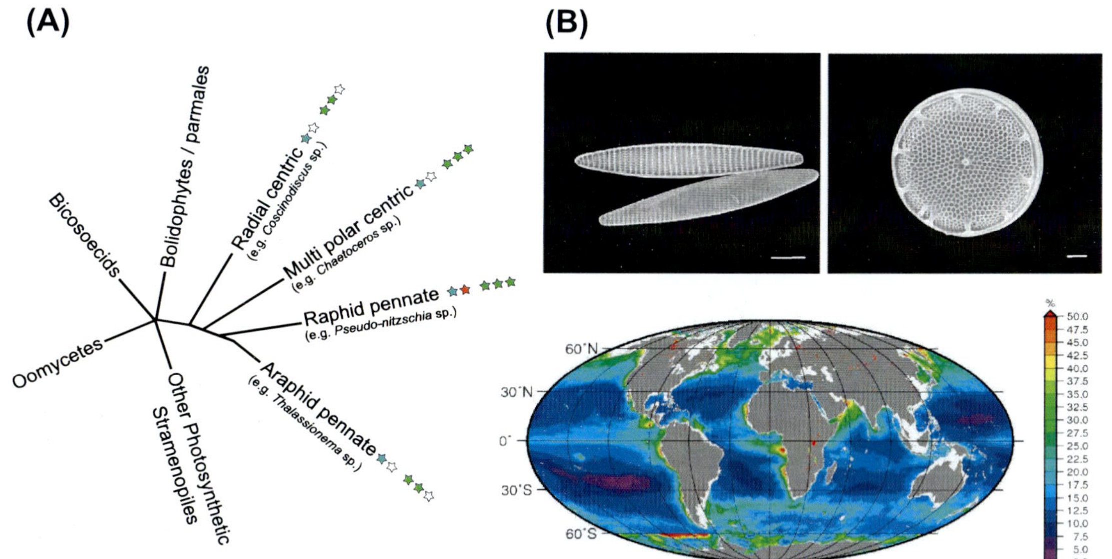

Figure 1.5 A) Legend as Figure 2A, but for diatoms. (B) Pennate diatom *Fragilariopsis kerguelensis*, top valve: interior side, bottom valve: exterior side, scale bar 10 μm (by Marina Montresor) (top left), centric diatom *Thalassiosira tealata* showing valve exterior with 10 peripheral strutted processes, a central strutted process and a peripheral labiate process, scale bar 1 μm (by Diana Sarno) (top right), and diatom pigment concentration estimates across the world oceans. *(adapted from Liu* et al.*, 2009)*

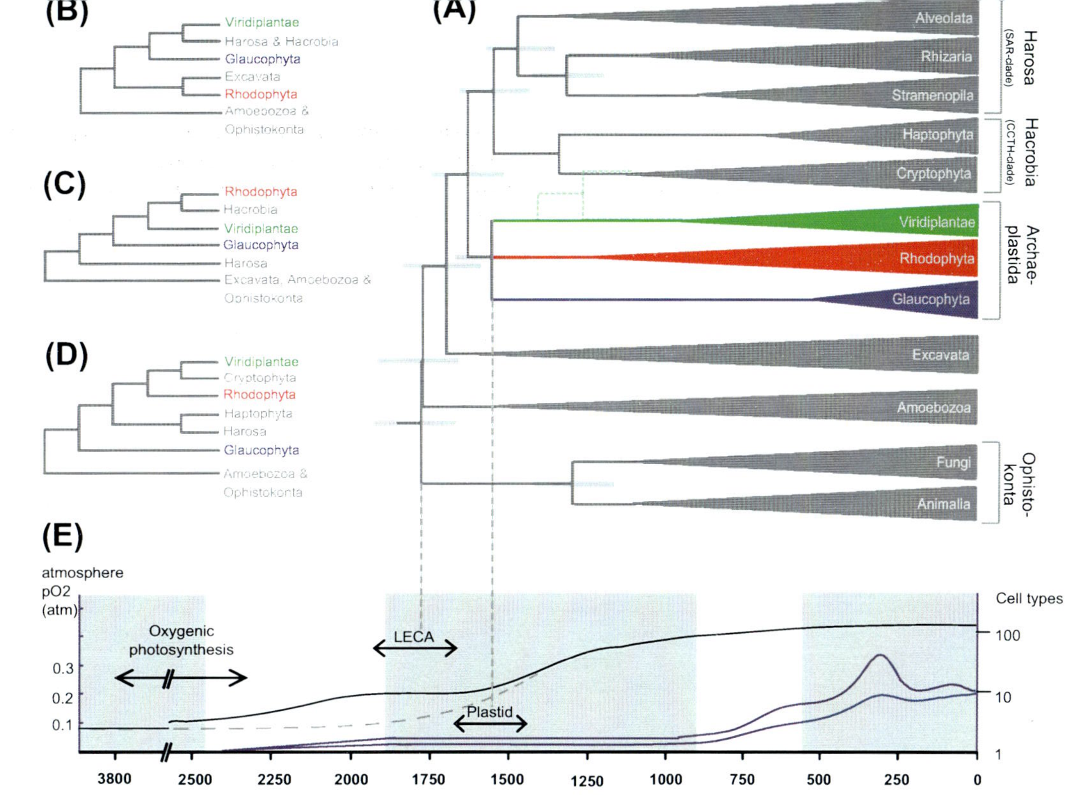

Figure 2.1 Relationships of Archaeplastida with main eukaryotic lineages and correlation between the rise in atmospheric oxygen and the evolution of organismal complexity. (A) Time-calibrated tree of extant eukaryotes (after Parfrey *et al.*, 2011). The tree topology is adjusted for the current uncertainty with respect to the branching order within the Archaeplastida. The dotted green line denotes the sister relationship between Viridiplantae and Cryptophyta in the analysis of Parfrey *et al.* (2011). Nodes are at mean divergence times and gray bars represent 95% highest probability density of node age. (B–D) Alternative topologies suggested by, respectively, Nozaki *et al.* (2009), Hampl *et al.* (2009) and Baurain *et al.* (2010). (E) Atmospheric partial oxygen pressure (blue lines) and cellular complexity (black line) (after Holland, 2006, and Hedges *et al.*, 2004). Blue lines denote maximum and minimum estimates of atmospheric O_2 partial pressure, respectively. Cellular complexity is defined as number of cell types. The black dashed line shows a more conservative interpretation of cellular complexity in the Proterozoic. The alternation of gray and white periods denotes the five different stages in oxygenation of the atmosphere according to Holland (2006). LECA: last eukaryotic common ancestor.

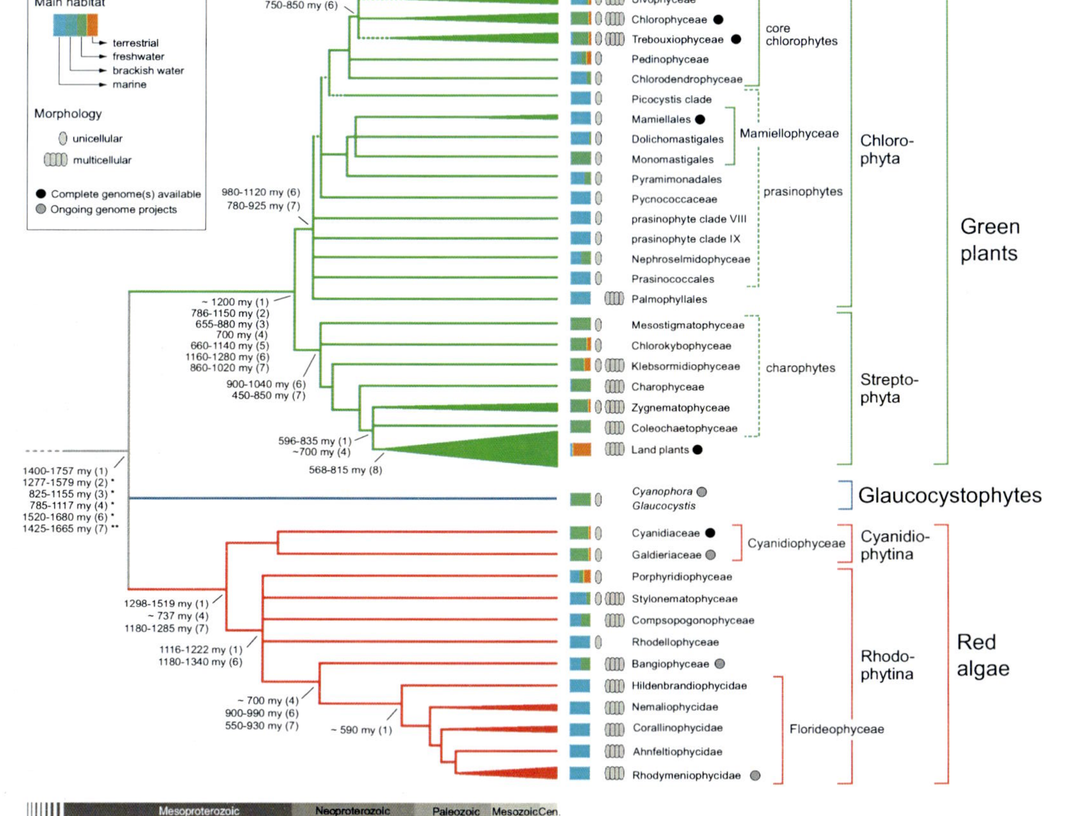

Figure 2.2 Relationships among major groups of Archaeplastida based on published molecular phylogenetic data (Leliaert *et al.*, 2012; Yoon *et al.*, 2010) with indication of main habitats and morphology. Divergence times are based on molecular clock studies calibrated with the fossil record: (1) Yoon *et al.* (2004), (2) Hedges *et al.* (2004), (3) Douzery, Snell, Bapteste, Delsuc, and Philippe (2004), (4) Berney and Pawlowski (2006), (5) Roger and Hug (2006) (r8s-PL method), (6) Herron *et al.* (2009) and (7) Parfrey *et al.* (2011). *excluding glaucophytes; **including cryptomonads.

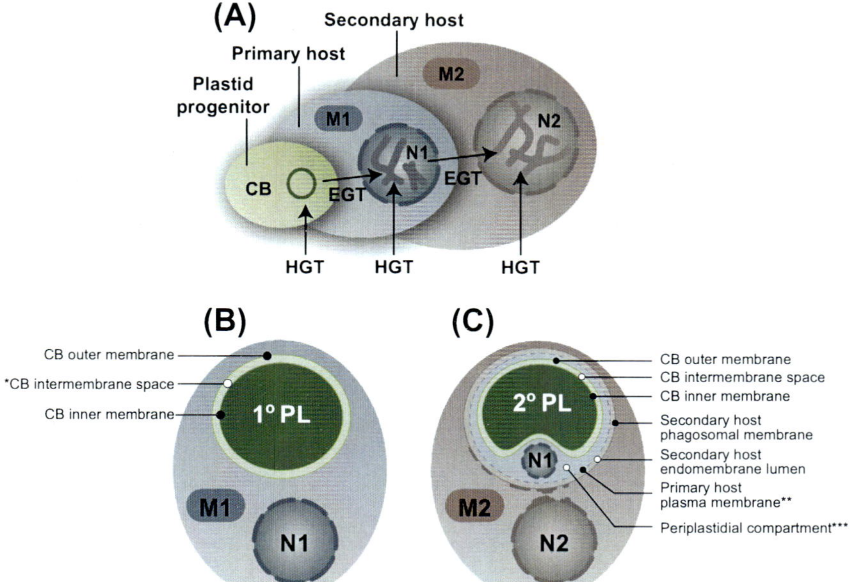

Figure 3.1 Plastid evolution by primary and secondary endosymbiosis. (A) Diagram showing gene flow in photosynthetic eukaryotes, beginning with endosymbiotic gene transfer (EGT) from the cyanobacterial (CB) progenitor of the plastid to the primary host nucleus (N1). In secondary endosymbiosis, another round of EGT occurs, in this case from the primary host nucleus to that of the secondary host (N2). Horizontal gene transfer (HGT) can also impact genome evolution at any stage. Gene transfers involving the mitochondria (M) of the primary and secondary hosts are omitted for simplicity. (B) Generic primary plastid-bearing alga. The primary plastids (PL) of red, green and glaucophyte algae are surrounded by two membranes. In the case of glaucophytes, the intermembrane space (*) contains a layer of peptidoglycan, presumably inherited from the cyanobacterial plastid progenitor. (C) A generic secondary plastid-bearing alga. Depending on the organism, these plastids are surrounded by three or four membranes. In the case of the three-membrane plastids of euglenids and peridinin-containing dinoflagellates, the primary host plasma membrane (**) is believed to have been lost. The remnant cytosol of the secondary endosymbiont is referred to as the periplastidial compartment (***). In cryptophytes and chlorarachniophytes, the primary host nucleus (N1) persists in a remnant form referred to as a nucleomorph. Nucleomorphs have been lost from the periplastidial compartments of all other secondary plastid-bearing algae. Refer to text for further discussion.

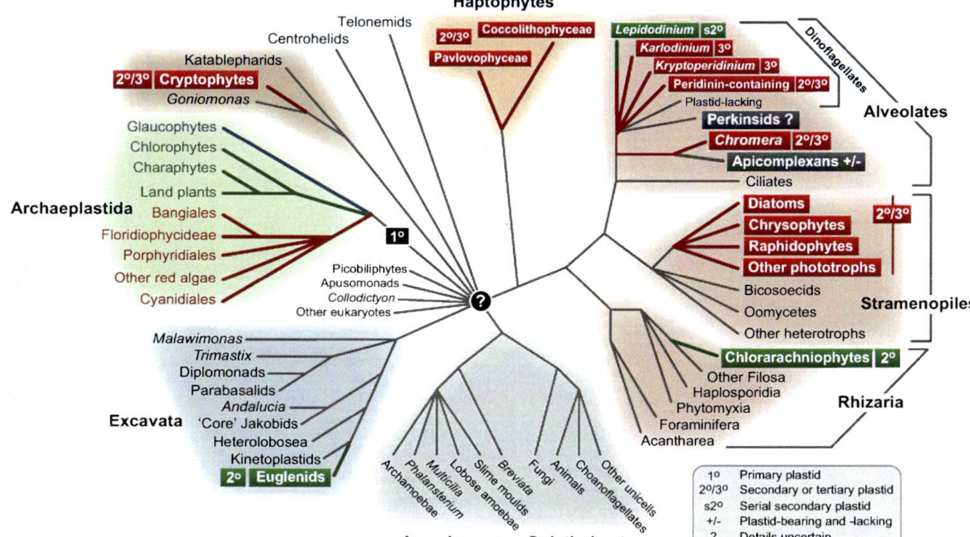

Figure 3.2 Distribution of plastids across the eukaryotic tree of life. The tree topology represents a synthesis of published data as of March 2012. The tree is unrooted and branch lengths are not to scale. Primary plastids are believed to have evolved from cyanobacteria in a common ancestor shared by red algae, green algae and glaucophyte algae. Subsequent secondary endosymbiotic events resulted in the spread of both red and green algal plastids to other eukaryotic lineages, including chlorarachniophytes and euglenids. Based on our current understanding of the higher order relationships between the major eukaryotic groups, secondary plastids of red algal ancestry appear to have spread by tertiary endosymbiosis. The donor and recipient lineages in such endosymbioses are not known. Green and red boxes indicate lineages harbouring green- and red-algal-derived plastids, respectively. Some apicomplexans have non-photosynthetic plastids and perkinsids appear to as well. Refer to text for discussion and references.

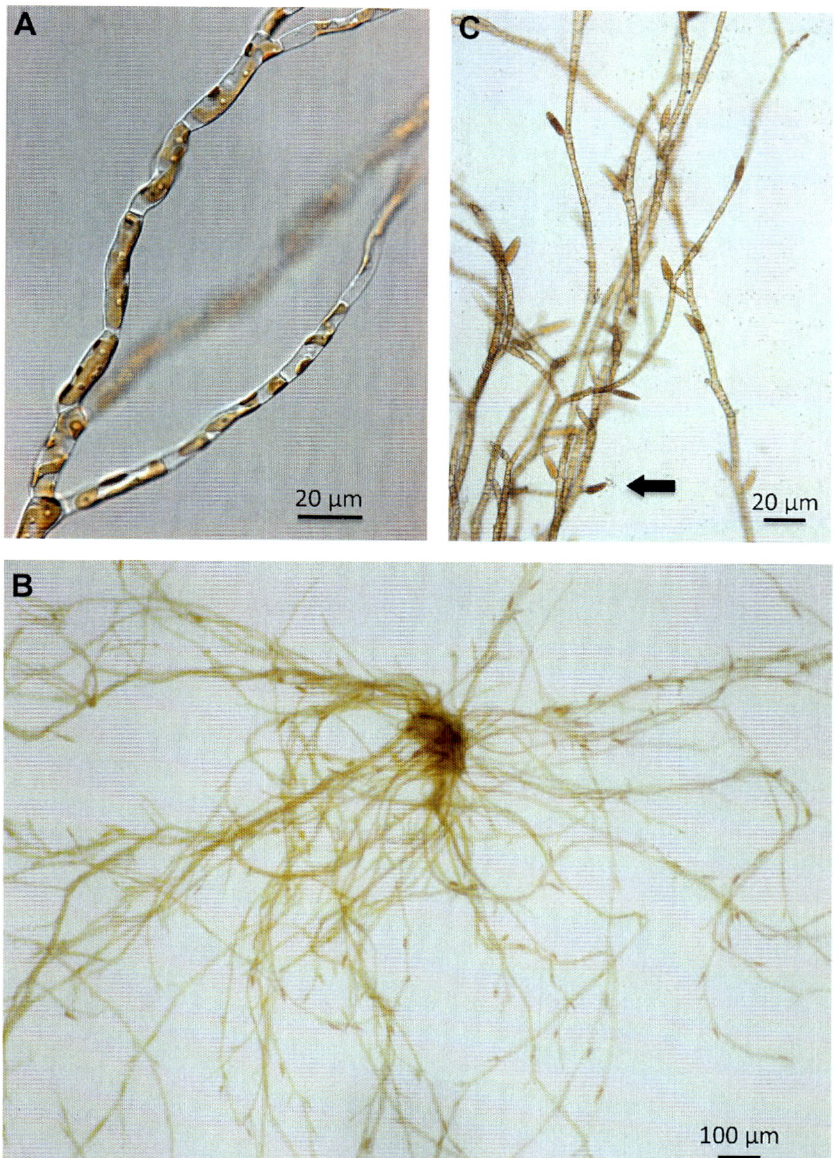

Figure 5.1 The filamentous brown alga *Ectocarpus*. Photographs are of the strain Ec32, which was used to obtain the complete genome sequence. (A) partheno-sporophyte filaments, (B) gametophyte filaments bearing plurilocular gametangia, (C) mature gametophyte filaments releasing gametes from plurilocular gametangia (arrow).

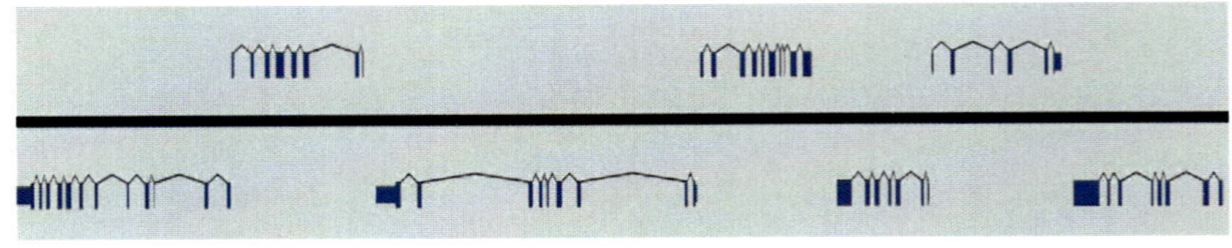

Figure 5.2 Diagram showing the alternating pattern of gene organization typical of many regions of the *Ectocarpus* genome. Coding exons are shown as dark blue bars and untranslated regions as light blue bars. Lines joining exons represent introns. Genes above the black line are transcribed from left to right and genes below the black line are transcribed from right to left.

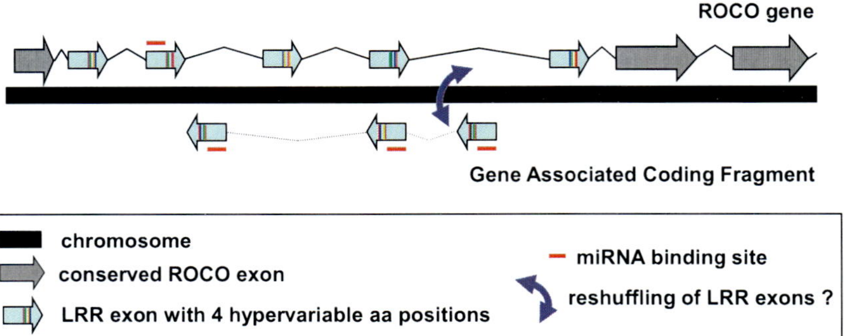

Figure 5.4 Gene structure and hypothetical post-transcriptional regulation features of the ROCO family. *Ectocarpus* ROCO proteins contain LRR domains in the N-terminal region. Each LRR motive is specified by precisely delineated, 24-codon exon (blue arrows). Each LRR exon contains four hypervariable codons corresponding to amino acids that are exposed at the protein surface and are believed to dictate the ligand-binding specificity of the LRR domain (multicolour vertical bars in each LRR exon). These amino acid residues evolve under positive selection, suggesting a potential role of ROCO proteins in pathogen detection or immune reactions. Sequences resembling the LRR-encoding exons are also present in the introns but located on the opposite strand of the DNA. It is possible that these sequences are integrated into ROCO gene sequences as new exons following intragenic recombination events (curved blue arrow). Phylogenetic analysis also suggests that recombination occurs between members of the ROCO family, leading to swapping of LRR exons (not shown here). Moreover, the more conserved N-terminal regions of some LRR exons are predicted to be targets of miRNAs (red bars), suggesting that shuffling of LRR exons might also modulate post-transcriptional regulation. It is not yet known whether the LRR-encoding sequences on the opposite strand are transcribed but if so they could serve as alternative targets for ROCO-directed miRNAs and hence potentially modulate ROCO gene expression.

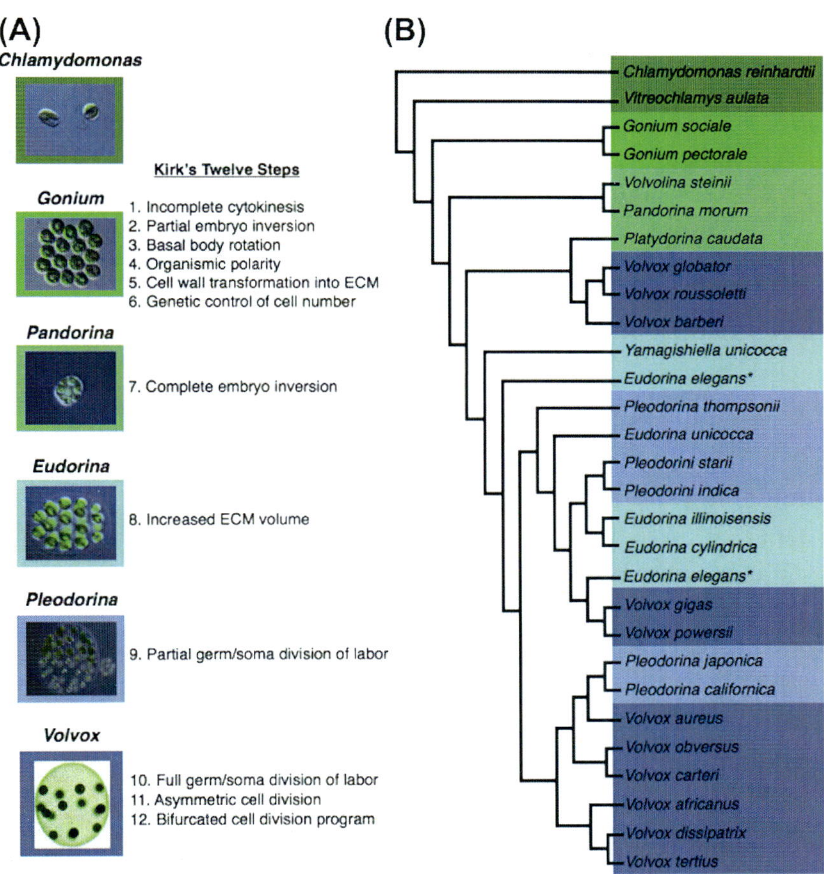

Figure 6.2 The morphology and phylogeny of the Volvocales. (A) Micrographs of Volvocine species and their relationship to each of Kirk's 12 steps of multicellular evolution (Kirk, 2005). (B) Phylogeny of selected Volvocales adapted from (Kirk, 2005; Nozaki, 2003). Species are color-coded based on morphology as indicated in panel (A). Asterisks indicate two isolates of *Eudorina elegans* that subsequently were reclassified as distinct species (Nozaki, 2003; Yamada, Miyaji, & Nozaki, 2008).

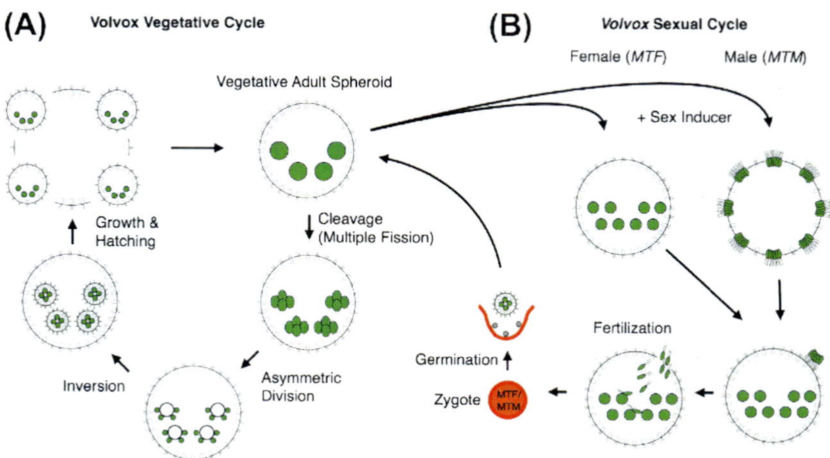

Figure 6.5 Vegetative and sexual developmental in *Volvox*. (A) The *Volvox* vegetative life cycle for males and females is identical and begins with a mature adult (upper right) whose gonidia (large green cells) undergo cleavage to begin embryogenesis. During the sixth cleavage cycle, asymmetric division occurs and leads to production of 16 large anterior cells that are destined to form the germ cells in the next generation. After a total of 12 cleavage cycles, the ~2000-celled embryo undergoes inversion to place the gonidial precursors inside the spheroid and the flagella of the somatic cells pointing outside in their final adult configuration. After a period of growth, the daughter spheroids hatch and continue to grow and mature into adults that can restart the vegetative cycle. (B) Sexual development begins when immature gonidia are exposed to sex-inducer protein. Their subsequent embryogenesis is then altered to produce egg-bearing females or sperm-packet-bearing males (Ferris *et al.*, 2010; Hallmann, Godl, Wenzl, & Sumper, 1998). Females contain 32–48 eggs and ~2000 somatic cells, while males contain 128 somatic cells and 128 packets of 64–128 sperm. Sperm packets are released whereupon they swim as a single unit until they encounter a sexual female. Upon interacting with a sexual female, the sperm packet dissolves into individual cells that enter the female ECM through a fertilization pore and swim until an individual sperm encounters and fuses with an egg to form a diploid zygote. When a *Volvox* zygospore germinates, only one of the four meiotic products survives and differentiates into a new vegetative spheroid.

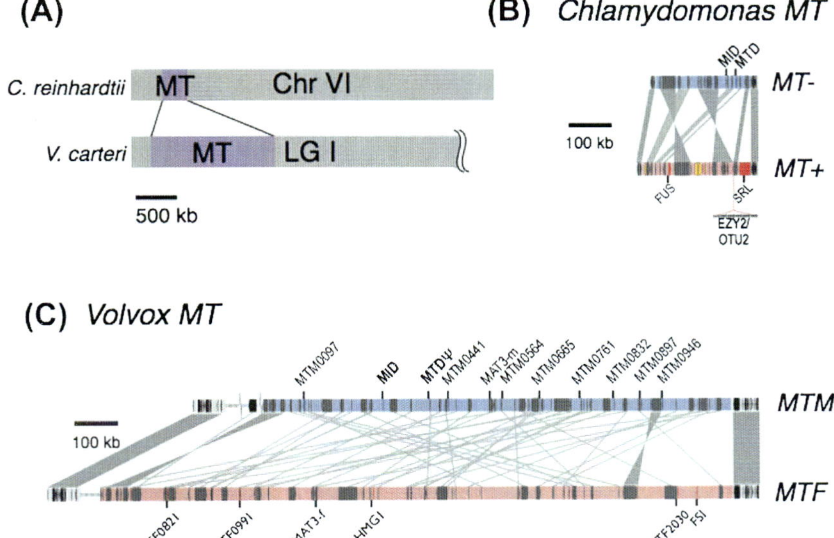

Figure 6.6 The mating-type loci of *Chlamydomonas* and *Volvox*. (A) The mating-type locus for both species is near the telomere of a syntenic chromosome (chromosome 6 in *Chlamydomonas* and linkage group I of *Volvox*). (B) *MT+* and *MT−* mating haplotypes of *Chlamydomonas*. Rearrangements between the two haplotypes are indicated by gray shading. Locations of the sex determining genes *MID* and *MTD1* and the sex limited gene gamete fusion gene FUS1 are shown. Also indicated are the *EZY2/OTU2* regions that may be important for uniparental chloroplast inheritance (Ferris *et al.*, 2002; Goodenough *et al.*, 2007). (C) The *Volvox* male and female mating-type loci are about six times larger than *Chlamydomonas MT* and contain very little syntenic gene order between haplotypes. Several novel genes are shown as well as genes encoding homologs *MID*, *MTD* and the retinoblastoma tumour suppressor homolog, *MAT3*. Figure adapted from Ferris *et al.* (2010) and Umen (2011).

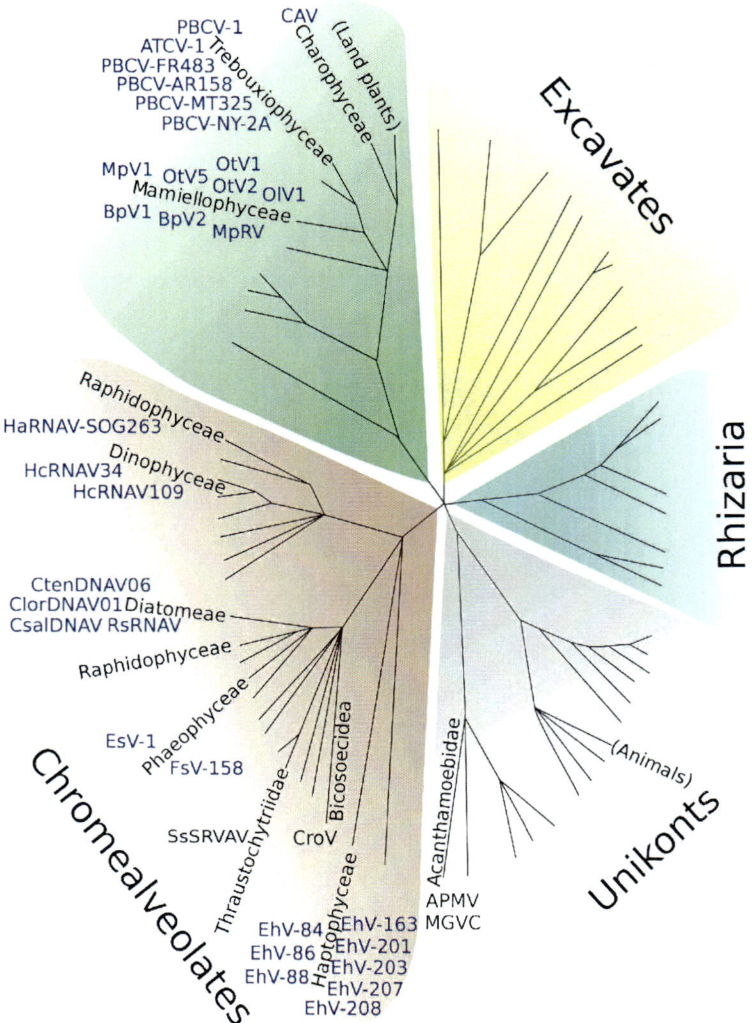

Figure 9.1 Algal viruses whose genomes have been sequenced. Known genomes of viruses infecting photosynthetic algae (text in blue, please refer to Table 9.1 for the names of hosts and viruses) lie mainly in two of the five eukaryotic kingdoms (coloured backgrounds) of life shown (most Unikonts, grey background, do not carry plastids, so their viruses are not included in this review). Only taxa with viruses mentioned in this review are labelled. Many lineages in Rhizaria, Metazoans (Unikonts) and Excavates can form symbioses with photosynthetic organisms (Johnson *et al.*, 2011). The Treouxiophyceaean algae infected by PBCV are usually symbionts of *Paramecium bursaria* (Alveolata) in nature. The positions of land plants and animals are shown for reference (tree simplified from Keeling *et al.*, 2005).